Improvement of Plant Production in the Era of Climate Change

Current trends in population growth suggest that global food production is unlikely to meet future demands under projected climate change scenarios unless the pace of plant improvement is accelerated. Plant production is facing many challenges due to changing environmental conditions and the growing demand for new plant-derived materials. These challenges come at a time when plant science is making significant progress in understanding the basic processes of plant growth and development. Major abiotic stresses like drought, heat, cold and salinity often cause a range of morphological, physiological, biochemical, and molecular changes affecting plant growth, development, and productivity; so sustainable food production poses a serious challenge to much of the world, particularly in emerging countries. This underscores the urgent need to find better ways to translate new advances in plant science into concrete successes in agricultural production. In order to overcome the negative effects of abiotic stress and to maintain food security in the face of these challenges, new, improved, and resilient plant varieties, contemporary breeding techniques, and a deep understanding of the mechanisms for offsetting harmful climate change are undoubtedly necessary. In this context, *Improvement of Plant Production in the Era of Climate Change* is a guide to the most advanced techniques that help in understanding plant response to abiotic stress, leading to new horizons and the strategy for the current translation studies application to overall solution to create a powerful production and crop improvement in such an adverse environment.

FEATURES

- Provides a state-of-the-art description of the physiological, biochemical, and molecular-level understanding of abiotic stress in plants.
- Courses taught in universities from basics to advanced level in field of plant physiology, molecular genetics, and bioinformatics will use this book.
- Focuses on climatic extremes and their management for plant protection and production, which is great threat to future generation and food security.
- Understanding of new techniques pointed out in this book will open the possibility of genetic engineering in crop plants with the concomitant improved stress tolerance.
- Addressing factors that are threatening future food production and providing potential solutions to these factors.
- Written by a diverse group of internationally famed scholars, this book adds new horizons in the field of abiotic stress tolerance.

Footprints of Climate Variability on Plant Diversity

Series Editor: Shah Fahad

Improvement of Plant Production in the Era of Climate Change

Edited by
Shah Fahad
University of Haripur, Pakistan
Muhammad Adnan
University of Swabi, Pakistan
Shah Saud
Northeast Agricultural University, China

CRC Press
Taylor & Francis Group
Boca Raton London New York

CRC Press is an imprint of the
Taylor & Francis Group, an **informa** business

Cover image: ParabolStudio/Shutterstock

First edition published 2023
by CRC Press
6000 Broken Sound Parkway NW, Suite 300, Boca Raton, FL 33487-2742

and by CRC Press
2 Park Square, Milton Park, Abingdon, Oxon, OX14 4RN

Library of Congress Cataloging-in-Publication Data
Names: Fahad, Shah (Assistant professor of agriculture), editor. | Adnan, Muhammad (Lecturer in agriculture), editor. | Saud, Shah, editor.
Title: Improvement of plant production in the era of climate change / Shah Fahad, Muhammad Adnan, Shah Saud.
Other titles: Footprints of climate variability on plant diversity.
Description: First edition. | Boca Raton, FL : CRC Press, 2022. | Series: Footprints of climate variability on plant diversity | Includes bibliographical references and index.
Identifiers: LCCN 2022001680 (print) | LCCN 2022001681 (ebook) | ISBN 9781032260716 (hardback) | ISBN 9781032260723 (paperback) | ISBN 9781003286417 (ebook)
Subjects: LCSH: Agriculture--Climatic factors. | Plants--Effect of stress on. | Sustainable agriculture. | Crops and climate.
Classification: LCC S600.5 .I48 2022 (print) | LCC S600.5 (ebook) | DDC 630.2/515--dc23/eng/20220218
LC record available at https://lccn.loc.gov/2022001680
LC ebook record available at https://lccn.loc.gov/2022001681

ISBN: 978-1-032-26071-6 (hbk)
ISBN: 978-1-032-26072-3 (pbk)
ISBN: 978-1-003-28641-7 (ebk)

DOI: 10.1201/9781003286417

Typeset in Times
by SPi Technologies India Pvt Ltd (Straive)

Contents

Acknowledgements

Words are bound and knowledge is limited to praise ALLAH, the Instant and Sustaining Source of all Mercy and Kindness, and the Sustainer of the Worlds. My greatest and ultimate gratitude is due to ALLAH (Subhanahu wa Taqadus). I thank ALLAH with all my humility, for everything that I can think of. His generous blessing and exaltation succeeded my thoughts and thrived my ambition to have the cherished fruit of my modest efforts in the form of this piece of literature from the blooming spring of blossoming knowledge. May ALLAH forgive my failings and weaknesses, strengthen and enliven my faith in HIM, and endow me with knowledge and wisdom. All praises and respects are for **Holy Prophet Muhammad** *Salle Allah Alleh Wassalam,* the greatest educator, the everlasting source of guidance and knowledge for humanity. He taught the principles of morality and eternal values and enabled us to recognize our Creator. I have a deep sense of obligation to my parents, my brothers, sisters, and son. Their unconditional love, care, and confidence in my abilities helped me achieve this milestone in my life. For this and much more, I am forever in their debt. It is to them that I dedicate this book. In this arduous time, I also appreciate the patience and serenity of my wife, who brought joy to my life in so many different ways. It is indeed on account of her affections and prayers that I was able to achieve something in my life.

Shah Fahad

About the Editors

Dr Shah Fahad is a Professor in the Hainan Key Laboratory for Sustainable Utilization of Tropical Bioresource, College of Tropical Crops, Hainan University, Haikou, China. He obtained his PhD in Agronomy from Huazhong Agriculture University, China, in 2015. After doing his postdoctoral research in Agronomy at the Huazhong Agriculture University (2015–17), he accepted the position of Assistant Professor at the University of Haripur. He has published over 332 peer-reviewed papers (impact factor 1050.18) with more than 300 research and 32 review articles, on important aspects of climate change, plant physiology and breeding, plant nutrition, plant stress responses and tolerance mechanisms, and exogenous chemical priming–induced abiotic stress tolerance. He has also contributed 58 book chapters to various book editions published by Springer, Wiley-Blackwell, and Elsevier. He has edited 17 book volumes, including this one, published by CRC press, Springer, and Intech Open. He won Young Rice International Scientist award, Distinguish scholar award, Top young investigator award, Dr Shah Jahan Agriculture award (PAS) in 2014, 2015, 2019, and 2021, respectively. He won 15 projects from international and national donor agencies. Dr. Shah Fahad name figured twice among the top 2 per cent scientists in a global list compiled by Stanford University, USA. Dr Shah Fahad name is also included in the distinguish list of 2021, Highly Cited Researchers announced by the Clarivate/Web of Sciences. He has worked and is presently continuing on a wide range of topics, including climate change, greenhouse emission gases, abiotic stresses tolerance, roles of phytohormones and their interactions in abiotic stress responses, heavy metals, regulation of nutrient transport processes.

Dr Muhammad Adnan is a lecturer in the Department of Agriculture at the University of Swabi (UOS), Pakistan. He completed his PhD (soil fertility and microbiology) from the Department of Soil and Environmental Sciences (SES) the University of Agriculture Peshawar, Pakistan, and Department of Plant, Soil and Microbial Sciences, Michigan State University, USA. He has received his MSc and BSc (Hons) in Soil and Environmental Sciences, from Department of SES the University of Agriculture, Peshawar-Pakistan.

Dr Shah Saud is currently working as an Assistant Professor in the College of Life Science, Linyi University, Linyi, China. He received his PhD in turf grasses (Horticulture) from Northeast Agricultural University, Harbin, China. He is currently working as a postdoctorate researcher in Department of Horticulture, Northeast Agricultural University, and Harbin, China. Dr Shah Saud has published over 160 research publications in peer-reviewed journals. He has edited 5 books and written 35 book chapters on important aspects of plant physiology, plant stress responses, and environmental problems in relation to agricultural plants. According to Scopus®, Dr. Shah Saud's publications have received roughly 3,000 citations with an h-index of 34.

Contributors

Muhammad Adnan
Department of Agriculture
University of Swabi
Pakistan

Gulnaz Afzal
Department of Zoology
The Islamia University of Bahawalpur
Bahawalpur, Pakistan

Muhammad Afzaal
Pakistan Meteorological Department
National Agromet Centre (NAMC)
Islamabad, Pakistan

Farooq Ahmad
Department of Zoology
The Islamia University of Bahawalpur
Bahawalpur, Pakistan

Ishfaq Ahmad
Centre for Climate Research and Development
COMSATS University
Islamabad, Pakistan

Ashfaq Ahmed
Department of Agronomy
University of Agriculture
Faisalabad, Pakistan

Saeed Ahmed
Department of Agronomy
MNS-University of Agriculture
Multan, Pakistan

Shakeel Ahmed
Department of Agronomy
Faculty of Agricultural Sciences and Technology (FAST)
Bahauddin Zakariya University
Multan, Pakistan

Muhammad Sohail Akram
Department of Botany
Faculty of Life Sciences
Government College University
Faisalabad, Pakistan

Duraid K. A. Al-Taey
Department of Horticulture
Faculty of Agriculture
Al-Qasim Green University
Baghdad, Iraq

Wafa'a A. Al-Taisan
Department of Biology
College of Science
Imam Abdulrahman Bin Faisal University
Dammam, Saudi Arabia

Abdelrazzaq Al-Tawaha
Department of Crop Science
Faculty of Agriculture
University Putra Malaysia
Selangor, Malaysia

Abdel Rahman M. Al-Tawaha
Department of Biological Sciences
Al Hussein Bin Talal University
Ma'an, Jordan

Ibrahim M. AlRawashdeh
Department of Biological Sciences
Al Hussein Bin Talal University
Ma'an, Jordan

Mohammad Nur Alam
Bangladesh Wheat and Maize Research Institute
Nashipur, Bangladesh

Madeeha Alamzeb
Department of Agronomy
The University of Agriculture Peshawar
Peshawar, Pakistan

Kawsar Ali
Department of Agriculture
Abdul Wali Khan University
Mardan, Pakistan

Amanullah
Department of Agronomy
The University of Agriculture
Peshawar, Pakistan

Shazma Anwar
Department of Agronomy
The University of Agriculture Peshawar
Peshawar, Pakistan

Muhammad Arif
Department of Agronomy
The University of Agriculture
Peshawar, Pakistan

Celaleddin Barutçular
Department of Field Crops
Faculty of Agriculture
University of Cukurova
Adana, Turkey

Zhulong Chan
Key Laboratory of Horticultural Plant Biology
Ministry of Education, College of Horticulture and Forestry Sciences
Huazhong Agricultural University
Wuhan, China

Hassan Javed Chaudhary
Department of Plant Sciences
Faculty of Biological Sciences
Quaid-i-Azam University
Islamabad, Pakistan

Fatih Çiğ
Department of Field Crops
Faculty of Agriculture
Siirt University
Siirt, Turkey

Murat Erman
Department of Field Crops
Faculty of Agriculture
Siirt University
Siirt, Turkey

Mabrouk Elsabagh
Department of Nutrition and Clinical Nutrition
Faculty of Veterinary Medicine
Kafr El-sheikh University
Kafr El-Sheikh, Egypt

Shah Fahad
Hainan Key Laboratory for Sustainable Utilization of Tropical Bioresource
College of Tropical Crops
Hainan University
Haikou, China
and
Department of Agronomy
The University of Haripur
Khyber Pakhtunkhwa, Pakistan

Abdul Ghaffar
Department of Agronomy
MNS-University of Agriculture
Multan, Pakistan

Muhammad Abdul Ghafar
Department of Zoology
The Islamia University of Bahawalpur
Bahawalpur, Pakistan

Ntsomboh-Ntsefong Godswill
Laboratory of Genetics & Plant Improvement and Plant Physiology Improvement Unit
Department of Plant Biology
Faculty of Science
University of Yaounde
Yaoundé, Cameroon

Ghulam Haider
Department of Agronomy
MNS-University of Agriculture
Multan, Pakistan

Bibi Haleema
Agricultural Research Institute
Tarnab, Peshawar

Mirza Hasanuzzaman
Department of Agronomy
Faculty of Agriculture
Sher-e-Bangla Agricultural University
Dhaka, Bangladesh

Zhang Hengheng
State Key Laboratory of Cotton Biology
Institute of Cotton Research of CAAS
Anyang, P.R. China

Akbar Hossain
Bangladesh Wheat and Maize Research Institute
Nashipur, Bangladesh

Gui Huiping
State Key Laboratory of Cotton Biology
Institute of Cotton Research of CAAS
Anyang, P.R. China

Jamshad Hussain
U.S.-Pakistan Centre for Advanced Studies in Agriculture and Food Security
University of Agriculture
Faisalabad, Pakistan

Muhammad Hussaan
Department of Botany
Faculty of Life Sciences
Government College University
Faisalabad, Pakistan

Imran
Department of Agronomy
The University of Agriculture Peshawar
Peshawar, Pakistan

Inamullah
Department of Agronomy
The University of Agriculture Peshawar
Peshawar, Pakistan

Asif Iqbal
State Key Laboratory of Cotton Biology
Institute of Cotton Research of CAAS
Anyang, P.R. China

Mazhar Iqbal
Department of Botany
Shaheed Benazir Bhutto University Sheringal Dir (U)
Pakistan

Muhammad Aamir Iqbal
Department of Agronomy
University of Poonch Rawalakot (AJK)
Pakistan

Naeem Iqbal
Department of Botany
Faculty of Life Sciences
Government College University
Faisalabad, Pakistan

Mohammad Sohidul Islam
Department of Agronomy
Hajee Mohammad Danesh Science and Technology University
Basherhat, Bangladesh

Nusrat Jahan
Department of Botany
Faculty of Life Sciences
Aligarh Muslim University
Aligarh, India

Talha Jan
Department of Agronomy
The University of Agriculture
Peshawar, Pakistan

Muhammad Tariq Javed
Department of Botany
Faculty of Life Sciences
Government College University
Faisalabad, Pakistan

Emine Karademir
Department of Field Crops
Faculty of Agriculture
Siirt University
Siirt, Turkey

Shah Khalid
Department of Agronomy
The University of Agriculture Peshawar
Peshawar, Pakistan

Asad Ali Khan
Department of Agronomy
The University of Agriculture Peshawar
Peshawar, Pakistan

Bushra Khan
Department of Soil Environmental Science
University of Agriculture
Peshawar, Pakistan

Imtiaz Khan
Department of Weed Science
The University of Agriculture
Peshawar, Pakistan

Hamdan Ali Khan
Department of Agronomy
The University of Agriculture Peshawar
Peshawar, Pakistan

Mehmood Alam Khan
Department of Agronomy
MNS-University of Agriculture
Multan, Pakistan

Muhammad Ishfaq Khan
Department of Weed Science
The University of Agriculture
Peshawar, Pakistan

Sabaz Ali Khan
Department of Environmental Sciences
COMSATS, Institute of Information Technology
Abbottabad, Pakistan

Hamid Majeed
Department of Nutrition and Clinical Nutrition
Faculty of Veterinary Medicine
Kafr El-sheikh University
Kafr El-Sheikh, Egypt

G. Malanga
Universidad de Buenos Aires
Facultad de Farmacia y Bioquímica, Fisicoquímica
Buenos Aires, Argentina

Ashutosh Kumar Mall
ICAR-Indian Institute of Sugarcane Research
Lucknow, India

David McNeil
University of Tasmania LPO, Sandy Bay
Tasmania, Australia

Ishaq Ahmad Mian
Department of Soil Environmental Science
University of Agriculture
Peshawar, Pakistan

Varucha Misra
ICAR-Indian Institute of Sugarcane Research
Lucknow, India

Muhammad Mubeen
Department of Environmental Sciences
COMSATS University Islamabad
Vehari Campus, Pakistan

Sahar Mumtaz
Department of Botany
Division of Science and Technology
University of Education
Lahore, Pakistan

Fazal Munsif
Department of Agronomy
The University of Agriculture
Peshawar, Pakistan

Wajid Nasim
Department of Agronomy
University College of Agriculture & Environmental Sciences
Islamia University of Bahawalpur
Bahawalpur, Pakistan

Nargis Naz
Department of Botany
The Islamia University of Bahawalpur
Bahawalpur, Pakistan

Niamatullah
Department of Human Nutrition
The University of Agriculture
Peshawar, Pakistan

Pang Nianchang
State Key Laboratory of Cotton Biology
Institute of Cotton Research of CAAS
Anyang, P.R. China

Nabeel Khan Niazi
Institute of Soil and Environmental Sciences
University of Agriculture Faisalabad
Faisalabad, Pakistan

Sytar Oksana
Institute of Agricultural Research for Development (IRAD)
Division of Valorization and Innovation
Yaounde, Cameroon

Yahia A. Othman
Department of Horticulture and Crop Science
The University of Jordan
Amman, Jordan

Brajendra Parmar
ICAR-Indian Institute of Rice Research
Soil Science
Hyderabad, India

Swapan Kumar Paul
Department of Agronomy
Bangladesh Agricultural University
Mymensingh, Bangladesh

C. P. Prajapati
ICAR-Indian Institute of Sugarcane Research
Lucknow, India

S. Puntarulo
Universidad de Buenos Aires
Facultad de Farmacia y Bioquímica, Fisicoquímica
Buenos Aires, Argentina

Ali M. Qaisi
Department of Pharmaceutical Sciences
School of Pharmacy
University of Jordan
Amman, Jordan

Dong Qiang
State Key Laboratory of Cotton Biology
Institute of Cotton Research of CAAS
Anyang, P.R. China

Muhammad Habib Ur Rahman
Department of Agronomy
MNS-University of Agriculture
Multan, Pakistan

Fahad Rasul
Department of Agronomy
University of Agriculture
Faisalabad, Pakistan

Abdur Rauf
Department of Chemistry
University of Swabi
Anbar, Pakistan

Muhammad Riaz
Department of Environmental Sciences and Engineering
Government College University Faisalabad
Faisalabad, Pakistan

E. Robello
Universidad de Buenos Aires
Facultad de Farmacia y Bioquímica, Fisicoquímica
Buenos Aires, Argentina

Ayman EL Sabagh
Department of Agronomy
Faculty of Agriculture
Kafrelsheikh University, Egypt

Muhammad Saeed
Department of Agronomy
The University of Agriculture Peshawar
Peshawar, Pakistan

Rokayya Sami
Department of Nutrition and Food Science
Taif University
Taif, Kingdom of Saudi Arabia

Jeyabalan Sangeetha
Department of Environmental Science
Central University of Kerala
Kasaragod, India

Muhammad Hamzah Saleem
MOA Key Laboratory of Crop Ecophysiology and Farming System Core in the Middle Reaches of the Yangtze River
College of Plant Science and Technology
Hauzhong Agricultural University
Wuhan, China

Shah Saud
College of Life Science
Linyi University
Linyi, China

Azizullah Shah
Pakistan Agriculture Research Council Islamabad
Islamabad, Pakistan

Mohd Abas Shah
ICAR-Central Potato Research Station
Jalandhar, India

Shaheen Shah
Department of Agronomy
The University of Agriculture Peshawar
Peshawar, Pakistan

Muhammad Shahid
Department of Bioinformatics and Biotechnology
Government College University
Faisalabad, Pakistan

S. Solomon
CSA University of Agriculture & Technology
Kanpur, India

Meizhen Song
State Key Laboratory of Cotton Biology
Institute of Cotton Research of CAAS
Anyang, P.R. China

Kashif Tanwir
Department of Botany
Faculty of Life Sciences
Government College University
Faisalabad, Pakistan

Devarajan Thangadurai
Department of Botany
Karnatak University
Dharwad, India

Allah Wasaya
College of Agriculture
Bahauddin Zakariya University
Bahadur Sub-Campus Layyah, Pakistan

Wang Xiangru
State Key Laboratory of Cotton Biology
Institute of Cotton Research of CAAS
Anyang, P.R. China

Zhang Xiling
State Key Laboratory of Cotton Biology
Institute of Cotton Research of CAAS
Anyang, P.R. China

Refat A. Youssef
Soils and Water Use Department
National Research Centre
Dokki, Egypt

Muhammad Zamin
Department of Agriculture
The University of Swabi
Swabi, Pakistan

Sayeda Sadaf Zehra
Department of Botany
The Islamia University of Bahawalpur
Bahawalpur, Pakistan

1

Adaptations in Cropping System and Pattern for Sustainable Crops Production under Climate Change Scenarios

Abdul Ghaffar
MNS University of Agriculture, Multan, Pakistan

Muhammad Habib Ur Rahman
MNS University of Agriculture, Multan, Pakistan; University of Bonn, Bonn, Germany

Saeed Ahmed
MNS University of Agriculture, Multan, Pakistan

Ghulam Haider
MNS University of Agriculture, Multan, Pakistan

Ishfaq Ahmad
COMSATS University Islamabad, Pakistan

Mehmood Alam Khan
MNS-University of Agriculture, Multan, Pakistan

Muhammad Afzaal
National Agromet Centre (NAMC), Islamabad, Pakistan

Shakeel Ahmed
Bahauddin Zakariya University, Multan, Pakistan

Shah Fahad
Hainan University, Haikou, China; The University of Haripur, Haripur, Pakistan

Jamshad Hussain and Ashfaq Ahmed
University of Agriculture – UAF, Faisalabad, Pakistan

CONTENTS

DOI: 10.1201/9781003286417-1

1.1 Introduction

1.1.1 What Are Cropping System and Cropping Pattern

1.1.1.1 Concept of Cropping System and Cropping Pattern

Cropping system, an important component of a farming system, represents a cropping pattern used on a farm and its interaction with farm resources, other farm enterprises, and available technology, which determine their makeup. Cropping system comprises all components required for the production of a particular crop and the interrelationships between them and the environment. *Cropping system* is defined as the order in which the crops are cultivated on a piece of land over a fixed period or the way in which different crops are grown. In the cropping system, sometimes a number of crops are grown together or they are grown separately at short intervals in the same field. When annual crops are considered, a cropping system usually means the combination of crops grown within a given year.

Cropping pattern is the pattern of crops for a given piece of land or the proportion of area under various crops at a point of time in a unit area, and it indicates the yearly sequence and spatial arrangements of crops and fallows in an area (Table 1.1).

1.1.2 Classification of Cropping System

Depending on the resources and technology available, different types of cropping systems are adopted on farms, which are as below.

1.1.2.1 Monocropping or Monoculture

In this system, only one crop is grown on farm land year after year. Corn, soybeans, cotton, millet, sunhemp, soybean, alfalfa, flax, and wheat are the common crops often grown using monocropping techniques (Table 1.2).

TABLE 1.1

Differences between Cropping Pattern and Cropping System

Cropping System	Cropping Pattern
The cropping patterns used on a farm and their interaction with farm resources, other farm enterprises, and available technology which determine their makeup.	The yearly sequence and spatial arrangement of crops or of crops and fallows on a given area.
Cropping pattern and its management help to derive benefits from a given resource base under specific environmental conditions.	Crop rotation practised by a majority of farmers in a given area or locality.
Pattern of crops taken up for a given piece of land, or order in which crops are cultivated on a piece of land over a fixed period, associated with soil, management practices such as tillage, manuring, and irrigation.	Type and management of crops in time and space.
Cropping system is a critical aspect in developing an effective ecological farming system to manage and organize crops so that they best utilize the available resources (soil, air, sunlight, water, labour, equipment).	Cropping pattern differs from region to region.

TABLE 1.2

Advantages and Disadvantages of Monocropping System

Advantages	Disadvantages
Specialized production	Declining biodiversity
Technological advances	Dangerous for bees
High efficiency	High risk of harvest loss
Greater yields of some produce	Fossil fuel dependent
Simpler to manage	Not climate smart
Higher earnings	Overproduction of commodity crops
	Soil degradation
	Pest problems

1.1.2.2 Multiple Cropping

Multiple cropping is defined as 'growing of two or more than two crops in the same field in the same time in a year.' It is not a new technique; it has been practised in Egypt (300 B.C.), Central America, South America, and Indonesia (Table 1.3).

1.1.2.3 Intercropping

In this system, farmers grow two or more crops simultaneously on the same field in one calendar year, e.g. sugarcane + oil seed crop and setaria + red gram (Tables 1.4 and 1.5).

TABLE 1.3

Advantages and Disadvantages of Multiple Cropping Systems

Advantages	Disadvantages
It is a better source of soil utilization.	The survival of pests becomes easy.
Increased yield per unit of land.	Pests can easily shift from one crop to another crop.
Increased export and foreign exchange.	Problem of weed management.
Costs of input decrease as compared to individual crop growing cost.	Implementation of new technology is difficult.
Reduce pest and disease attack.	Competition effect can cause yield loss.
Different types of products can be produced at a time.	

TABLE 1.4

Advantages and Disadvantages of Intercropping Systems

Advantages	Disadvantages
Intercropping gives additional yield and income/unit area than sole cropping.	Yield decreases as the crops differ in their competitive abilities.
It acts as an insurance against failure of crops in abnormal year.	Management of intercrops having different cultural practices seems to be a difficult task.
Intercrops maintain the soil fertility as the nutrient uptake is made from both layers of soil.	Improved implements cannot be used efficiently
Reduction in soil runoff and controls weeds.	Higher amount of fertilizer or irrigation water cannot be utilized properly as the component crops vary in their response of these resources.
Intercrops provide shade and support to the other crop.	Harvesting is difficult.
Intercropping system utilizes resources efficiently and their productivity is increased.	
Intercropping with cash crops is highly profitable.	

TABLE 1.5

Types of Intercropping

Mixed Cropping	Row Intercropping	Strip Intercropping	Relay Intercropping
Growing two or more crops simultaneously with no distinct row arrangement.	Growing two or more crops simultaneously where one or more crops are planted in rows.	Growing two or more crops simultaneously in different strips wide enough to permit independent cultivation but narrow enough for the crops to interact agronomically.	A second crop is planted after the first crop has reached its reproductive stage but before it is ready for harvest.
Examples of mixed intercropping are wheat, gram, mustard.	Examples of row intercropping are sunflower, potato, dry bean.	Examples of strip intercropping are strips of wheat, corn, soybean.	Examples of relay intercropping are rice-cauliflower, onion-gourd.

1.1.3 Principles of Cropping System

Each cropping system has its own features, but it has to follow the following basic principles.

i. Choose crops that complement each other.
ii. Choose crops and cropping rotation which utilize available resources efficiently
iii. Choose crops and cropping system that maintain and enhance soil fertility
iv. Choose crops which have a diversity of growth cycles
v. Choose a diverse species of crops

1.1.4 Objectives of Cropping System

The objective of any cropping system is the efficient utilization of all resources, viz. land, water, and solar radiation; maintaining stability in production; and obtaining higher net returns. The efficiency can be measured by the quantity of produce obtained per unit resource used in a given time. The scope or objectives of any cropping system are the following:

i. To increase crop diversity with noncereals such as oilseed and legume crops
ii. To intensify rotations (i.e. less fallow)

iii. To maintain or increase soil quality
iv. To reduce or eliminate tillage
v. To sustain economic profitability for farmers

1.1.5 Significance for Crop Production

1.1.5.1 Features and Significances of Cropping System

Cropping system contains many features, which indicate its importance towards increasing sustainable crop production under current environmental conditions. The most important features of any cropping system should be

i. Improve soil structure and enhance soil tilth through appropriate measures of soil surface management.
ii. Control soil erosion through runoff management and ground cover establishment.
iii. Replenish nutrients harvested in crops, from trees, and by animals through a judicious combination of inorganic fertilizers and organic amendments
iv. Encourage carbon sequestration through the maximum use of conservation tillage practices
v. Improve production from existing lands through sustainable utilization of resources

1.1.6 Soil Fertility Management

Soil fertility management has become a global issue in a sustainable manner which is attributed to current climate change scenarios and lack of proper cropping system. Soil nutrients are depleted during crop growth, and must be renewed or replaced in order to get maximum income from the particular piece of land. In intensive cropping system, soil fertility management has become a complex problem because of residual effect of nutrients applied to the previous crops, possible effect of legumes in the system, complementary and competitive interaction from the component crops and influence of crop residues left in the soil. In sequence cropping system, system productivity can be increased with the application of phosphorus (P) along with nitrogen (N), and further increased with use of N, P, and potassium (K). Application of N at recommended dose is advocated to each of the crop in the cropping system. The management of P in cropping system needs careful adjustment of P fertilizer dose taking into the account type of fertilizer, soil characteristics, and their yield level; extent of P removal; and growing environment. In cropping system – involving wheat, fertilizer P dose to kharif crop can be reduced if preceding wheat has received P in adequate amounts. The removal of K in proportion to N is very high in cropping systems, particularly those involving cereal and fodder crops. It is important to apply K fertilizer at recommended dose to maintain soil fertility. Among the secondary nutrients, sulphur (S) application is necessary if cropping system includes oilseed crops particularly to sesamum–mustard, soybean–safflower, and groundnut–mustard cropping sequence. On the other hand, zinc (Zn) deficiency is the most common as nearly 50% soils of intensive cultivated areas suffer from Zn deficiency. Use of organics prevents Zn deficiency in intensive cropping system under normal soils (Table 1.6).

TABLE 1.6

Various Types of Biofertilizers, Targeted Crops, and Their Rate of Application to the Soil

Name of Biofertilizer	Targeted Crops	Dose	Nutrient Fix (kg ha^{-1})	Nutrient Elements
Rhizobium	Pulses and legumes, oilseed	1:100	50–100	N
PSB	All crops	1:100	30–35	P
Azotobacter	Rice, sugarcane, vegetables, etc.	1:100	20–40	N
Azospirillum	Maize, sugarcane, sorghum, and millet	1:100	20–30	N

Source: Ray et al. 2015.

1.1.7 Water Management

Under current climate change scenarios, the maintenance of water has become a big issue, and it must be maintained within a wide range throughout the growing period for the better development of crop plants. The selection of crops should be done according to requirement of crop and soil structure, which can be a sustainable approach towards increasing crop production under current climate change scenarios. In both of these cases, the choice of crop is the most important sustainable approach as when there is shortage of water, then crop with low water requirement should be used in cropping system instead of crop with high water requirement. Hence, there is a need to reshuffle the cropping systems in order to maintain crop production under current climate change scenarios. Rice–groundnut cropping system is the most suitable cropping system instead of rice–rice cropping system under current conditions of water scarcity. Under limited water supply, however, rice–chickpea–green gram and rice–mustard–green gram are more suitable cropping systems with high water-use efficiency. On the other hand, agronomic practices such as planting of crops on ridges, proper irrigation scheduling, and rain water harvesting, etc. are the best water management strategies.

1.1.8 Efficient Cropping System

The cropping system which utilizes farm resources, farm enterprises, and farm technology in a sustainable fashion is called *efficient cropping system.* It depends upon farm resources, farm enterprises, and farm technology because farm is an organized economical unit. The farm resources include land, labour, water, capital, and infrastructure. When land is limited, intensive cropping should be adapted to fully utilize available water and labour. When sufficient and cheap labour is available, vegetable crops should be included in the cropping systems as they require more labour. Capital-intensive crops like sugarcane, banana, etc. find a space in the cropping system when capital is not a constraint. In low rainfall regions (750 mm/annum) monocropping is followed and when rainfall is more than 750 mm, intercropping is practised. When there is sufficient irrigation water available, triple and quadruple cropping should be adopted in order to utilize irrigation water sufficiently. When other climatic factors are not limiting, farm enterprises like dairy, poultry, etc. also influence the type of cropping system. When the farm enterprises include dairy, cropping system should contain fodder crops. Components change in cropping system also takes place with the developments of technology. All above-mentioned cropping systems are efficient, as these cropping systems are adjusted in order to utilize all farm resources, enterprises, and technology efficiently.

1.1.9 Cropping Systems of the World

1.1.9.1 Major Cropping Systems of the World

All types of cropping systems, i.e. monocropping, multiple cropping, and intercropping, have been practised across the world. The practice of multiple cropping systems is considered high as compared to monocropping systems. In some areas of the world, intercropping is meeting an increasing demand of food, which is attributed to its more production per unit area. Major cropping systems which are largely practised across the world include rice–wheat, cotton–wheat, maize–wheat, soybean–safflower, sugarcane–pulses, sunflower–cotton, etc. On the other hand, most efficiently practised cropping systems across the world are listed in Table 1.7.

1.1.10 Cropping Systems in Asia

1.1.10.1 Major Cropping Systems in Asia

Asia is largest continent in the world with respect to area and population. There are various types of cropping systems practised in Asia. The most important and efficiently utilized cropping systems are enlisted in Table 1.8. The rice–wheat cropping system occupies about 18 million ha area in Asia, of which 75% are in the Indo-Gangetic Plains of India (10 million ha), Pakistan (2.2 million ha), Bangladesh

TABLE 1.7

Distribution of Most Efficient Cropping Systems in the World

Monocropping	Sequence Cropping	Intercropping
Maize	Rice–Wheat	Maize–Peanut
Wheat	Maize–Wheat	Wheat–Raya
Cotton	Cotton–Wheat	Sorghum–Red gram
Rice	Peanut–Wheat	Sugarcane–Beans
Sorghum	Millet–Mustard	Sugarcane–Potato
Groundnut	Sugarcane–Pulses	Toria–Gohbi sarson
Mustard	Soybean–Safflower	Pearl millet–Gram
Gram	Sesame–Chickpea	Sunflower–Cotton
Sugarcane	Sorghum–Gram	Maize–Peas
Soybean	Cotton–Peanut	Maize–Green gram
Cowpea	Cotton–Pigeon pea	Sorghum–Pulse
Chilli	Rice–Potato–Green gram	Cocunut–Marigold
Finger millet	Jute–Rice–Potato	Banana–Papaya–Coffee–Cacao
Red gram	Rice–Potato–Sesame	Corn–Bean–Squash
	Rice–Mustard–Maize	Forage sorghum with silage corn
	Cowpea–Mustard–Jute	Wheat–Gram–Mustard
	Rice–Potato–Groundnut	Maize–Pigeon pea
	Green gram–Maize–Wheat	Maize–Soybean
		Maize–Black gram
		Groundnut–Red gram

TABLE 1.8

Distribution of Most Efficient Cropping Systems in Asia

Monocropping	Sequence Cropping	Intercropping
Rice	Rice–Wheat	Wheat–Gram–Mustard
Wheat	Cotton–Wheat	Maize–Soybean
Cotton	Maize–Wheat	Pearl millet–Gram
Maize	Peanut–Wheat	Sugarcane–Beans
Sorghum	Millet–Mustard	Sugarcane–Potato
Groundnut	Cotton–Peanut	Toria–Gohbi sarson
Mustard	Rice–Maize	Pearl millet–Gram
Millet	Sesame–Chickpea	Sorghum–Redgram
Sugarcane	Sugarcane–Pulses	Sugarcane–Potato

(0.8 million ha), and Nepal (0.5 million ha) (Mishra and Singh, 2012). Both rice and wheat crops feed about 20% of the world's population (Mishra and Singh, 2012). Similarly, cotton–wheat, maize–wheat, peanut–wheat, rice–maize, rice–pulses, rice–rice, and maize–soybean are widely practised cropping systems in many parts, i.e. Pakistan, Nepal, Bangladesh, and India of South Asia.

1.1.11 Cropping System in Pakistan

1.1.11.1 Major Cropping Systems in Pakistan

There are three major cropping systems which are practised throughout the Pakistan. These include cotton–wheat, rice–wheat, and wheat–miscellaneous cropping systems. Rice–wheat cropping system is widely practised in northern areas, i.e. Gujranwala, Mandi Bahaudin, Sheikhupura, Okara, Narrowal,

TABLE 1.9

Major Cropping Systems – Their Distribution, Production, and Constraints in Pakistan

	Cropping System		
Parameter	**Rice–Wheat**	**Cotton–Wheat**	**Wheat–Miscellaneous**
Regions	Gujranwala, Mandi Bahaudin, Sheikhupura, Okara, Narrowal, Jhang, Hafizabad, Lahore, and some areas of Sindh	Multan, Vehari Muzaffargarh Rahim Yar Khan Bahawalpur areas of Punjab, and some areas of Sindh	Northern and southern areas of Punjab, Sindh, KPK, and Baluchistan
Production	Efficient cropping system High yield per unit area	Efficient cropping system High yield per unit area	Some are efficient and many are less efficient High production from less investment
Constraints	i) Increased pest population ii) Residue management iii) Late sowing of wheat iv) Creation of hardpan v) Fertility issue vi) Labour issue	i) Delayed planting of wheat ii) Improper crop nutrition iii) Weed infestation iv) Irregular rainfall	i) Weed management issue ii) Pest management issue iii) Improper crop nutrition iv) Intercultural operation issue v) Crop management issue

and Jhang of Pakistan. Cotton–wheat cropping system is widely practised in southern areas, i.e. Multan, Vehari, Muzaffargarh, Rahim Yar Khan, and Bahawalpur of Pakistan while wheat–miscellaneous cropping system is practised throughout Pakistan. The detail of major cropping systems of Pakistan is enlisted in Table 1.9.

1.2 Impact of Climate Change on Cropping System

1.2.1 Soil Health under Current Cropping System

1.2.1.1 Soil Moisture Contents

Under current climate change scenarios, soil moisture content has been fluctuating because of variability in climate variables. Such as variability in precipitation, temperature, and type of land use causes rapid changes in soil moisture content, since the timescale for response is usually within a few hours. Firstly, the decrease in atmospheric precipitation has resulted in a decrease in water infiltration (I) and water storage (S) in the soil, and plants' water supply. On the other hand, it causes surface runoff (R) in hilly lands with undulating surfaces, consequences of which are water erosion hazard and also the risk of wind erosion for dry surfaces. Ultimately, it reduces infiltration rate and groundwater recharge and increases water loss due to evaporation and transpiration with increasing temperature regime. Temperature increase results in greater evapotranspiration, leading to loss of water from the soil. The integral influence of climate–hydrology–vegetation–land-use changes are reflected by the field water balance and soil moisture regime. Under current cropping systems, the rotation of cover crops with main crops has been practised at a very low rate, which is the main reason behind high rate of evaporation.

1.2.2 Soil Organic Matter

Soil organic matter has been affected negatively under current climate variability and it will continue with climate variability. Arid regions of the globe have been affected more because of higher temperature resulting in higher decomposition of organic matter into the soil. Some researchers have opined that global warming and increasing CO_2 levels in the atmosphere can favor increased plant growth, which in turn could provide more organic matter for the soil. On the other hand a rise in air temperature and that of the soil is consistent with an increase in decomposition and loss of soil organic matter. Thus, there is significant curiosity in the fate of such carbon, particularly the extent to which soils and land use can be used to regulate the sequestration of carbon from the atmosphere or the loss of soil organic carbon to the atmosphere. The balance of opinion currently is that in the absence of mitigating action, losses through organic matter decomposition are likely to exceed levels gained from increased plant growth, thus adding to atmospheric CO_2 levels and the greenhouse gas effect and to lower levels of soil organic matter which is a similar finding as that of Beedlow and coworkers' (2004). Increased water-use efficiency under higher CO_2 conditions will lead to higher productivity, especially in water-limited systems; but the magnitude of the response will depend on other limiting factors such as soil nitrogen.

1.2.3 Soil Microbes

Climate variability has both direct and indirect effects on speeding up or slowing down of terrestrial microbial community composition and their functions. Death and disturbance; negative effect on extinct/shift due reduction in metabolic activity; reduction of biomass, diversity, and composition may lead to negative or positive results on its physiology and greenhouse gases emission. When temperature increases, microbial community structures are distorted and processes like respiration, fermentation, and methanogenesis are also accelerated. The impacts of climate change for biotic and abiotic components are the risk of injury, illness, death from the resulting heat waves; wildfires; intense storms; floods rises; extinction; natural disasters; extreme heat; poor air quality; drought; and spreading and emerging diseases. On the other hand, bacteria, fungi, algae, and archaea are considered as drivers of climate change. These microbes accelerate global warming through organic matter decomposition and increase in the flux of CO_2 in the atmosphere (Swati et al. 2014; Weiman 2015). Microbial decomposition of soil carbon is producing a positive feedback to rising global temperatures. Microbial biomass and enzymes are a powerful tool to stimulate warming because decomposed carbon-based organic matter efficiently releases toxic compounds to environment and prevents from climate change. Temperature directly affects enzyme activity and microbial physiological property. Efficiency of soil microorganisms in using carbon determines the soil carbon response to climate change (Allison et al. 2010; Steinweg et al. 2008). In conclusion, climate variability has reduced microbial concentration in the soil by causing injury, illness, and death.

1.3 Greenhouse Gases Emission from Current Cropping System and Patterns

1.3.1 Causes of Greenhouse Gases Emission

Greenhouse gases (GHGs) emission has increased significantly under current cropping systems due to human activities, natural factors, poor management practices of crop production, and industrial areas. The major greenhouse gases are carbon dioxide, methane, nitrous oxide, and the halocarbons.

i. Emission of carbon dioxide comes from the burning of fossil fuel, decomposition of plant material and organic matter, and deforestation (Davidson and Janssens 2006)
ii. Emission of methane comes from rice production system, fossil fuels production, distribution and combustion, landfills and waste, biomass burning, livestock farming (Singh et al. 2010)
iii. Emission of nitrous oxide comes from agriculture sector during fertilizer use, especially nitrogenous fertilizers and burning of fossil fuels (Sanford et al. 2012)
iv. Poor management practices of crop production such as poor land use, changes in land use, and deforestation are also important sources of greenhouse gases

1.3.2 Negative Impacts of Greenhouse Gases

The emissions of GHGs have resulted in acid rain and Earth's temperature rise (global warming). The destabilization of the global climate destroys natural ecosystem and increases natural disasters, such as violent storms, floods, droughts, etc. The acid rain and these natural disasters affect the agricultural productivity (Valasai et al. 2005). Arid regions are most affected areas due to the emission of greenhouse gases (climate variability).

i. Increase in temperature above the base, but not exceeding optimum temperature, generally leads to lower yield in cereal crop (Fahad et al. 2016a, 2016b, 2016c, 2016d); however, yield in root crops and grassland will increase. Crops should be adjusted according to environmental conditions prevailing to that area.
ii. Higher level of carbon dioxide increases the crop yield if sufficient water and nutrients are available. However, when there is insufficiency of water and nutrients, yields may be reduced.
iii. Elevated CO_2 can reduce protein and nitrogen content in alfalfa and soybean plants, resulting in a loss of quality. Reduced grain and forage quality can reduce the ability of pasture and rangeland to support grazing livestock.
iv. Due to higher temperature the rate of evaporation will increase, this will reduce the moisture availability and hence negatively affects crop yield.
v. According to CERES Crop Model, wheat yields could decline by up to 60% owing to heat stress.
vi. Various studies have discussed that the natural disasters, like droughts, floods, cyclones, hurricanes, typhoons are the main results due to increase in global warming and rapid climate change, which negatively affect crop production throughout the world (Adnan et al. 2018; Akram et al. 2018a, 2018b; Aziz et al. 2017a, 2017b; Rahman et al. 2017; Hafiz et al. 2016, 2019; Muhammad et al. 2019; Sajjad et al. 2019; Saud et al. 2013, 2014, 2016, 2017; Shah et al. 2013; Qamar-Ur et al. 2017; Wajid et al. 2017; Yang et al. 2017; Zahida et al. 2017; Depeng et al. 2018; Fahad and Bano 2012; Fahad et al. 2013, 2014a, 2014b, 2015a, 2015b, 2016a, 2016b, 2016c, 2016d, 2017, 2018, 2019a, 2019b).

1.3.3 Contribution of Green House Gases to Climate Variability

Greenhouse gases are the most noteworthy drivers of observed climate change since the mid-20th century. The atmospheric concentration of carbon dioxide (CO_2) has been increased from 278 ppm in 1750 to 405.5 ppm in 2017 by 146% and is presently increasing at the rate of 2.24 ppm y^{-1} (0.55%) (WMO 2018). The atmospheric concentration of methane (CH_4) has increased from 722 ppb in 1750 to 1,859 ppb in 2017 by 257% and is presently increasing at the rate of 0.38% or 6.9 ppb y^{-1} (IPCC 2018). The atmospheric concentration of nitrous oxide (N_2O) has increased from 270 ppb to 330 ppb at the rate of 0.27% or 0.93 ppb y^{-1}), which has already caused ~1°C (1.8°F) increase in global temperature since the Industrial Revolution (IPCC 2018) with dire consequences as exemplified by the increase in frequency of extreme events throughout the world. The contribution of GHGs towards greenhouse effect can be determined by the characteristics of that gas, its abundance, and any indirect effects it may cause.

1.3.4 Water-use Efficiency and Water Productivity of Cropping System

1.3.4.1 Concept of Wateruse Efficiency and Water Productivity

According to Briggs and Shantz, *water-use efficiency (WUE)* is a concept which shows relationship between plant productivity and water use. They introduced the term, WUE, as a measure of the amount of biomass produced per unit of water used by a plant. Water productivity is generally defined as crop yield per cubic metre of water consumed.

Water productivity with respect to crop production is referred to as crop water productivity (CWP), and is defined as the amount of crop produced per volume of water used.

1.3.4.2 Factors Affecting Water-use Efficiency and Water Productivity

Under current cropping systems, WUE and water productivity have decreased because of various factors. Climate change scenarios have great impact on WUE and water productivity under current cropping systems. Under current climate change scenarios, temperature has increased; hence, evapotranspiration has increased which negatively affects WUE. Soil depth and texture can make a significant difference in efficiency levels. The soil is highly impermeable and there is significantly leakage from the system. These factors account for an overall efficiency level of 70%.

Evaporation and evapotranspiration losses are associated with open canals, irrigated field, and crop growth. There is very low WUE in arid and semiarid regions because of high rate of evaporation and less rainfall. Other factors may include seepage, percolation, soil depth and texture, size of irrigated areas and methods for water application, design of irrigation structures and their operation, and maintenance and management skills.

1.3.4.3 Improvements Regarding Wateruse Efficiency and Water Productivity

There are a lot of issues associated with less WUE and water productivity, and it is necessary to tackle these issues because of decreased water resources under current climate change scenarios.

i. Lining of canals and water courses in order to reduce seepage, percolation, and evaporation losses in irrigated areas
ii. Efficient utilization of water harvesting and conserving strategies in rainfed areas
iii. Development of water-efficient crops (biological measures) under current climate change scenarios
iv. Readjustment of current cropping systems in order to enhance WUE and water productivity, e.g. rice–peanut cropping system should be adopted instead of rice–rice cropping system

Remember that a single approach cannot improve WUE and water productivity under water limited scenarios. Combining biological water-saving measures with engineering solutions and agronomic and soil manipulation can increase WUE and water productivity to some extent.

1.3.5 Fertilizer-use Efficiency in Current Cropping System

1.3.5.1 Concept of Fertilizer-use Efficiency

Efficiency in any system is an expression of obtainable output with the addition of unit amount of input. The NUE/FUE can be expressed in several ways. Mosier et al. described four indices to describe nutrient-use efficiency (NUE) as below Table 1.10.

The efficiency of fertilizer is low such as the efficiency of N = 30%–40% in rice and 50%–60% in other cereal crops, K = 60%–80%, P = 15%–20% and S = 812%. In case of micronutrients, the efficiency of most of them is below 5%.

1.3.5.2 Factors Affecting Fertilizer-use Efficiency

There are various factors and issues which are lowering the fertilizer use efficiency under current cropping systems. Some main issues related to low FUE are enlisted in Table 1.11.

TABLE 1.10

Description of Four Indices Which Describe Efficiency of Fertilizer Nitrogen (NUE)

Indices	Description
Partial factor productivity (PFP)	kg crop yield per kg input applied
Agronomic efficiency (AE)	kg crop yield increase per kg nutrient applied
Apparent recovery efficiency (RE)	kg nutrient taken up per kg nutrient applied
Physiological efficiency (PE)	kg yield increase per kg nutrient taken up

TABLE 1.11

Issues Related to Low Fertilizer Use Efficiency under Current Cropping Systems

Soil Fertility Issues	Agronomic Issues	Seed-Related Issues	Weather-Related Issues
Soil salinity, sodicity	Delayed sowing	Nonavailability of sufficient quantity of seeds	High-intensity rain leading to nutrient loss
Water logging	Low seed rate		
Lack of adequate soil testing facilities	Poor weed management	Lack of high-yielding varieties of crops	Abnormal high/low temperature
Low status of soil organic carbon	Insufficient tillage	Lack of more efficient nutrient using genotypes	
Inadequate P and K application	Large-scale monoculture		
Inappropriate time and method of fertilizer application	Inefficient irrigation		
Inadequate addition of organic manures	Inadequate plant protection		
	Plant competition		

TABLE 1.12

Universal Basic Scientific Principles behind the '4R' Strategy

Nutrient Management Strategy	Universal Basic Scientific Principles behind the Strategy
Right Source	i. Supply in plant available forms ii. Suit soil properties iii. Recognize synergisms among elements iv. Blend compatibility
Right Rate	i. Appropriately assess soil nutrient supply ii. Assess all available indigenous nutrient sources iii. Assess plant demand iv. Predict fertilizer-use efficiency
Right Time	i. Assess timing of crop uptake ii. Assess dynamics of soil nutrient supply iii. Recognize timing of weather factors
Right Place	i. Recognize root-soil dynamics ii. Manage spatial variability iii. Requirements of tillage system iv. Limit potential on-field transport

1.3.5.3 Improvements Regarding Fertilizer-use Efficiency

FUE can be increased by decreasing the losses of nutrients due to: (a) leaching, (b) volatilization, (c) immobilization, (d) chemical reaction between various components in the mixture, (e) change in capacity to supply nutrients, and (f) unfavourable effects associated with fertilizer application. The 4 'R' strategies are basic needs which can decrease the nutrient losses and increase FUE under current climate change scenarios (Table 1.12).

1.4 Biodiversity under Climate Change

1.4.1 Level of Biodiversity

The level of biodiversity can be defined as the total number of genetic characteristics in the genetic makeup of a species. Biodiversity has a vital role in the maintenance of ecosystem because it performs lot of functions, i.e. maintain soil fertility, cycle nutrients, deliver clean water to streams and rivers, act as buffer against pests and diseases, pollinate plants, etc. Climate change is affecting the habitats of several species, which has decreased level biodiversity under current cropping systems. There are many species which have been affected physiologically, which is attributed to their vulnerability to temperature variability under current environmental conditions. Climate change is considered a major global threat, which has several negative impacts on biodiversity and natural ecosystems. All levels of biodiversity from organism to biome have been affected due to climate change.

Several studies have shown that a large number of species have become extinct due to climate change (Loarie et al. 2009; Bellard et al. 2012). For example, Monzón et al. (2011) have studied effects of biodiversity level and shown that 19 species have become extinct due to climate change. On the other hand, International Union for Conservation of Nature Red List of Threatened Species has predicted that 4,161 species are being threatened by climate variability, of which 29% are due to temperature extremes, 28% are due to drought, and 33% due to climate change-induced habitat shifts and alteration (IUCN 2016).

1.4.2 Threats to Biodiversity

Under current cropping systems, biodiversity is declining because of the following main threats to biodiversity: deforestation, desertification, pollution of marine environment, increasing wildlife trade, climate change, and loss of habitat due to human activities, etc. Deforestation and desertification are the direct causes of extinction and loss of biodiversity.

Forest ecosystems contain 80% of the world's terrestrial biodiversity and are the main sources of wood fibre and biomass energy, as well as critical components of the global cycles of water, energy, and nutrient. The degradation of forests represents not only loss of biodiversity but also an enormous potential source of GHG emissions, which is the main cause of climate change.

The climate variability, i.e. increase in the amount of CO_2 in the air, affects the physiological functioning of plant and species composition. Mostly number of species is vulnerable to high temperature, which will increase by 1.7°C by 2100.

Wildlife trade is also considered a main threat to biodiversity depletion. Hence increasing wildlife trade has decreased the biodiversity under current cropping systems.

Pollution of the marine environment; overexploitation of resources (overhunting, overfishing, and overharvesting); and development of critical habitats such as wetlands, mangroves, and water-flow from poor land-use practices are major threats to the aquatic biodiversity.

Poor land management and conversion of forest land into agricultural land have led to desertification process, which directly affects habitats of biodiversity.

1.4.3 The Value of Biodiversity in Agriculture

Biodiversity is considered a sustainable approach for the proper functioning of the ecosystem and for delivering ecosystem services. Thus, the conservation of biodiversity in an agroecosystem makes agricultural production more sustainable and economically viable. Biodiversity performs several functions, i.e. pollination of crops, biological crop protection, protection of soils against erosion, nutrient cycling, control of water flow, and distribution and maintenance of proper structure and fertility of soils, etc. in an ecosystem and enhance the crop production under changing environmental conditions. Similarly, bio diversity makes the agriculture system more profitable business under current cropping systems. On the other hand, conservation of biodiversity might be the sustainable method which can decrease the excessive use of chemical pesticides, fertilizers, water and other agricultural inputs in order to increase the crop production and meet global food demand.

1.4.4 Biodiversity Management

Anthropogenic climate change is threatening biodiversity and the sustained provision of ecosystem services. Hence, there is need to manage the biodiversity in a sustainable fashion under current climate change scenarios. Biodiversity can be saved and managed properly through the conservation of natural habitats, i.e. local coastlines, waterways, wetlands, grassy woodlands, rural roadsides, bush and native grasslands, etc. of biodiversity. Similarly, there is need of additional research and action towards reducing the negative impacts of climate variability on biodiversity and increasing synergy of biodiversity conservation and sustainable use with climate change mitigation and adaptation activities.

1.5 Weeds Flora and Fauna under Current Cropping System

1.5.1 Cropping System and Weed Population

There has been an increased geographical distribution of weed population under current cropping system, which is attributed to variations in climate variables, i.e. increased CO_2 level in the atmosphere. It is estimated that weed population has caused 34% crop production losses across the world. The variations in the crop sequence create differences in resource competition, allelopathic interference, soil disturbances, and mechanical damage to develop an unsuitable ecology for weeds, which may prevent the proliferation of any particular weed. For instance, the rice–wheat cropping system does not favour wild oat ecologically (Walia 2006), which is a serious weed in non-paddy wheat cropping system. Various cultural practices, i.e. crop rotation, tillage systems, application of agrochemicals and other agricultural practices affect the soil seed bank and weed flora.

In some cropping systems, weed population causes more problems, especially in organic farming, where application of synthetic agrochemicals is not allowed. One study has shown that when a farm was converted from conventional to organic farming, the composition of weed seed in the soil seed bank hardly changed after six years (Ahmad et al. 2015). Different crop rotations (crops with different life cycles) play an important role in the reduction of weed seed bank and weed population, i.e. wheat–maize and wheat–sugar beet (Koocheki et al. 2009). In South Asia, *Phalaris minor* has caused severe damage, especially to rice–wheat cropping system.

1.5.2 Harmful Aspects of Weeds Flora

Increased weed population has become a main threat to crop production under cropping systems because weeds compete with the main crop for light, water, nutrients, and space. Inadequate crop rotation and repeated use of same kinds of herbicides have increased weed population. Climate variability is considered another factor which increases the weed growth and development through carbon dioxide fertilization under current cropping systems. It is estimated that weeds and other harmful organisms can compete for light, water, and nutrients and can potentially reduce 10%–40% crop yield on average. Several weeds are considered hosts of many insect pests and pathogens of main crop, which adversely affect crop growth and development. On the other hand, weeds have allelopathic impacts on main crop, hence crop growth is reduced. Weeds are reducing the quality of crop produce. For instance, when winter grains are threshed, weed seeds of wild mustard and sweet clover mix into the main crop. Weeds growing into the water channels have been seen to reduce the water-use efficiency under current cropping systems through both competing for water and causing hindrance to water flow. Similarly, weeds cause hindrance during the use of tillage implements for intercultural operations. It is concluded that weeds are decreasing the crop productivity and increasing the cost of production under current cropping systems.

1.6 Threats and Challenges Due to Insect Pests and Diseases

1.6.1 Estimation of Crop Yield Losses

Climate variability has increased the crop losses because of increased incidences of insect pests and diseases on all crops under current cropping system, especially maize, rice, and wheat across the world (Deutsch et al. 2018). This study has also shown that 10%–25% yields of wheat, rice, and maize are lost to insect pests due to per degree Celsius of warming. It is estimated that 40%–50% of all crop yields are lost to pests, crop diseases, or postharvest losses across the developing world. In United States, it accounts for 20%–25 %. Insect pests, pathogens, and weeds are causing one-third of crop losses due to climate variability globally (Riegler 2018). Globally, an increase of 1°C in temperature can decrease maize and rice yield by 7.4% and 3.2%, respectively (Parmesan et al. 2018). Pests, weeds, and diseases have directly and indirect impacted on crop production because of temperature and precipitation fluctuations across the world (Kistner et al. 2018). Variations in climate variables, e.g. increase in temperature, have decreased wheat crop production by increasing the aphid population and weed infestation (Tian et al. 2019). Climate variability is associated with the geographical distribution of hosts, their insect pests, and pathogens and increases the impacts by both the native and invasive pest species (Ramsfield et al. 2016). On the other hand, continuous cropping system, e.g. rice–wheat cropping system, has improved the conditions for the buildup of plant diseases (leaf diseases and kernel bunt in rice and wheat, respectively). In conclusion, climate variability has decreased the crop production through the provision of favourable environment to insect pests and pathogens under current cropping system.

1.6.2 Estimation of Crop Quality Losses

Climate variability has increased the infestation of insect pests and diseases on the crops because it makes the environmental favourable for the growth and reproduction of insects pests and pathogens across the world. It causes crop losses both quantitatively and qualitatively. Under current cropping systems, crop quality losses are considered more because of increased precipitation, temperature, and carbon dioxide concentration (Rao et al. 2017). Climate variability has led to an increase in viral diseases on wheat crops in Australia. Brown and Funk (2008) discovered that the marmorated stink bug was responsible for the yield and quality losses of maize, which are increasing with climate variability under current cropping system (Cissel et al. 2015). Mostly crop quality losses have been found associated with the horticultural crops, which causes the reduction in nutrition, germination rate, and weight (Yuya et al. 2009).

1.6.3 Pesticide Resistance

Pesticide resistance has become a main threat and challenge to current cropping systems under climate change scenarios across the world (Sparks and Nauen 2015). Climate variables, i.e. temperature, precipitation, and carbon dioxide concentration etc., negatively affect the performance of pesticides through directly altering pesticide metabolism and translocation and indirectly increasing pesticide detoxification in host plants, which reduces the pesticide availability (Matzrafi 2019). Climate variability has made the environmental conditions favourable for the proliferation of insect pests and hence pests become resistant to pesticides through evolving physiological changes.

1.7 Climate Change Scenarios

1.7.1 Future Climate Scenarios across the World

1.7.1.1 Estimation of Future Climate Scenarios

The estimation of future climate change scenarios is difficult due to uncertainty in the projections of climate change. This uncertainty in projections might be due to future anthropogenic and natural forces; incomplete understanding and imprecise models of the climate system; and also due to the existence of internal climate variability. Greenhouse gas (GHG) concentrations continue to rise, we expect to see future changes to the climate system that are greater than those already observed and attributed to human activities. It is possible to understand future climate change using models and to use models to characterize outcomes and uncertainties under specific assumptions about future forcing scenarios (Table 1.13).

TABLE 1.13

Projected Climate Change Scenarios

Climate Variable	Projected Climate Scenarios
Temperature rise	Increase. IPCC projects global mean temperature to increase by 1.5°C by 2030–2052. Regional changes will vary. High latitude and inland areas will likely have more warming than global average.
Precipitation change	Direction generally uncertain. Very high latitude areas and equatorial regions could see an increase. Mediterranean region may see a decrease. Change in other areas is generally uncertain, but more information can be obtained by careful examination of model output.
Global sea level rise	Mean sea level rise will accelerate. Global sea level will rise by about 0.1 m by 2100.
Drought	Increase in most mid-continental areas during summer.
Flooding intensity	Increase in many areas across the globe.

Source: IPCC 2018.

1.7.2 Types of Climate Change Scenarios

Climate change scenarios can be divided into following main categories.

i. **Arbitrary climate change scenarios**
Arbitrary climate change scenarios are the variations in main climate variables selected to analyze the sensitivity of a system to possible changes in climate. These are often identical annual changes in variables, such as temperature and precipitation. An example is combinations of 1°C, 2°C, and 4°C increases in temperature combined with no change and increases and decreases of 10% and 20% in precipitation. Different changes can be assumed for different seasons. These scenarios are most useful for analyzing the sensitivity of systems to changes in individual variables and combined changes. The advantages of these scenarios are easy to create and apply and can represent broad range of potential changes in climate.

ii. **Analogue climate change scenarios**
Analogue climate change scenarios can be predicted from historical instrumental records of climate and from pale climate reconstructions. The instrumental record will often be a complete multidecadal record of often daily or subdaily weather variables. The advantage of these data is that they will be recorded at each observation station and thus could provide better information on regional distribution of climate than many climate models. Another advantage of these scenarios is to capture the climate variability with high spatial and temporal resolution.

Pale climate reconstructions can reflect a wider range of climate conditions than instrumental record can go hundreds or even thousands of years back in time. Their advantage is that by covering a longer period of time than the instrumental record they may include more variance of mean climate conditions.

iii. **Climate model–based scenarios**
A climate model provides mathematical representation of the climate variables. The model can enable us to simulate how global and regional climates may change as a result of anthropogenic influences on the climate. These models can predict climate variability at global and regional scale. Global climate models include from simple, one-dimensional models (MAGICC) to more complex models (GCMs). GCMs can model the atmosphere and oceans, and their interactions with land surfaces.

1.7.3 Future Land-use Scenarios

Land-use activities are the also major driving forces behind the current climate change scenarios across the world. Land-use and land cover variability can modify the surface fluxes of heat and water vapors.

This alteration in the fluxes will affect the atmospheric boundary layer and will provide energy for the initiation of thunderstorms. These thunderstorms will transport heat, moisture, and wind energy to higher latitudes. Thunderstorms occur over a relatively small percentage of Earth's surface; a change in their spatial patterns would be expected to have global climate consequences. Vegetation and soil typically act as carbon sink and both store carbon dioxide that is produced through the process of photosynthesis. On the other hand, the conversion of forest to agricultural land leads to local temperature increases comparable to that simulated as being due to the radiative effect of the addition of carbon dioxide to the atmosphere.

In a region, type of vegetation can cause the climate variability through known effect of reflection of sunlight back into the space. For instance, the change from a dark forest to grassland makes the surface lighter, and causes it to reflect more sunlight: an increase in albedo. Humans change the land surface mainly to create more agricultural land. Since the preindustrial era, albedo increased due to land-use change, which has a cooling effect on the planet.

Conventional tillage practices are also responsible for the emission of GHGs into the atmosphere, which will increase mean surface temperature and other climate variables. The forest and accurate land-use measures can reduce net carbon emissions equivalent of 10%–20% of projected fossil fuel emissions through 2050.

1.7.4 Future Water-use Scenarios

Global precipitation is increasing, which is attributed to increased global mean surface temperature, and will continue to increase until the mean surface temperature will increase. The relationship between global precipitation and global temperature is approximately linear. According to the assumptions of the concentration-driven RCPs, global mean precipitation will increase by 1%–3%/1°C for scenarios except RCP2.6 for which rate of rise is 0.5%–4%/1°C at the end of the 21st century. The general pattern of change indicates that high latitude land masses are likely to experience greater amounts of precipitation due to the increased specific humidity of the warmer troposphere and increased transport of water vapour from the tropics by the end of this century under the RCP8.5 scenario. Many mid-latitude and subtropical arid and semiarid regions will likely experience less precipitation and many moist mid-latitude regions will likely experience more precipitation by the end of this century under the RCP8.5 scenario.

Annual surface evaporation is projected to increase as global temperatures rise over most of the ocean and is projected to change over land following a similar pattern as precipitation.

Decreases in annual runoff are likely in parts of southern Europe, the Middle East, and southern Africa by the end of the 21st century under the RCP8.5 scenario. Increases in annual runoff are likely in the high northern latitudes corresponding to large increases in winter and spring precipitation by the end of the 21st century under the RCP8.5 scenario.

1.7.5 Projection of Change in Temperature

Global mean temperature is rising which is attributed to the emission of GHGs, and it will continue to rise until the emission of GHGs. Temperature change will not be regionally uniform. According to the Intergovernmental Panel on Climate Change (IPCC), global mean temperature will increase to 1.5°C between 2030 and 2050 if it continues to increase at the current rate (IPCC 2018). According to the assumptions of the concentration-driven RCPs, global mean temperature will be 0.3°C to 1.7°C (RCP2.6), 1.1°C to 2.6°C (RCP4.5), 1.4°C to 3.1°C (RCP6.0), 2.6°C to 4.8°C (RCP8.5). Hot extremes of different frequency, duration, and magnitude along with heat stress can be expected. On the other hand, occasional cold winter extremes will continue to occur (Collins et al. 2013).

1.7.6 Projection of Change in Sea Level

An average global increase in sea level by 0.1 m by 2100 has been projected, which is because of increased GHG concentrations (IPCC 2018). Sea level rise is not uniform around the globe, and its magnitude and rate of the rise depend upon the emission pathways. The increase in sea level will vary from region to region which will be due to regionally varying rates of thermal expansion of the oceans and regional differences in atmospheric circulation. SIMCLIM Sea Level Generator developed at the International Global Change Institute (IGCI) is software which has the ability to generate future scenarios of sea level change in the world.

Future sea level scenarios will have a number of negative impacts on an ecosystem. These negative impacts might be changes in surface water quality and groundwater characteristics, increased flood risk and potential loss of life, higher storm-surge flooding, impacts on agriculture and aquaculture through decline in soil and water quality, increased coastal erosion, and inhibition of primary production processes.

1.7.7 Projection of Change in Greenhouse Gases

GHGs from human activities are the most significant drivers of observed climate change since the mid 20th century. The level of GHGs is increasing, which is attributed to various factors such as burning of fossil fuels, deforestation, and emission of nitrous oxide and methane from agriculture sector. According to the IPCC, limited reductions of about 35% of methane and black carbon by 2050 relative to 2010 have been reported. The increasing demand on bioenergy will increase emissions of nitrous oxide (IPCC 2018).

1.8 Climate Change Scenario in Pakistan and Cropping System

1.8.1 Factors Contributing Climate Change Scenarios

In Pakistan, climate variability has become a main threat to the crop production under current cropping systems. There are various factors, i.e. burning of fossil fuels, desertification, deforestation, increasing livestock farming, excessive use of foliar sprays and synthetic fertilizers, and emission of GHGs, which are contributing towards climate change scenarios in Pakistan. The burning of fossil fuels is considered a main factor which has added a large quantity of carbon dioxide – about 40% higher – than at the beginning of industrialization. Similarly, other GHGs, i.e. nitrous oxide and methane, have been emitted in small quantities and these gases are trapping heat more effectively. Deforestation is the second main factor that leads to carbon emission and loss of habitat for large number of plants and animal species. Under current cropping system, excessive application of foliar spray and synthetic fertilizers is increasing the concentration of GHGs in the atmosphere.

1.8.2 Impact of Present Climate Change Scenarios on Crop Production

Climate variability has caused many social, environmental, and economic impacts under current cropping systems. Climate variability has driven many weather extreme events, i.e. drought, flooding, and attack of insect pests on crops (McDowell et al. 2018). In Pakistan, flooding and drought are main threats to crop production during summer months. Every year, thousands of hectares are damaged because of flooding. It reduces the soil fertility level because excessive water causes runoff of nutrients from the soil surface. The arid and semiarid regions of Pakistan are under threat because these regions are already facing water shortage and high temperature. Drought and heat stress affect adversely various stages of pulses crop growth such as flowering, pod development, and pod filling, which are most sensitive to drought and heat stress (Sehgal et al. 2019). In Pakistan, increased temperature is reducing the sunflower production because it reduces the sunflower duration to remain green that finally leads to less leaf area index (LAI), achene yield, and total dry matter production (Awais et al. 2018). Climate variability has increased the intensity of rainfall which causes great damage to younger plants, promotes ripening, grain lodging in standing crop, and depletes soil fertility. It can also affect negatively on the yield and quality of fruits and vegetables. High temperature is favourable for the outbreak of insect pests because pests have become more active and have an expanded geographical range. Thus excessive use of pesticides along with other agricultural inputs has increased the cost of production under current cropping systems. Similarly, high temperature and increased precipitation cause the spread of diseases because wet vegetation promotes the germination of spores and the proliferation of fungi and bacteria, which negatively affects crop production. On the other hand, the population of nematodes and roundworms has increased in the soil because of increased soil moisture under current environmental conditions. Climate variability has caused severe postharvest damages of agricultural products because of high concentration of mycotoxin which is favoured by high humidity and temperature. Climate variability is negatively affecting the cotton yield and quality (Amin et al. 2018). In conclusion, climate variability is reducing the crop yield and quality, and making the crop production costlier.

1.8.3 Projection of Future Climate Change Scenarios

According to the report of Asian Development Bank, the increase in temperature will be as high as 4.38°C in Pakistan and regional changes in temperature will be different. In both summer and winter, the rise in temperature will be higher in southern areas of Pakistan. The level of precipitation will be more in summer and low in winter in southern areas of Pakistan. Winter precipitation is projected to decrease particularly in the southern parts of the basin with greater warming in the winter than in other seasons of the year. There will be more number of rainy days, an increase in rainfall intensity, and high volume of rainfall.

According to Pakistan Meteorological Department, the increase in temperature will be 3°C–5°C (RCP4.5) and 4°C–6°C under Representative Concentration Pathways (RCP)4.5 and RCP8.5, respectively, by the end of 2050. RCP8.5 also indicates a more abrupt increase in temperature in the region after 2060.

According to model, snow-covered areas of Pakistan show a larger increase in mean temperature compared to central and southern regions under both RCP scenarios. However, RCP8.5 shows a more abrupt increase in temperature in the region after 2060. In northern areas, the temperature increase will be 10°C–12°C whereas RCP4.5 shows a similar increasing trend with less intensity (5°C–6°C). Similarly, RCP4.5 and 8.5 indicate that there will be increase in annual mean precipitation by 4 mm/day and 2–3 mm/day, respectively, until the end of 2050 and 21st century.

In Pakistan, it is difficult to project sea level rise by the end of 2100 because there is limited availability of data at country level (Chaudhry, 2017; Basir et al. 2015). In the past century, the rise in average mean sea level was 1.1 mm/year for Pakistan. At global and regional levels the sea level projections can be useful in determining the extent of risks of Pakistan in future. Pakistan is located in South Asia where sea level rise will be 0.2–0.6 m by the end of this century. In Pakistan, future sea level rise will most likely affect the low-lying coastal areas south of Karachi and Indus River delta and many other southern areas of Pakistan.

1.9 Adaptation Technologies and Strategies for Sustainable Production of Cropping System

1.9.1 Agronomic Management Strategies

1.9.1.1 Conservation Tillage

Conservation tillage may be defined as a broad set of practices with a goal leaving some crop residue (over 30%) on the soil's surface to increase water infiltration and reduce erosion (SSSA 1997). It is considered most important component of sustainable agriculture. Conservation tillage is an agricultural management approach that aims to minimize the frequency or intensity of tillage operations in an effort to promote certain economic and environmental benefits. With this technique, at least 30% of the soil surface is covered with crop residue/organic residue following planting (Dinnes 2004). Conservation tillage can be divided into the following major categories: reduced tillage, mulch tillage, strip tillage, ridge tillage, and no tillage systems (Tables 1.14 and 1.15).

1.9.1.2 Crop Rotation

Crop rotation is the practice of cyclically growing a sequence of different plant species on the same piece of land following a defined order of the crop succession with a fixed length (Stein and Steinmann 2018;

TABLE 1.14

Comparison of Various Types of Conservation Tillage

Mulch Tillage	Ridge Tillage	Strip Tillage	Reduced Tillage	Zero Tillage
Tillage that leave >30% residue.	Tillage that leave 30%–50% residue.	Involves tilling the soil only in narrow strips with the rest of the field left untilled (strip-till).	Tillage that leave 15%–30% residue.	Involves planting of crops directly into residue that hasn't been tilled at all.
Soil is disturbed prior to planting	Soil undisturbed from planting to harvest except for nutrient injection.	Soil is disturbed only in narrow strips	Soil coverage after planting	Planting or drilling in a narrow seedbed
Chisels, field cultivators, disks, sweeps, or blades are used.	Sweeps, disk opener, coulters are used.	Disk opener and row cleaner are used.	Chisel plow and field cultivators are used.	Coulters, row cleaners, disk openers are used.
Weed control is accomplished with herbicide.	Weed control is accomplished with herbicides and/or cultivation.	Weed control is accomplished with herbicide.	Weed control is accomplished with herbicide.	Weed control is also accomplished with pre- and postemergence herbicides.

TABLE 1.15

Advantages and Disadvantages of Conservation Tillage

Advantages	Disadvantages
Increases carbon sequestration.	Increases soil bulk density.
Improves soil water infiltration and reduces water erosion.	Decreases soil temperature.
Improves the stabilization of soil surface.	Specialized, expensive equipment is required
Reduces leaching of nutrients.	Much hand labour in the case of very small-scale growers.
Decreases evaporation and increases soil moisture retention.	Requires more herbicides and pesticides.
Reduces the cost of fossil fuel and the associated carbon emissions to the atmosphere.	A significant amount of time is required to detect changes.

Hadi et al. 2014; Leteinturier et al. 2006). The selection of optimal crop rotation is important for successful sustainable agriculture, e.g. cotton–gram, sugarcane–wheat, etc. Legumes are essential in any rotation and should comprise 30% to 50% of the crop land. A mixed cropping, pasture and livestock system is desirable or even essential for the success of sustainable agriculture.

Rotation of different crops can be used for the breaking of pest cycles and addition of extra nutrients into the soil. Crop rotations can build soil fertility; preserve the environment; control weeds, diseases, and insects; and add to crop and market diversity (Baldwin 2006) (Figure 1.1).

Crop rotation should be done in following ways;

i) Deep rooted crops should be rotated with shallow rooted crops, e.g. cotton, castor, pigeon pea–potato, lentil, green gram.
ii) Exhaustive crops should be rotated with restorative crops, e.g. potato, sorghum, sugarcane, castor–sunhemp, black gram, cowpea.
iii) Cover crops should be included in crop rotation.
iv) Dicot crops should be rotated by monocot crops such as mustard, potato–rice, wheat–sugarcane.

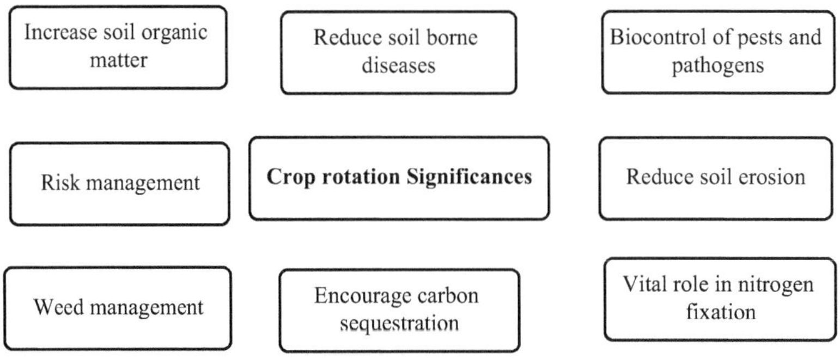

FIGURE 1.1 Significance of crop rotation in a particular cropping system.

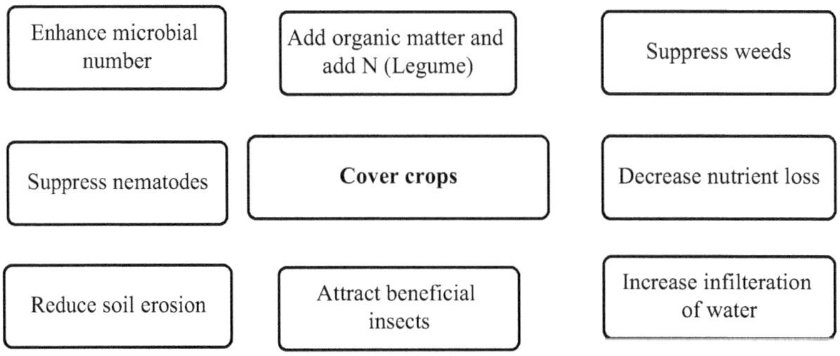

FIGURE 1.2 Significance of cover crops in a particular cropping system.

The rice–wheat crop rotation occupies about 18 million hectares area in Asia, of which 75% are in the Indo-Gangetic Plains of India (10 million ha), Pakistan (2.2 million ha), Bangladesh (0.8 million ha), and Nepal (0.5 million ha) (Mishra and Singh, 2012). Both rice and wheat crops feed about 20% of the world population (Shah et al. 2016; Mishra and Singh, 2012).

1.9.1.3 Cover Crops

A *cover crop* is planted to manage soil erosion, soil fertility, soil quality, water, weeds, pests, diseases, biodiversity, and wildlife in an agro ecosystem (shown in Figure 1.2). Under climate scenarios, the crop production is severely affected due to poor soil quality, weeds infestation, insect pest and disease attack due to climate variability. Among winter legumes, subterranean clover, crimson clover, field peas, and hairy vetch are the most important cover crops. Similarly, berseem clover, cowpeas, soybeans, velvet bean are summer annual legumes. Alfalfa, crown vetch, red clover, sweet clover and white clover are biennial and perennial legumes which are used as cover crops. These cover crops should be incorporated in current cropping systems in order to maintain crop production and to prevent soil from degradation, severe attack of insect pests and diseases, and infestation by weeds due to climate variability. The incorporation of cover crops in current cropping systems is considered a best strategy for sustainable production of cropping systems.

1.9.1.4 Change in Cropping Pattern and Calendar of Planting

The change in cropping pattern and crop calendar is considered a sustainable approach towards decreasing the negative impacts of climate variability and improving crop production. Cropping pattern involves changing the crops which are adversely affecting the soil health to crops having ability to improve soil health under current environmental conditions, such as adjustment of exhaustive crops with restorative crops (potato, sorghum, sugarcane, black gram, and cowpea). Similarly, change in calendar of planting is the best strategy that can prevent from many environmental risks (drought, insect pests attack) under changing environmental conditions, such as early planting of oilseed crops can prevent from flea beetle, early planting of corn is less susceptible to earworm and stem borer (Dosdall and Stevenson 2005).

1.9.1.5 Conservation of Soil and Water

Under rainfed areas, soil degradation is a big issue which not only depletes soil fertility but also makes the lowland areas unproductive. Because of climate variability, irregular rainfalls can cause severe damage to soil quality through soil erosion. Thus the conservation of soil and water has become a requisite of the farming community in a sustainable fashion. The most important soil and water conservation measures/approaches are:

Agronomic measures: These measures reduce the impact of raindrops through interception and increases infiltration rates. These also reduce the surface runoff and soil erosion. Strip cropping, mix cropping, intercropping, fallowing, mulching, contour ploughing, grazing management, and agroforestry are agronomic measures which can save soil and water for sustainable crop production under climate change scenarios especially under rainfed areas.

Biological measures: These measures provide protective cover of vegetation for water and soil conservation. These measures can prevent from splash erosion, reduce the velocity of surface runoff, and facilitate the accumulation of soil particles. The roots and organic matter stabilize the soil aggregates and increase infiltration rate. Vegetative strips, e.g. acacia line, protective bush land and reforestation, are biological measures. In biological measures, the incorporation of cover crops is considered an important approach towards reducing the erosion and maintaining soil fertility under current climatic conditions.

Physical measures: These measures provide physical structure for water and soil conservation. Stone terraces/Bunds, check dams, contour ditches, dams, retention reservoirs, planting pits, and grassed waterways are physical measures that prevent soil and water loss.

1.9.1.6 Use of Organic Fertilizers

The use of artificial fertilizers causes the decline in soil health, which adversely affects humans, animals, and plants. Nitrogenous fertilizers cause diseases in human beings and animals by polluting ground water. Organic fertilizers can perform various functions that cannot be performed by manmade artificial fertilizers such as improvement in chemical properties of soil (Chrispaul et al. 2010). Therefore, several farmers have adopted extensive use of organic sources of nutrient for crop production such as green manures, animal dung, composted materials, and domestic waste due to high prices of synthetic fertilizers (Fening et al. 2010). The gradual increase in soil organic matter and nutrients supply are the major roles of organic resources in soil (Bhattacharyya et al. 2009) and comparison between organic and synthetic fertilizers has been shown in Table 1.16. Farm yard manure, compost, poultry manure, crop residues, and other natural resources are widely used organic sources of nutrients across the world. Organic sources of nutrients such as biochar, farmyard manure, etc. decrease GHG emission and enhance the carbon sequestration, which helps to reduce the global warming (Lehmann et al. 2003). Under current climate change scenarios, it is necessary to encourage the use of organic sources of nutrients in order to sustain environment and improve crop production. Comparison of organic and synthetic fertilizers has been listed in Table 1.16.

1.9.1.7 Integrated Pest Management

Integrated pest management (IPM) in crops is an important strategy for insect pest control, which plays an important role for sustainable agriculture (Wu et al. 2009). IPM is a pest management system which includes cultural, mechanical, biological, and chemical methods used to reduce pest population and minimize their economic damages (Table 1.17).

TABLE 1.16

Comparison of Organic and Synthetic Fertilizers

Organic Fertilizers	Synthetic Fertilizers
Derived from byproducts of living organisms	Derived from the petroleum industry
Nutrient levels can fluctuate and are generally lower than synthetic fertilizers	Nutrient levels can be formulated for specific needs
Micronutrients are typically present, which are essential for plants	Generally don't contain micronutrient
Soil microbes release the nutrients	Water releases the nutrients
Release rate is slow	Release rate is quick and short
Plant growth over time	Rapid plant growth
Require fewer applications	Require more frequent application
Reduce the chances of leaching or runoff	More chances of leaching of nutrients

TABLE 1.17

Description of IPM

Cultural Methods	Biological Methods	Mechanical Methods	Chemical Methods
Suppression of pest population by making the environment less favourable for survival, growth, and reproduction of pest species.	Suppression of pest population by using biological agents and natural products.	Suppression of pest population by using mechanical means.	Suppression of pest population by using synthetic pesticides.
It includes the selection of crop field, variety, crop rotation, deep ploughing, and time of planting.	It includes the predators and parasites such as *Trichogramma* species, dragonfly and ladybird beetle.	It includes the physical means such as various types of traps (sex pheromones).	It includes the seed treatment of fungicides and foliar spray of pesticides.
Crop rotation and early planting of crops are the best strategies for the suppression of soil-borne diseases such as root-knot nematodes which is a well-known constraint in cotton production across the world (Starr et al. 2005; Draz 2009).	In India, the use of neem products/neem-based pesticides and release of *Trichoderma* species can be used a sustainable strategy which has decreased insect pests (Singh et al. 2008).	Sex pheromone traps and light traps are successfully used mechanical methods in insect pest management across the world.	Various studies have shown that seed treatment is a sustainable strategy for the control of insect pests and diseases across the world such as cottonseed treatment with neonicotinoid (North et al. 2017; de Medeiros et al. 2015).

IPM system consists of four general guidelines:

i) To analyze the pest status of each potentially damaging organism to establish economic thresholds
ii) To devise strategies to increase biotic factors to control pest reproduction
iii) Use the best combination of natural enemies, resistant plant varieties, and environmental modification to control pest outbreaks with minimum ecological disruption
iv) To check systematically and regularly insects, diseases, and weed population.

Under current climate change scenarios, the only use of synthetic products for the control of insect pests is risky because it contributes to the emission of GHGs. Therefore, sustainable pest management of insect pests has become necessary for the control of pest population in order to mitigate the risks of climate change and maintenance of natural biodiversity. Integrated pest management (IPM) in crops is considered a best sustainable pest management strategy across the world.

1.10 Breeding and Biotechnology Strategies

1.10.1 Development of Drought-tolerant Varieties

The cropping system including crops which have more water requirement are more prone to drought stress. This cropping system has to adjust in order to improve crop production under current climate change scenarios, but it is a temporary solution. E.g. rice–rice is an efficient cropping system for total yield, but it consumes large amount of water especially in summer. If water is scarce in summer instead of rice, groundnut is used in the cropping system. The best strategy is to develop water-saving and drought-resistant crops to cope with the water shortage and increasing demand for food production in the future. Different drought tolerant varieties of mungbean have been developed for the sustainable production of pulses across the world (Reyes et al. 2017).

1.10.2 Development of Disease-resistant Varieties

Diseases infestation has been observed more in sequence cropping due to continuous cropping. One way to avoid crop damage is to reshuffle cropping system and provide awareness to farmers about crop rotation which is considered the best approach towards reducing infestation of pathogens. Some pathogens that cause diseases can survive in the soil from year to year in one form or the other, usually as sclerotia, spores, or hyphae. Rotation of host crops with nonhost crops prevents the buildup of large populations of pathogens. Similarly, development of disease-resistant varieties is considered a sustainable approach towards crop production under changing climate scenarios. Varieties of alfalfa, common bean, soybean, cowpea, peppers, and tomato are available with resistance to certain root-knot nematode species.

1.10.3 Development of Insect Pests–resistant Varieties

Insect pest infestation is found more in sequence cropping due to continuous cropping, as such a cropping system provides shelter and food to the insect pests of previous crops. In such situations, the development of insect pests–resistant varieties becomes necessary in order to fulfil the increasing food demands of ever-increasing population under current environmental conditions. Various pest-resistant varieties of annual crops have been developed, e.g. Bt maize is resistant to maize weevil (Gassmann et al. 2011).

1.10.4 Development of Short-duration Varieties

Development of short-duration varieties is the most important approach, which can prevent from many damages such as from insect pests attack, drought stress, etc. Various short-duration varieties of rice, chickpea, and cassava have been developed with resilience to disease, cold, and other such problems (Iftekharuddaula et al. 2016; Gowda et al. 2015). In Bangladesh, early maturing submergence-tolerant varieties of rice have been developed in order to avoid incidences of cold and diseases later in the season (Iftekharuddaula et al. 2016). RRI dhan33 is an early-maturing rainfed lowland rice variety with 118-day growth duration. Similarly, different early maturity varieties of chickpea have been developed for sustainable crop production.

1.11 Precision Farming and Precise Application of Resources

Precision farming is conceptualized by a system approach to reorganize the total system of agriculture towards a low-input, high-efficiency, sustainable agriculture (Cook and Bramley 1998). It has been used for the evaluation of how the small pieces of land or plots have different production environments and yields in Asia (Farid et al. 2013; Chandel et al. 2014; Shivanna and Nagendrappa 2014). Precision farming has become the requisite of agriculture because of global warming; climate change; reduction in the water resources and soil quality; health hazards due to the excessive application of insecticides, pesticides, and fertilizers to meet the needs of the growing population. These issues can be handled by using fewer inputs, eliminating health hazards, making the environment safer by lowering the use of insecticides, pesticides, and fertilizers. It becomes possible when we know about the potential of specific land and its composition of all nutrients necessary for a better yield and soil water requirements.

Under current climate change scenarios, this modern technology can be successfully applied in farming by using computers, global positioning systems (GPS), telecommunications, variable rate technology farm implements, geographic information systems (GIS), machine guidance, and remote sensing (Zhang et al. 2002). All these equipment coordinate with one another and make agricultural practices very easy and more input responsive. With the help of these modern technologies, precision farming is capable of collecting more comprehensive data on production variability in both space and time. This variability in the crop field can be responded positively in proper time which is considered a major property of precision farming. For the precisely application of available resources, precision farming combines the following technologies.

1.11.1 Variable Rate Technology

Variable rate application (VRA) in precision farming is an area of technology that focuses on the automated application of materials to a given field. The technology in which the materials are applied is based on data that is collected by sensors, maps, and GPS. These materials may be fertilizers, chemicals, and seeds, and hence this technology helps in the accurate application of these materials to the field.

Map-based VRA adjusts the application rate based on an electronic map. With the help of GPS and prescription maps, variable rate technology can measure the field position and hence the concentration of agricultural inputs required by the crop is changed as the applicator moves through the crop field. There is no need of maps and positioning system in sensor-based variable rate applications. Sensors on the applicator measure soil properties or crop characteristics 'on the go.' Based on this continuous availability of information, a control system estimates the input requirements of the soil or plants and transfers the information to a controller which makes sure the availability of the inputs to the location measured by the sensor. This variable rate technology can be a sustainable approach towards enhancing the crop production and decreasing the cost of production due to accurate application of all agricultural inputs to the fields. Spray nozzles having sensors shut down when there is no vegetation under the nozzles and it reduces the wastage of fertilizers and pesticides. It decreases the bad environmental impacts; input used like fertilizers, pesticides, insecticides, etc.; and cost of production (Hedley 2015).

1.11.2 Global Positioning System Soil Sampling

Global Positioning System (GPS) is a satellite network revolving around the globe. This system operates by determining reflecting GPS signals from Earth's surface. It requires satellites of GPS which provide radar transmitter and GPS receiver. The position of radar transmitters and GPS receiver should be co-located, but the angle of incidence may vary over the selected points of soil but in general it is above 30 degrees (Privette et al. 2011). Mapping and determining variations in soil properties of a field needs accurate information of position from where samples were taken. In GPS soil sampling, grid method of sampling is used and location of sampling point should be accurate to develop soil data layer.

Autosteer and sensors with GPS system ensure the automatic application of agricultural practices so farmers can also work at night in the fields. There is precise seeding in the fields by the seed drill controlled by the GPS system. Maps of the soil give all information about the inputs to be supplied in the soil; hence, farmers can apply nutrients more precisely and timely and can reduce wastage of nutrients (Srinivasan 2006; McBratney et al. 2005).

1.11.3 Geographical Information System

Geographical Information Systems (GIS) is able to manage large data of soil. GIS technology has been successfully used for the management of cotton and many other crops in China and many other countries of the world. In China, GIS was used for the evaluation of hail disaster risk management to the cotton crop. Under climate change scenarios, organic matter content and nutrient status of soil are very important for sustainable production (Senthurpandian et al. 2010).

GIS can produce soil fertility status spatially as well as temporarily for site-specific management of resources (nutrients, water) for sustainable crop production (Patil et al. 2011). For this purpose GIS technology can be used for evaluation of soil variability (Sarmah and Astec 2018). GIS has enabled researchers and other scientific organizations to study the soil properties through spatial visualization using vector and raster maps (Glowienka et al. 2016).

1.11.4 Computer based Applications

There are many kinds of information technologies, including radars, mobile/telephones, FAX machines, computers, and satellites, etc. in the world. These technologies contribute many forms of information systems such as systems for information reclamation and systems that help us to solve problems or make decisions. An information retrieval system (IRS) is an environment of people, technologies, and procedures (software) that help find data, information, and knowledge resources that can be located in a particular library or, for that matter, anywhere they exist. Information about available resources is acquired, stored, searched, and retrieved when it is needed. These IT-based technologies are useful for the cultivation of crops. The amount of water sprinkled in a balanced quantity is also computerized. The production capacity of the farm has been increased owing to use of computer in agriculture field. There are lesser losses due to work being monitored by computer. By using computer in traditional field like agricultural field, we can increase the productivity and minimize the errors.

1.11.5 Remote Sensing Technology

Remote sensing is the process of detecting and monitoring the physical characteristics of an area by measuring its reflected and emitted radiation at a distance from the targeted area. These radiations travel from substance/source directly through space vacuum or are reflected indirectly and captured by specific sensors. There are different reflectance properties of different substances such as rock, soils, water, vegetation, etc. Hence, remote sensing can be used for the characterization of soil attributes (Grunwald et al. 2014). The nature and shape of spectral reflectance curve of a soil depends on chemical and physical properties of soil (Wadodkar et al. 2014). Satellite imagery provided by remote sensing can be used for the evaluation of requirements of different inputs.

With the help of remote sensing, soil fertility maps preparation has been made for the spatial valuation of soil fertility in sense of nutrient deficiency. Remote sensing can be used satisfactorily for studying soil properties such as soil crusting, salinization, texture, moisture, and in mineralogy to classify the soils.

1.11.6 Crop Simulation Models

Various Crop Simulation Models (CSMs) can be used for the simulation of growth, development, and yield of crops for different weather and soil conditions and management practices (Hoogenboom et al. 2004). CSMs are the primary tools available to assess the potential consequences of climate change on crop production and informative adaptive strategies in agriculture risk management. Several studies have been carried out on the use of crop models across the world. Under current climate change scenarios, application of crop models has become necessary due to climate variability. Various crop models, DSSAT, APSIM, CROPGRO, EPIC, and CERES, are available that can be used for the prediction of crop yield under future climate. Hence, these crop models can help in the development of virtual cultivars according to predicted future climate scenarios. These virtual cultivars can perform well under future climate conditions because crop models can predict crop yield as a function of weather parameters, soil conditions, and management practices. Hence current cropping systems can be adjusted on the basis of variations in weather parameters and soil conditions.

1.12 Application of Decision Support Systems for Sustainable Crops Production

Computer systems that provide users with support to analyze complex information and help to make decisions are called decision support systems (DSSs). These are information systems with a specific function to help people with problem solving to some extent and decision-making process. This system consists of a collection of people, procedures, software, and databases with a purpose. In such systems, computer is the most important technology.

1.12.1 Decision Support System in Crop Productivity Improvement

Decision support system (DSS) has been widely used for the better management of agricultural operations and activities across the world. Crop productivity can be estimated with reliable results using DSS. 'Crop Environment Resource Synthesis (CERES) – Wheat' is a part of DSS that has been successfully used to simulate the crop growth and development of wheat under variable climatic, water, and nitrogen levels in semiarid and subtropical regions in India. Cropping System Model (CSM) named as 'DSSAT-CSMCERES-Wheat 4.0' has been used for the estimation of crop yield, evapotranspiration, crop water productivity (CWP), and irrigation water productivity (IWP). In India, CROPGRO – Soybean model has been used successfully to simulate climatic change, growth, and yield of soybean under climate change scenarios. Different decision-support systems can be used for analyzing the crop productivity under current and future climate change scenarios in order to sustain crop productivity and input resources.

1.12.2 Decision Support System in Crop Water Requirements Management

DSS can be applied for the efficient management of water resources and crop water requirements. Crop water requirements depend on factors such as evaporation, evapotranspiration, meteorological factors such as solar radiation, air temperature, humidity, wind speed, etc. CROPWAT is a DSS which has been used to decide irrigation scheduling for different crop patterns and to calculate crop water requirements across the world. CROPWAT 8.0 has been also used in the assessment of reference evapotranspiration under temperature conditions and best results have been obtained when applied to simulate different crop water requirements as per the need of crop under different planting dates and probable canal water supplies. Similarly, DSS acts as a helpful tool for the estimation of spatial variation of climatic water balance, probabilistic monthly monsoon rainfall, and mapping of cold periods in the specific region.

1.12.3 Decision Support System in Irrigation Scheduling

Irrigation scheduling is one of the most important operations that can be used for the sustainable utilization of water with the help of DSS. In India, a DSS tool has been applied for the Simulation of Water and SALT (SWASALT) under irrigated areas and semiarid regions. This model has been successfully used to prevent farm water logging and soil salinization because of canal irrigation. Similarly, this model can also be used to calculate Water Management Response Indicators (WMRI). This indicator has the ability to optimize the on-farm irrigation schedule by minimizing the percolation losses of ground water for different soil types. Different models can be used for the efficient utilization of water on the farm, and these models include canal flow model, water balance model, GIS-based integrated model of rainfall and groundwater flow model. These models have been successfully used to increase crop production for different cropping patterns under a variety of climatic conditions.

1.12.4 Decision Support System in Advisory System

DSS is considered a successful tool in advisory system because it has helped to improve farm productivity by delivering high-quality farm-specific agroexpert decisions in a timely manner across the world. DSS can be used for the better management of crops, as it helps in taking decisions from sowing to harvesting, which reduces the cost of production and increases farm productivity as well as quality of agricultural commodities. Various DSS tools can be used for proper nutrient and pest management advice for grape farms through mobile phones such as 'mKRISHITM'. DSS can be used for the sustainable IPM by creating continuous awareness among farmers and in turn to improve crop productivity which is considered an important operation in agriculture sector.

1.13 Conclusion

The production of current cropping systems is decreasing, which is attributed to climate change scenarios and less adaptation of climate-resilient technologies and strategies across the world. Climate-resilient technologies and strategies and reshuffling of cropping systems on the basis of climate change scenarios can increase the production of current cropping systems under climate change scenarios. These climate-resilient technologies and strategies can be categorized into agronomic management strategies, breeding and biotechnology strategies, decision support systems, and modern concepts and tools. Agronomic management strategies, i.e. conservation tillage, crop rotation, cover crops, use of organic amendments, integrated nutrient management (INM), integrated pest management (IPM), etc. can mitigate the risks of climate variability and also improve production of current cropping systems under climate change scenarios. Breeding and biotechnology strategies, i.e. development of disease-resistant varieties, insect pests–resistant varieties, drought-resistant varieties, and short-duration varieties can improve production of current cropping systems and prevent crop yield and quality losses from climate-driven severe attack of insect pests, diseases, and drought stress. Decision support system (DSS) can be a useful tool for decision making and precisely application of agriculture resources in different cropping systems. Modern concepts and technologies, i.e. global positioning system (GPS), geographical information system (GIS), remote sensing (RS), variable rate technology (VAT), crop simulation models (CSM), can be used to assess and manage the variability efficiently prevailing in different cropping systems under climate change scenarios. In conclusion, there is need to adopt all-above discussed climate-resilient technologies and strategies for the sustainable production of current cropping systems; conservation of soil, water, biodiversity, and other agricultural resources; and improving the environmental quality. On the other hand, there is also need to readjust the current cropping systems according to the climate change scenarios.

REFERENCES

Adnan M, Zahir S, Fahad S, Arif M, Mukhtar A, Imtiaz AK, Ishaq AM, Abdul B, Hidayat U, Muhammad A, Inayat-Ur R, Saud S, Muhammad ZI, Yousaf J, Amanullah Hafiz MH, Wajid N (2018) Phosphate-solubilizing bacteria nullify the antagonistic effect of soil calcification on bioavailability of phosphorus in alkaline soils. *Sci Rep* 8:4339. https://doi.org/10.1038/s41598-018-22653-7

Ahmad K, Adnan M, Khan MA, Hussain Z, Junaid K, Saleem N, Ali M, Basir A (2015) Bioactive neem leaf powder enhances the shelf life of stored mungbean grains and extends protection from pulse beetle. *Pak J Weed Sci Res* 21(1):71–81.

Akram R, Turan V, Hammad HM, Ahmad S, Hussain S, Hasnain A, Maqbool MM, Rehmani MIA, Rasool A, Masood N, Mahmood F, Mubeen M, Sultana SR, Fahad S, Amanet K, Saleem M, Abbas Y, Akhtar HM, Waseem F, Murtaza R, Amin A, Zahoor SA, ul Din MS, Nasim W (2018a) Fate of organic and inorganic pollutants in paddy soils. In: Hashmi MZ, Varma A (ed) *Environmental pollution of paddy soils, soil biology*. Springer International Publishing Ag, Cham, Switzerland, pp. 197–214.

Akram R, Turan V, Wahid A, Ijaz M, Shahid MA, Kaleem S, Hafeez A, Maqbool MM, Chaudhary HJ, Munis, MFH, Mubeen M, Sadiq N, Murtaza R, Kazmi DH, Ali S, Khan N, Sultana SR, Fahad S, Amin A, Nasim W (2018b) Paddy land pollutants and their role in climate change. In: Hashmi MZ, Varma A (ed) *Environmental pollution of paddy soils, soil biology*. Springer International Publishing Ag, Cham, Switzerland, pp. 113–124.

Allison SD, Wallenstein MD, Bradford MA (2010) Soil carbon response to warming dependent on microbial physiology. *Nat Geo Sci* 3:336–340.

Amin A, Nasim W, Mubeen M, Ahmad A, Nadeem M, Urich P, Fahad S, Ahmad S, Wajid A, Tabassum F, Hammad HM (2018) Simulated CSM-CROPGRO-cotton yield under projected future climate by SimCLIM for southern Punjab Pakistan. *Agric Syst* 167:213–222.

Awais M, Wajid A, Saleem MF, Nasim W, Ahmad A, Raza MAS, Bashir MU, Mubeen M, Hammad HM, Rahman MHU, Saeed U (2018) Potential impacts of climate change and adaptation strategies for sunflower in Pakistan. *Envir Sci Pollut Res* 25(14):13719–13730.

Aziz K, Daniel KYT, Fazal M, Muhammad ZA, Farooq S, Fan W, Fahad S, Ruiyang Z (2017a) Nitrogen nutrition in cotton and control strategies for greenhouse gas emissions: A review. *Environ Sci Pollut Res* 24:23471–23487. https://doi.org/10.1007/s11356-017-0131-y

Aziz K, Daniel KYT, Muhammad ZA, Honghai L, Shahbaz AT, Mir A, Fahad S (2017b) Nitrogen fertility and abiotic stresses management in cotton crop: A review. *Environ Sci Pollut Res* 24:14551–14566. https://doi.org/10.1007/s11356-017-8920-x

Baldwin RE (2006) Globalisation: The great unbundling (s) (No. BOOK). Economic Council of Finland.

Basir A, Ali R, Alam M, Shah AS, Khilwat A, Adnan M, Ibrahim M, Rehman I, Adnan T (2015) Potential of wheat (*Tritium aestivum* L.) Advanced lines for yield and yield attributes under different planting dates in Peshawar Valley. *Am Eurasian J Agric Environ Sci* 15(12):2484–2488.

Beedlow PA, Tingey DT, Phillips DL, Hogsett WE, Olszyk DM (2004) Rising atmospheric CO_2 and carbon sequestration in forests. *Front Ecol Environ* 2:315–322.

Bellard C, Bertelsmeier C, Leadley P, Thuiller W, Courchamp F (2012) Impacts of climate change on the future of biodiversity. *Ecol Lett* 15:365–377.

Bhattacharyya R, Prakash V, Kundu S, Srivastva AK, Gupta HS, Mitra S (2009) Long term effects of fertilization on carbon and nitrogen sequestration and aggregate associated carbon and nitrogen in the Indian sub-Himalayas. *Nutr Cycl Agroecosyst* 86:1–16.

Brown ME, Funk CC (2008) Food security under climate change. *Science* 319(5863):580–581.

Chandel NS, Agrawal KN, Tripathi H, Garg SK (2014) Development of yield maps in wheat using yield monitor. *Bhartiya Krishi Anusandhan Patrika* 29 (3):111–115.

Chaudhry QUZ (2017) *Climate change profile of Pakistan*. Asian Development Bank.

Chrispaul M, David MM, Joseph AO (2010) Samuel VO (2010) Effective microorganism and their influence on growth and yield of pigweed (Amaranthusdubians). *J Agric Biol Sci* 5:17–22.

Cissel WJ, Mason CE, Whalen J, Hough-Goldstein J, Hooks CR (2015) Effects of brown marmorated stink bug (Hemiptera: Pentatomidae) feeding injury on sweet corn yield and quality. *J Econ Entom* 108(3):1065–1071.

Collins M, Knutti R, Arblaster J, Dufresne JL, Fichefet T, Friedlingstein P, Gao X, Gutowski WJ, Johns T, Krinner G, Shongwe M (2013) Long-term climate change: Projections, commitments and irreversibility 1029–1136.

Cook SE, Bramley RG (1998) Precision agriculture – Opportunities, benefits and pitfalls. *Aust J Exp Agric* 38:753–763.

Davidson EA, Janssens IA (2006) Temperature sensitivity of soil carbon decomposition and feedbacks to climate change. *Nature* 440:165–173.

de Medeiros FHVD, Souza RMD, Ferro HM, Zanotto E, Machado JDC (2015) Screening of endospore-forming bacteria for cotton seed treatment against bacterial blight and damping-off. *Adv Plants Agric Res* 2(4):56–61.

Depeng W, Fahad S, Saud S, Muhammad K, Aziz K, Mohammad NK, Hafiz MH, Wajid N (2018) Morphological acclimation to agronomic manipulation in leaf dispersion and orientation to promote "Ideotype" breeding: evidence from 3D visual modeling of "super" rice (*Oryza sativa* L.). *Plant Physiol Biochem*. https://doi.org/10.1016/j.plaphy.2018.11.010

Deutsch CA, Tewksbury JJ, Tigchelaar M, Battisti DS, Merrill SC, Huey RB, Naylor RL (2018) Increase in crop losses to insect pests in a warming climate. *Science* 361(6405):916–919.

Dinnes DL (2004) *Assessment of practices to reduce nitrogen and potassium non-point source pollution of Iowa's surface waters*, Iowa Department of National resources, Des Moines, LA.

Dosdall LM, Stevenson FC (2005) Managing flea beetles (Phyllotreta spp.) (Coleoptera: Chrysomelidae) in canola with seeding date, plant density, and seed treatment. *Agron J* 97:1570–1578.

Draz KA (2009) *Cotton Pests*. Faculty of Agriculture (Damanhor), Alexandira University.

Fahad S, Bajwa AA, Nazir U, Anjum SA, Farooq A, Zohaib A, Sadia S, Nasim W, Adkins S, Saud S, Ihsan MZ, Alharby H, Wu C, Wang D, Huang J (2017) Crop production under drought and heat stress: plant responses and management options. *Front Plant Sci* 8:1147. https://doi.org/10.3389/fpls.2017.01147

Fahad S, Bano A (2012) Effect of salicylic acid on physiological and biochemical characterization of maize grown in saline area. *Pak J Bot* 44:1433–1438.

Fahad S, Chen Y, Saud S, Wang K, Xiong D, Chen C, Wu C, Shah F, Nie L, Huang J (2013) Ultraviolet radiation effect on photosynthetic pigments, biochemical attributes, antioxidant enzyme activity and hormonal contents of wheat. *J Food Agri Environ* 11(3&4):1635–1641.

Fahad S, Hussain S, Bano A, Saud S, Hassan S, Shan D, Khan FA, Khan F, Chen Y, Wu C, Tabassum MA, Chun MX, Afzal M, Jan A, Jan MT, Huang J (2014a) Potential role of phytohormones and plant growth-promoting rhizobacteria in abiotic stresses: Consequences for changing environment. *Environ Sci Pollut Res* 22(7):4907–4921. https://doi.org/10.1007/s11356-014-3754-2

Fahad S, Hussain S, Matloob A, Khan FA, Khaliq A, Saud S, Hassan S, Shan D, Khan F, Ullah N, Faiq M, Khan MR, Tareen AK, Khan A, Ullah A, Ullah N, Huang J (2014b) Phytohormones and plant responses to salinity stress: a review. *Plant Growth Regul* 75(2):391–404. https://doi.org/10.1007/s10725-014-0013-y

Fahad S, Hussain S, Saud S, Hassan S, Chauhan DS, Khan F et al. (2016a) Responses of rapid viscoanalyzer profile and other rice grain qualities to exogenously applied plant growth regulators under high day and high night temperatures. *PLoS One* 11(7):e0159590. https://doi.org/10.1371/journal.pone.0159590

Fahad S, Hussain S, Saud S, Hassan S, Ihsan Z, Shah AN, Wu C, Yousaf M, Nasim W, Alharby H, Alghabari F, Huang J (2016c) Exogenously applied plant growth regulators enhance the morphophysiological growth and yield of rice under high temperature. *Front Plant Sci* 7:1250. https://doi.org/10.3389/fpls.2016.01250

Fahad S, Hussain S, Saud S, Hassan S, Tanveer M, Ihsan MZ, Shah AN, Ullah A, Nasrullah KF, Ullah S, Alharby HNW, Wu C, Huang J (2016d) A combined application of biochar and phosphorus alleviates heat-induced adversities on physiological, agronomical and quality attributes of rice. *Plant Physiol Biochem* 103:191–198.

Fahad S, Hussain S, Saud S, Khan F, Hassan S, Jr, Nasim W, Arif M, Wang F, Huang J (2016b) Exogenously applied plant growth regulators affect heat-stressed rice pollens. *J Agron Crop Sci* 202:139–150.

Fahad S, Hussain S, Saud S, Tanveer M, Bajwa AA, Hassan S, Shah AN, Ullah A, Wu C, Khan FA, Shah F, Ullah S, Chen Y, Huang J (2015a) A biochar application protects rice pollen from high-temperature stress. *Plant Physiol Biochem* 96:281–287.

Fahad S, Muhammad ZI, Abdul K, Ihsanullah D, Saud S, Saleh A, Wajid N, Muhammad A, Imtiaz AK, Chao W, Depeng W, Jianliang H (2018) Consequences of high temperature under changing climate optima for rice pollen characteristics-concepts and perspectives. *Archives Agron Soil Sci.* https://doi.org/10.1080/03650340.2018.1443213

Fahad S, Nie L, Chen Y, Wu C, Xiong D, Saud S, Hongyan L, Cui K, Huang J (2015b) Crop plant hormones and environmental stress. *Sustain Agric Rev* 15:371–400.

Fahad S, Adnan M, Hassan S, Saud S, Hussain S, Wu C, Wang D, Hakeem KR, Alharby HF, Turan V, Khan MA, Huang J (2019b) Rice responses and tolerance to high temperature. In: Hasanuzzaman M, Fujita M, Nahar K, Biswas JK (ed) *Advances in rice research for abiotic stress tolerance*. Woodhead Publ Ltd, Abington, Cambridge, UK, pp. 201–224.

Fahad S, Rehman A, Shahzad B, Tanveer M, Saud S, Kamran M, Ihtisham M, Khan SU, Turan V, Rahman MHU. (2019a) Rice responses and tolerance to metal/metalloid toxicity. In: Hasanuzzaman M, Fujita M, Nahar K, Biswas JK (ed) *Advances in rice research for abiotic stress tolerance*. Woodhead Publ Ltd, Abington, Cambridge, UK, pp. 299–312.

Farid HU, Bakhsh A, Ahmad N, Ahmad A (2013) Evaluation of management zones for site-specific application of crop inputs. *Pak J Life Social Sci* 11(1):29–35.

Fening JO, Ewusi-Mensah N, Safo EY (2010) Improving the fertilizer value of cattle manure for sustaining small holder crop production in Ghana. *J Agro* 9:92–101.

Gassmann AJ, Petzold-Maxwell JL, Keweshan RS, Dunbar MW (2011) Field-evolved resistance to Bt maize by western corn rootworm. *PloS one* 6(7):22629.

Glowienka E, Michalowska K, Pekala A, Hejmanowska B (2016) *October application of GIS and remote sensing techniques in multitemporal analyses of soil properties in the Foreland of the Carpathians.* In *IOP Conference Series: Earth and Environmental Science* (Vol. 44, No. 5, p. 052044). IOP Publishing.

Gowda CL, Chaturvedi SK, Gaur P M, Sameer Kumar CV, Jukanti AK (2015) Pulses research and development strategies for India.

Grunwald S, Vasques GM, Rivero RG (2014) Fusion of soil and remote sensing data to model soil properties. *Adv Agron.*

Rahman MHU, Ashfaq A, Aftab W, Manzoor H, Fahd R, Wajid I, Md. Aminul I, Vakhtang S, Muhammad A, Asmat U, Abdul W, Syeda RS, Shah S, Shahbaz K, Fahad S, Manzoor H, Saddam H, Wajid N (2017)

Application of CSM-CROPGRO-Cotton model for cultivars and optimum planting dates: evaluation in changing semi arid climate. *Field Crops Res.* http://dx.doi.org/10.1016/j.fcr.2017.07.007

Hadi F, Rahman AU, Ibrar M, Dastagir G, Arif M, Naveed K, Adnan M (2014) Weed diversity with special reference to their ethnomedicinal uses in wheat and maize at Rech valley, Hindokush Range, Chitral, Pakistan. *Pak J Weed Sci Res* 20(3):335–346.

Hafiz MH, Muhammad A, Farhat A, Hafiz FB, Saeed AQ, Muhammad M, Fahad S, Muhammad A (2019) Environmental factors affecting the frequency of road traffic accidents: A case study of sub-urban area of Pakistan. *Environ Sci Pollut Res* https://doi.org/10.1007/s11356-019-04752-8

Hafiz MH, Wajid F, Farhat A, Fahad S, Shafqat S, Wajid N, Hafiz FB (2016) Maize plant nitrogen uptake dynamics at limited irrigation water and nitrogen. *Environ Sci Pollut Res* 24(3):2549–2557. https://doi.org/10.1007/s11356-016-8031-0

Hedley C (2015) The role of precision agriculture for improved nutrient management on farms. *J Sci Food Agric* 95(1):12–19

Hoogenboom GW, Jeffrey W, Messina CD (2004) From genome to crop: Integration through simulation modeling. *Field Crops Res* 90:145–163.

Iftekharuddaula KM, Ahmed HU, Ghosal S, Amin A, Moni ZR, Ray BP, Barman HN, Siddique MA, Collard BC, Septiningsih EM (2016) Development of early maturing submergence-tolerant rice varieties for Bangladesh. *Field Crops Res* 190:44–53.

IUCN (International Union for Conservation of Nature) (2016) *The IUCN Red List of Threatened Species.* Gland, Switzerland: IUCN.

IPCC (Intergovernmental Panel on Climate Change) (2018) Global warming of 1.5°C. Summary for Policy Makers. Switzerland: World Meteorological Organization, United Nations Environment Program, and Intergovernmental Panel on Climate Change.

Kistner E, Kellner O, Andresen J, Todey D, Morton LW (2018) Vulnerability of specialty crops to short-term climatic variability and adaptation strategies in the Midwestern USA. *Clim Change* 146(1–2):145–158.

Koocheki A, Nassiri M, Alimoradi L, Ghorbani R (2009) Effect of cropping systems and crop rotations on weeds. *Agron Sust Develop* 29(2):401–408.

Lehmann J, Silva JPD, Steiner C, Nehls T, Zech W, Glaser B (2003) Nutrient availability and leaching in an archaeological Anthrosol and a Ferralsol in the Central Amazon basin: fertilizer, manure and charcoal amendments. *Plant Soil* 249:343–357.

Leteinturier B, Herman JL, De Longueville F, Quintin L, Oger R (2006) Adaptation of a crop sequence indicator based on a land parcel management system. *Agric Ecosystems Environ* 112(4):324–334.

Loarie SR, Duffy PB, Hamilton H, Asner GP, Field CB, Ackerly DD (2009) The velocity of climate change. *Nature* 462:1052–1055.

Matzrafi M (2019) Climate change exacerbates pest damage through reduced pesticide efficacy. *Pest Manage Sci* 75(1):9–13.

McBratney A, Whelan B, Ancev T, Bouma J (2005) Future directions of precision agriculture. *Precis Agric* 6:7–23.

McDowell NG, Michaletz ST, Bennett KE, Solander KC, Xu C, Maxwell RM, Middleton RS (2018) Predicting chronic climate-driven disturbances and their mitigation. *Trends Ecol Evo* 33(1):15–27.

Mishra JS, Singh VP (2012) Tillage and weed control effects on productivity of a dry seeded rice–wheat system on a Vertisol in Central India. *Soil Till Res* 123:11–20.

Monzón J, Moyer-Horner L, Palamar MB (2011) Climate change and species range dynamics in protected areas. *Bioscience* 61(10):752–761.

Muhammad Z, Abdul MK, Abdul MS, Kenneth BM, Muhammad S, Shahen S, Ibadullah J, Fahad S (2019) Performance of *Aeluropus lagopoides* (mangrove grass) ecotypes, a potential turfgrass, under high saline conditions. *Environ Sci Pollut Res.* https://doi.org/10.1007/s11356-019-04838-3

North JH, Gore J, Catchot AL, Stewart SD, Lorenz GM, Musser FR, Cook DR, Kerns DL, Dodds DM (2017) Value of neonicotinoid insecticide seed treatments in mid-south cotton (*Gossypium hirsutum* [Malvales: Malvaceae]) production systems. *J Economic Entomol* 111(1):10–15.

Parmesan C, Hanley ME, Singer MC (2018) Model vs experiment to predict crop losses.

Patil SS, Patil VC, Al-Gaadi KA. (2011) Spatial variability in fertility status of surface soils. *World App Sci J* 14:1020–1024.

Privette CV, Khalilian A, Torres O, Katzberg S (2011) Remote sensing of environment utilizing space-based GPS technology to determine hydrological properties of soils. *Remote Sens Environ* 115(12):3582–3586.

Qamar-Uz Z, Zubair A, Muhammad Y, Muhammad ZI, Abdul K, Fahad S, Safder B, Ramzani PMA, Muhammad N (2017) Zinc biofortification in rice: Leveraging agriculture to moderate hidden hunger in developing countries. *Arch Agron Soil Sci* 64:147–161. https://doi.org/10.1080/03650340.2017.1338343

Ramsfield TD, Bentz BJ, Faccoli M, Jactel H, Brockerhoff EG (2016) Forest health in a changing world: Effects of globalization and climate change on forest insect and pathogen impacts. *Forestry* 89(3):245–252.

Rao NK, Dulloo ME, Engels JM. (2017) A review of factors that influence the production of quality seed for long-term conservation in genebanks. *Genetic Res Crop Evol* 64(5):1061–1074.

Ray SK, Chatterjee D, Saha S, Kamble KH, Deka BC, Kendra V, Jharnapani MA. (2015) Guide for Soil Nutrient Management with special reference to Wokha. ICAR Research Complex for NEH Region, Nagaland Centre.

Reyes JAO, Ocampo ET, Bartolome MEPL, Sabanal AQC, Maghirang RG (2017) Towards the development of mung bean varieties with high and stable yields for post-rice drought conditions. *Philippine J Crop Sci* 42(1): 119–120.

Riegler M (2018) Insect threats to food security. *Science* 361(6405):846–846.

Sajjad H, Muhammad M, Ashfaq A, Waseem A, Hafiz MH, Mazhar A, Nasir M, Asad A, Hafiz UF, Syeda RS, Fahad S, Depeng W, Wajid N (2019) Using GIS tools to detect the land use/land cover changes during forty years in Lodhran district of Pakistan. *Environ Sci Pollut Res*. https://doi.org/10.1007/s11356-019-06072-3

Sanford RA, Wagner DD, Cu QW, Chee-Sanford J, Thomas SH, et al. (2012) Unexpected non-denitrifier nitrous oxide reductase gene diversity and abundance in soils. *Proc Natl Acad Sci* 109:19709–19714.

Sarmah K, Astec A (2018). Role of GIS Based Technologies in Sustainable Agriculture Resource Planning & Management Using Spatial Decision Support Approach (1):30–34.

Saud S, Chen Y, Fahad S, Hussain S, Na L, Xin L, Alhussien SA (2016) Silicate application increases the photosynthesis and its associated metabolic activities in Kentucky bluegrass under drought stress and post-drought recovery. *Environ Sci Pollut Res* 23(17):17647–17655. https://doi.org/10.1007/s11356-016-6957-x

Saud S, Chen Y, Long B, Fahad S, Sadiq A (2013) The different impact on the growth of cool season turf grass under the various conditions on salinity and draught stress. *Int J Agric Sci Res* 3:77–84.

Saud S, Fahad S, Yajun C, Ihsan MZ, Hammad HM, Nasim W, Amanullah Jr, Arif M, Alharby H (2017) Effects of nitrogen supply on water stress and recovery mechanisms in Kentucky Bluegrass plants. *Front Plant Sci* 8:983. https://doi.org/10.3389/fpls.2017.00983

Saud S, Li X, Chen Y, Zhang L, Fahad S, Hussain S, Sadiq A, Chen Y (2014) Silicon application increases drought tolerance of Kentucky bluegrass by improving plant water relations and morph physiological functions. *SciWorld J* 2014:1–10. https://doi.org/10.1155/2014/368694

Sehgal A, Sita K, Bhandari K, Kumar S, Jitendra Kumar PV, Prasad V, Siddique KHM, Nayyar H. (2019) Influence of drought and heat stress, applied independently or in combination during seed development, on qualitative and quantitative aspects of seeds of lentil (Lens culinaris Medikus) genotypes, differing in drought sensitivity. *Plant cell Environ* 42(1):198–211.

Senthurpandian VK, Jayaganesh S, Srinivas S, Palani N, Muraleedharan N (2010) Application of geographic information system to fertility management of tea soils of Anamallais. *Asian J Earth Sci* 3(3):36–141.

Shah B, Khan IA, Khan A, Din MMU, Adnan M, Junaid K, Shah SRA, Zaman M, Ahmad N, Akbar R, Fayaz W, Rahman IU (2016) Determination of physio-morphic basis of resistance in different maize cultivars against insect pests. *J Ento Zool Studies* 4(1):317–321.

Shah F, Lixiao N, Kehui C, Tariq S, Wei W, Chang C, Liyang Z, Farhan A, Fahad S, Huang J (2013) Rice grain yield and component responses to near 2°C of warming. *Field Crop Res* 157:98–110.

Shivanna A M, Nagendrappa G (2014) Chemical analysis of soil samples to evaluate the soil fertility status of selected command areas of three tanks in Tiptur Taluk of Karnataka, India. *IOSR J App Chem* 7(11):1–5.

Singh A, Vasisht A K, Kumar R, Das D k (2008) Adoption of integrated pest management practices in paddy and cotton: a case study in Haryana and Punjab. *Agric Economics Res Rev* 21:221.

Singh BK, Bardgett RD, Smith P, Dave SR (2010) Microorganisms and climate change: terrestrial feedbacks and mitigation options. *Nat Rev Microbiol* 8:779–790.

Sparks TC, Nauen R (2015) Mode of action classification and insecticide resistance management. *Pestic Biochem Physiol* 121:122–128.

Srinivasan A (2006) *Handbook of precision agriculture: principles and applications*; Bringhamton: New York, NY, 3–18.

SSSA (Soil Science Society of America) (1997) Glossary of Soil Science Terms, 1996. Soil Sci Soc Am Madison WI.

Starr JL, Carneiro RG, Ruano O (2005) Nematode parasites of cotton and other tropical fibre crops. 733–750 in M. Luc, R. A. Sikora, and J. Bridge (eds). *Plant parasitic nematodes in sub-tropical and tropical agriculture*, 2nd ed. Wallingford: CAB International.

Stein S, Steinmann HH (2018) Identifying crop rotation practice by the typification of crop sequence patterns for arable farming systems–A case study from Central Europe. *Eur J Agron* 92:30–40.

Steinweg JM, Plante AF, Conant RT, Paul EA, Tanaka DL (2008) Patterns of substrate utilization during long-term incubations at different temperatures. *Soil Biol Biochem* 40:2722–2728.

Swati T, Ramesh S, Shaily J (2014) Effect of climate change on plant-microbe interaction: an overview. *Eur J Mol Biotechnol* 5:149–156.

Tian B, Yu Z, Pei Y, Zhang, Z, Siemann E, Wan S, Ding J (2019) Elevated temperature reduces wheat grain yield by increasing pests and decreasing soil mutualists. *Pest Manag Sci* 75(2):466–475.

Valasai G D, Harijan K, Uqaili M S, Memon H R (2005) *Impact of greenhouse gases on agricultural productivity in Pakistan*. In *Proceedings of the first international conference on environmentally sustainable development* 1–3.

Wadodkar M R, Ravishankar T, Joshi K (2014) Application of remote sensing techniques for soil fertility assessment.

Wajid N, Ashfaq A, Asad A, Muhammad T, Muhammad A, Muhammad S, Khawar J, Ghulam MS, Syeda RS, Hafiz MH, Muhammad IAR, Muhammad ZH, Rahman MHU, Veysel T, Fahad S, Suad S, Aziz K, Shahzad A (2017) Radiation efficiency and nitrogen fertilizer impacts on sunflower crop in contrasting environments of Punjab. *Pakistan Environ Sci Pollut Res* 25:1822–1836. https://doi.org/10.1007/s11356-017-0592-z

Walia U A (2006) *Weed Ecology and Weed Competition*. Kalyani Publishers, India 61–70.

Weiman S (2015) Microbes help to drive global carbon cycling and climate change. *Microbe Mag* 10:233–238.

WMO (World Meteorological Organization) (2018) The state of greenhouse gases in the atmosphere based on global observations through 2017. Greenhouse Gas Bulletin 8.

Wu K, Lu Y, Wang Z (2009) Advance in integrated pest management of crops in China. *Chinese Bulletin Entom* 46(6):831–836.

Yang Z, Zhang Z, Zhang T, Fahad S, Cui K, Nie L, Peng S, Huang J (2017) The effect of season-long temperature increases on rice cultivars grown in the central and southern regions of China. *Front Plant Sci* 8:1908. https://doi.org/10.3389/fpls.2017.01908

Yuya A I, Tadesse A, Azerefegne F, Tefera T (2009) Efficacy of combining Niger seed oil with malathion 5% dust formulation on maize against the maize weevil, *Sitophilus zeamais* (Coleoptera: Curculionidae). *J Stored Prod Res* 45:67–70.

Zahida Z, Hafiz FB, Zulfiqar AS, Ghulam MS, Fahad S, Muhammad RA, Hafiz MH, Wajid N, Muhammad S (2017) Effect of water management and silicon on germination, growth, phosphorus and arsenic uptake in rice. *Ecotoxicol Environ Saf* 144:11–18.

Zhang N, Wang M, Wang N (2002) Precision agriculture – A worldwide overview Comp. *Elec Agric* 36(2–3):113–132.

2

Extreme Low Temperature and Plant Life

Muhammad Arif and Talha Jan
The University of Agriculture, Peshawar, Pakistan

Muhammad Riaz
Government College University Faisalabad, Faisalabad, Pakistan

Fazal Munsif
The University of Agriculture, Peshawar, Pakistan

Shah Fahad
Hainan University, Haikou, China; The University of Haripur, Haripur, Pakistan

Niamatullah
The University of Agriculture, Peshawar, Pakistan

Muhammad Adnan
University of Swabi, Pakistan

Sahar Mumtaz
University of Education, Lahore, Pakistan

Bibi Haleema
Agricultural Research Institute, Peshawar, Pakistan

Amanullah
The University of Agriculture, Peshawar, Pakistan

Kawsar Ali
Abdul Wali Khan University, Mardan, Pakistan

Fahad Rasul
University of Agriculture, Faisalabad, Pakistan

Muhammad Ishfaq Khan
The University of Agriculture, Peshawar, Pakistan

Muhammad Zamin
University of Swabi, Pakistan

Ishaq Ahmad Mian
University of Agriculture, Peshawar, Pakistan

Bushra Khan
University of Peshawar, Peshawar, Pakistan

Imtiaz Khan
The University of Agriculture, Peshawar, Pakistan

Shah Saud
Linyi University, Linyi, China

DOI: 10.1201/9781003286417-2

CONTENTS

2.1 Introduction

Cold stress or low temperature stress is a matter of routine for plants of polar and other cold areas of the world. However, with shifting seasons some nonpolar crop species such as rice and sugarcane are experiencing cold stress and resulting injury. Cold stress comprises of freezing and chilling stress. Under chilling stress conditions, plant injury occurs without ice formation in plant tissues whereas under freezing stress ice formation occurs in plant tissues. Low temperature range of 2°C to 10°C is enough to cause chilling injury, but few susceptible species such as rice and sugarcane can experience chilling injury at temperature as high as 15°C (Thomashow 1999). Chilling stress alters cell membrane structure from flexible liquid-crystal to a solid-gel phase, thus distorting normal cellular structure and affecting cell functions. Also the accumulation of reactive oxygen species and other toxic metabolites occur at low temperatures. The kinetics of several metabolic and physiological processes are thermodynamically hampered by low temperature or chilling stress (Ruelland et al. 2009). It results in decreased rate and uniformity of seed emergence, reduced seedling vigour after germination, and delayed plants' morphological processes, incurring yield reductions (Ruelland et al. 2009). Plants are categorized as tolerant and susceptible on the basis of survivability under low temperature conditions. Plants in polar regions are tolerant to cold stress, whereas tropical and subtropical plant and crop species are very much susceptible to cold stress.

2.2 Effects of Low Temperature Stress

2.2.1 Germination

Seeds require optimum temperature for germination and emergence. Temperature below optimum severely impairs germination and results in the loss of seed vigour (Wang et al. 2016). The kinetics of various metabolic and physiological functions in seed are thermodynamically limited due to chilling stress and the seed often loses its ability to germinate (Ruelland et al. 2009). It has been known to cause alterations in DNA, RNA, and protein structures; membrane damage; reduced respiration; and less ATP production, ultimately causing loss in seed germination and vigour. Negative effects of low temperature or chilling stress have been reported on various crops such as cotton, rice (Wang et al. 2016), potato, tomato (Starck et al. 2000), sugarcane, and muskmelons (Wang et al. 2004).

2.2.2 Growth

Growth restrictions are usually posed due to disastrous effects of chilling on plant leaves. This includes decreased leaf cells production with prolonged cell cycle, leaf necrosis, delayed leaf development, leaf wilting, and increased pathogen and disease incidence (Rymen et al. 2007). Chilling causes decrease in leaf area due to decreased rate of cell division and cell elongation, as well as slow leaf initiation rate, which results in lesser biomass production and assimilation (Adnan et al. 2016; Lukatkin et al. 2012). Chilling stress also poses severe threat to root growth by changing root morphology, and decreasing root length and biomass (Cutforth et al. 1986). This effect has consequences in the form of reduced exploration for water and nutrients due to decreased volume of root system.

2.2.3 Yield and Yield-Related Attributes

Reduced growth due to chilling stress ultimately causes reduction in yield and yield attributes. Besides reduction in growth rate, chilling stress has detrimental impact on plants' reproductive system, which ultimately causes reduction in yield. It has been observed that chilling stress may prove more severe during reproductive stage than pre- and postreproductive stages. Disastrous effects of chilling stress on crop yield have been observed in many cultivated crops. Barley productivity was severely impaired due to chilling stress (Shah et al. 2016). In chickpea, chilling stress caused flower abortion and poor pod setting leading to severe yield reduction. Cold temperature is held accountable for 30%–40% reduction in rice yields because of spikelets degeneration, spikelets sterility, and panicle deformation (Andaya and Mackill 2003). In other grain crops, chilling stress during reproductive stage may cause flower abscission, pollen tube distortion, pollen sterility, ovule abortion, poor fruit set, and ultimately reduced yields. In wheat, chilling stress during stem elongation stopped internode extension, caused spikelet mortality, and lowered biomass accumulation (Ibrahim et al. 2016; Whaley et al. 2004).

2.2.4 Photosynthesis and Gaseous Exchange

Chilling stress has been proved detrimental for photosynthesis and gaseous exchange. Low photosynthetic rate under chilling stress has been attributed to low stomatal conductance; reduced chloroplast development, quantum efficiency, and yield; as well as restricted assimilate transport (Sowiński et al. 2005). Plant photosynthetic function is reduced due to overexcitation of thylakoid membranes which enhance photoinhibition process. Moreover, chilling slows down the rate of CO_2 fixation and restricts the production of NADPC supplement available to accept electrons from the electron transport chain. This is yet another reason for reduced photosynthesis accompanied by slowing down of the rate of CO_2 fixation (Wise 1995). All these events result in reduced overall photosynthetic efficiency of plants.

2.2.5 Water Relations

Plant water relations is a set of important plant attributes, such as root hydraulic conductance, leaf water potential, stomatal conductance, transpiration rate, relative water content, leaf water potential, transpiration rate, stomatal conductance, as well as leaf and canopy temperature. The main mechanism in work that affects plant water relations is reduced transpiration which badly affects root hydraulic function. Transpiration is reduced due to development of vapour pressure deficit between leaf surface and ambient atmosphere. This causes a decline in leaf water potential and subsequent loss of turgor pressure (Aroca et al. 2003). This trend was obvious in sugarcane where leaf water potential was reduced due to the onset of chilling stress. In maize chilling stress caused shoot water due to decreased transpiration and root water uptake. The ultimate consequence of chilling stress related to water relations is chilling-induced wilting in plants.

2.2.6 Nutrient Relations

One among many reasons that cause decline in plant productivity is disturbed nutrient relations under chilling stress conditions, although the effect varies among plant species and growth stages, nutrient availability, and duration of stress period. The physiochemical and microbial processes that modify the plant–nutrient relationship are impaired due to low temperature. Plan root system shows poor performance and hence reduces the uptake of several nutrients including nitrogen (N), phosphorus (P), and potassium (K). Reasons behind low nutrient relations under chilling stress are numerous including low hydraulic conductance, reduced root length, thicker root axis, and poor root branching. The effect was obvious in several plant species including nutrient exhaustive crops such as maize (Hund et al. 2007).

2.2.7 Oxidative Stress

Several studies have held chilling stress responsible for the overproduction of reactive oxygen species (ROS) such as singlet oxygen ($1O_2$), hydroxyl radicals (OH^-), superoxide radicals (O_2^-), and hydrogen peroxide (H_2O_2). Excess production of ROS causes damage to DNA and RNA, protein oxidation and membrane lipid peroxidation, and even plant death (Apel and Hirt 2004). They cause destruction at cellular level by reacting with almost every cell constituent (Sattler et al. 2000). Studies have shown that chilling stress severely impaired the enzymatic activities and disturbance in membrane properties causing tissue necrosis, which pointed towards the occurrence of oxidative damage (Chinnusamy et al. 2007). In fact, under chilling stress plant becomes more susceptible to oxidative stress, which is originally caused by other stresses such as drought and heat stress. This was obvious from the effect of chilling stress on maize leaves which was more pronounced on bundle sheath proteins as compared to mesophyll cells. Overproduction of ROS under chilling stress causes serious cellular damage by rapidly reacting with essential cellular structures (Sattler et al. 2000).

2.3 Avoidance and Tolerance Mechanism

2.3.1 Plants Adaptation to Cold Stress

Certain plant species have evolved morphological traits to cope with extended cold periods (Guy et al. 2008). These plant species generally have short plant stature with high root-to-shoot ratio and have smaller leaf surface area. They are usually slow growing with C3 photosynthetic apparatus and store assimilated sugars in underground parts. These plants get a short mild growing season, so they store reserves for that period and rapidly mobilize those reserves when temperatures are relatively high. Phenological occurrences in these plants are strictly controlled by their genes (Guy et al. 2008).

2.3.2 Avoidance and Tolerance

Chilling stress avoidance by plants is achieved by preventing sensitive tissues from freezing. Some plant species accumulate heat in their bodies during day time and dissipate it during night when temperature is downf; some avoid cold temperature by undergoing seed or organ dormancy, while others make use of leaves to protect their growing meristems. Some plants undergo a phenomenon called *supercooling*. In this process the formation of ice nucleators is inhibited preventing ice nucleation even at temperatures as low as −40°C (Wisniewski and Fuller 1999). By doing so, plants' internal processes are not affected by the presence of ice around their bodies.

Coping with low temperature stress is also sought at molecular level through the production of cryoprotectant compounds like soluble sugars (stachyose, raffinose, trehalose, saccharose), sugar alcohols (ribitol, sorbitol, inositol), and low-molecular-weight nitrogenous compounds (glycine, betaine, proline). In another study it was found that the production of phenolic compounds (Griffith and Yaish 2004) and putrescine content can help plants survive chilling stress. Among specific crop species, *Pringlea antiscorbutica* was reported to have increased levels of polyamines (putrescine and agmatine), which was deemed as marker of chilling tolerance in this crop (Hummel et al. 2004).

2.4 Summary and Conclusion

Another form of extreme temperature, chilling also has detrimental effects on plant life. It causes damage to seed cell membrane, reduces respiration, and causes reduction in ATP production, ultimately causing loss in seed germination and vigour. The rates of cell division and cell elongation are decreased under low temperature stress resulting in decreased leaf area. Chilling stress affects plant photosynthetic function by damaging chloroplast and thylakoid membranes. The rate of CO_2 fixation is also reduced under chilling stress. All these events lead to lesser biomass production and assimilation. Chilling stress may prove even more severe during reproductive stages. The nutrient balance and uptake efficiency of plant roots are hampered due to root damage, reduced root length, thicker root axis, and poor root branching, and hence the uptake of several nutrients including N, P, and K is reduced under chilling stress. Like heat stress, chilling stress also promotes the production of ROS, and plant suffers oxidative stress which causes severe cellular damages by reacting with cell constituents.

Indeed plants suffer from both these stresses; however, some plant species have adopted many avoidance and tolerance mechanisms to avoid and tolerate these extreme stresses. Avoidance mechanisms for heat stress include transpirational cooling, changing leaf orientation, and alteration of membrane lipid compositions. Other mechanisms include maintaining water balance through closure of stomata, increased stomatal densities, and evolving larger xylem vessels. For coping with low temperature stress, plants undergo alteration at molecular level by production of cryoprotectant compounds like soluble sugars, sugar alcohols, and low-molecular-weight nitrogenous compounds. Since extreme temperatures cannot be controlled anthropogenically, genetic breeding for the production of tolerant varieties is perhaps the ultimate need of the hour, second only to climate change mitigation strategies.

REFERENCES

Adnan M, Shah Z, Saleem N, Basir A, Rahman IU, Ullah H, Ibrahim M, Shah JA, Muhammad AK, Ali Shah SR (2016). Isolation and evaluation of summer legumes Rhizobia as PGPR. *Pure App Bio* 5(1):127–133.

Andaya V, Mackill D (2003) Mapping of QTLs associated with cold tolerance during the vegetative stage in rice. *J Exp Bot* 54:2579–2585.

Apel K, Hirt H (2004) Reactive oxygen species: metabolism, oxidative stress, and signal transduction. *Annu Rev Plant Biol* 55:373–399.

Aroca R, Vernieri P, Irigoyen JJ, Sánchez-Dıaz M, Tognoni F, Pardossi A (2003) Involvement of abscisic acid in leaf and root of maize (*Zea mays* L.) in avoiding chilling-induced water stress. *Plant Sci* 165:671–679.

Chinnusamy V, Zhu J, Zhu J-K (2007) Cold stress regulation of gene expression in plants. *Trends Plant Sci* 12:444–451.

Cutforth H, Shaykewich C, Cho C (1986) Effect of soil water and temperature on corn (*Zea mays* L.) root growth during emergence. *Can J Soil Sci* 66:51–58.

Griffith M, Yaish MW (2004) Antifreeze proteins in overwintering plants: a tale of two activities. *Trends Plant Sci* 9:399–405.

Guy C, Kaplan F, Kopka J, Selbig J, Hincha DK (2008) Metabolomics of temperature stress. *Physiol Plantarum* 132:220–235.

Hummel I, El Amrani A, Gouesbet G, Hennion F, Couée I (2004) Involvement of polyamines in the interacting effects of low temperature and mineral supply on *Pringlea antiscorbutica* (*Kerguelen cabbage*) seedlings. *J Exp Bot* 55:1125–1134.

Hund A, Richner W, Soldati A, Fracheboud Y, Stamp P (2007) Root morphology and photosynthetic performance of maize inbred lines at low temperature. *Eur J Agron* 27:52–61.

Ibrahim M, Jamal Y, Basir A, Adnan M, Rahman IU, Khan IA, Attaullah (2016) Response of Sesame (*Sesamum indicum* L.) to various levels of nitrogen and phosphorus in agro-climatic condition of Peshawar. *Pure Appl Biol* 5(1):121–126.

Lukatkin AS, Brazaityte A, Bobinas C, Duchovskis P (2012) Chilling injury in chilling-sensitive plants: a review. *Agriculture* 99:111–124.

Ruelland E, Vaultier M-N, Zachowski A, Hurry V (2009) Cold signalling and cold acclimation in plants. *Adv Bot Res* 49:35–150.

Rymen B, Fiorani F, Kartal F, Vandepoele K, Inzé D, Beemster GT (2007) Cold nights impair leaf growth and cell cycle progression in maize through transcriptional changes of cell cycle genes. *Plant Physiol* 143:1429–1438.

Sattler U, Calsou P, Boiteux S, Salles B (2000) Detection of oxidative base DNA damage by a new biochemical assay. *Arch Biochem Biophys* 376:26–33.

Shah B, Khan IA, Akbar R, Fayaz W, Zaman M, Din MMU, Ahmad N, Adnan M, Junaid K, Shah SRA, Rahman IU (2016) Population dynamics of corn flea beetle and fourspotted leaf beetle on different maize cultivars at Peshawar. *J Ento Zool Studies* 4(1):308–311.

Sowiński P, Rudzińska-Langwald A, Adamczyk J, Kubica I, Fronk J (2005) Recovery of maize seedling growth, development and photosynthetic efficiency after initial growth at low temperature. *J Plant Physiol* 162:67–80.

Starck Z, Niemyska B, Bogdan J, Tawalbeh RA (2000) Response of tomato plants to chilling stress in association with nutrient or phosphorus starvation. *Plant Soil* 226:99–106.

Thomashow MF (1999) Plant cold acclimation: freezing tolerance genes and regulatory mechanisms. *Ann Rev Plant Biol* 50:571–599.

Wang W et al. (2016) Pre-sowing seed treatments in direct-seeded early rice: consequences for emergence, seedling growth and associated metabolic events under chilling stress. *Sci Rep* 6:19637.

Wang W, Vinocur B, Shoseyov O, Altman A (2004) Role of plant heat-shock proteins and molecular chaperones in the abiotic stress response. *Trends Plant Sci* 9:244–252.

Whaley J, Kirby E, Spink JH, Foulkes M, Sparkes D (2004) Frost damage to winter wheat in the UK: the effect of plant population density. *E J Agron* 21:105–115.

Wise RR (1995) Chilling-enhanced photooxidation: the production, action and study of reactive oxygen species produced during chilling in the light. *Photosynthesis Res* 45:79–97.

Wisniewski M, Fuller M (1999) Ice nucleation and deep supercooling in plants: new insights using infrared thermography. In: *Cold-adapted organisms*. Springer-Verlag, Berlin, pp. 105–118.

3

Heat-Induced Oxidative Stress in Plants: Consequences and Survival Mechanisms

Akbar Hossain and Mohammad Nur Alam
Bangladesh Wheat and Maize Research Institute, Nashipur, Bangladesh

Zhulong Chan
Huazhong Agricultural University, Wuhan, China

Ayman EL Sabagh
Kafrelsheikh University, Egypt

Shah Fahad
College of Tropical Crops, Hainan University, Haikou, China;
The University of Haripur, Haripur, Pakistan

Mirza Hasanuzzaman
Sher-e-Bangla Agricultural University, Bangladesh

CONTENTS

3.1 Introduction

The plants are sessile organisms facing the challenges of adverse environments, and as needed they change the physiological pathways, signal transductions, and genotypic arrangements/combinations to minimize the harmful effects of harsh environments (Barkla et al. 2013; Xu et al. 2016; Antoniou et al. 2017; Alam et al. 2018a; Deepranjan et al. 2021; Haider et al. 2021; Huang et al. 2021; Ikram et al. 2021; Jabborova et al. 2021; Khadim et al. 2021a, 2021b; Manzer et al. 2021; Muzammal et al. 2021; Abdul et al. 2021a, 2021b; Ashfaq et al. 2021; Amjad et al. 2021; Atif et al. 2021; Athar et al. 2021; Adnan et al.

DOI: 10.1201/9781003286417-3

2018a, 2018b, 2019; Akram et al. 2018a, 2018b; Aziz et al. 2017a, 2017b; Chang et al. 2021; Chen et al. 2021; Emre et al. 2021; Habib Ur et al. 2017; Hafiz et al. 2016, 2019; Ghulam et al. 2021; Guofu et al. 2021; Hafeez et al. 2021; Khan et al. 2021; Kamarn et al. 2017; Muhammad et al. 2019; Safi et al. 2021; Sajjad et al. 2019; Saud et al. 2013, 2014, 2016, 2017, 2020; Shah et al. 2013; Qamar-Ur et al. 2017; Hamza et al. 2021; Irfan et al. 2021; Wajid et al. 2017; Yang et al. 2017; Zahida et al. 2017; Depeng et al. 2018; Hussain et al. 2020; Hafiz et al. 2020a, 2020b; Shafi et al. 2020; Wahid et al. 2020; Subhan et al. 2020; Zafar-ul-Hye et al. 2020a, 2020b; Zafar-ul-Hye et al. 2021; Adnan et al. 2020; Ilyas et al. 2020; Saleem et al. 2020a, 2020b, 2020c; Rehman et al. 2020; Farhana et al. 2020; Wu et al. 2020; Mubeen et al. 2020; Farhana et al. 2020; Jan et al. 2019; Wu et al. 2019; Ahmad et al. 2019; Baseer et al. 2019; Hafiz et al. 2018; Tariq et al. 2018; Fahad and Bano 2012; Fahad et al. 2013, 2014a, 2014b, 2015a, 2015b, 2016a, 2016b, 2016c, 2016d, 2017, 2018a, 2018b, 2019a, 2019b, 2020, 2021a, 2021b, 2021c, 2021d, 2021e, 2021f; Hesham and Fahad 2020; Farhat et al. 2020). Major abiotic stresses such as heat, salinity, and drought have become the greatest threats to crop production due to the changing climate (Cheeseman 2016; Iqra et al. 2020; Akbar et al. 2020; Mahar et al. 2020; Noor et al. 2020; Bayram et al. 2020; Amanullah 2017, 2018a, 2018b; Amanullah et al. 2020, 2021; Rashid et al. 2020; Arif et al. 2020; Amir et al. 2020; Saman et al. 2020; Muhammad et al. 2020; Md Jakir and Allah 2020; Mahmood Ul et al. 2021; Farah et al. 2020; Sadam et al. 2020; Unsar et al. 2020; Fazli et al. 2020; Md. Enamul et al. 2020; Gopakumar et al. 2020; Zia-ur-Rehman 2020; Ayman et al. 2020; Mohammad et al. 2020a, 2020b; Senol 2020; Amjad et al. 2020; Ibrar et al. 2020; Sajid et al. 2020; Muhammad et al. 2021; Sidra et al. 2021; Zahir et al. 2021; Sahrish et al. 2022). To alleviate the detrimental effects of abiotic stresses, plants inherently try to survive through developing a complex signalling pathway by the commanding of responsive genes (Quan et al. 2016; Zaki et al. 2015; Hu et al. 2016a; Li et al. 2017; Alam et al. 2018b). The roles of those genes are well-defined, as they signal to plant cells to produce antioxidants and hormones, and undertake enzymatic and metabolic activities to help plants survive against stressors. Those genes also genetically modify/manipulate as per stress-specific environment (Moshelion and Altman 2015; Sagor et al. 2016). Among the abiotic stressors, high-temperature stress has a great effect on plants through changing the modulation of gene expressions, physiological and metabolic activities (Antoniou et al. 2017; Yin et al. 2017; Alam et al. 2018b). Again heat stress damages photosynthetic machineries and metabolic pathways and creates oxidative stress which breaks DNA/RNA bases, molecules of amino acid, fatty acid, carbohydrate products, ultimately stunting plant growth (Alam et al. 2018a; Sun et al. 2018). Moreover, phytohormones (abscisic acid, salicylic acid, jasmonic acid, melatonin, brassinosteroids, etc.) and small molecule compounds (calcium, nitric oxide, etc.) have been experimentally evidenced to mitigate the adverse effects of heat stress on plants (Wei et al. 2015; Hasan et al. 2015; Carvalho et al. 2016; Ali et al. 2016a; Edel et al. 2017). During heat condition, plants send the signal to activate downstream genes to produce much protein to mitigate the detrimental effects of heat stress on them (Sun et al. 2018). Through such ways, plants try to survive in adverse environmental conditions. The chapter concentrates on harmful effects of high temperature–induced stress on plants, plants self-survivability, and exogenous molecules like melatonin, brassinosteroids, etc., and also ways to protect plants under heat stresses are described.

3.2 Detrimental Effects of Heat-induced Physiological Stress on Plants

High-temperature stress (HS) is the foremost abiotic stress, which is much detrimental to phenotypic, physiological, and genotypic expressions of plants (Ashwani et al. 2010; Fahad et al. 2014a, 2014b, 2015a, 2015b, 2016a, 2016b, 2016c, 2016d, 2017, 2018a, 2019a, 2019b). Its severity across the globe is increasing day by day due to the continuously rising ambient temperature. The Intergovernmental Panel on Climate Change (IPCC 2007) estimated that air temperature across the globe is predicted to rise by 0.2°C every decade, and will increase by 1.8°C–4.0°C by the year 2100 than the current level. Therefore, it is confirmed that heat stress has become the greatest threat to sustainable agricultural production in the future. Heat stress causes diverse and continuous alterations in plants' growth, reproduction, physicobiochemical processes, leading to reduction in yield (Table 3.1) (Mittler and Blumwald 2010; McClung and Davis 2010; Fahad et al. 2017; Shah et al. 2016; Fahad et al. 2014a, 2014b, 2015a, 2015b, Fahad et al.

TABLE 3.1
Adverse Effect of Heat Stress in Different Growth Stages of Plants

Plant Species	Levels of Heat	Growth Stage	Major Consequences	References
Tall fescue (*Festuca arundinacea* Schreb.)	Optimum 22°C /18°C (day/night) and high (35°C /30°C) (day/night)	Seedling stage	• Decreased SOD and CAT in P234881 accession, but increased POD compared to their respective control. • Increased CAT, POD, and MDA in P578718 accession, compared to their respective control.	Bi et al. (2016)
	Heat stress: 40°C /36°C for 8 days (day/night) Acclimated temperature: (34°C for 4 h and 22°C for 20 h	Tillering stage	• Heat shock protein (HSP) genes, *LMW-HSP* and *HMW-HSP,* exhibited transcriptional memory for their higher transcript abundance during one or more subsequent stresses in preacclimated plants. • Leaf organic acids (citric acid, malic acid, threonic acid, etc.), sugars (sucrose, glucose, talose, glucoheptose, etc.), amino acids (serine, proline, etc.), and one fatty acid (butanoic acid) were found more in preacclimated plants .	Hu et al. (2015)
Rice (*Oryza sativa* L.)	Above 33°C, 10 days	Heading stage	• Reduced the number of pollen grains and spikelet fertility	Hurkman et al. (2009)
	32°C (night)	Reproductive stage	• Decreased spikelet fertility, grain length, width and weight, and yield	Suwa et al. (2010)
	25°C –42.5°C	Vegetative growth stage	• Decreased in the CO_2 assimilation rate	Djanaguiraman et al. (2011)
Wheat (*Triticum aestivum* L.)	32°C /24°C (day/night), 24 h	Spikelet initiation stage	• Increased spikelet sterility, and reduced grain yield	Saitoh (2008)
	37°C /28°C (day/night), 20 days	Grain filling and maturity stages	• Shortened the duration of grain filling and maturity stages, and decreased in kernel weight and yield.	Rahman et al. (2009)
	30°C /25°C day/night	From 60 DAS to maturity stage	• Reduced leaf size, shortened the periods of days to booting, heading, anthesis and maturity; drastic reduction of number of grains/spike; and finally reduced grain size and yield.	Djanaguiraman et al. (2010)
	38°C, 24 and 48 h	Seedling stage	• Decreased Chl and relative water content, diminished antioxidative capacity	Hasanuzzaman et al. (2013a)
	Avg. 34°C /29 °C (day/night)	Growth and Maturity stages	• Reduced tillers, leaf area, dry matter and grain yield	Alam et al. (2014)
Sorghum (*Hordeum vulgare*)	0°C /30°C (day/night)	65 DAS to maturity stage	• Decreased Chl content, photosystem II, photochemistry, net photosynthetic rate (Pn) and antioxidant enzyme activity, and increased ROS content, and thylakoid membrane damage, finally reduced yield.	Mohammed and Tarpley (2010)
Maize (*Zea mays*)	35°C /27°C (day/night), 14 days	Reproductive stage	• Reduced ear expansion, particularly suppression of cob extensibility by impairing hemicelluloses and cellulose synthesis through the reduction of photosynthate supply.	Yin et al. (2010)
Soybean (*Glycine max*)	38°C /28°C (day/night), 14 days	Flowering stage	• Decreased the leaf Pn and stomatal conductance, increased thicknesses of the palisade and spongy layers, damaged plasma membrane, chloroplast membrane, and thylakoid membranes, distorted mitochondrial membranes, cristae and matrix.	Tan et al. (2011)

DAS, Days after sowing.

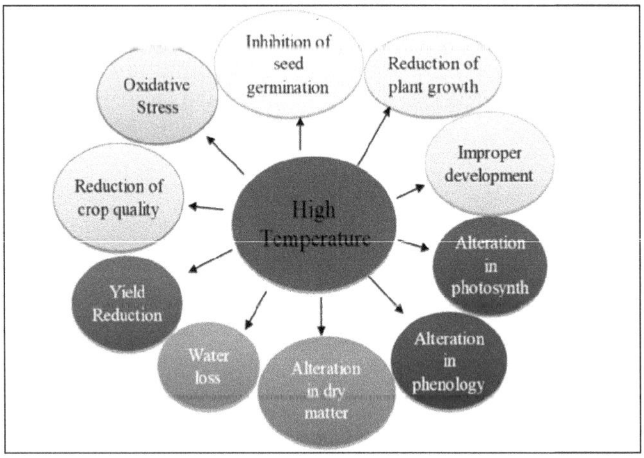

FIGURE 3.1 High temperature stress consequences in the plant cells (Adapted from Hasanuzzaman et al. 2013a).

2016a, 2016b, 2016c, 2016d, 2018b, 2019a, 2019b; Hasanuzzaman et al. 2013a; Figure 3.1). The plant, particularly, as sessile organism can't move to favourable environments during unsuitable conditions like HS (Lobell and Asner 2003; Lobell et al. 1980; Lobell and Field 2007). The adverse effects of elevated temperature on cereals vary with the timing, duration, sternness, as well as plant species/cultivars/variety (Fahad et al. 2016b). For example, C4 plants are more capable of surviving under stress condition than C3 plants.

During HS the twigs and leaves of affected plants show symptoms like sunburn, leaves senescence, growth inhibition, and discoloration of fruits and leaves (Ismail and Hall 1999; Vollenweider and Gunthardt-Goerg 2005) and heat stress also decreases the number of spikes and florets per plant, and the seed-setting (Prasad et al. 2006; Fahad et al. 2016b). The insides of floret, anthers, and pollens are much susceptible to high temperature than the ovules (Fahad et al. 2016b). Increased temperature can lessen seed germination, and thus cause poor germination and seedling establishment. A substantial reduction in the intermodal length and biomass accumulation, along with early leaf senescence in cereals, including wheat under heat stress, was discovered by Ebrahim et al. (1998).

Under heat stress, maize and sugarcane exhibited a significant reduction in the growth and net assimilation rate (Wahid and Close 2007). The major effects of the heat stress were observed in tomato, as it affected fertilization, meiosis process, and growth of fertilized embryo, ultimately causing a substantial reduction in the yield (Camejo et al. 2005). The grain weight and the total number of grains in wheat under elevated temperatures were reduced substantially (Ferris et al. 1998). The tillering of rice was very sensitive to elevated night temperature (Fahad et al. 2016a). High temperature during the reproductive phase can cause a substantial yield reduction of major cereals in the temperate regions. Since heat stress reduces the protein, oil, and starch content markedly, the quantity and quality of the final product in cereals and oilseed crops deteriorate (Wilhelm et al. 1999; Maestri et al. 2002; Ali et al. 2016b). The grain weight of a rice cultivar is almost stable in a stress-free environment (Mohammed and Tarpley 2010), but under high night temperature, rice grain production per unit area is reduced due to a decrease in individual grain weight (Fahad et al. 2016a). High temperature reduces substantial yield production in peanut (*Arachis hypogea* L.) and common beans (*Phaseolus vulgaris* L.) (Vara Parasad et al. 1999; Rainey and Griffiths 2005).

Considering the heat-induced oxidative stress, plants produce excess reactive oxygen species (ROS: H_2O_2, OH^-)/reactive nitrogen species (RNS: N_2O, NO), which lead to oxidative damage in plants' cell (Hasanuzzaman et al. 2012, 2013b). Disproportion between the systemic expression of ROS and RNS, and also a biological system's ability to readily depollute the reactive intermediates or to renovate the resulting damages is called *oxidative stress*. Under stress condition, instabilities of cells components cause toxic effects through the production of peroxides and free radicals that lead to damage of the cells'

TABLE 3.2

Heat-Induced Oxidative Damage in Plant Cells Due to Production of Higher Concentration of ROS

Lipid	Protein	DNA
• Chain breakage • Increase in membrane fluidity and permeability	• Site-specific amino acid modification • Fragmentation of the peptide chain • Aggregation of crosslinked reaction products • Altered electric charge • Enzyme inactivation • Increase susceptibility of proteins of proteolysis	• Deoxyribose oxidation • Strand breakage • Removal of nucleotides • Modification of bases • DNA-protein crosslinks

Source: (Hasanuzzaman et al. 2013a).

proteins (amino acids, peptide bonds), lipids (saturated and unsaturated fatty acids), strands breakage of DNA/RNA, exclusion of nucleotides, alteration of bases, DNA-protein crosslink, etc. (Table 3.2). Under HS the cell membranes' permeability, stability of various proteins, RNA species, cytoskeleton structures, the efficiency of enzymatic reactions in the plants' cell are generally altered, which ultimately creates metabolic imbalances in plants' cells (Pagamas and Nawata 2008; Ruelland and Zachowski 2010; Ahuja et al. 2010; Suzuki et al. 2011, 2012). Some common effects of heat stress on plants have been pointed out below in Table 3.3.

TABLE 3.3

Reported Levels of Melatonin in Plants

Common Name	Scientific Name	Family	Concentration (ng/g)	References
Seeds				
Milk thistle	*Silybum marianum*	Asteraceae	2	Reiter and Tan (2002)
Poppy	*Popaver somniferum*	Papaveraceae	6	
Anise	*Pimpinela anisum*	Apiaceae	7	
Coriander	*Coriandrum sativum*	Apiaceae	7	
Celery	*Apium graveolens*	Apiaceae	7	
Flax	*Linum usitatissimum*	Linaceae	12	
Green cardamom	*Elettaria cardamomum*	Zingiberaceae	15	
Alfalfa	*Medicago sativa*	Fabaceae	16	
Fennel	*Foeniculum vulgare*	Apiaceae (Umbelliferae)	28	
Sunflower	*Helianthus annuus*	Asteraceae	29	
Fenugreek	*Trigonella foenum-graecum*	Fabaceae	43	
Wolf berry	*Lycium barbarum*	Solanaceae	103	
Black mustard	*Brassica nigra*	Brassicaceae	129	
White mustard	*Brassica hirta*	Brassicaceae	189	
Barley	*Hordeum vulgare* L.	Poaceae	0.4, 0.87	Hattori et al. (1995); Badria (2002); Wang et al. (2017)
Rice (different varieties)	*Oryza sativa japonica* L.	Poaceae	1, 1.50, 1–201	
Oat	*Avena sativa* L.	Poaceae	2	
Corn (different varieties)	*Zea mays* L.	Poaceae	2, 1.88, 11–2034	
Tall fescue	*Festuca arundinacea*	Poaceae	5	Hattori et al. (1995)
Huang-qin	*Scutellaria biacalensis*	Lamiaceae	7	Manchester et al. (2000)
Almond	*Prunus amygdalus* Batsch	Rosaceae	39	

(*Continued*)

TABLE 3.3 (CONTINUED)

Reported Levels of Melatonin in Plants

Common Name	Scientific Name	Family	Concentration (ng/g)	References
Germinated Seeds				
Alfalfa	*Medicago sativa*	Fabaceae	0.133	Aguilera et al. (2015)
Lentil	*Lens sculenta*	Fabaceae	0.217	
Mung bean	*Vigna radiata*	Fabaceae	0.166	
Onion	*Allium cepa*	Amaryllidaceae	0.302	
Broccoli	*Brassica oleraceae*	Brassicaceae	0.439	
Red cabbage	*Brassica oleraceae capitata rubra*	Brassicaceae	0.857	
Radish	*Raphanus sativus japonicum, rambo, sinicum rosae*	Brassicaceae	0.536	
Leaves				
St. John's wort	*Hypericum perforatum*	Hypericaceae	1750	Murch and Saxena (2006)
Fever few, gold	*Tanacetum parthenium*	Asteraceae	1920	
Fever few, green	*Tanacetum parthenium*	Asteraceae	2450	
Morning glory	*Pharbitis choisy*	Convolvulaceae	0.0005	Van Tassel et al. (2001)
Chinese cabbage	*Brassica chinensis*	Brassicaceae	0.10	Hattori et al. (1995); Badria (2002)
Cabbage	*Brassica oleracea* L. var. *capitata*	Brassicaceae	0.10, 0.30	
Tomato (Transgenic)	*Solanum lycopersicum* L.	Solanaceae	45	Wang et al. (2017)
Lupin	*Lupinus albus* L.	Fabaceae	75.6	Arnao and Hernández-Ruiz (2013)
Shoots				
Morning glory	*Pharbitis choisy*	Convolvulaceae	0.004	Van Tassel et al. (2001)
Asparagus	*Asparagus officinalis* L.	Asparagaceae	0.01	Hattori et al. (1995)
Red pigweed	*Chenopodium rubrum* L.	Chenopodiaceae	0.20	Kolář et al. (1997)
Flowers				
St. John's wort	*Hypericum perforatum*	Hypericaceae	4390	Murch and Saxena (2006)

It is well-documented that under stress conditions, ROS act as a cellular messenger in redox signalling (Hasanuzzaman et al. 2012, 2013b; Fahad et al. 2014a, 2014b, 2015a, 2015b, 2016a, 2016b, 2016c, 2016d, 2017, 2018a, 2019a, 2019b). Plant cell components such as chloroplasts, mitochondria, cell wall, plasma membrane, endoplasmic reticulum, and microbodies are responsible for the production of ROS/RNS under adverse conditions (Zahoor et al. 2016; Hasanuzzaman et al. 2013b; Figure 3.2), while the enzymes such as NADH/NAD oxidase, xanthine oxidase, lipoxygenase, cyclooxygenase, etc. motivated or stimulated by biotic/abiotic stresses for converting molecular oxygen (O_2) into superoxide anion radical (O_2^-) in the cells of affected plants which is the principal producer of ROS (H_2O_2, OH^-) and RNS (N_2O, NO) (Reddy et al. 2004; Almeselmani et al. 2006; Sharma et al. 2012).

Under stress conditions (including heat stress), numerous physiological and molecular processes and pathways such as regulation of the cellular redox status, production of antioxidants and hormones, stimulation of calcium and ROS signalling, and induction of defensive genes are involved (Peleg and Blumwald 2011; Hasanuzzaman et al. 2010, 2012, 2013b). Also several metabolic alterations including overproduction of ROS, i.e. hydrogen peroxide (H_2O_2), superoxide radical (O_2^-), singlet oxygen (O_1^-), hydroxyl radical (OH^-), etc.; RNS i.e. nitrous oxide (N_2O), nitric oxide (NO), peroxynitrite ($ONOO^-$) etc.; malondialdehyde (MDA; Figure 3.3), photoinhibition, denaturation of some proteins such as chloroplast

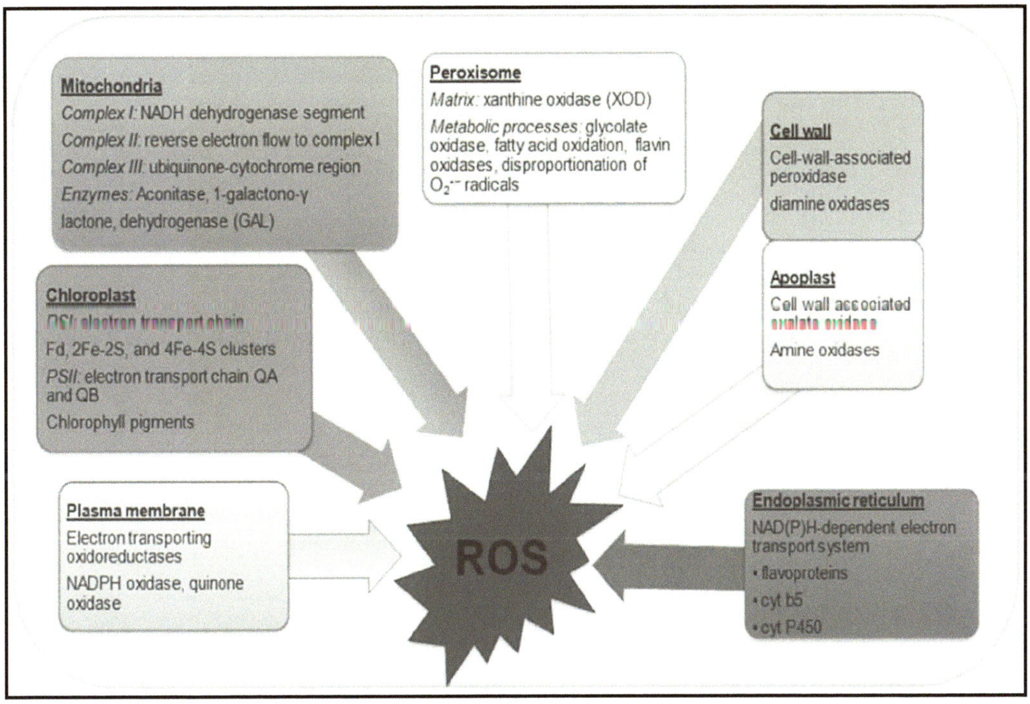

FIGURE 3.2 Production of ROS in the plant cells due to stress induced oxidative stress (adapted from Hasanuzzaman et al. 2013a, 2013b).

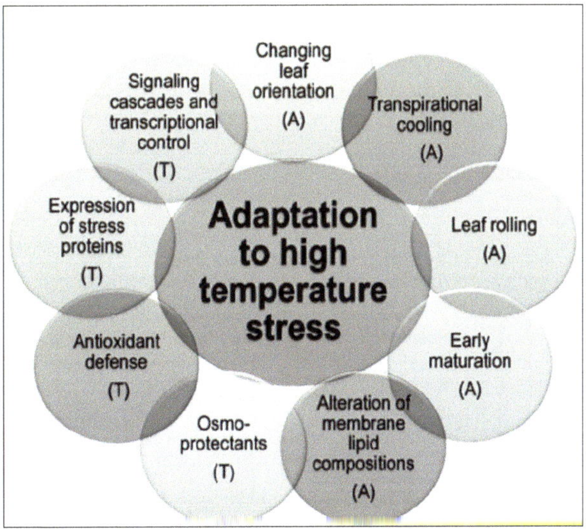

FIGURE 3.3 Survival mechanisms of plants under high temperature stress. A: Avoidance, T: Tolerance (Hasanuzzaman et al. 2013a).

proteins, oxidation and misfolding of protein, damage to biofilm structure and functions, inhibition in protein synthesis, and inhibition of photosynthesis by damaging structural organization of photosystem II (PSII) under heat stress condition (Salvucci and Crafts-Brandner 2004; Lazar et al. 2013). Under heat stress, the oxidative degradation of lipids occurs in plant cells. In that case, free radicals 'steal' electrons from the lipids and release them in cell membranes, resulting in cell damage (Figure 3.4).

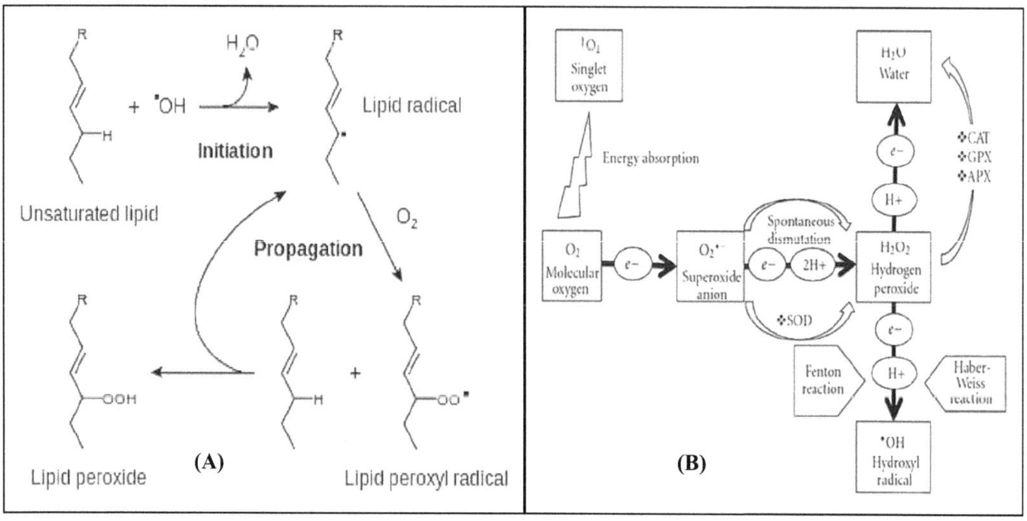

FIGURE 3.4 (A) Process to production of lipid peroxidation in plant cells under stress condition; (B) Schematic representation of generation of reactive oxygen species (ROS) in plants (adapted from Sharma et al. 2012).

3.3 Adaptation Mechanisms of Plants to High Temperature

Although, plants unceasingly fight to survive against various environmental stresses, including heat stress, to some extent plants are able to tolerate HS by altering the physiological changes in their cells, and frequently by changing signal pathways for metabolism, particularly through generating compatible solutes/antioxidants, which are able to maintain cell turgor by osmotic adjustment, organize proteins and cellular structures, and also modify the antioxidant system to reestablish the cellular redox balance and homeostasis (Valliyodan and Nguyen 2006; Munns and Tester 2008; Janska et al. 2010; Adnan et al. 2016; Fahad et al. 2014a, 2014b, 2015a, 2015b, 2016a, 2016b, 2016c, 2016d, 2017, 2018a, 2019a, 2019b; Hasanuzzaman et al. 2010, 2013b; Figure 3.5).

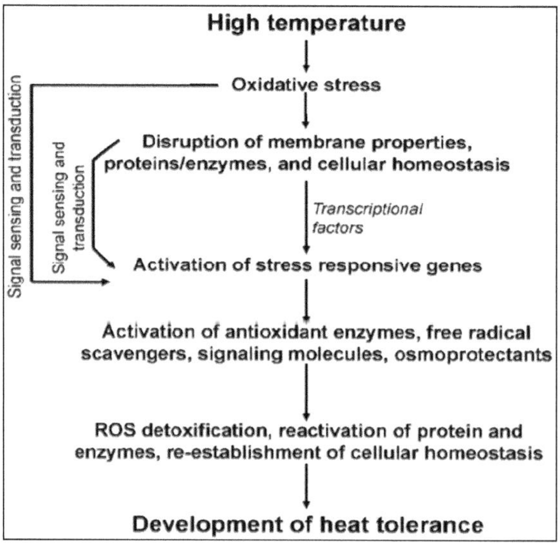

FIGURE 3.5 Heat-induced signalling transduction mechanism and development of heat tolerance in plants (Hasanuzzaman et al. 2013a).

At the molecular level, HS causes the modifications in expression level of genes involved in direct protection of plant cells (Chinnusamy et al. 2007; Shinozaki and Yamaguchi-Shinozaki 2007), including those involving biosynthesis of osmoprotectants, detoxifying enzymes, transporters, and regulatory proteins (Semenov and Halford 2009; Krasensky and Jonak 2012). In such condition (HS), the modification of physiological and biochemical processes by the changes of gene expression gradually lead to the development of heat tolerance in the form of acclimation, or in the ideal case, adaptation (Hasanuzzaman et al. 2010, 2013b; Figure 3.5).

3.4 Management through Exogenous Melatonin and 2,4-Epibrassinolide against Heat Stress

Phytohormones play a vital role in response of plants to high temperature (Ahammed et al. 2013; Fahad et al. 2014a, 2014b). For example, elevated temperature persuades Abscisic acid (ABA) synthesis that leads to ABA-mediated stomatal closure, which in fact ultimately reduces transpiration losses (Carvalho et al. 2016). Salicylic acid is well known as an abiotic stress-responsive hormone, which can also promote photosynthesis under heat stress by influencing various physiological processes and biochemical reactions (Raskin 1992). Jasmonic acid (JA) or jasmonate is one of the most important cellular regulators that critically control a spectrum of physiological processes, and mediate stress responses (Kazan 2015). Exogenous application of JA, as well as an increase of its endogenous level by the overexpression of JA biosynthetic gene, confers thermotolerance in *Arabidopsis thaliana* (Clarke et al. 2009). Furthermore, the biosynthesis of plant metabolites, such as soluble sugars, amino acids, organic acids, and lipids is also affected by high temperature; nevertheless, they are also crucial for plant adaptation to high-temperature stress (Carvalho et al. 2016).

In recent times, exogenous applications of protectants in the form of osmoprotectants (proline, Pro; glycine betaine, GB; trehalose, Tre, etc.), phytohormones (melatonin, brassinosteroids, BR; 24-epibrassinolide, 24-EBL; 28-homobrassinolide, 28-HBL; abscisic acid, ABA; gibberellic acid, GA; jasmonic acid, JA; salicylic acid, SA; etc.), signalling molecules (e.g. nitric oxide, NO), polyamines (putrescine, Put; spermidine, Spd, and spermine, Spm), trace elements (selenium, Se; silicon, Si; etc.) and nutrients (nitrogen, N; phosphorus, P; potassium, K; calcium, Ca; etc.) have been found to be effective in mitigating HS-induced damage in plants (Barnabás et al. 2008; Sirhindi et al. 2009; Waraich et al. 2012; Arora et al. 2012; Hasanuzzaman et al. 2012, 2013a, 2013b; Shi et al. 2015; Fan et al. 2015; Xu et al. 2016; Yadava et al. 2016; Hu et al. 2016a, 2016b; Wu et al. 2017). Among these protectants, how melatonin (Figure 3.6) and 24-EBL protect plants under heat stress is discussed under the following subheading:

3.4.1 Melatonin

Melatonin (MT) is a ubiquitous, amphiphilic, and pleiotropic signalling molecule (Ceraulo et al. 1999; Xu et al. 2016) that provides physiological protection to plant and animal kingdoms against diverse environmental stresses, and has numerous cellular and physiological functions in animals, fungi, bacteria, many

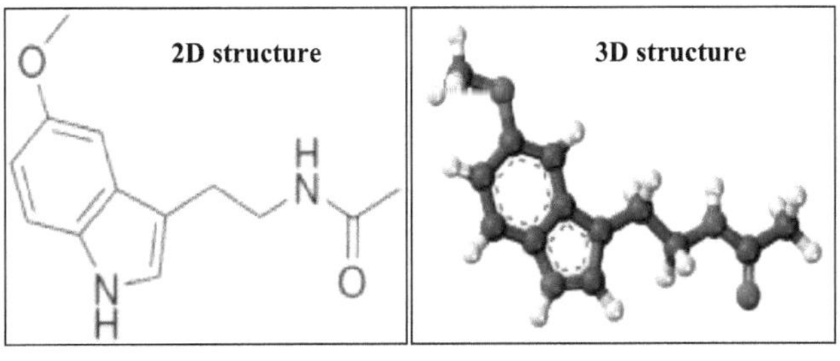

FIGURE 3.6 Chemical formula: $C_{13}H_{16}N_2O_2$; chemical structure: N-acetyl-5-methoxytryptamine (melatonin) and molecular mass: 232.278 g mol^{-1} of melatonin (adapted from NCBI 2019a).

TABLE 3.4
Exogenous Application of Melatonin or Modulation of Exogenous Melatonin Regulates Plant Abiotic Stress Tolerance

SI No.	Concentration of Melatonin or Transgene	Plant Species	Changes and Mechanisms	References
1	5μM and 20μM	*Arabidopsis thaliana*	Expression of HSF and HSP	Shi et al. (2015)
2	20μM	Tomato (*Solanum lycopersicum*)	HSP abundance and expression	Xu et al. (2016)
3	Overexpression of tomato *ASMT*	Tomato (*S. lycopersicum*)	HSP abundance and expression	Xu et al. (2016)
4	0.3μM, 1μM, 6μM, 12μM, 30μM, 60μM, 90 μM	*Phacelia tanacetifolia*	Seed germination	Tiryaki and Keles (2012)

SNAT: serotonin N-acetyltransferase; ASMT: N-acetylserotonin methyltransferase.

plants and crops, including cucumber, wheat, mungbean, sunflower, grape, soybean, tomato (Szafranska et al. 2013; Meng et al. 2014; Mukherjee et al. 2014; Turk et al. 2014; Liu et al. 2015; Wei et al. 2015) and most popular beverages including coffee, tea, wine, and beer (Mercolini et al. 2008; Maldonado et al. 2009; Iriti et al. 2010; Stege et al. 2010; Table 3.4 and Figure 3.6).

3.4.1.1 Functions of Melatonin

Melatonin serves as an important signal molecule during plant developmental processes and multiple abiotic stresses, and provides physiological protection against a variety of environmental stresses, such as heat, chilling, drought, salt, heavy metals, and ultraviolet radiation (Hasan et al. 2015; Reiter et al. 2015; Shi et al. 2015; Li et al. 2016; Xu et al. 2016; Antoniou et al. 2017). It can freely cross cell membranes and distribute to any aqueous compartment, including cytosol, nucleus, and mitochondria (Menendez-Pelaez et al. 1993). In plants like kelps and other algae, the distribution of MT between cytoplasts/symplasts, vacuoles, and apoplasts was described (Hardeland 1999; Hardeland and Poeggeler 2003; Hardeland et al. 2007). MT has an amphiphilic attribute, which indicates that membrane receptors are not necessarily required for mediating its functions. Nonreceptor-mediated actions of melatonin have been identified. For example, it can scavenge free radicals and work as an antioxidant (Reiter et al. 2008; Xu et al. 2010; Um and Kwon 2010; Romero et al. 2010).

3.4.1.2 The Roles of Melatonin in Plants

MT effectively scavenges a variety of ROS/RNS, and protects cells, tissues, and organisms from oxidative stress (Arnao and Hernández-Ruiz 2006; Tan et al. 2007; Jung et al. 2010; Kaur et al. 2010; Laothong et al. 2010; Nopparat et al. 2010; Shi et al. 2015; Figure 3.7).

Since the formation of ROS/RNS and their metabolisms are basic mechanisms related to cellular biology and physiopathology in all aerobic organisms, it is deduced that the initial function of melatonin in organisms is to serve as an antioxidant with the other functions being acquired during evolution (Hardeland et al. 1995; Tan et al. 2010). The receptor-dependent and receptor-independent activities of melatonin are attributable to its highly pleiotropic nature. Tan et al. (2000) hypothesized that MT might be the first line of defence to protect plants against harsh environments, such as cold, heat, drought, and air or soil pollution. This hypothesis was subsequently supported by other reports (Murch et al. 2009; Tan et al. 2010). To elucidate the roles of MT in plants, Tan et al. (2000) reported that when plants were moved from room temperature to a cold (4°C) environment, MT content was significantly elevated. The observation of elevated melatonin in plants from alpine and the Mediterranean environments exposed to intense UV radiation led to the discovery of large natural temperature variations (Hardeland et al. 2007; Murch et al. 2009; Tan et al. 2010). Wei et al. (2015) experimentally proved that exogenous MT improved salt and drought tolerance of plants, demonstrating the significant potential of this signalling and pleiotropic

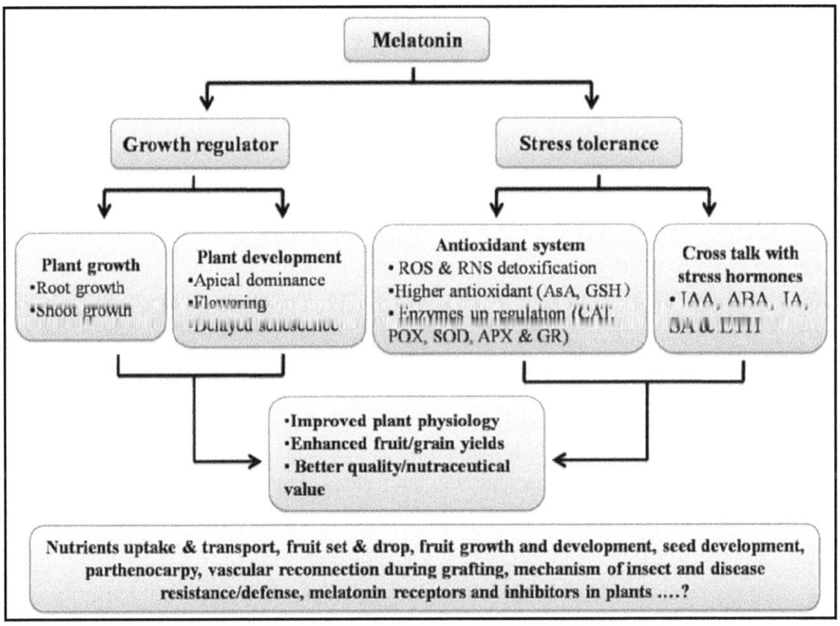

FIGURE 3.7 Functions of melatonin in plants process. ROS, Reactive oxygen species; RNS, Reactive nitrogen species; AsA, Ascorbic acid; GSH, Glutathione; CAT, Catalase; POX, Peroxidase; SOD, Superoxide dismutase; APX, Ascorbate peroxidase; GR, Glutathione reductase; IAA, Indole-3-acetic acid; ABA, Abscisic acid; JA, Jasmonic acid; SA, Salicylic acid; and ETH, Ethylene (Adapted from Nawaz et al. 2016).

molecule for improving field crops like soybean. In bermudagrass (*Cynodon dactylon* L.), exogenous melatonin improved salt-, drought-, and cold-stress tolerance when compared to control. MT-treated plants exhibited lower ROS burst, EL, cell damage, higher plant height-biomass; an increase in levels of amino acids, organic acids, sugars, and alcohols than untreated plants, indicating that MT affected the carbohydrate and nitrogen metabolisms which were the key solutes involved in the osmotic stress response (Shi et al. 2015). Recently, similar data were observed in *Citrus aurantium* seedlings by Kostopoulou et al. (2015); in roots, seedlings, and cotyledons of sunflower by Mukherjee et al. (2014); in cucumber seeds by Zhang et al. and in *Chara australis* cell cultures by Beilby et al. (2015) under salt stress; and also in grape (*Vitis vinifera*) cuttings under water-deficient stress by Meng et al. (2014), in cucumber seedlings under HS by Zhao et al. and in tomato plants under alkaline stress by Liu et al. (2015). Under salt stress, apple seedlings (*Malus hupehensis*) pretreated with melatonin showed higher shoot height, leaf number, chlorophyll content, and lower EL compared with control treatment. It is noticeable that H_2O_2 levels were halved, ROS-metabolizing enzyme levels (ascorbate peroxidase, catalase, and peroxidase activities) were increased, and Na^+ and K+ transporters (NHX1 and AKT1) were upregulated, which would be helpful for plants to alleviate saline-induced inhibition (Li et al. 2012).

Abiotic and biotic stressors provoke an increase in endogenous melatonin levels through the upregulation of MT biosynthetic genes. Both abiotic and biotic stress effects are mediated by an oxidation burst, ROS induction being the first cellular signal. Stressors act as negative effectors in many cellular and physiological processes, such as photosynthesis and membranes' integrity. Endogenous melatonin can change the expression of many genes and regulatory factors that attenuate or reverse the negative effects of biotic/abiotic stressors on physiological processes, acting as a positive effector against stress. Also, MT can act directly as a free radical scavenger (direct antioxidant) on reactive oxygen and reactive nitrogen species, lipid peroxides, and toxic chemicals, controlling relevant aspects such as membrane integrity and the proper functioning of the redox network (Arnao and Hernández-Ruiz 2015).

Shi et al. (2015) summarized that the transcript levels of class A1 heat shock factors (*HSFA1*s), which serve as the master regulators of heat stress responses were significantly upregulated in exogenous

MT-treated *Arabidopsis* seedlings under heat stress. They also observed that exogenous melatonin-enhanced thermotolerance was largely alleviated in *HSFA1s* quadruple knockout (QK) mutants and *HSFA1s*-activated transcripts of heat-responsive genes, including *HSFA2*. Heat stress–associated 32 (*HSA32*), heat shock protein 90 (*HSP90*), and heat shock protein 101 (*HSP101*), largely contributed to MT-mediated thermotolerance.

3.5 Future Prospects of Melatonin

1. Some successful examples of transgenic plants having introduced genes from vertebrates have shown substantially elevated concentrations of melatonin. In the future, additional transgenic crops will likely be produced, and the altered biosynthesis of melatonin in these plants may be used as a tool to induce resistance against biotic and abiotic stresses leading to increased crop yields. Also, considering the importance and therapeutic value of melatonin for humans, the pharmaceutical industry should capitalize on the potential human benefits of related pharmaceutical preparations for humans, animals, and plants. Presumably, in the future, we will see fruits and vegetables with higher levels of melatonin produced by the combination of conventional and modern breeding approaches. Few reports are available related to the possible role of melatonin in helping to control diseases and insects; this area of research should be aggressively explored with a definition of the specific defence mechanisms. This information could lead to the use of melatonin on a commercial scale. Melatonin enhances the phytoremediation capacity of plants, but further studies are required. Whether melatonin improves the phytoremediation capacity of the hyperaccumulator plant species would be important to document, and if so mechanism involved would require definition.
2. It has been widely reported that application of melatonin promotes root growth, but the roles of melatonin in nutrient uptake still need to be investigated; to date, no reports have been published clarifying the interactions between melatonin and nutrient uptake and transport. Similarly, very limited information is available on the response of foliar applications of melatonin to its absorption and modification of plant growth and development.

3.6 2,4-Epibrassinolide

Brassinosteroids (BRs; Figure 3.8) is a group of about 40 different polyhydroxy steroids and recognized as the sixth class plant hormone. BRs were firstly isolated by Mitchell et al. (1970) when they observed the promotion in stem elongation and cell division by the treatment of organic extracts of rapeseed pollen grains (*Brassica napus* L.) (Mitchell et al. 1970). 24-Epibrassinolide (EBL) and 28-homobrassinolide

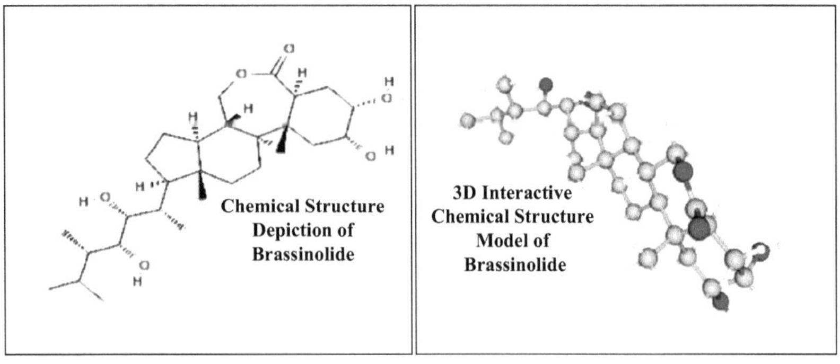

FIGURE 3.8 Chemical structure depiction, 3D interactive chemical structure model, chemical formula: $C_{28}H_{48}O_6$, molar mass: 480.69 g mol^{-1} of Brassinolide (adapted from NCBI 2019b).

(HBL) are the two most active forms of BRs that are also available commercially in the market. EBL was extracted from *Aegle marmelos* (L.) correa leaves (Bael/Bel/Bengal quince; plant of the Rutaceae family), and evaluated for the antigen toxicity against maleic hydrazide (MH)-induced genotoxicity in garlic (*Allium cepa* L.) chromosomal aberration. The percentage of chromosomal aberrations induced by maleic hydrazide (0.01%) declined significantly with EBL treatment (Sondhi et al. 2008).

3.6.1 Functions of Brassinosteroids

BRs (Figure 3.8) regulated various aspects of plant growth and development, including cell elongation, photomorphogenesis, xylem differentiation, and seed germination (Sasse 2003; Fahad et al. 2014a, 2014b), as well as adaptation to abiotic and biotic environmental stresses (Divi and Krishna 2009; Yadava et al. 2016). Clouse and Sasse (1998) showed that EBL promoted cell expansion and cell elongation. BRs were involved in numerous plant physiological and biological activities as phytohormone (Deng et al. 2007; Xue et al. 2009). BRs were also reported to counteract both abiotic and biotic stresses in plants (Xia et al. 2008; Sharma et al. 2008). Brassinolide is one kind of BR shown to have biological activity. Kim et al. (1990) extracted 'Brassinolide' from scots pine tree (*Pinussil verstris*) using chromatography-mass spectrometry, and they stated that it works as plant growth regulator. Nemhauser et al. (2004) found that EBL worked with auxin to expand and elongate the cell. EBL promoted the vascular differentiation and BR signal transduction (Caño-Delgado et al. 2004), and was essential for pollen elongation and pollen tube formation (Hewitt et al. 1985). They also observed that EBL could provide some protection to plants during chilling and drought stresses.

3.6.2 The Roles of 2,4-Epibrassinolide/Brassinosteroids in Plants

Clouse and Sasse (1998) *stated that t*he variety developed by tissue culture faced too much acceleration of senescence than the mutant developed by BR treatment. This indicated that BRs are also biologically active molecules. Stress-protective properties of BRs (especially epibrassinolide [EBL] and homobrassinolide [HBL]) were reported by various researchers in plants against a number of stresses like chilling (Dhaubhadel et al. 1999), salt (Özdemir et al. 2004; Houimli et al. 2010; Liu et al. 2013), heat (Dhaubhadel et al. 2002), and heavy metals (Bajguz 2010; Janeczko et al. 2005; Arora et al. 2012). Under different cobalt toxicity conditions, Arora et al. (2012) sprayed (EBL) as foliar application with the concentration of 1.0µM, 0.01µM, and 0.1µM on 15 days old seedlings of *Brassica juncea* L. The stress generated by cobalt ions were alleviated, and the activities of antioxidative enzymes (SOD, CAT, POD, GR, APOX, MDHAR, and DHAR) and protein content were also elevated considerably in plant leaves treated with EBL compared with control. Among them, 0.01µM concentration was the most effective treatment. The influence of EBL on the antioxidative defence system of plants under oxidative stress was studied by Cao et al. (2005); Hayat et al. (2007a, 2007b); and Sirhindi et al. (2009). The changes in the activities of antioxidative enzymes by EBL differed with plant species and stress conditions (Almeida et al. 2005; Hayat et al. 2007a, 2007b).

BR treatment improved tolerance to high temperature in *Brassica juncea* L. (Kumar et al. 2010). The ability of HBL to confer resistance to stress in *B. juncea* L. was discovered by Kumar (2012). Application of EBL and HBL showed the protective roles on the shoot and root growth, and physiological changes in seedlings of *B. juncea* L. under seasonal stress (Sirhindi et al. 2009). Application of BRs to cucumbers demonstrated to increase the metabolism and removal of pesticides, which could be beneficial for reducing the human ingestion of residual pesticides from nonorganically grown vegetables (Anuradha and Rao 2003). After application of EBL on tomato (*Lycopersicon esculentum* L.) under salt stress, the concentrations of chlorophyll a and chlorophyll b were elevated, and thus pigmentation was increased as well (Kang et al., 2004). Under salinity, BR-treated rice seeds considerably restored the pigments compared to nontreated plants (El-Meleigy et al. 2004).

EBL and HBL were applied on 10-day-old seedlings of *B. juncea* L. (RCM 619) under field conditions. Exogenous of EBL and HBL significantly ameliorated the total protein content, CAT, and SOD activities as compared to control treatment. The PPO activity was significantly enhanced on application of 10^{-8}M HBL as compared to other concentrations of EBL and HBL. HBL was more effective at germination stage in scavenging free radicals, which were produced in greater amount during germination from basic metabolic processes, whereas EBL was effective for the initial growth of seedlings as evidenced by the increase of shoot length (Sirhindi et al. 2009). EBL spraying on pepper seedlings grown under salinity stress exhibited higher biomass and Chl content, while EL and proline concentration was decreased considerably compared to control seedlings (Houimli et al. 2010). Liu et al. (2014) sprayed 150mM NaCl and different concentrations of EBL (10^{-6}M, 10^{-8}M, 10^{-10}M, 10^{-12}M) at the four-leaf stage of canola for 15 days. The result showed that 10^{-10}M concentration was the optimal treatment, and used in a subsequent experiment.

Further, the study exhibited that EBL promoted shoot growth and increased osmotic adjustment ability in all plant parts, especially in younger leaves and roots due to increase of free amino acid content in upper leaves, soluble sugars in middle leaves, organic acids and proline in lower leaves, all of these compounds in roots, as well as essential inorganic ions. Na^+ and Cl^- were sharply increased in different plant parts under salt stress, but EBL reduced their accumulation. EBL improved the uptake of K^+, Ca^{2+}, Mg^{2+}, and NO_3^- in roots, which were mainly transported to upper leaves, while NO_3^- was mainly transported to middle leaves. Thus, EBL improved ion homeostasis of K^+/Na^+, Ca^{2+}/Na^+, Mg^{2+}/Na^+, and NO_3^-/Cl, especially in younger leaves and roots. Yadava et al. (2016) applied EBL externally on maize grown at growth chamber under high temperature (48°C; RH 50%) and exposed the plants to heat stress for 3, 6, 9, 24, and 48 h. They observed that exogenous EBL arrested protein degradation and enhanced cell membrane stability as compared to control. The biochemical activities of antioxidant enzymes, such as CAT, SOD, and POX were found to be dynamically and variably modulated with post EBL treatment (Figure 3.9).

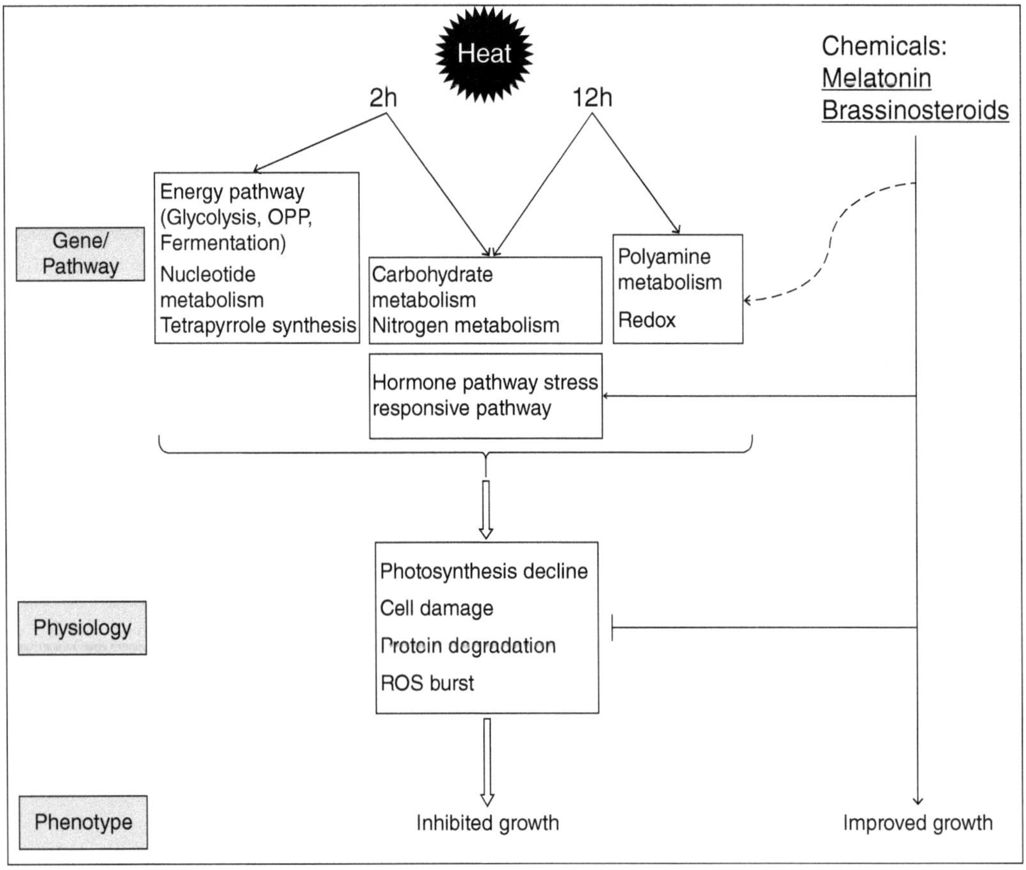

FIGURE 3.9 A model depicts the inhibitory effects of heat stress on the growth of tall fescue and improved growth by exogenous melatonin (MT) and 24-epibrassinolide (EBL) (adapted from Alam et al. 2018b).

3.6.3 Changes of Hormone Pathway Related Genes of the Plant to Escape Heat Stress

Many hormone-related genes, including BR, MT, auxin (IAA), and abscisic acid (ABA) showed significant changes after heat stress treatment. Genes involved in BR, MT, and IAA biosynthesis and signalling pathways were mainly downregulated, while the majority of genes in the ABA pathway were upregulated (Fahad et al. 2014a, 2014b).

Heat stress affects nucleotide, carbohydrate, nitrogen, and polyamine metabolisms; redox, hormone, and energy pathways; and regulates the expression of stress-responsive genes, resulting in the inhibition of photosynthesis, the increases of cell damages and ROS contents, degradation of proteins and other macromolecules. The ultimate result, the growth of tall fescue was inhibited. However, exogenous application of chemicals like MT and EBL modulated the expressions of stress-responsive genes, as well as the genes involved in hormone biosynthesis and signalling pathways, and thus alleviated the inhibitory effects of heat stress, leading to improvement in the growth of tall fescue turfgrass (Alam et al. 2018a, 2018b).

3.7 Conclusion

Heat stress extensively changed, ROS, RNS, antioxidant enzymes, sugar, salt, cation, anion, amino acid, especially transcriptome of plants. The degradation pathways of DNA, RNA, and protein were activated by heat stress. After heat, the genes involved in energy metabolism were upregulated to minimize photosynthesis damages caused by heat stress. *HSF*, *HSP*, *HSC*, and hormone-related genes were modulated shortly after heat, indicating self-adjustment of the plants in response to heat stress. Short-term heat stress activated nucleotide metabolism and tetrapyrrole synthesis, while long-term heat stress regulated polyamine metabolism and redox pathways. In addition, exogenous application of MT, EBL, and other osmoprotectants improved plant tolerance to heat stress as evidenced by the decrease of ROS level, EL and MDA contents, and the increase of total chlorophyll content and antioxidant enzyme activities. Therefore, heat stress inhibited plant growth while MT, EBL, SA, JA, etc. alleviated the inhibitory effects of heat stress in the plants resulting in treated plants showing comparatively better growth than untreated seedlings.

REFERENCES

Abdul S, Muhammad AA, Shabir H, Hesham A, El E, Sajjad H, Niaz A, Abdul G, RZ Sayyed, Fahad S, Subhan D, Rahul D (2021a) Zinc nutrition and arbuscular mycorrhizal symbiosis effects on maize (*Zea mays* L.) growth and productivity. *J Saudi Society Agric Sci* https://doi.org/10.1016/j.sjbs.2021.06.096

Abdul S, Muhammad AA, Subhan D, Niaz A, Fahad S, Rahul D, Mohammad JA, Omaima N, Muhammad Habib Ur R, Bernard RG (2021b) Effect of arbuscular mycorrhizal fungi on the physiological functioning of maizeunder zinc-deficient soils. *Sci Rep* 11:18468.

Adnan M, Fahad S, Muhammad Z, Shahen S, Ishaq AM, Subhan D, Zafar-ul-Hye M, Martin LB, Raja MMN, Beena S, Saud S, Imran A, Zhen Y, Martin B, Jiri H, Rahul D (2020) Coupling phosphate-solubilizing bacteria with phosphorus supplements improve maize phosphorus acquisition and growth under lime induced salinity stress. *Plants* 9(900). https://doi.org/10.3390/plants9070900

Adnan M, Fahad S, Khan IA, Saeed M, Ihsan MZ, Saud S, Riaz M, Wang D, Wu C (2019) Integration of poultry manure and phosphate solubilizing bacteria improved availability of Ca bound P in calcareous soils. *3 Biotech* 9(10):360.

Adnan M, Shah Z, Sharif M, Rahman H (2018b) Liming induces carbon dioxide (CO_2) emission in PSB inoculated alkaline soil supplemented with different phosphorus sources. *Environ Sci Poll Res* 25(10):9501–9509.

Adnan M, Zahir S, Fahad S, Arif M, Mukhtar A, Imtiaz AK, Ishaq AM, Abdul B, Hidayat U, Muhammad A, Inayat-Ur R, Saud S, Muhammad ZI, Yousaf J, Amanullah, Hafiz MH, Wajid N (2018a) Phosphate-solubilizing bacteria nullify the antagonistic effect of soil calcification on bioavailability of phosphorus in alkaline soils. *Sci Rep* 8:4339. https://doi.org/10.1038/s41598-018-22653-7

Adnan M, Shah Z, Arif M, Khan MJ, Mian IA, Sharif M, Alam M, Basir A, Ullah H, Rahman I, Saleem N (2016) Impact of rhizobial inoculum and inorganic fertilizers on nutrients (NPK) availability and uptake in wheat crop. *Can J Soil Sci* 96:169–176.

Aguilera Y, Herrera T, Benítez V, Arribas SM, López de Pablo AL, Esteban RM, Martín-Cabrejas MA (2015) Estimation of scavenging capacity of melatonin and other antioxidants: contribution and evaluation in germinated seeds. *Food Chem* 170:203–211.

Ahammed GJ, Ruan YP, Zhou J, Xia XJ, Shi K, Zhou YH, Yu, JQ (2013) Brassinosteroid alleviates polychlorinated biphenyls-induced oxidative stress by enhancing antioxidant enzymes activity in tomato. *Chemosphere* 90:2645–2653.

Ahmad S, Kamran M, Ding R, Meng X, Wang H, Ahmad I, Fahad S, Han Q (2019) Exogenous melatonin confers drought stress by promoting plant growth, photosynthetic capacity and antioxidant defense system of maize seedlings. *PeerJ* 7:e7793. http://doi.org/10.7717/peerj.7793

Ahuja I, de Vos RCH, Bones AM, Hall RD (2010) Plant molecular stress responses face climate change. *Trends Plant Sci* 15:664–674.

Akbar H, Timothy JK, Jagadish T, Golam M, Apurbo KC, Muhammad F, Rajan B, Fahad S, Hasanuzzaman M (2020) Agricultural land degradation: processes and problems undermining future food security. In: Fahad S, Hasanuzzaman M, Alam M, Ullah H, Saeed M, Khan AK, Adnan M (ed) *Environment, climate, plant and vegetation growth*. Springer Publ Ltd, Springer Nature Switzerland AG. Part of Springer Nature, pp. 17–62. https://doi.org/10.1007/978-3-030-49732-3

Akram R, Turan V, Hammad HM, Ahmad S, Hussain S, Hasnain A, Maqbool MM, Rehmani MIA, Rasool A, Masood N, Mahmood F, Mubeen M, Sultana SR, Fahad S, Amanet K, Saleem M, Abbas Y, Akhtar HM, Waseem F, Murtaza R, Amin A, Zahoor SA, ul Din MS, Nasim W (2018a) Fate of organic and inorganic pollutants in paddy soils. In: Hashmi MZ, Varma A (ed) *Environmental pollution of paddy soils, soil biology*. Springer International Publishing Ag, Cham, Switzerland, pp. 197–214.

Akram R, Turan V, Wahid A, Ijaz M, Shahid MA, Kaleem S, Hafeez A, Maqbool MM, Chaudhary HJ, Munis, MFH, Mubeen M, Sadiq N, Murtaza R, Kazmi DH, Ali S, Khan N, Sultana SR, Fahad S, Amin A, Nasim W (2018b) Paddy land pollutants and their role in climate change. In: Hashmi MZ, Varma A (ed) *Environmental pollution of paddy soils, soil biology*. Springer International Publishing Ag, Cham, Switzerland, pp. 113–124.

Alam MN, Bodruzzaman M, Hossain MM, Sadekuzzaman M (2014) Growth performance of spring wheat under heat stress conditions. *Int J Agron Agric Res* 4(6):91–103.

Alam MN, Wang Y, Zhulong Chan Z (2018a) Physiological and biochemical analyses reveal drought tolerance in cool-season tall fescue (*Festuca arundinacea*) turf grass with the application of melatonin. *Crop Pasture Sci* 69(10):1041–1049. https://doi.org/10.1071/CP18394

Alam MN, Zhang L, Yang L, Islam MR, Liu Y, Luo H, Yang P, Qingfeng Wang Q, Chan Z (2018b) Transcriptomic profiling of tall fescue in response to heat stress and improved thermotolerance by melatonin and 24-epibrassinolide. *BMC Genom* 19:224–237. https://doi.org/10.1186/s12864-018-4588-y

Ali A, Imtiyaz M, Adnan M, Arshad M, Rahman IU, Jamal Y, Muhammad H, Saleem N, Rahman Z (2016a) Effect of Zinc Activities on Shoot, Root Biomass and Phosphorus Uptake in Wheat Gynotypes. *Am Eurasian J Agric Environ Sci* 16(1):204–208.

Ali M, Ullah Z, Mian IA, Khan M, Khan N, Adnan M, Saeed M (2016b) Response of maize to nitrogen levels and seed priming. *Pure Appl Biol* 5(3):578–587.

Almeida JM, Fidalgo F, Confraria A, Santos A, Pires H, Santos I (2005) Effect of hydrogen peroxide on catalase gene expiration, isoform activities and levels in leaves of potato sprayed with homobrassinolide and ultrastructure changes in mesophyll cells. *Funct Plant Biol* 32:707–720.

Almeselmani M, Deshmukh PS, Sairam RK, Kushwaha SR, Singh TP (2006) Protective role of antioxidant enzymes under high temperature stress. *Plant Sci* 171(3):382–388.

Amanullah, Muhammad I, Haider N, Shah K, Manzoor A, Asim M, Saif U, Izhar A, Fahad S, Adnan M et al. (2021) Integrated foliar nutrients application improve wheat (*Triticum Aestivum* L.) productivity under calcareous soils in drylands. *Commun Soil Sci Plant Anal.* http://dx.doi.org/10.1080/00103624.2021.1956521

Amanullah, FS (eds.) (2017) Rice - technology and production. IntechOpen Croatia 2017. http://dx.doi.org/10.5772/64480

Amanullah, FS (eds.) (2018a) Corn-Production and human Health in changing climate. IntechOpen United Kingdom 2018. http://dx.doi.org/10.5772/intechopen.74074

Amanullah, FS (eds.) (2018b) Nitrogen in agriculture – updates. IntechOpen Croatia 2018. http://dx.doi.org/10.5772/65846

Amanullah, SK, Imran, HAK, Muhammad A, Abdel RA, Muhammad A, Fahad S, Azizullah S, Brajendra P (2020) Effects of climate change on irrigation water quality. In: Fahad S, Hasanuzzaman M, Alam

M, Ullah H, Saeed M, Khan AK, Adnan M (ed), *Environment, climate, plant and vegetation growth.* Springer Publ Ltd, Springer Nature Switzerland AG. Part of Springer Nature. pp. 123–132. https://doi.org/10.1007/978-3-030-49732-3

Amir M, Muhammad A, Allah B, Sevgi Ç, Haroon ZK, Muhammad A, Emre A (2020) Bio fortification under climate change: the fight between quality and quantity. In: Fahad S, Hasanuzzaman M, Alam M, Ullah H, Saeed M, Khan AK, Adnan M (eds), *Environment, climate, plant and vegetation growth.* Springer Publ Ltd, Springer Nature Switzerland AG. Part of Springer Nature. pp. 173–228. https://doi.org/10.1007/978-3-030-49732-3

Amjad I, Muhammad H, Farooq S, Anwar H (2020) Role of plant bioactives in sustainable agriculture. In: Fahad S, Hasanuzzaman M, Alam M, Ullah H, Saeed M, Khan AK, Adnan M (ed) *Environment, climate, plant and vegetation growth.* Springer Publ Ltd, Springer Nature Switzerland AG. Part of Springer Nature. pp. 591–606. https://doi.org/10.1007/978-3-030-49732-3

Amjad SF, Mansoora N, Din IU, Khalid IR, Jatoi GH, Murtaza G, Yaseen S, Naz M, Danish S, Fahad S et al. (2021) Application of zinc fertilizer and mycorrhizal inoculation on physio- biochemical parameters of wheat grown under water-stressed environment. *Sustainability* 13:11007. https://doi.org/10.3390/su131911007

Antoniou C, Chatzimichail G, Xenofontos R, Pavlou JJ, Panagiotou E, Christou A, Fotopoulos V (2017) Melatonin systemically ameliorates drought stress-induced damage in *Medicago sativa* plants by modulating nitro-oxidative homeostasis and proline metabolism. *J Pineal Res* 62:e12401. https://doi.org/10.1111/jpi.12401

Anuradha S, Rao SSR (2003) Application of brassinosteroids to rice seeds (*Oryza sativa* L.) reduced the impact of salt stress on growth, prevented photosynthetic pigment loss and increased nitrate reductase activity. *Plant Growth Regul* 40:29–32.

Arif M, Talha J, Muhammad R, Fahad S, Muhammad A, Amanullah, Kawsar A, Ishaq AM, Bushra K, Fahd R (2020) Biochar; a remedy for climate change. In: Fahad S, Hasanuzzaman M, Alam M, Ullah H, Saeed M, Khan AK, Adnan M (eds), *Environment, climate, plant and vegetation growth.* Springer Publ Ltd, Springer Nature Switzerland AG. Part of Springer Nature. pp. 151–172. https://doi.org/10.1007/978-3-030-49732-3

Arnao MB and Hernández-Ruiz J (2006) The physiological function of melatonin in plants. *Plant Signal Behav* 1(3):89–95.

Arnao MB, Hernández-Ruiz J (2013) Growth conditions determine different melatonin levels in *Lupinus albus* L. *J Pineal Res* 55:149–155.

Arnao MB, Hernández-Ruiz J (2015) Functions of melatonin in plants: a review. *J Pineal Res* 59:133–150.

Arora P, Bhardwaj R, Kanwar MK (2012) Effect of 24-epibrassinolide on growth, protein content and antioxidative defense system of *Brassica juncea* L. subjected to cobalt ion toxicity. *Acta Physiol Plant* 34(5):2007–2017.

Ashfaq AR, Uzma Y, Niaz A, Muhammad AA, Fahad S, Haider S, Tayebeh Z, Subhan D, Süleyman T, Hesham AELE, Pramila T, Jamal MA, Sulaiman AA, Rahul D (2021) Toxicity of Cadmium and Nickel in the context of applied activated carbon biochar for improvement in soil fertility. *Saudi Society Agric Sci.* https://doi.org/10.1016/j.sjbs.2021.09.035

Ashwani P, Sopory SK, Bohnert HJ, Govindjee (2010) Abiotic stress adaptation in plants: physiological, molecular and genomic foundation. Publisher: Springer, Dordrecht, Part of Springer Nature. 526 p. Print ISBN: 978-90-481-3111-2. https://doi.org/10.1007/978-90-481-3112-9.

Athar M, Masood IA, Sana S, Ahmed M, Xiukang W, Sajid F, Sher AK, Habib A, Faran M, Zafar H, Farhana G, Fahad S (2021) Bio-diesel production of sunflower through sulphur management in a semi-arid subtropical environment. *Environ Sci Pollution Res.* https://doi.org/10.1007/s11356-021-16688-z

Atif B, Hesham A, Fahad S (2021) Biochar coupling with phosphorus fertilization modifies antioxidant activity, osmolyte accumulation and reactive oxygen species synthesis in the leaves and xylem sap of rice cultivars under high-temperature stress. *Physiol Mol Biol Plants.* https://doi.org/10.1007/s12298-021-01062-7

Ayman EL, Hossain A, Barutçular C, Iqbal MA, Islam MS, Fahad S, Sytar O, Çig F, Meena RS, Erman M (2020) Consequences of salinity stress on the quality of crops and its mitigation strategies for sustainable crop production: an outlook of arid and semi-arid regions. In: Fahad S, Hasanuzzaman M, Alam M, Ullah H, Saeed M, Khan AK, Adnan M (ed) *Environment, climate, plant and vegetation growth.* Springer Publ Ltd, Springer Nature Switzerland AG. Part of Springer Nature. pp. 503–534. https://doi.org/10.1007/978-3-030-49732-3

Aziz K, Daniel KYT, Fazal M, Muhammad ZA, Farooq S, Fan W, Fahad S, Ruiyang Z (2017a) Nitrogen nutrition in cotton and control strategies for greenhouse gas emissions: a review. *Environ Sci Pollut Res* 24:23471–23487. https://doi.org/10.1007/s11356-017-0131-y

Aziz K, Daniel KYT, Muhammad ZA, Honghai L, Shahbaz AT, Mir A, Fahad S (2017b) Nitrogen fertility and abiotic stresses management in cotton crop: a review. *Environ Sci Pollut Res* 24:14551–14566. https://doi.org/10.1007/s11356-017-8920-x

Badria FA (2002) Melatonin, serotonin, and tryptamine in some Egyptian food and medicinal plants. *J Med Food* 5(3):153–157.

Bajguz A (2010) Metabolism of brassinosteroids in plants. *Plant Physiol Biochem* 45:95–107.

Barkla BJ, Vera-Estrella R, Pantoja O (2013) Progress and challenges for abiotic stress proteomics of crop plants. *Proteomics* 3:1801–1815.

Barnabás B, Jäger K, Fehér A (2008) The effect of drought and heat stress on reproductive processes in cereals. *Plant Cell Environ* 31:11–38.

Baseer M, Adnan M, Fazal M, Fahad S, Muhammad S, Fazli W, Muhammad A, Jr. Amanullah, DW, Saud S, Muhammad N, Muhammad Z, Fazli S, Beena S, Mian AR, Ishaq AM (2019) Substituting urea by organic wastes for improving maize yield in alkaline soil. *J Plant Nutr.* https://doi.org/10.1080/01904167.2019.1659344

Bayram AY, Seher Ö, Nazlican A (2020) Climate change forecasting and modeling for the year of 2050. In: Fahad S, Hasanuzzaman M, Alam M, Ullah H, Saeed M, Khan AK, Adnan M (ed), *Environment, climate, plant and vegetation growth.* Springer Publ Ltd, Springer Nature Switzerland AG. Part of Springer Nature. pp. 109–122. https://doi.org/10.1007/978-3-030-49732-3

Beilby MJ, Al Khazaaly S, Bisson MA (2015) Salinity-induced noise in membrane potential of Characeae Chara australis: effect of exogenous melatonin. *J Membrane Biol* 248:93–102.

Bi A, Fan J, Hu Z, Wang G, Amombo E, Fu J, Hu T (2016) Differential acclimation of enzymatic antioxidant metabolism and photosystem II photochemistry in tall fescue under drought and heat and the combined stresses. *Front Plant Sci* 7:453. https://doi.org/10.3389/fpls.2016.00453

Camejo D, Rodriguez P, Morales MA et al. (2005) High temperature effects on photosynthetic activity of two tomato cultivars with different heat susceptibility. *J Plant Physiol* 162:281–289.

Caño-Delgado A, Yin Y, Yu C, Vafeados D, Mora-García S, Cheng JC, Nam KH, Li J, Chory J (2004) BRL1 and BRL3 are novel brassinosteroids receptors that function in vascular differentiation in Arabidopsis. *Develop* 131(21):5341–5351.

Cao S, Xu Q, Cao Y, Qian K, An K, Zhu Y, Binzeng H, Zhao H, Kuai B (2005) Loss-of-function mutations in DET2 gene lead to an enhanced resistance to oxidative stress in Arabidopsis. *Physiol Plant* 123: 57–66.

Carvalho LC, Coito JL, Gonçalves EF, Chaves MM, Amâncio S (2016) Diffrential physiological response of the grapevine varieties Touriga Nacional and Trincadeira to combined heat, drought and light stresses. *Plant Biol* 18:101–111.

Ceraulo L, Ferrugia M, Tesoriere L, Segreto S, Livrea MA, Liveri VT (1999) Interactions of melatonin with membrane models: portioning of melatonin in AOT and lecithin reversed micelles. *J Pineal Res* 26: 108–112.

Chang W, Qiujuan J, Evgenios A, Haitao L, Gezi L, Jingjing Z, Fahad S, Ying J (2021) Hormetic effects of zinc on growth and antioxidant defense system of wheat plants. *Sci Total Environ.* https://doi.org/10.1016/j.scitotenv.2021.150992

Cheeseman J (2016) Food security in the face of salinity, drought, climate change, and population growth. In: Khan M, Ozturk M, Gul B, Ahmed MZ (ed) *Halophytes for food security in dry lands.* San Diego: Academic Press, pp. 111–123,

Chen Y, Guo Z, Dong L, Fu Z, Zheng Q, Zhang G, Qin L, Sun X, Shi Z, Fahad S, Xie F, Saud S (2021) Turf performance and physiological responses of native Poa species to summer stress in Northeast China. *PeerJ* 9:e12252 http://doi.org/10.7717/peerj.12252

Chinnusamy V, Zhu J, Zhou T, Zhu JK (2007) Small RNAs: big role in abiotic stress tolerance of plants. In: Jenks MA, Hasegawa PM, Jain SM (eds) *Advances in molecular breeding toward drought and salt tolerant crops.* Springer: Dordrecht, The Netherland, pp. 223–260.

Clarke SM, Cristescu SM, Miersch O, Harren FJ, Wasternack C, Mur LA (2009) Jasmonates act with salicylic acid to confer basal thermotolerance in *Arabidopsis thaliana. New Phytol* 182:175–187.

Clouse SD and Sasse JM (1998) Brassinosteroids: essential regulators of plant growth and development. *Annu Rev Plant Physiol Plant Mol Biol* 49:427–451.

Deepranjan S, Ardith S, Siva D, Sonam S, Shikha, MP, Amitava R, Sayyed RZ, Abdul G, Mohammad JA, Subhan D, Fahad S, Rahul D (2021) Optimizing nutrient use efficiency, productivity, energetics, and economics of red cabbage following mineral fertilization and biopriming with compatible rhizosphere microbes. *Sci Rep* 11:15680.

Deng Z, Zhang X, Tang W, Oses-Prieto JA, Suzuki N, Gendron JM, Chen H, Guan S, Chalkley RJ, Peterman TK, Burlingame AL (2007) A Proteomics study of Brassinosteroid response in Arabidopsis. *Mol Cell Proteomics* 6:2058–2071.

Depeng W, Fahad S, Saud S, Muhammad K, Aziz K, Mohammad NK, Hafiz MH, Wajid N (2018) Morphological acclimation to agronomic manipulation in leaf dispersion and orientation to promote "Ideotype" breeding: evidence from 3D visual modeling of "super" rice (*Oryza sativa* L.). *Plant Physiol Biochem*. https://doi.org/10.1016/j.plaphy.2018.11.010

Dhaubhadel S, Browning KS, Gallie DR, Krishna P (2002) Brassinosteroid functions to protect the translational machinery and heat shock protein synthesis following thermal stress. *Plant J* 29:68–69.

Dhaubhadel S, Chaudhary S, Dobinson KF, Krishna P (1999) Treatment with 24-epibrassinolide (a brassinosteroid) increases the basic thermotolerance of *Brassica napus* and tomato seedlings. *Plant Mol Biol* 40:332–342.

Divi UK, Krishna P (2009) Brassinosteroids confer stress tolerance. In: Hirt H (ed) *Plant stress biology: genomics goes systems biology*. Wiley-VCH, Weinheim, pp. 119–135.

Djanaguiraman M, Prasad PVV, Al-Khatib K (2011) Ethylene perception inhibitor 1-MCP decreases oxidative damage of leaves through enhanced antioxidant defense mechanisms in soybean plants grown under high temperature stress. *Environ Exp Bot* 71:215–223.

Djanaguiraman M, Prasad PVV, Seppanen M (2010) Selenium protects sorghum leaves from oxidative damage under high temperature stress by enhancing antioxidant defense system. *Plant Physiol Biochem* 48:999–1007.

Ebrahim MK, Zingsheim O, El-Shourbagy MN, Moore PH, Komor E (1998) Growth and sugar storage in sugarcane grown at temperature below and above optimum. *J Plant Physiol* 153:593–602.

Edel KH, Marchadier E, Brownlee C, Kudla J, Hetherington AM (2017) The evolution of calcium-based signaling in plants. *Curr Biol* 27(13):R667–R679. https://doi.org/10.1016/j.cub.2017.05.020

El-Meleigy EA, Gabr MF, Mohamed FH, Ismail MA (2004) Responses to NaCl salinity of tomato cultivated and breeding lines differing in salt tolerance in callus cultures. *Int J Agric Biol* 1560–8530.

Emre B, Ömer SU, Martín LB, Andre D, Fahad S, Rahul D, Muhammad Z-ul-H, Ghulam SH, Subhan D (2021) Studying soil erosion by evaluating changes in physico-chemical properties of soils under different land-use types. *J Saudi Society Agric Sci* 20:190–197.

Fahad S, Abdul B, Adnan M (eds) (2018b) Global wheat production. IntechOpen United Kingdom 2018. http://dx.doi.org/10.5772/intechopen.72559

Fahad S, Bajwa AA, Nazir U, Anjum SA, Farooq A, Zohaib A, Sadia S, Nasim W, Adkins S, Saud S, Ihsan MZ, Alharby H, Wu C, Wang D, Huang J (2017) Crop production under drought and heat stress: plant responses and management options. *Front Plant Sci* 8:1147. https://doi.org/10.3389/fpls.2017.01147

Fahad S, Bano A (2012) Effect of salicylic acid on physiological and biochemical characterization of maize grown in saline area. *Pak J Bot* 44:1433–1438.

Fahad S, Chen Y, Saud S, Wang K, Xiong D, Chen C, Wu C, Shah F, Nie L, Huang J (2013) Ultraviolet radiation effect on photosynthetic pigments, biochemical attributes, antioxidant enzyme activity and hormonal contents of wheat. *J Food Agri Environ* 11(3&4):1635–1641.

Fahad S, Hussain S, Bano A, Saud S, Hassan S, Shan D, Khan FA, Khan F, Chen Y, Wu C, Tabassum MA, Chun MX, Afzal M, Jan A, Jan MT, Huang J (2014a) Potential role of phytohormones and plant growth-promoting rhizobacteria in abiotic stresses: consequences for changing environment. *Environ Sci Pollut Res* 22(7):4907–4921. https://doi.org/10.1007/s11356-014-3754-2

Fahad S, Hussain S, Matloob A, Khan FA, Khaliq A, Saud S, Hassan S, Shan D, Khan F, Ullah N, Faiq M, Khan MR, Tareen AK, Khan A, Ullah A, Ullah N, Huang J (2014b) Phytohormones and plant responses to salinity stress: a review. *Plant Growth Regul* 75(2):391–404. https://doi.org/10.1007/s10725-014-0013-y

Fahad S, Hussain S, Saud S, Hassan S, Chauhan BS, Khan F et al (2016a) Responses of rapid viscoanalyzer profile and other rice grain qualities to exogenously applied plant growth regulators under high day and high night temperatures. *PLoS One* 11(7):e0159590. https://doi.org/10.1371/journal.pone.0159590

Fahad S, Hussain S, Saud S, Hassan S, Ihsan Z, Shah AN, Wu C, Yousaf M, Nasim W, Alharby H, Alghabari F, Huang J (2016c) Exogenously applied plant growth regulators enhance the morphophysiological growth and yield of rice under high temperature. *Front Plant Sci* 7:1250. https://doi.org/10.3389/fpls.2016.01250

Fahad S, Hussain S, Saud S, Hassan S, Tanveer M, Ihsan MZ, Shah AN, Ullah A, Nasrullah KF, Ullah S, Alharby HNW, Wu C, Huang J (2016d) A combined application of biochar and phosphorus alleviates heat-induced adversities on physiological, agronomical and quality attributes of rice. *Plant Physiol Biochem* 103:191–198.

Fahad S, Hussain S, Saud S, Khan F, Hassan S, Jr, Nasim W, Arif M, Wang F, Huang J (2016b) Exogenously applied plant growth regulators affect heat-stressed rice pollens. *J Agron Crop Sci* 202:139–150.

Fahad S, Hussain S, Saud S, Tanveer M, Bajwa AA, Hassan S, Shah AN, Ullah A, Wu C, Khan FA, Shah F, Ullah S, Chen Y, Huang J (2015a) A biochar application protects rice pollen from high-temperature stress. *Plant Physiol Biochem* 96:281–287.

Fahad S, Muhammad ZI, Abdul K, Ihsanullah D, Saud S, Saleh A, Wajid N, Muhammad A, Imtiaz AK, Chao W, Depeng W, Jianliang H (2018a) Consequences of high temperature under changing climate optima for rice pollen characteristics-concepts and perspectives. *Archives Agron Soil Sci.* http://dx.doi.org/10.1080/03650340.2018.1443213

Fahad S, Nie L, Chen Y, Wu C, Xiong D, Saud S, Hongyan L, Cui K, Huang J (2015b) Crop plant hormones and environmental stress. *Sustain Agric Rev* 15:371–400.

Fahad S, Saud S, Yajun C, Chao W, Depeng W (eds.) (2021f) Abiotic stress in plants. IntechOpen United Kingdom 2021. http://dx.doi.org/10.5772/intechopen.91549

Fahad S, Hasanuzzaman M, Alam M, Ullah H, Saeed M, Ali Khan I, Adnan M (eds.) (2020) *Environment, climate, plant and vegetation growth.* Springer Nature Switzerland AG 2020. DOI: https://doi.org/10.1007/978-3-030-49732-3

Fahad S, Adnan M, Hassan S, Saud S, Hussain S, Wu C, Wang D, Hakeem KR, Alharby HF, Turan V, Khan MA, Huang J (2019b) Rice responses and tolerance to high temperature. In: Hasanuzzaman M, Fujita M, Nahar K, Biswas JK (ed), *Advances in rice research for abiotic stress tolerance.* Woodhead Publ Ltd, Abington, Cambridge, UK, pp. 201–224.

Fahad S, Rehman A, Shahzad B, Tanveer M, Saud S, Kamran M, Ihtisham M, Khan SU, Turan V, Rahman MHU (2019a) Rice responses and tolerance to metal/metalloid toxicity. In: Hasanuzzaman M, Fujita M, Nahar K, Biswas JK (eds), *Advances in rice research for abiotic stress tolerance.* Woodhead Publ Ltd, Abington, Cambridge, UK, pp. 299–312.

Fahad S, Sönmez O, Saud S, Wang D, Wu C, Adnan M, Arif M, Amanullah (eds.) (2021e) Engineering tolerance in crop plants against abiotic stress. In: *Footprints of climate variability on plant diversity*, 1st edn. CRC Press, Boca Raton.

Fahad S, Sonmez O, Saud S, Wang D, Wu C, Adnan M, Turan V (eds.) (2021b) Climate change and plants: biodiversity, growth and interactions. In: *Footprints of climate variability on plant diversity*, 1st edn. CRC Press, Boca Raton.

Fahad S, Sonmez O, Saud S, Wang D, Wu C, Adnan M, Turan V (eds.) (2021c) Developing climate resilient crops: improving global food security and safety. In: *Footprints of climate variability on plant diversity*, 1st edn CRC Press, Boca Raton.

Fahad S, Sönmez O, Saud S, Wang D, Wu C, Adnan M, Turan V (eds.) (2021a) Plant growth regulators for climate-smart agriculture. In: *Footprints of climate variability on plant diversity*, 1st edn. CRC Press, Boca Raton, FL.

Fahad S, Sönmez O, Turan V, Adnan M, Saud S, Wu C, Wang D (eds.) (2021d) Sustainable soil and land management and climate change. In: *Footprints of climate variability on plant diversity*, 1st edn. CRC Press, Boca Raton.

Fan J, Chen K, Amombo E, Hu Z, Chen L, Fu J (2015) Physiological and molecular mechanism of nitric oxide (NO) involved in bermudagrass response to cold stress. *PLoS One* 10 (7): e0132991.

Farah R, Muhammad R, Muhammad SA, Tahira Y, Muhammad AA, Maryam A, Shafaqat A, Rashid M, Muhammad R, Qaiser H, Afia Z, Muhammad AA, Muhammad A, Fahad S (2020) Alternative and non-conventional soil and crop management strategies for increasing water use efficiency. In: Fahad S, Hasanuzzaman M, Alam M, Ullah H, Saeed M, Khan AK, Adnan M (ed) *Environment, climate, plant and vegetation growth.* Springer Publ Ltd, Springer Nature Switzerland AG. Part of Springer Nature, pp. 323–338. https://doi.org/10.1007/978-3-030-49732-3

Farhana G, Ishfaq A, Muhammad A, Dawood J, Fahad S, Xiuling L, Depeng W, Muhammad F, Muhammad F, Syed AS (2020) Use of crop growth model to simulate the impact of climate change on yield of various wheat cultivars under different agro-environmental conditions in Khyber Pakhtunkhwa, Pakistan. *Arabian J Geosci* 13:112 https://doi.org/10.1007/s12517-020-5118-1

Farhat A, Hafiz MH, Wajid I, Aitazaz AF, Hafiz FB, Zahida Z, Fahad S, Wajid F, Artemi C (2020) A review of soil carbon dynamics resulting from agricultural practices. *J Environ Manage* 268:110319.

Fazli W, Muhmmad S, Amjad A, Fahad S, Muhammad A, Muhammad N, Ishaq AM, Imtiaz AK, Mukhtar A, Muhammad S, Muhammad I, Rafi U, Haroon I, Muhammad A (2020) Plant-microbes interactions and functions in changing climate. In: Fahad S, Hasanuzzaman M, Alam M, Ullah H, Saeed M, Khan AK, Adnan M (ed) *Environment, climate, plant and vegetation growth.* Springer Publ Ltd, Springer Nature Switzerland AG. Part of Springer Nature. pp. 397–420. https://doi.org/10.1007/978-3-030-49732-3

Ferris R, Ellis RH, Wheeler TR, Hadley P (1998) Effect of high temperature stress at anthesis on grain yield and biomass of field grown crops of wheat. *Ann Bot* 82:631–639.

Ghulam M, Muhammad AA, Donald LS, Sajid M, Muhammad FQ, Niaz A, Ateeq Ur R, Shakeel A, Sajjad H, Muhammad A, Summia M, Aqib HAK, Fahad S, Rahul D, Mazhar I, Timothy DS (2021) Formalin fumigation and steaming of various composts differentially influence the nutrient release, growth and yield of muskmelon (*Cucumis melo* L.). *Sci Rep* 11:21057.

Gopakumar L, Bernard NO, Donato V (2020) Soil microarthropods and nutrient cycling. In: Fahad S, Hasanuzzaman M, Alam M, Ullah H, Saeed M, Khan AK, Adnan M (eds), *Environment, climate, plant and vegetation growth.* Springer Publ Ltd, Springer Nature Switzerland AG. Part of Springer Nature. pp. 453–472. https://doi.org/10.1007/978-3-030-49732-3

Guofu L, Zhenjian B, Fahad S, Guowen C, Zhixin X, Hao G, Dandan L, Yulong L, Bing L, Guoxu J, Saud S (2021) Compositional and structural changes in soil microbial communities in response to straw mulching and plant revegetation in an abandoned artificial pasture in Northeast China. *Glob Ecol Conserv* 31:e01871.

Habib Ur R, Ashfaq A, Aftab W, Manzoor H, Fahd R, Wajid I, Md. Aminul I, Vakhtang S, Muhammad A, Asmat U, Abdul W, Syeda RS, Shah S, Shahbaz K, Fahad S, Manzoor H, Saddam H, Wajid N (2017) Application of CSM-CROPGRO-Cotton model for cultivars and optimum planting dates: evaluation in changing semi-arid climate. *Field Crops Res.* http://dx.doi.org/10.1016/j.fcr.2017.07.007

Hafeez M, Farman U, Muhammad MK, Xiaowei L, Zhijun Z, Sakhawat S, Muhammad I, Mohammed AA, Mandela F-G, Nicolas D, Muzammal R, Fahad S, Yaobin L (2021) Metabolic-based insecticide resistance mechanism and ecofriendly approaches for controlling of beet armyworm *Spodoptera exigua*: a review. *Environ Sci Pollution Res.* https://doi.org/10.1007/s11356-021-16974-w

Hafiz MH, Abdul K, Farhat A, Wajid F, Fahad S , Muhammad A, Ghulam MS, Wajid N, Muhammad M, Hafiz FB (2020b) Comparative effects of organic and inorganic fertilizers on soil organic carbon and wheat productivity under arid region, *Commun Soil Sci Plant Anal.* http://dx.doi.org/10.1080/00103624.2020.1763385

Hafiz MH, Farhat A, Ashfaq A, Hafiz FB, Wajid F, Carol Jo W, Fahad S, Gerrit H (2020a) Predicting Kernel growth of maize under controlled water and nitrogen applications. *Int J Plant Prod.* https://doi.org/10.1007/s42106-020-00110-8

Hafiz MH, Farhat A, Shafqat S, Fahad S, Artemi C, Wajid F, Chaves CB, Wajid N, Muhammad M, Hafiz FB (2018) Offsetting land degradation through nitrogen and water management during maize cultivation under arid conditions. *Land Degrad Dev* 1–10. http://dx.doi.org/10.1002/ldr.2933

Hafiz MH, Muhammad A, Farhat A, Hafiz FB, Saeed AQ, Muhammad M, Fahad S, Muhammad A (2019) Environmental factors affecting the frequency of road traffic accidents: a case study of sub-urban area of Pakistan. *Environ Sci Pollut Res.* https://doi.org/10.1007/s11356-019-04752-8

Hafiz MH, Wajid F, Farhat A, Fahad S, Shafqat S, Wajid N, Hafiz FB (2016) Maize plant nitrogen uptake dynamics at limited irrigation water and nitrogen. *Environ Sci Pollut Res* 24(3):2549–2557. https://doi.org/10.1007/s11356-016-8031-0

Haider SA, Lalarukh I, Amjad SF, Mansoora N, Naz M, Naeem M, Bukhari SA, Shahbaz M, Ali SA, Marfo TD, Subhan D, Rahul D, Fahad S (2021) Drought stress alleviation by potassium-nitrate-containing chitosan/montmorillonite microparticles confers changes in *Spinacia oleracea* L. *Sustain* 13:9903. https://doi.org/10.3390/su13179903

Hamza SM, Xiukang W, Sajjad A, Sadia Z, Muhammad N, Adnan M, Fahad S et al. (2021) Interactive effects of gibberellic acid and NPK on morpho-physio-biochemical traits and organic acid exudation pattern in coriander (*Coriandrum sativum* L.) grown in soil artificially spiked with boron. *Plant Physiol Biochem* 167:884–900.

Hardeland R (1999) Melatonin and 5-methoxytryptamine in nonmetazoans. *Reprod Nutr Dev* 39:399–408.

Hardeland R, Balzer I, Poeggeler B, Fuhrberg B, Una H, Behrmann G, Wolf R, Meyer TJ, Reiter RJ (1995) On the primary functions of melatonin in evolution: mediation of photoperiodic signals in a unicell, photooxidation, and scavenging of free radicals. *J Pineal Res* 18:104–111.

Hardeland R, Pandi-Perumal SR, Poeggeler B (2007) Melatonin in plants-focus on a vertebrate night hormone with cytoprotective properties. *Funct Plant Sci Biotechnol* 1:32–45.
Hardeland R, Poeggeler B (2003) Non-vertebrate melatonin. *J Pineal Res* 34:233–241.
Hasan M, Ahammed GJ, Yin L, Shi K, Xia X, Zhou Y, Yu J, Zhou J (2015) Melatonin mitigates cadmium phytotoxicity through modulation of phytochelatins biosynthesis, vacuolar sequestration, and antioxidant potential in *Solanum lycopersicum* L. *Front Plant Sci* 6:601.
Hasanuzzaman M, Hossain MA, da Silva JAT, Fujita M (2012) Plant responses and tolerance to abiotic oxidative stress: an Antioxidant defense is a key factor. In Bandi V, Shanker AK, Shanker C, Mandapaka M (eds) *Crop stress and its management: perspectives and strategies*. Springer: Berlin, Germany, pp. 261–316.
Hasanuzzaman M, Hossain MA, Fujita M (2010) Physiological and biochemical mechanisms of nitric oxide induced abiotic stress tolerance in plants. *Am J Plant Physiol* 5:295–324.
Hasanuzzaman M, Nahar K, Alam MM, Roychowdhury R, Fujita M (2013a) Physiological, biochemical, and molecular mechanisms of heat stress tolerance in plants. *Int J Mol Sci* 14:9643–9684.
Hasanuzzaman M, Nahar K, Fujita M (2013b) Extreme temperatures, oxidative stress and antioxidant defense in plants. In: Vahdati K, Leslie C (eds) *Abiotic stress—plant responses and applications in agriculture*. IntechOpen, pp. 169–205. http://dx.doi.org/10.5772/54833
Hattori A, Migitaka H, Iigo M, Itoh M, Yamamoto K, Ohtani-Kaneko R, Hara M, Suzuki T, Reiter RJ (1995) Identification of melatonin in plants and its effects on plasma melatonin levels and binding to melatonin receptors in vertebrates. *Biochem Mol Biol Int* 35:627–634.
Hayat S, Ali B, Hasan S, Ahmad A (2007a) Effect of 28-homobrassinolide on salinity-induced changes in *Brassica juncea*. *Turkish J Biol* 31:141–146.
Hayat S, Ali B, Hasan S, Ahmad A (2007b) Brassinosteroids enhanced antioxidants under cadmium stress in *Brassica juncea*. *Environ Exp Bot* 60(1):33–41.
Hesham FA, Fahad S (2020) Melatonin application enhances biochar efficiency for drought tolerance in maize varieties: modifications in physio-biochemical machinery. *Agron J* 112(4):1–22.
Hewitt FR, Hough T, O'Neill P, Sasse JM, Williams EG, Rowan KS (1985) Effect of brassinolide and other growth regulators on the germination and growth of pollen tubes of Prunus avium using a multiple hanging drop assay. *Aust J Plant Physiol* 12:201–211.
Houimli SIM, Denden M, Mouhandes BD (2010) Effects of 24-epibrassinolide on growth, chlorophyll, electrolyte leakage and proline by pepper plants under NaCl-stress. *EurAsian J Biosci* 4:96–104.
Hu L, Zhang Z, Xiang Z, Yang Z (2016b) Exogenous application of citric acid ameliorates the adverse effect of heat stress in tall fescue (*Lolium arundinaceum*). *Front Plant Sci* 7:179.
Hu T, Liu SQ, Amombo E, Fu JM (2015) Stress memory induced rearrangements of HSP transcription, photosystem II photochemistry and metabolism of tall fescue (*Festuca arundinacea* Schreb.) in response to high-temperature stress. *Front Plant Sci* 16(6):403. https://doi.org/10.3389/fpls.2015.00403
Hu Z, Fan J, Xie Y, Amombo E, Liu A, Gitau MM, Khaldun AB, Chen L, Fu J (2016a) Comparative photosynthetic and metabolic analyses reveal mechanism of improved cold stress tolerance in bermudagrass by exogenous melatonin. *Plant Physiol Biochem* 100:94–104.
Huang L-Y, Li X-X, Zhang Y-B, Fahad S, Wang F (2021) dep1 improves rice grain yield and nitrogen use efficiency simultaneously by enhancing nitrogen and dry matter translocation. *J Integrative Agri.* http://dx.doi.org/10.1016/S2095-3119(21)63795-4
Hurkman WJ, Vensel WH, Tanaka CK, Whitehand L, Altenbach SB (2009) Effect of high temperature on albumin and globulin accumulation in the endosperm proteome of the developing wheat grain. *J Cereal Sci* 49:12–23.
Hussain MA, Fahad S, Rahat S, Muhammad FJ, Muhammad M, Qasid A, Ali A, Husain A, Nooral A, Babatope SA, Changbao S, Liya G, Ibrar A, Zhanmei J, Juncai H (2020) Multifunctional role of brassinosteroid and its analogues in plants. *Plant Growth Regul.* https://doi.org/10.1007/s10725-020-00647-8
Ibrar K, Aneela R, Khola Z, Urooba N, Sana B, Rabia S, Ishtiaq H, Mujaddad Ur R, Salvatore M (2020) Microbes and environment: global warming reverting the frozen zombies. In: Fahad S, Hasanuzzaman M, Alam M, Ullah H, Saeed M, Khan AK, Adnan M (eds), *Environment, climate, plant and vegetation growth*. Springer Publ Ltd, Springer Nature Switzerland AG. Part of Springer Nature, pp. 607–634. https://doi.org/10.1007/978-3-030-49732-3
Ikram U, Khadim D, Muhammad T, Muhammad S, Fahad S (2021) Gibberellic acid and urease inhibitor optimize nitrogen uptake and yield of maize at varying nitrogen levels under changing climate. *Environ Sci Pollution Res.* https://doi.org/10.1007/s11356-021-16049-w

Ilyas M, Mohammad N, Nadeem K, Ali H, Aamir HK, Kashif H, Fahad S, Aziz K, Abid U, (2020) Drought tolerance strategies in plants: a mechanistic approach. *J Plant Growth Regulation.* https://doi.org/10.1007/s00344-020-10174-5

IPCC (Intergovernmental Panel on Climate Change) (2007) Climate change-The physical science basis. In Contribution of Working Group I to the Fourth Assessment Report of the Intergovernmental Panel on Climate Change; Cambridge University Press: Cambridge, UK, 2007.

Iqra M, Amna B, Shakeel I, Fatima K, Sehrish L, Hamza A, Fahad S (2020) Carbon cycle in response to global warming. In: Fahad S, Hasanuzzaman M, Alam M, Ullah H, Saeed M, Khan AK, Adnan M (eds), *Environment, climate, plant and vegetation growth.* Springer Publ Ltd, Springer Nature Switzerland AG. Part of Springer Nature. pp. 1–16. https://doi.org/10.1007/978-3-030-49732-3

Irfan M, Muhammad M, Muhammad JK, Khadim MD, Dost M, Ishaq AM, Waqas A, Fahad S, Saud S et al. (2021) Heavy metals immobilization and improvement in maize (*Zea mays* L.) growth amended with biochar and compost. *Sci Rep* 11:18416.

Iriti M, Varoni EM, Vitalini S (2010) Melatonin in traditional Mediterranean diets. *J Pineal Res* 49:101–105.

Ismail AM, Hall AE (1999) Reproductive-stage heat tolerance, leaf membrane thermostability and plant morphology in cowpea. *Crop Sci* 39:1762–1768.

Jabborova D, Sulaymanov K, Sayyed RZ, Alotaibi SH, Enakiev Y, Azimov A, Jabbarov Z, Ansari MJ, Fahad S, Danish S et al. (2021) Mineral fertilizers improves the quality of turmeric and soil. *Sustain* 13:9437. https://doi.org/10.3390/su13169437

Jan M, Muhammad A-H, Adnan NS, Muhammad Y, Javaid I, Xiuling L, Depeng W, Fahad S, (2019) Modulation in growth, gas exchange, and antioxidant activities of salt-stressed rice (*Oryza sativa* L.) genotypes by zinc fertilization. *Arabian J Geosci* 12:775. https://doi.org/10.1007/s12517-019-4939-2

Janeczko A, Koscielniak J, Pilipowicz M, Szarek-Lukaszewska G, Skoczowski A (2005) Protection of winter rape photosystem 2 by 24-epibrassinolide under cadmium stress. *Photosynthetica* 43:293–298.

Janska A, Marsik P, Zelenkova S, Ovesna J (2010) Cold stress and acclimation: what is important for metabolic adjustment? *Plant Biol* 12:395–405.

Jung KH, Hong SW, Zheng HM, Lee HS, Lee H, Lee DH, Lee SY, Hong SS (2010) Melatonin ameliorates cerulein-induced pancreatitis by the modulation of nuclear erythroid 2-related factor 2 and nuclear factor-kappa B in rats. *J Pineal Res* 48:239–250.

Kamarn M, Wenwen C, Irshad A, Xiangping M, Xudong Z, Wennan S, Junzhi C, Shakeel A, Fahad S, Qingfang H, Tiening L (2017) Effect of paclobutrazol, a potential growth regulator on stalk mechanical strength, lignin accumulation and its relation with lodging resistance of maize. *Plant Growth Regul* 84:317–332. https://doi.org/10.1007/s10725-017-0342-8

Kang SA, Marjavaara PJ, Crane BR (2004) Electron transfer between cytochrome c and cytochome c peroxidase in single crystals. *J Am Chem Soc* 126(35):10836–10837.

Kaur C, Sivakumar V, Ling EA (2010) Melatonin protects periventricular white matter from damage due to hypoxia. *J Pineal Res* 48:185–193.

Kazan K (2015) Diverse roles of jasmonates and ethylene in abiotic stress tolerance. *Trends Plant Sci* 20:219–229.

Khadim D, Fahad S, Jahangir MMR, Iqbal M, Syed SA, Shah AK, Ishaq AM, Rahul D et al. (2021a) Biochar and urease inhibitor mitigate NH_3 and N_2O emissions and improve wheat yield in a urea fertilized alkaline soil. *Sci Rep* 11:17413.

Khadim D, Saif-Ur-R, FS, Syed SA, Shah AK et al. (2021b) Influence of variable biochar concentration on yield-scaled nitrous oxide emissions, Wheat yield and nitrogen use efficiency. *Sci Rep* 11:16774.

Khan MMH, Niaz A, Umber G, Muqarrab A, Muhammad AA, Muhammad I, Shabir H, Shah F, Vibhor A, Shams HA-H, Reham A, Syed MBA, Nadiyah MA, Ali TKZ, Subhan D, Rahul D (2021) Synchronization of Boron application methods and rates is environmentally friendly approach to improve quality attributes of *Mangifera indica* L. On sustainable basis. *Saudi J Bio Sci.* https://doi.org/10.1016/j.sjbs.2021.10.036

Kim SK, Abe H, Little CHA, Pharis RP (1990) Identification of two brassinosteroids from the cambial region of Scots pine (*Pinus silverstris*) by gas chromatography-mass spectrometry, after detection using a dwarf rice lamina inclination bioassay. *Plant Physiol* 94:1709–1713.

Kolář J, Macháčková I, Eder J, Prinsen E, Van Dongen W, Van Onckelen H, Illnerová H (1997) Melatonin: occurrence and daily rhythm in *Chenopodium rubrum. Phytochem* 44(8):1407–1413.

Kostopoulou Z, Therios I, Roumeliotis E, Kanellis AK, Molassiotis A (2015) Melatonin combined with ascorbic acid provides salt adaptation in *Citrus aurantium* L. seedlings. *Plant Physiol Biochem* 86:155–165.

Krasensky J, Jonak C (2012) Drought, salt, and temperature stress-induced metabolic rearrangements and regulatory networks. *J Exp Bot* 63(4):1593–1608.

Kumar M, Sirhindi G, Bhardwaj R, Kumar S, Jain G (2010) Effect of exogenous H_2O_2 on antioxidant enzymes of *Brassica juncea* L. seedlings in relation to 24-epibrassinolide under chilling stress. *Indian J Biochem Biophys* 47:378–382.

Kumar S, Sirhindi G, Bhardwaj R, Kumar M, Arora P (2012) Role of 24-epibrassinolide in amelioration of high temperature stress through antioxidant defense system in *Brassica juncea* L. *Plant Stress* 6: 550–558.

Laothong U, Pinlaor P, Hiraku Y (2010) Protective effect of melatonin against Opisthorchisviverrini-induced oxidative and nitrosative DNA damage and liver injury in hamsters. *J Pineal Res* 49:271–282.

Lazar D, Murch SJ, Beilby MJ, Al Khazaaly S (2013) Exogenous melatonin affects photosynthesis in characeae *Chara australis*. *Plant Signal Behav* 8(3):e23279.

Li H, Hu T, Amombo E, Fu J (2017) Transcriptome profilings of two tall fescue (*Festuca arundinacea*) cultivars in response to lead (Pb) stress. *BMC Genom* 18:145.

Li Q, Zhao XQ, Chang AK, Zhang QM, Bai FW (2012) Ethanol-induced yeast flocculation directed by the promoter of TPS1 encoding trehalose-6-phosphate synthase 1 for efficient ethanol production. *Metab Eng* 14(1):1–8. https://doi.org/10.1016/j.ymben.2011.12.003

Li X, Tan D-X, Jiang D, Liu F (2016) Melatonin enhances cold tolerance in drought-primed wild-type and abscisic acid-deficient mutant barley. *J Pineal Res* 61:328–339.

Liu J, Gao H, Wang X, Zheng Q, Wang C, Wang X, Wang Q (2013) Effects of 24-epibrassinolide on plant growth, osmotic regulation and ion homeostasis of salt-stressed canola. *Plant Biol* 16:440–450.

Liu J, Wang QL, Chang Q, Han LN, Pei GL (2014) Isocitrate lyase is required for urediniospore germination of *Puccinia striiformis* f. sp. tritici. *Mol Biol Repor* 41(12):7797–7806.

Liu N, Jin Z, Wang S, Gong B, Wen D, Wang X, Wei M, Shi Q (2015) Sodic alkaline stress mitigation with exogenous melatonin involves reactive oxygen metabolism and ion homeostasis in tomato. *Sci Hort* 181:18–25.

Lobell DB, Asner GP (2003) Climate and management contributions to recent trends in U.S. agricultural yields. *Science* 299.

Lobell DB, Field CB (2007) Global scale climate–crop yield relationships and the impacts of recent warming. *Environ Res Lett* 2:014002. https://doi.org/10.1088/1748-9326/2/1/014002

Lobell DB, Schlenker W, Costa-Roberts, J (1980) Climate trends and global crop production since. *Science* 333:616–620.

Maestri E, Klueva N, Perrotta C, Gulli M, Nguyen HT, Marmiroli N (2002) Molecular genetics of heat tolerance and heat shock proteins in cereals. *Plant Mol Biol* 48:667–681.

Mahar A, Amjad A, Altaf HL, Fazli W, Ronghua L, Muhammad A, Fahad S, Muhammad A, Rafiullah, Imtiaz AK, Zengqiang Z (2020) Promising technologies for Cd-contaminated soils: drawbacks and possibilities. In: Fahad S, Hasanuzzaman M, Alam M, Ullah H, Saeed M, Khan AK, Adnan M (eds), *Environment, climate, plant and vegetation growth.* Springer Publ Ltd, Springer Nature Switzerland AG. Part of Springer Nature, pp. 63–92. https://doi.org/10.1007/978-3-030-49732-3

Mahmood Ul H, Tassaduq R, Chandni I, Adnan A, Muhammad A, Muhammad MA, Muhammad H-R, Mehmood AN, Alam S, Fahad S (2021) Linking plants functioning to adaptive responses under heat stress conditions: a mechanistic review. *J Plant Growth Regul.* https://doi.org/10.1007/s00344-021-10493-1

Maldonado MD, Moreno H, Calvo JR (2009) Melatonin present in beer contributes to increase the levels of melatonin and antioxidant capacity of the human serum. *Clinical Nut* 28:188–191.

Manchester LC, Tan DX, Reiter RJ, Park W, Monis K, Qi W (2000) High levels of melatonin in the seeds of edible plants: possible function in germ tissue protection. *Life Sci* 67:3023–3029.

Manzer HS, Saud A, Soumya M, Abdullah A, Qasi DA, Bander MA, Hayssam MA, Hazem MK, Fahad S, Vishnu DR, Om PN (2021) Molybdenum and hydrogen sulfide synergistically mitigate arsenic toxicity by modulating defense system, nitrogen and cysteine assimilation in faba bean (*Vicia faba* L.) seedlings. *Environ Pollut* 290:117953.

McClung CR, Davis SJ (2010) Ambient thermometers in plants: from physiological outputs towards mechanisms of thermal sensing. *Curr Biol* 20:1086–1092.

Md Jakir H, Allah B (2020) Development and applications of transplastomic plants; a way towards eco-friendly agriculture. In: Fahad S, Hasanuzzaman M, Alam M, Ullah H, Saeed M, Khan AK, Adnan M (eds), *Environment, climate, plant and vegetation growth.* Springer Publ Ltd, Springer Nature Switzerland AG. Part of Springer Nature. pp. 285–322. https://doi.org/10.1007/978-3-030-49732-3

Md. Enamul H, Shoeb AZM, Mallik AH, Fahad S, Kamruzzaman MM, Akib J, Nayyer S, Mehedi KM, Swati AS, Md. Yeamin A, Most SS (2020) Measuring vulnerability to environmental hazards: qualitative to quantitative. In: Fahad S, Hasanuzzaman M, Alam M, Ullah H, Saeed M, Khan AK, Adnan M (eds), *Environment, climate, plant and vegetation growth.* Springer Publ Ltd, Springer Nature Switzerland AG. Part of Springer Nature, pp. 421–452. https://doi.org/10.1007/978-3-030-49732-3

Menendez-Pelaez A, Poeggeler B, Reiter RJ, Barlow-Walden L, Pablos MI, Tan DX (1993) Nuclear localization of melatonin in different mammalian tissues: immunocytochemical and radioimmunoassay evidence. *J Cell Biochem* 53:373–382.

Meng JF, Xu TF, Wang ZZ, Fang YL, Xi ZM, Zhang ZW (2014) The ameliorative effects of exogenous melatonin on grape cuttings under water-deficient stress: antioxidant metabolites, leaf anatomy, and chloroplast morphology. *J Pineal Res* 57:200–212.

Mercolini L, Addolorata Saracino M, Bugamelli F, Ferranti A, Malaguti M, Hrelia S, Raggi MA (2008) HPLC-F analysis of melatonin and resveratrol isomers in wine using an SPE procedure. *J Sep Sci* 31:1007–1014.

Mitchell JW, Mandava N, Worley JF, Plimmer JR, Smith MV (1970) Brassins-a new family of plant hormones from rape pollen. *Nature* 225(5237):1065–1066.

Mittler R, Blumwald E (2010) Genetic engineering for modern agriculture: challenges and perspectives. *Annu Rev Plant Biol* 61:443–462.

Mohammad IA-W, Abdelazeem S, Munir A, Khalid E, Adel RAU (2020b) Extent of climate change in Saudi Arabia and its impacts on agriculture: a case study from Qassim region. In: Fahad S, Hasanuzzaman M, Alam M, Ullah H, Saeed M, Khan AK, Adnan M (eds), *Environment, climate, plant and vegetation growth.* Springer Publ Ltd, Springer Nature Switzerland AG. Part of Springer Nature, pp. 635–658. https://doi.org/10.1007/978-3-030-49732-3

Mohammad IA-W, Munir A, Adel RAU, Mutair A, Muhammad IR (2020a) Advances in pyrolytic technologies with improved carbon capture and storage to combat climate change. In: Fahad S, Hasanuzzaman M, Alam M, Ullah H, Saeed M, Khan AK, Adnan M (eds), *Environment, climate, plant and vegetation growth.* Springer Publ Ltd, Springer Nature Switzerland AG. Part of Springer Nature, pp. 535–576. https://doi.org/10.1007/978-3-030-49732-3

Mohammed AR, Tarpley L (2010) Effects of high night temperature and spikelet position on yield-related parameters of rice (*Oryza sativa* L.) plants. *Eur J Agron* 33:117–123.

Moshelion M, Altman A (2015) Current challenges and future perspectives of plant and agricultural biotechnology. *Trends Biotechnol* 33:337–342.

Mubeen M, Ashfaq A, Hafiz MH, Muhammad A, Hafiz UF, Mazhar S, Muhammad SD, Asad A, Amjed A, Fahad S, Wajid N (2020) Evaluating the climate change impact on water use efficiency of cotton-wheat in semi-arid conditions using DSSAT model. *J Water Clim Change.* https://doi.org/10.2166/wcc.2019.179/622035/jwc2019179.pdf

Muhammad N, Muqarrab A, Khurram S, Fiaz A, Fahim N, Muhammad A, Shazia A, Omaima N, Sulaiman AA, Fahad S, Subhan D, Rahul D (2021) Kaolin and Jasmonic acid improved cotton productivity under water stress conditions. *J Saudi Soc Agric Sci* 28:6606–6614. https://doi.org/10.1016/j.sjbs.2021.07.043

Muhammad TQ, Amna F, Amna B, Barira Z, Xitong Z, Ling-Ling C (2020) Effectiveness of conventional crop improvement strategies vs. omics. In: Fahad S, Hasanuzzaman M, Alam M, Ullah H, Saeed M, Khan AK, Adnan M (eds) *Environment, climate, plant and vegetation growth.* Springer Publ Ltd, Springer Nature Switzerland AG. Part of Springer Nature. pp. 253–284. https://doi.org/10.1007/978-3-030-49732-3

Muhammad Z, Abdul MK, Abdul MS, Kenneth BM, Muhammad S, Shahen S, Ibadullah J, Fahad S (2019) Performance of *Aeluropus lagopoides* (mangrove grass) ecotypes, a potential turfgrass, under high saline conditions. *Environ Sci Pollut Res.* https://doi.org/10.1007/s11356-019-04838-3

Mukherjee S, David A, Yadav S, Baluška F, Bhatla SC (2014) Salt stress-induced seedling growth inhibition coincides with differential distribution of serotonin and melatonin in sunflower seedling roots and cotyledons. *Physiol Plant* 152:714–728.

Munns R, Tester M (2008) Mechanisms of salinity tolerance. *Ann Rev Plant Biol* 59:651–681.

Murch SJ, Alan AR, Cao J, Saxena PK (2009) Melatonin and serotonin in flowers and fruits of *Datura metel* L. *J Pineal Res* 47:277–283.

Murch SJ, Saxena PK (2006) A melatonin-rich germplasm line of St John's wort (*Hypericum perforatum* L.). *J Pineal Res* 41(3):284–287.

Muzammal R, Fahad S, Guanghui D, Xia C, Yang Y, Kailei T, Lijun L, Fei-Hu L, Gang D (2021) Evaluation of hemp (*Cannabis sativa* L.) as an industrial crop: a review. *Environ Sci Pollution Res.* https://doi.org/10.1007/s11356-021-16264-5

Nawaz MA, Huang Y, Bie Z, Ahmed W, Reiter RJ, Niu M, Hameed S (2016) Melatonin: current status and future perspectives in plant science. *Front Plant Sci* 6:1230. https://doi.org/10.3389/fpls.2015.01230

NCBI (National Center for Biotechnology Information) (2019a) PubChem Database. Melatonin, CID=896, https://pubchem.ncbi.nlm.nih.gov/compound/Melatonin. Accessed July 13 2019.

NCBI (National Center for Biotechnology Information) (2019b) PubChem Database. Brassinolide, CID=115196, https://pubchem.ncbi.nlm.nih.gov/compound/Brassinolide. Accessed July 13 2019.

Nemhauser JL, Mockler TC, Chory J (2004) Interdependency of brassinosteroid and auxin signaling in Arabidopsis. *PLoS Biol* 2(9): e258. https://doi.org/10.1371/journal.pbio.0020258

Noor M, Naveed Ur R, Ajmal J, Fahad S, Muhammad A, Fazli W, Saud S, Hassan S (2020) Climate change and costal plant lives. In: Fahad S, Hasanuzzaman M, Alam M, Ullah H, Saeed M, Khan AK, Adnan M (eds) *Environment, climate, plant and vegetation growth.* Springer Publ Ltd, Springer Nature Switzerland AG. Part of Springer Nature, pp. 93–108. https://doi.org/10.1007/978-3-030-49732-3

Nopparat C, Porter JE, Ebadi M, Govitrapong P (2010) The mechanism for the neuroprotective effect of melatonin again stmethamphetamine induced autophagy. *J Pineal Res* 49:382–389.

Özdemir F, Bor M, Demiral T, Turkan I (2004) Effect of 24-epibrassinolide on seed germination, seedling growth, lipid peroxidation, proline content and antioxidative system of rice (*Oryza sativa*) under salinity stress. *Plant Growth Regul* 42:203–211.

Pagamas P, Nawata E (2008) Sensitive stages of fruit and seed development of chili pepper (*Capsicum annuum* L. var. Shishito) exposed to high-temperature stress. *Sci Hort* 117:21–25.

Peleg Z, Blumwald E (2011) Hormone balance and abiotic stress tolerance in crop plants. *Curr Opin Plant Biol* 14:290–295.

Prasad PVV, Boote KJ, Allen LH (2006) Adverse high temperature on pollen viability, seed-set, seed yield and harvest index of grain-sorghum [*Sorghum bicolor* (L.) Moench] are more severe at elevated carbon dioxide due to higher tissue temperature. *Agric For Meteorol* 139:237–251.

Qamar-Uz Z, Zubair A, Muhammad Y, Muhammad ZI, Abdul K, Fahad S, Safder B, Ramzani PMA, Muhammad N (2017) Zinc biofortification in rice: leveraging agriculture to moderate hidden hunger in developing countries. *Arch Agron Soil Sci* 64:147–161. https://doi.org/10.1080/03650340.2017.1338343

Quan W, Liu X, Wang H, Chan Z (2016) Comparative physiological and transcriptional analyses of two contrasting drought tolerant alfalfa varieties. *Front Plant Sci* 6:1256.

Rahman MA, Chikushi J, Yoshida S, Karim AJMS (2009) Growth and yield components of wheat genotypes exposed to high temperature stress under control environment. *Bangladesh J Agric Res* 34:361–372.

Rainey K, Griffiths P (2005) Evaluation of *Phaseolus acutifolius* A. Gray plant introductions under high temperatures in a controlled environment. *Genet Resour Crop Evol* 52:117–120.

Rashid M, Qaiser H, Khalid SK, Mohammad IA-W, Zhang A, Muhammad A, Shahzada SI, Rukhsanda A, Ghulam AS, Shahzada MM, Sarosh A, Muhammad FQ (2020) Prospects of biochar in alkaline soils to mitigate climate change. In: Fahad S, Hasanuzzaman M, Alam M, Ullah H, Saeed M, Khan AK, Adnan M (eds) *Environment, climate, plant and vegetation growth.* Springer Publ Ltd, Springer Nature Switzerland AG. Part of Springer Nature, pp. 133–150. https://doi.org/10.1007/978-3-030-49732-3

Raskin I (1992) Role of salicylic acid in plants. *Ann Rev Plant Biol* 43:439–463.

Reddy AR, Chaitanya KV, Vivekanandan M (2004) Drought-induced responses of photosynthesis and antioxidant metabolism in higher plants. *J Plant Physiol* 161(11):1189–1202.

Rehman M, Fahad S, Saleem MH, Hafeez M, Muhammad HR, Liu F, Deng G (2020) Red light optimized physiological traits and enhanced the growth of ramie (*Boehmeria nivea* L.). *Photosynthetica* 58 (4):922–931.

Reiter R, Tan DX, Zhou Z, Cruz M, Fuentes-Broto L, Galano A (2015) Phytomelatonin: assisting plants to survive and thrive. *Molecules* 20:7396–7437.

Reiter RJ, Tan DX (2002) Melatonin: an antioxidant in edible plants. *Ann New York Acad Sci* 957(1):341–344.

Reiter RJ, Tan DX, Jou MJ, Korkmaz A, Manchester LC, Paredes SD (2008) Biogenic amines in the reduction of oxidative stress: melatonin and its metabolites. *Neuroendocrinol Lett* 29:391–398.

Romero A, Egea J, Garcia AG, Lopez MG (2010) Synergistic neuroprotective effect of combined low concentrations of galantamine and melatonin against oxidative stress in SH-SY5Y neuroblastoma cells. *J Pineal Res* 49:141–148.

Ruelland E, Zachowski A (2010) How plants sense temperature. *Environ Exp Bot* 69:225–232.

Sadam M, Muhammad TQ, Ghulam M, Muhammad SK, Faiz AJ (2020) Role of biotechnology in climate resilient agriculture. In: Fahad S, Hasanuzzaman M, Alam M, Ullah H, Saeed M, Khan AK, Adnan M (eds)

Environment, climate, plant and vegetation growth. Springer Publ Ltd, Springer Nature Switzerland AG. Part of Springer Nature. pp. 339–366. https://doi.org/10.1007/978-3-030-49732-3

Safi UK, Ullah F, Mehmood S, Fahad S, Ahmad Rahi A, Althobaiti F et al. (2021) Antimicrobial, antioxidant and cytotoxic properties of *Chenopodium glaucum* L. *PLoS ONE* 16(10): e0255502. https://doi.org/10.1371/journal.pone.0255502

Sagor GH, Berberich T, Tanaka S, Nishiyama M, Kanayama Y, Kojima S, Muramoto K, Kusano T (2016) A novel strategy to produce sweeter tomato fruits with high sugar contents by fruit-specific expression of a single bZIP transcription factor gene. *Plant Biotech J* 14(4):1116–1126.

Sahrish N, Shakeel A, Ghulam A, Zartash F, Sajjad H, Mukhtar A, Muhammad AK, Ahmad K, Fahad S, Wajid N, Sezai E, Carol Jo W, Gerrit H (2022) Modeling the impact of climate warming on potato phenology. *European J AgroN* 132:126404.

Saitoh H (2008) *Ecological and Physiology of Vegetable*. Nousangyoson Bunka Kyoukai, Tokyo, Japan.

Sajid H, Jie H, Jing H, Shakeel A, Satyabrata N, Sumera A, Awais S, Chunquan Z, Lianfeng Z, Xiaochuang C, Qianyu J, Junhua Z (2020) Rice production under climate change: adaptations and mitigating strategies. In: Fahad S, Hasanuzzaman M, Alam M, Ullah H, Saeed M, Khan AK, Adnan M (eds) *Environment, climate, plant and vegetation growth*. Springer Publ Ltd, Springer Nature Switzerland AG. Part of Springer Nature, pp. 659–686. https://doi.org/10.1007/978-3-030-49732-3

Sajjad H, Muhammad M, Ashfaq A, Waseem A, Hafiz MH, Mazhar A, Nasir M, Asad A, Hafiz UF, Syeda RS, Fahad S, Depeng W, Wajid N (2019) Using GIS tools to detect the land use/land cover changes during forty years in Lodhran district of Pakistan. *Environ Sci Pollut Res*. https://doi.org/10.1007/s11356-019-06072-3

Saleem MH, Fahad S, Adnan M, Mohsin A, Muhammad SR, Muhammad K, Qurban A, Inas AH, Parashuram B, Mubassir A, Reem MH (2020a) Foliar application of gibberellic acid endorsed phytoextraction of copper and alleviates oxidative stress in jute (*Corchorus capsularis* L.) plant grown in highly copper-contaminated soil of China. *Environ Sci Pollution Res*. https://doi.org/10.1007/s11356-020-09764-3

Saleem MH, Fahad S, Shahid UK, Mairaj D, Abid U, Ayman ELS, Akbar H, Analía L, Lijun L (2020c) Copper-induced oxidative stress, initiation of antioxidants and phytoremediation potential of flax (*Linum usitatissimum* L.) seedlings grown under the mixing of two different soils of China. *Environ Sci Poll Res*. https://doi.org/10.1007/s11356-019-07264-7

Saleem MH, Rehman M, Fahad S, Tung SA, Iqbal N, Hassan A, Ayub A, Wahid MA, Shaukat S, Liu L, Deng G (2020b) Leaf gas exchange, oxidative stress, and physiological attributes of rapeseed (*Brassica napus* L.) grown under different light-emitting diodes. *Photosynthetica* 58(3):836–845.

Salvucci ME, Crafts-Brandner SJ (2004) Inhibition of photosynthesis by heat stress: the activation state of Rubisco as a limiting factor in photosynthesis. *Plant Physiol* 120:179–186.

Saman S, Amna B, Bani A, Muhammad TQ, Rana MA, Muhammad SK (2020) QTL Mapping for abiotic stresses in cereals. In: Fahad S, Hasanuzzaman M, Alam M, Ullah H, Saeed M, Khan AK, Adnan M (eds), *Environment, climate, plant and vegetation growth*. Springer Publ Ltd, Springer Nature Switzerland AG. Part of Springer Nature, pp. 229–252. https://doi.org/10.1007/978-3-030-49732-3

Sasse JM (2003) Physiological actions of brassinosteroids: an update. *J Plant Growth Regul* 22:276–288.

Saud S, Chen Y, Fahad S, Hussain S, Na L, Xin L, Alhussien SA (2016) Silicate application increases the photosynthesis and its associated metabolic activities in Kentucky bluegrass under drought stress and post-drought recovery. *Environ Sci Pollut Res* 23(17):17647–17655. https://doi.org/10.1007/s11356-016-6957-x

Saud S, Chen Y, Long B, Fahad S, Sadiq A (2013) The different impact on the growth of cool season turf grass under the various conditions on salinity and drought stress. *Int J Agric Sci Res* 3:77–84.

Saud S, Fahad S, Cui G, Chen Y, Anwar S (2020) Determining nitrogen isotopes discrimination under drought stress on enzymatic activities, nitrogen isotope abundance and water contents of Kentucky bluegrass. *Sci Rep* 10:6415. https://doi.org/10.1038/s41598-020-63548-w

Saud S, Fahad S, Yajun C, Ihsan MZ, Hammad HM, Nasim W, Amanullah Jr, Arif M and Alharby H (2017) Effects of Nitrogen Supply on Water Stress and Recovery Mechanisms in Kentucky Bluegrass. *Plants Front Plant Sci* 8:983. http://dx.doi.org/10.3389/fpls.2017.00983

Saud S, Li X, Chen Y, Zhang L, Fahad S, Hussain S, Sadiq A, Chen Y (2014) Silicon application increases drought tolerance of Kentucky bluegrass by improving plant water relations and morph physiological functions. *SciWorld J* 2014:1–10. https://doi.org/10.1155/2014/368694

Semenov MA, Halford NG (2009) Identifying target traits and molecular mechanisms for wheat breeding under a changing climate. *J Exp Bot* 60:2791–2804.

Senol C (2020) The effects of climate change on human behaviors. In: Fahad S, Hasanuzzaman M, Alam M, Ullah H, Saeed M, Khan AK, Adnan M (eds) *Environment, climate, plant and vegetation growth*. Springer Publ Ltd, Springer Nature Switzerland AG. Part of Springer Nature, pp. 577–590. https://doi.org/10.1007/978-3-030-49732-3

Shafi MI, Adnan M, Fahad S, Fazli W, Ahsan K, Zhen Y, Subhan D, Zafar-ul-Hye M, Martin B, Rahul D (2020) Application of single superphosphate with humic acid improves the growth, yield and phosphorus uptake of wheat (*Triticum aestivum* L.) in calcareous soil. *Agron* 10:1224. http://dx.doi.org/10.3390/agronomy10091224

Shah B, Khan IA, Khan A, Din MMU, Junaid K, Adnan M, Shah SRA, Zaman M, Ahmad N, Akbar R, Fayaz W, Rahman IU (2016) Correlation between proximate chemical composition and insect pests of maize cultivars in Peshawar. *J Ento Zool Studies* 4(1):312–316.

Shah F, Lixiao N, Kehui C, Tariq S, Wei W, Chang C, Liyang Z, Farhan A, Fahad S, Huang J (2013) Rice grain yield and component responses to near 2°C of warming. *Field Crop Res* 157:98–110.

Sharma P, Bhardwaj R, Arora N, Arora HK, Kumar A (2008) Effects of 28-homobrassinolide on nickel uptake, protein content and antioxidative defence system in *Brassica juncea*. *Biol Plant* 52(4):767–770.

Sharma P, Jha AB, Dubey RS, Pessarakli M (2012) Reactive oxygen species, oxidative damage, and antioxidative defense mechanism in plants under tressful conditions. *J Bot* 2012:26.

Shi H, Jiang C, Ye T, Tan DX, Reiter RJ, Zhang H, Liu R, Chan Z (2015) Comparative physiological, metabolomic, and transcriptomic analyses reveal mechanisms of improved abiotic stress resistance in bermudagrass (*Cynodon dactylon* (L). Pers.) by exogenous melatonin. *J Exp Bot* 66:681–694.

Shinozaki K, Yamaguchi-Shinozaki K (2007) Gene networks involved in drought stress response and tolerance. *J Exp Bot* 58:221–227.

Sidra K, Javed I, Subhan D, Allah B, Syed IUSB, Fatma B, Khaled DA, Fahad S, Omaima N, Ali TKZ, Rahul D (2021) Physio-chemical characterization of indigenous agricultural waste materials for the development of potting media. *J Saudi Society Agricultural Sci* https://doi.org/10.1016/j.sjbs.2021.08.058

Sirhindi G, Kumar S, Bhardwaj R, Kumar M (2009) Effects of 24 epibrassinolide and 28-homobrassinolide on the growth and antioxidant enzyme activities in the seedlings of *Brassica juncea* L. *Physiol Mol Biol Plants* 15:335–341.

Sondhi N, Bhardwaj R, Kaur S, Singh B, Kumar N (2008) Isolation of 24-epibrassinolide from leaves of "Aeglemarmelos" and evaluation of its antigenotoxicity potential employing Allium cepa chromosomal aberration assay. *Plant Growth Regulator* 54(3):217–224.

Stege PW, Sombra LL, Messina G, Martinez LD, Silva MF (2010) Determination of melatonin in wine and plant extracts by capillary electrochromatography with immobilized carboxylic multi-walled carbon nanotubes as stationary phase. *Electrophoresis* 31:2242–2248.

Subhan D, Zafar-ul-Hye M, Fahad S, Saud S, Martin B, Tereza H, Rahul D (2020) Drought Stress Alleviation by ACC Deaminase Producing *Achromobacter xylosoxidans* and *Enterobacter cloacae*, with and without Timber Waste Biochar in Maize. *Sustain* 12(6286). http://dx.doi.org/10.3390/su12156286

Sun X, Wang P, Jia X, Huo L, Che R, Ma F (2018) Improvement of drought tolerance by overexpressing MdATG18a is mediated by modified antioxidant system and activated autophagy in transgenic apple. *Plant Biotech J* 16(2):545–557.

Suwa R, Hakata H, Hara H, El-Shemy HA, Adu-Gyamfi JJ, Nguyen NT, Kanai S, Lightfoot DA, Mohapatra PK, Fujita K (2010) High temperature effects on photosynthetic partitioning and sugar metabolism during ear expansion in maize (*Zea mays* L.) genotypes. *Plant Physiol Biochem* 48:124–130.

Suzuki N, Koussevitzky S, Mittler R, Miller G (2012) ROS and redox signaling in the response of plants to abiotic stress. *Plant Cell Environ* 35:259–270.

Suzuki N, Miller G, Morales J, Shulaev V, Torres MA, Mittler R (2011) Respiratory burst oxidases: the engines of ROS signaling. *Curr Opin Plant Biol* 14:691–699. https://doi.org/10.1016/j.pbi.2011.07.014

Szafranska K, Glinska S, Janas KM (2013) Ameliorative effect of melatonin on meristematic cells of chilled and re-warmed *Vigna radiata* roots. *Biol Plant* 57:91–96.

Tan DX, Hardeland R, Manchester LC, Paredes SD, Korkmaz A, Sainz RM, Mayo JC, Fuentes-Broto L, Reiter RJ (2010) The changing biological roles of melatonin during evolution: from an antioxidant to signals of darkness, sexual selection and fitness. *Biol Rev* 85:607–623.

Tan DX, Manchester LC, Reiter RJ, Qi WB, Karbownik M, Calvo JR (2000) Significance of melatonin in antioxidative defense system: reactions and products. *Biol Signals Recept* 9:137–159. http://dx.doi.org/10.1159/00001463

Tan DX, Manchester LC, Terron MP, Flores LJ, Reiter RJ (2007) One molecule, many derivatives: a never-ending interaction of melatonin with reactive oxygen and nitrogen species? *J Pineal Res* 42:28–42.

Tan W, Wei Meng Q, Brestic M, Olsovska K, Yang X (2011) Photosynthesis is improved by exogenous calcium in heat-stressed tobacco plants. *J Plant Physiol* 168:2063–2071.

Tariq M, Ahmad S, Fahad S, Abbas G, Hussain S, Fatima Z, Nasim W, Mubeen M, Ur Rehman MH, Khan MA, Adnan M (2018) The impact of climate warming and crop management on phenology of sunflower-based cropping systems in Punjab, *Pakistan. Agri Forest Met* 15(256):270–282.

Tiryaki I, Keles H (2012) Reversal of the inhibitory effect of light and high temperature on germination of *Phacelia tanacetifolia* seeds by melatonin. *J Pineal Res* 52:332–339.

Turk H, Erdal S, Genisel M, Atici O, Demir Y, Yanmis D (2014) The regulatory effect of melatonin on physiological, biochemical and molecular parameters in cold-stressed wheat seedlings. *Plant Growth Regul* 74:139–152.

Um HJ, Kwon TK (2010) Protective effect of melatonin on oxaliplatin induced apoptosis through sustained Mcl-1 expression and antioxidant action in renal carcinoma Caki cells. *J Pineal Res* 49:283–290.

Unsar N-U, Muhammad R, Syed HMB, Asad S, Mirza AQ, Naeem I, Muhammad HR, Fahad S, Shafqat S (2020) Insect pests of cotton crop and management under climate change scenarios. In: Fahad S, Hasanuzzaman M, Alam M, Ullah H, Saeed M, Khan AK, Adnan M (eds), *Environment, climate, plant and vegetation growth.* Springer Publ Ltd, Springer Nature Switzerland AG. Part of Springer Nature, pp. 367–396. https://doi.org/10.1007/978-3-030-49732-3

Valliyodan B, Nguyen HT (2006) Understanding regulatory networks and engineering for enhanced drought tolerance in plants. *Curr Opin Plant Biol* 9:189–195.

Van Tassel DL, Roberts N, Lewy A, O'Neill SD (2001) Melatonin in plant organs. *J Pineal Res* 31(1):8–15.

Vara Parasad PV, Craufurd PQ, Summerfield RJ (1999) Fruit number in relation to pollen production and viability in groundnut exposed to short episodes of heat stress. *Ann Bot* 84:381–386.

Vollenweider P, Gunthardt-Goerg MS (2005) Diagnosis of abiotic and biotic stress factors using the visible symptoms in foliage. *Environ Pollut* 137:455–465.

Wahid A, Close TJ (2007) Expression of dehydrins under heat stress and their relationship with water relations of sugarcane leaves. *Biol Plant* 51:104–109.

Wahid F, Fahad S, Subhan D, Adnan M,, Zhen Y, Saud S, Manzer HS, Martin B, Tereza H, Rahul D (2020) Sustainable management with mycorrhizae and phosphate solubilizing bacteria for enhanced phosphorus uptake in calcareous soils. *Agriculture* 10(334). http://dx.doi.org/10.3390/agriculture10080334

Wajid N, Ashfaq A, Asad A, Muhammad T, Muhammad A, Muhammad S, Khawar J, Ghulam MS, Syeda RS, Hafiz MH, Muhammad IAR, Muhammad ZH, Muhammad HR, Veysel T, Fahad S, Suad S, Aziz K, Shahzad A (2017) Radiation efficiency and nitrogen fertilizer impacts on sunflower crop in contrasting environments of Punjab. *Pakistan Environ Sci Pollut Res* 25:1822–1836. https://doi.org/10.1007/s11356-017-0592-z

Wang Y, Reiter RJ, Chan Z (2017) Phytomelatonin: a universal abiotic stress regulator. *J Exp Bot* 69(5):963–974.

Waraich EA, Ahmad R, Halim A, Aziz T (2012) Alleviation of temperature stress by nutrient management in crop plants: a review. *J Soil Sci Plant Nut* 12:221–244.

Wei W, Li QT, Chu YN, Reiter RJ, Yu XM, Zhu DH, Zhang WK, Ma B, Lin Q, Zhang JS, Chen SY (2015) Melatonin enhances plant growth and abiotic stress tolerance in soybean plants. *J Exp Bot* 66:695–707.

Wilhelm EP, Mullen RE, Keeling PL, Singletary GW (1999) Heat stress during grain filling in maize: effects of kernel growth and metabolism. *Crop Sci* 39:1733–1741.

Wu C, Kehui C, She T, Ganghua L, Shaohua W, Fahad S, Lixiao N, Jianliang H, Shaobing P, Yanfeng D (2020) Intensified pollination and fertilization ameliorate heat injury in rice (*Oryza sativa* L.) during the flowering stage. *Field Crops Res* 252:107795.

Wu C, Tang S, Li G, Wang S, Fahad S, Ding Y (2019) Roles of phytohormone changes in the grain yield of rice plants exposed to heat: a review. *PeerJ* 7:e7792. http://dx.doi.org/10.7717/peerj.7792

Wu W, Zhang Q, Ervin EH, Yang Z, Zhang X (2017) Physiological mechanism of enhancing salt stress tolerance of perennial ryegrass by 24-Epibrassinolide. *Front Plant Sci* 8:1017. https://doi.org/10.3389/fpls.2017.01017

Xia XJ, Zhang Y, Wu JX, Wang JT, Zhou YH, Shi K, Yu YL, Yu JQ (2008) Brassinosteroids promote metabolism of pesticides in cucumber. *J Agric Food Chem* 57(18):8406–8413.
Xu SC, He MD, Zhong M, Zhang YW, Wang Y, Yang L, Yang J, Yu ZP, Zhou Z (2010) Melatonin protects against nickel-induced neurotoxicity in vitro by reducing oxidative stress and maintaining mitochondrial function. *J Pineal Res* 49:86–94.
Xu W, Cai SY, Zhang Y, Wang Y, Ahammed GJ, Xia XJ, Shi K, Zhou YH, Yu JQ, Reiter RJ, Zhou J (2016) Melatonin enhances thermotolerance by promoting cellular protein protection in tomato plants. *J Pineal Res* 61(4):457–469.
Xue LW, Du JB, Yang H, Xu F, Yuan S, Lin HH (2009) Brassinosteroids counteract abscisic acid in germination and growth of Arabidopsis. *Zeitschrift für Naturforschung C* 64(3–4):225–230.
Yadava P, Kaushal J, Gautam A, Parmar H, Singh I (2016) Physiological and biochemical effects of 24-epibrassinolide on heat-stress adaptation in maize (*Zea mays* L.). *Nat Sci* 8:171–179.
Yang Z, Zhang Z, Zhang T, Fahad S, Cui K, Nie L, Peng S, Huang J (2017) The effect of season-long temperature increases on rice cultivars grown in the central and southern regions of China. *Front Plant Sci* 8:1908. https://doi.org/10.3389/fpls.2017.01908
Yin M, Wang Y, Zhang L, Li J, Quan W, Yang L, Wang Q, Chan Z (2017) The Arabidopsis Cys2/His2 zinc finger transcription factor ZAT18 is a positive regulator of plant tolerance to drought stress. *J Exp Bot* 68(11):2991–3005.
Yin Y, Li S, Liao W, Lu Q, Wen X, Lu C (2010) Photosystem II photochemistry, photoinhibition, and the xanthophyll cycle in heat-stressed rice leaves. *J Plant Physiol* 167:959–966.
Zafar-ul-Hye M, Muhammad TH, Muhammad A, Fahad S, Martin B, Tereza D, Rahul D, Subhan D (2020b) Potential role of compost mixed biochar with rhizobacteria in mitigating lead toxicity in spinach. *Scientific Rep* 10:12159. https://doi.org/10.1038/s41598-020-69183-9.
Zafar-ul-Hye M, Muhammad N, Subhan D, Fahad S, Rahul D, Mazhar A, Ashfaq AR, Martin B, Ji˘rí H, Zahid HT, Muhammad N (2020a) Alleviation of cadmium adverse effects by improving nutrients uptake in bitter gourd through cadmium tolerant rhizobacteria. *Environment* 7 (54). http://dx.doi.org/10.3390/environments7080054
Zafar-ul-Hye M, Akbar MN, Iftikhar Y, Abbas M, Zahid A, Fahad S, Datta R, Ali M, Elgorban AM, Ansari MJ et al. (2021) Rhizobacteria inoculation and Caffeic acid alleviated drought stress in lentil plants. *Sustain* 13:9603. https://doi.org/10.3390/su13179603
Zahida Z, Hafiz FB, Zulfiqar AS, Ghulam MS, Fahad S, Muhammad RA, Hafiz MH, Wajid N, Muhammad S (2017) Effect of water management and silicon on germination, growth, phosphorus and arsenic uptake in rice. *Ecotoxicol Environ Saf* 144:11–18.
Zahir SM, Zheng HG, Ala Ud D, Amjad A, Ata Ur R, Kashif J, Shah F, Saud S, Adnan M, Fazli W, Saud A, Manzer HS, Shamsher A, Wajid N, Hafiz MH, Fahad S (2021) Synthesis of silver nanoparticles using Plantago lanceolata extract and assessing their antibacterial and antioxidant activities. *Sci Rep* 11:20754.
Zahoor M, Khan N, Ali M, Saeed M, Ullah Z, Adnan M, Ahmad B (2016) Integrated effect of organic waste and NPK fertilizers on nutrients uptake in potato crop and soil fertility. *Pure Appl Biol* 5(3):601–607.
Zaki AB, Ahmad N, Khan IA, Shah B, Khan A, Rasheed MT, Adnan M, Junaid K, Huma Z, Ahmed S (2015) Adulticidal efficacy of *Azadirachtaindica* (neem tree), *Sesamumindicum*(til) and *Pinussabinaena* (pine tree) extracts against *Aedesaegypti*under laboratory conditions. *J Ento Zool Stud* 3(6):112–116.
Zia-Ur-Rehman M (2020) Environment, climate change and biodiversity. In: Fahad S, Hasanuzzaman M, Alam M, Ullah H, Saeed M, Khan AK, Adnan M (ed) *Environment, climate, plant and vegetation growth.* Springer Publ Ltd, Springer Nature Switzerland AG. Part of Springer Nature, pp. 473–502. https://doi.org/10.1007/978-3-030-49732-3

4

The Incidence of Heat Stress on the Quality of Food Crops in the Era of the Changing Climate: An Overview

Ayman EL Sabagh
Kafrelsheikh University, Egypt; Siirt University, Turkey

Celaleddin Barutçular
University of Cukurova, Adana, Turkey

Akbar Hossain
Bangladesh Wheat and Maize Research Institute, Dinajpur, Bangladesh

Fatih Çiğ, Murat Erman, and Emine Karademir
Siirt University, Turkey

Mohammad Sohidul Islam
Hajee Mohammad Danesh Science and Technology University, Bangladesh

Swapan Kumar Paul
Bangladesh Agricultural University, Mymensingh, Bangladesh

Allah Wasaya
Bahauddin Zakariya University, Layyah, Pakistan

Muhammad Aamir Iqbal
University of Poonch Rawalakot (AJK), Pakistan

Sytar Oksana
Institute of Agricultural Research for Development (IRAD), Yaounde, Cameroon

Ntsomboh-Ntsefong Godswill
University of Yaounde 1, Cameroon; Kiev National University of Taras Shevchenko, Kyiv, Ukraine

Shah Fahad
Hainan University, Haikou, China; The University of Haripur, Haripur, Pakistan

Muhammad Mubeen and Hamid Majeed
COMSATS University Islamabad, Vehari, Pakistan

Wajid Nasim
University College of Agriculture & Environmental Sciences, Bahawalpur, Pakistan

Mabrouk Elsabagh
Kafr El-Sheikh University, Kafr El-Sheikh, Egypt

DOI: 10.1201/9781003286417-4

CONTENTS

4.1 Introduction

Heat stress is one of the major abiotic constraints confronted to sustainable crop production especially in developing countries, as it influences the growth, physiology, and metabolism of plants (Lunde et al. 2007; Hossain et al. 2013; Barutcular et al. 2017; Hossain et al. 2018; Yildirim et al. 2018; Fahad et al. 2014a, 2014b, 2015a, 2015b, 2016a, 2016b, 2016c, 2016d, 2018, 2019, 2020, 2021a, 2021b, 2021c, 2021d, 2021e). Field crop production, as well as food security, is directly affected by global warming (Iqbal et al. 2019a, 2019b; Ainsworth and Ort 2010). Crops tend to produce the maximum yields as per varietal potential when grown near their optimal temperature. In the context of climate change, the temperature is often higher than the optimum due to environmental fluctuations; hence, there is the probability of the crops and especially grain being exposed to extended periods of supra-optimal temperatures (Monjardino et al. 2007; Iqbal et al. 2018). The heat has multiple negative impacts on crop yields, including reducing leaf photosynthesis and enhancing leaf senescence rates. More critically for yield determination, however, are the reported effects of decreasing grain number when heat stress occurs before or around anthesis and reduced grain weight when it occurs during grain filling. Exposure to high temperature usually results in a reduction in chlorophyll biosynthesis (Dutta et al. 2009). Reduction in the activities of source and sink takes place under heat stress, which greatly affects the growth and ultimately the economic yield of food crops (Taiz and Zeiger 2006; Iqbal et al. 2018a).

Adverse environmental conditions during critical crop growth stages have been identified as a major constraint to crop productivity as well as grain quality (Jiang and Huang 2001; Iqbal et al. 2018b; Adnan et al. 2018a, 2018b; Akram et al. 2018a, 2018b; Aziz et al. 2017a; Habib Ur et al. 2017; Hafiz et al. 2016, 2019; Kaman et al. 2017; Muhammad et al. 2019, 2019; Sajjad et al. 2019, 2019; Saud et al. 2013, 2014, 2016, 2017; Fatima et al. 2016; Shah et al. 2013; Qamar-Uz et al. 2017, 2017; Wajid et al. 2017, 2017; Yang et al. 2017; Zahida et al. 2017; Depeng et al. 2018; Fahad and Bano 2012; Fahad et al. 2017; Fahad et al. 2013, 2014a, 2014b, 2015a, 2015b, 2016a, 2016b, 2016c, 2016d, 2018, 2019). High-temperature stress accelerates faster foliage senescence, poor synthesis of assimilates, and reduced translocation of photosynthates to the developing grain (Al-Khatib and Paulsen 1984; Iqbal et al. 2018c). Heat stress adversely affects the ability of legumes to retain their reproductive organs and achieve satisfactory pod set. Heat stress reduced the yield of the faba-bean (*Vicia faba* L.) due to gametophyte damage and consequent failure of fertilization (Bishop et al. 2016; Iqbal et al. 2019a). Grain protein content and gluten quality are the two most important parameters determining the wheat quality (Motzo et al. 2007). It has been reported that high-temperature stress during the grain-filling stage reduces grain-filling duration that ultimately reduces the grain quality (Wardlaw and Moncur 1995), thus accelerating cell death, and an earlier harvest maturity (Altenbach et al. 2011). The reproductive phase of development is the most sensitive stage to heat stress in several crops including wheat, chickpea, maize, and sorghum. The reproductive processes involving pollen and stigma viability, pollination anthesis, pollen tube growth, and early embryo development are particularly vulnerable to heat stress. However, in general, male reproductive tissues are much more sensitive at all stages of development to heat stress than female reproductive tissues (Lamaoui et al. 2018). Other studies revealed significant increase of protein content in the wheat flour due to the adverse effect of heat stress (Daniel and Triboi 2000; Balla and Veisz 2007; Labuschagne et al. 2009; Hrušková and Švec 2009).

Scientists have been trying to overcome the problem of environmental stresses by employing different strategies such as the application of various organic compatible solutes that effectively take part in plant stress tolerance including proline (Pro), glycine betaine (GB), trehalose (Tre), and several others (Ali et al. 2017; Arshad et al. 2016). The exogenous application of proline as the foliar application has been employed to enhance stress tolerance in several crops (Athar et al. 2008).

However, very little information is available on the relative importance of the influence of the interaction of genotype and environmental parameters (especially temperature) on grain quality in the Mediterranean region (Rharrabti et al. 2003). Therefore, evaluation of the grain quality of crops under heat stress, as well as assessing the dose of exogenous application of osmoprotectants that could improve quality attributes of grain by mitigating stress effects, is of great importance. Scientists need to address: (i) how climate change is increasing heat stress on crops, (ii) enhance the understanding of heat stress–specific adverse effects on various crops parameters such as yield and yield components, (iii) update heat stress–induced yield reduction statistics of crops, (iv) specifically review yield (grain) deterioration indices under heat stress, and (v) review some common management strategies associated with specific response to heat stress. This review highlights the basis of heat stress effects on crop yield, quality, and management.

4.2 The Adverse Effects of Heat Stress on the Quality of Food Crops

Heat stress, defined as a rise in temperature beyond threshold levels sufficient to cause irreversible damage, greatly limits growth and yield of crop plants when it occurs transiently or continually. Different researchers have previously studied the impacts of heat stress on different crops like potato, rice, maize, and soybean (Lafta and Lorenzen 1995; Wahid et al. 2007, 2007; Peng et al. 2004; Barutcular et al. 2016; Yildirim et al. 2018; Fahad et al. 2014a, 2014b, 2015a, 2015b, 2016a, 2016b, 2016c, 2016d, 2018, 2019).

4.3 The Heat Stress Impacts on the Quality of Major Cereals

The heat stress produces reactive oxygen species (ROS) that cause oxidative damage to cellular membranes, reduce photosynthesis, and ultimately reduce the yield of different cereal crops (Larkindale and Knight 2002; Vacca et al. 2004; Fahad et al. 2014a, 2014b, 2015a, 2015b, 2016a, 2016b, 2016c, 2016d, 2018, 2019; Hussain et al. 2016). Heat induces oxidative damages to cause a reduction in the translocation of photosynthates and abnormal growth of cells as reported by Berry and Bjorkman (1980). The photosynthesis process is also affected by heat stress in different cereal crops like wheat and corn (Tewari and Tripathy 1998; Mohanty and Mohanty 1988). Crops grown in heat stress areas produce low yields and substandard quality seeds (Wardlaw et al. 2002). All plant growth stages are affected by the heat stress, but plant maturity phase shows more sensitivity against heat stress (Wahid et al. 2007, 2007). Heat stress and water stress are interlinked in plants; with increased transpiration rates, plants require more water to mitigate these stresses. Heat stress affects seed germination, panicle formation in cereals like rice, seed growth, yield, and seed quality (Rizhsky et al. 2004; Roberts 1988). Plant photosynthesis varies for different crops as temperatures change. It was shown that for viticulture photosynthesis depends upon the high temperature and its optimum range is about 30°C–40°C, but when the temperature crosses this optimum limit at >40°C leaf photosynthetic activity stops (Sadras 2007). Moreover, heat stress inhibits the activity of the enzyme ribulose bisphosphate carboxylase that is one of the major enzymes required in photosynthesis to fix carbon dioxide (Salvucci and Crafts-Brandner 2004; Lea and Leegood 1999). High temperature influences mitochondrial activity, which leads to injured plants producing low quality of seeds especially in cereals like wheat and rice (Gifford 2003). Under heat stress conditions, the plants' transpiration increases until a point when water becomes unavailable for the assimilation of photosynthates, and stomata's closing, and the sign of wilt can be observed (Byrla et al. 2001).

4.3.1 Impacts of Heat Stress on Wheat

Wheat is an important cereal crop and its growth is adversely affected due to temperature stress (Pipolo et al. 2004). During the grain filling stage, high temperatures reduce the plants' photosynthesis, resulting in a decrease of photosynthates translocated to the grains, leading to lower yield and grains of inferior quality (Machado and Paulsen 2001). Previously, Rehman et al. (2009) found that at 5°C higher temperature compared to the different variants 15°C/10°C, 20°C /15°C and 25°C/20°C of control variants in the greenhouse has adversely affected the growth, morphological and yield traits of wheat. High temperature also disrupted the cellular mechanisms, leading to dwarf plants and small kernel size (Demirevska-Kepova et al. 2005; Tewolde et al. 2006). Hot weather ultimately leads to drought. Drought stress influences wheat during all developmental stages, but grain formation and the reproductive stage are the most critical ones (Pradhan et al. 2012). Wheat yield was decreased from 30% to 1% during the mild drought stress at postanthesis while this reduction increased up to 92% in case of prolonged mild drought stress at flowering and grain formation (De Oliveira et al. 2013). Castro et al. (2007) revealed that high temperatures (35°C) during ripening stage drastically impacts seed traits and protein content during the grain filling stage along with reducing thousand seed weight. Mohammadi et al. (2006) studied the effects of high temperature on spike traits and reported that various spike traits such as spike length, the number of grains, thousand-grain weight were reduced because of heat stress. Heat stress was recorded to have a significant impact on grain formation of wheat. The experiment with growth chamber treatment involved 2 d with high apex temperatures (33°C/30°C day/night) followed by 4 d with the temperature treatment 3°C during the night (2100–0700), above 30°C from 1100 to 1700, and 20°C for the rest of the day found reduction of seed number in field pea.

4.3.2 Impacts of Heat Stress on Maize

Maize (*Zea mays* L.) is the world's largest cereal crop, playing a major role in guaranteeing global food security (Abdelaal et al. 2017; Adnan et al. 2018a, 2018b; Molla et al. 2019). According to Tao et al. (2016a, 2016b), the major obstacle to increase maize productivity in the past three decades in China is heat stress (HTS) during grain filling (Tao et al. 2016a, 2016b). In the temperate climate, heat stress decreases the grain filling rate of spring maize and reduces kernel weight (Tao et al. 2013; Mayer et al. 2014). Summer maize has an adequate average daily temperature of 24°C–26°C (63–92 d after emerging) at grain filling stage. A number of grains have been reported to be reduced with a rise in temperature at an average daily temperature of 26°C–45°C (Li et al. 2003). Photosynthesis is the most sensitive and easily suppressed symptom of hot weather compared to any other symptoms of stress. Ben-Asher et al. (2008) report that in maize at high temperatures of 35°C/30°C (day/night) and 40°C/35°C (day/night), the net photosynthetic rate (Pn) was 50%–60% lower compared with that at low temperatures of 25°C/20°C (day/night) and 30°C/25°C (day/night). The lower photosynthesis rate decreases the biomass production and photosynthates available to be translocated to the grains (Wahid et al. 2007, 2007).

4.3.3 Impacts of Heat Stress on Rice

Rice (*Oryza sativa* L.) is one of the essential cereals developed and grown over the world (Anis et al. 2019; Fahad et al. 2016a, 2016b, 2016c, 2016d, 2018, 2019). Although it has been utilized as a model plant for several years, the development reactions of rice to high temperature are still inadequately comprehended

(Nagai and Makino 2009). Most rice is at present developed in areas where current temperatures are near to optimum for rice growth. Rice yield has been assessed to have decreased by 41% before the 21st century due to the increase of environmental temperature (Ceccarelli et al. 2010).

The heat stress affects all critical growth stages like germination, tillering, spike formation, and maturity of rice (Mittler and Blumwald 2010). Heat stress at the flowering and grain-filling stages seriously affect the spikelet fertility and grain quality, but its flowering phase is more sensitive to heat stress, with temperature ranges between 32°C and 36°C, causing sterility in spikelets. Heat stress inhibits pollen development on the stigma that ultimately results in kernels' abortion and reduced yield (Jagadish et al. 2010). The development of the male reproductive organ is highly sensitive to variation in temperatures at anthesis leading to high sterility, impaired fertility and grain filling (Liu et al. 2013). Heat stress at grain filling stage decreases the number of spikelets, grain weight, and yield (Prasad et al. 2006; Jagadish et al. 2007; Ahmed et al. 2015). The heat stress reduces the seed weight and yield, and leads to accumulation of carbohydrates like starch and glucose in the seeds of rice under elevated carbon dioxide (Hussain et al. 1999).

4.3.4 Effect of High Temperature on Quality of Potato

The potato crop is the fourth main food crop in the world after maize, rice, and wheat. High temperatures during the growing season cause an array of changes in potato plants, which affect its development and may lead to a drastic reduction in economic yield. In natural conditions, drought and heat stress are two different types of abiotic stresses that occur in the field simultaneously or separately, in the case of irrigation use on potato plantations (Rykaczewska 2015). The elevated temperature had a profound impact on leaf and tuber transcript expression with large number of transcripts displaying a rhythmic oscillation at the higher growth temperature (Hancock et al. 2014).

4.3.5 Effect of High Temperature on Quality of Tomatoes

High temperature affects the quality of several genotypes of processing tomatoes. The quality parameters include colour, carotenoid concentration, acidity, pH, and content of total soluble solids. Fruits exposed to either 27.5°C or 30°C during ripening showed a significant decrease in carotenoid concentration and colour score as compared with fruits kept at 25°C. Soluble solid content did not change and acidity increased with high temperatures. Among the individual carotenoids, although lycopene concentration decreased, a more pronounced decrease occurred in the concentration of its precursors, phytoene and phytofluene (Yakir and DeNiro 1990). The temperature has a significant influence on many aspects of growth and development in tomato. Temperature significantly affects the partitioning of assimilates between the vegetative and generative parts. At higher temperatures, trusses appear faster and therefore, initially, there are more fruits on a plant at a higher temperature. Lower fruit yield of tomato genotypes under high temperature is mainly due to limiting carbohydrate supply. Effects of high temperature were more pronounced at flowering stage compared to the preflowering stage (Islam 2011).

4.4 Impacts of Heat Stress on the Quality of Crops Products

4.4.1 Effects on the Protein and Amino Acid Contents of Wheat

Heat stress limits global crop production, and the situation is projected to become worse due to higher temperatures caused by global warming. The extreme temperature during grain filling has been identified as a major source of variation in wheat flour quality characteristics (Barutcular et al. 2016). The reduction in the duration of the grain-filling period due to heat stress has demonstrated a shortened duration of glutenin synthesis, which in turn reduced dough strength, protein accumulation, etc. (Labuschagne et al. 2009). Instead of the proteins that play an active role in the biosynthesis and metabolism, storage proteins and those involved in protecting the plants against biotic and abiotic stresses are accumulated preferentially. The responses of specific proteins depend on the temperature at the early or middle phases of grain filling (Hurkman et al. 2009). At the same time, other studies have shown that protein level in bread wheat flour increases significantly under heat stress influence (Balla and Veisz 2007).

Yield and quality of wheat are fully influenced by different environmental factors, particularly temperature, salinity, and drought (Out et al. 2018; Yassin et al. 2019a, 2019b; Jahan et al. 2019). High temperature and drought occur most frequently after heading, during the grain-filling period, causing stress that influences the quality of the wheat kernels. In wheat, it was observed in several studies that heat stress during grain filling decreased viscosity of the flour.

4.4.2 Effects on the Protein and Amino Acid Content in Rice

The exposure of rice crops to high temperature reduced the crops' vegetative growth duration with lower accumulated carbohydrate partitioning towards seeds, along with the increase of the protein and ash content (Samol et al. 2015). The critical period of high temperature for rice quality is between 6 and 12 days after flowering (Tashiro and Wardlaw 1991). Decreased grain weight, reduced grain filling, the higher percentage of white chalky rice and milky white rice are common effects of heat stress exposure during ripening stage (Kobata et al. 2004; Samol et al. 2015). In addition, the high temperature also causes a serious reduction in grain size and amylose content (Yamakawa et al. 2007; Samol et al. 2015). In rice, several studies observed that heat stress during grain filling decreases the peak and breakdown viscosity of the flour and the deterioration in food quality is more severe when plants are exposed to high temperature (Liu et al. 2013).

4.4.3 Effects on the Starch, Protein, and Amino Acid Content in Maize

Heat stress decreases grain yield and accelerates grain filling rate, increases starch content, starch granule size, crystallinity, and iodine binding capacity, as well as decreases the protein content in maize, thus making the grain less tender and prone to retrogradation (Yang et al. 2015). High temperature accelerates physiological and biochemical reactions in grains, thus increasing the grain filling rate (Cairns et al. 2013). The amount of soluble sugar transferred to starch was revealed to gradually decrease as grain filling proceeded. A study revealed that starch content was increased by heat stress at fresh waxy maize varieties (Yang et al. 2015). On the contrary, Wang and Frei described that starch content was decreased by heat stress in maize, wheat, and barley. Thitisaksakul et al. (2012) reported that high temperature during grain filling increased the mean starch granule size and that the small granules were more easily affected than large ones.

4.5 Effects of High Temperature on Seed Oil and Fatty Acids

Oil content and oil composition (individual fatty acids) of oilseed crops are affected by geneics (cultivars, genotypes within cultivars) and environment variables, especially drought and heat stress (Bellaloui et al. 2013). The instability of conventional cultivars and mid-oleic acid germplasm across environment created a challenge, mainly due to temperature changes that influence the enzymes controlling the biosynthesis of soybean seed fatty acids, especially at seed-fill stage (Bachlava and Cardinal 2009). Therefore, heat stress is a major environmental factor that affects total oil and alters fatty acid composition (Bellaloui et al. 2013).

Heat stress during the reproductive stages in sunflower often negatively influences the seed fertilization, seed number, rate and duration of grain filling period, embryo growth, seed weight, and seed oil characteristics (Prasad et al. 2008). It has been reported that constant high temperatures of ≥35°C for seven consecutive days rapidly decreased the triacylglycerol accumulation and strongly modified the fatty acid composition in the poststress period. The period from 12 to 19 daa (days after anthesis) showed the greatest sensitivity to heat stress regarding embryo and grain weight responses, whereas the period of greatest sensitivity for oil quality was from 19 to 26 daa (Rondanini et al. 2003). Temperature, especially more fluctuations of day and night temperatures, is the main factor affecting oil quality (Izquierdo et al. 2013).

Constant high temperatures during the growing season or brief periods of heat stress during seed development modify oil concentration and fatty acid composition in traditional sunflower and may lead to reduction in the commercial oil quality (van der Merwe et al. 2015). The relative proportions of fatty acids in traditional sunflower are strongly influenced by the temperature during seed development. In controlled environment studies, high temperatures during seed development have been found to cause a decrease in linoleic acid and a corresponding increase in oleic acid percentages in the oil (van der Merwe et al. 2015). The increment of palmitic and stearic acid (StAs) due to high temperature was in agreement with findings of Rondanini et al. (2003). Conversely, Martínez-Force et al. (1998) reported that the StAs concentration progressively decreased with an increase in temperature in sunflower. An earlier study reported by Izquierdo et al. (2013) revealed that plant genotypes which have high oleic acid were less sensitive to temperature stress than low oleic and linoleic acid content genotypes. Similarly, Bellaloui et al. (2013) found that concentrations of linolenic acids (LAs) in soybean were decreased under higher temperature in seed developing period, whereas oleic acid concentrations were increased. Sunflower hybrids with a low concentration of oleic acid potentials were more sensitive to temperature, whereas hybrids with higher oleic acid potentials were tolerant to heat stress (Lagravère et al. 2004).

There is a correlation between late sowing dates, and oil and protein content of cotton. Wei et al. (2017) reported that when cotton seeds were sown later than the normal dates, oil and protein content in seeds decreased significantly as compared with the optimum sowing. The changing amplitude of embryo oil content of cotton seed was larger than that of embryo protein content, suggesting that embryo oil accumulation was more sensitive than protein accumulation in response to sowing date under heat stress influence and solar radiation, which are the primary factors for oil accumulation (Chen et al. 2014). A high temperature that remains above 25°C during plant seed growing is associated with a reduction in seed oil concentration (Weiss 2000). However, when grain filling is delayed due to late sowing dates, reductions in oil concentration are observed when low temperatures and low radiation prevail during grain filling (Leon et al. 2003). The cause of the reduction in oil concentration could be attributed to a shortening of the grain-filling period at heat stress (Connor and Hall 1997). High temperatures during seed maturation have been found to reduce the rate and duration of oil deposition in the grain (Rondanini et al. 2003). Wilcox and Cavins (1992) found that palmitic acid (PA) concentrations in the grains of soybean decreased, whereas StAs increased when sown late. They also found that the linolenic acid (LA) concentration in the grains of soybean was more sensitive to change in temperature due to late sowing than other fatty acids.

Seed filling is the most critical growth stage which gets affected by environmental stresses, particularly by temperature variations. Oil begins to accumulate in developing seeds at 15–20 days after flowering (DAF) and heat stress at this stage negatively affects oil accumulation and composition. Protein and oil deposition in soybean seeds mostly occurred in 20 to 40 DAF and continued up to 70 DAF (Rose 1988). However, when sown late, the protein concentration in seeds increased while oil concentration decreased. The fatty acid composition was strongly affected by air temperature, and the oleic acid percentage increased as the temperature increased at the pod-filling period (Wolf et al. 1982). The high oleic acid and low LA concentrations in the planting at a temperature near 21°C–24°C, and the low oleic acid and high LA in late planting, especially the June planting, could be due to the inverse relationship between these two constituents under different temperature regimes (Bellaloui et al. 2013).

4.6 Effect on Fibre Quality of the Cotton

The heat stress is a major environmental factor influencing growth, yield and quality of fibre crops. Cotton fibre quality was found to be less sensitive to high temperature at the boll development stage (Hake and Silvertooth 1990). Fibre rings of cotton are composed of cellulose and are fully dependent on the photosynthesis rate. And the rings consist of two cellulose layers (a thicker layer) which are formed during the day and another layer (more porous) which is laid down at night (Basra and Malik 1984). Another important indicator for international cotton classers and spinners is micronaire. It is one of the two most important fibre characteristics, which is the determinant of both fineness (linear density) and maturity of cotton fibre (degree of cell wall development) (Hake and Silvertooth 1990; Kakar et al. 2012).

However, under unfavourable conditions, the quality of cotton fibre, as well as yield, deteriorate, particularly due to a hot environment (heat stress) in the late sown condition through hampered photosynthesis (Kakar et al. 2012; Gormus et al. 2017a, 2017b; Omar et al. 2018).

Heat stress affects the yield and fibre quality, as it influences the growth of the seedlings, roots, vegetative parts, and also flowering and fruit formation (Singh et al. 2007). A strong negative correlation between high temperature and lint yield with quality of cotton was observed by Singh et al. (2007). Heat stress negatively influences plant growth, physiology, and ultimately lint yield (Sarwar et al. 2018). Fibre micronaire value increases due to high night temperatures (Krieg 2002) and shorter fibres resulted at high temperature (Meredith 2005). The highest influence of increasing temperature was an increase in fibre strength, but reduction in ginning percentage, fluctuations in fibre length, and less stable micronaire (Hesketh and Low 1968). Conversely, Emara (2012) and Samir et al. reported that sowing dates did not exhibit any significant effect on the fibre strength or fibre elongation percentages.

4.7 Effect on Forage Quality

Forage quality, especially the digestibility and protein content, determine the value of forage crop (Akdeniz et al. 2019). Crop growing season and plant developmental stages are important factors that influence the forage quality (Darby and Lauer 2002). Incidence of stress at any crop growth stage influences forage digestibility leading to the reduction of nutritive value of the feed (Turner et al. 2006). Mahmood et al. (2010) reported that heat stress reduces growth and alters the forage quality of maize. Germination, seed vigour, and seed weight of forage rape seeds also reduced on exposure to heat stress up to 30°C (Rashid et al. 2018). Similarly, Li et al. (2018) conducted chlorophyll fluorescence analysis and revealed that heat stress reduced the rate of photosynthesis, and this reduction was multiplied under Pi-deficient conditions. Thus, morphological development can be used to accurately predict the forage quality of any crop throughout the season (Mitchell and Moser 2000).

4.8 Management Strategies to Mitigate the Adverse Effects of Heat Stress

In response to temperature stress, various approaches have been proposed and are being used to mitigate the drastic impacts of heat stress leading to the adjustment of the cellular milieu and plant tolerance. In response to the stress signals, plants naturally have developed diverse mechanisms for combating and tolerating them. Various mechanisms are involved in cold and heat tolerance. Among the various methods to induce heat stress in plants, foliar application and/or presowing seed treatment with low concentrations of inorganic salts, osmoprotectants, signalling molecules (growth hormones), and oxidants (H_2O_2), as well as preconditioning of plants, are common approaches (Wahid et al. 2007; EL Sabagh et al. 2019a, 2019b, 2019c).

Enhancing tolerance to heat stress in crop plants is getting more important currently. The genetic improvements in combination with the proper cultural practices are considered important in managing the abiotic stresses (Wahid et al. 2007; Deepranjan et al. 2021; Haider et al. 2021; Huang et al. 2021; Ikram et al. 2021; Jabborova et al. 2021; Khadim et al. 2021a, 2021b; Manzer et al. 2021; Muzammal et al. 2021; Abdul et al. 2021a, 2021b; Ashfaq et al. 2021; Amjad et al. 2021; Atif et al. 2021; Athar et al. 2021; Adnan et al. 2018a, 2018b, 2019; Akram et al. 2018a, 2018b; Aziz et al. 2017a, 2017b; Chang et al. 2021; Chen et al. 2021; Emre et al. 2021; Habib Ur et al. 2017; Hafiz et al. 2016, 2019; Ghulam et al. 2021; Guofu et al. 2021; Hafeez et al. 2021; Khan et al. 2021; Kamarn et al. 2017; Muhammad et al. 2019; Safi et al. 2021; Sajjad et al. 2019; Saud et al. 2013, 2014, 2017, 2016, 2020; Shah et al. 2013; Qamar-Uz et al. 2017; Hamza et al. 2021; Irfan et al. 2021; Wajid et al. 2017; Yang et al. 2017; Zahida et al. 2017; Depeng et al. 2018; Hussain et al. 2020; Hafiz et al. 2020a, 2020b; Shafi et al. 2020; Wahid et al. 2020; Subhan et al. 2020; Zafar-ul-Hye et al. 2020a, 2020b; Zafar-ul-Hye et al. 2021; Adnan et al. 2020; Ilyas et al. 2020; Saleem et al. 2020a, 2020b, 2020c; Rehman et al. 2020; Farhat et al. 2020; Wu et al. 2020; Mubeen et al. 2020; Farhana et al. 2020; Jan et al. 2019; Wu et al. 2019; Ahmad et al. 2019; Baseer et al. 2019; Hafiz et al. 2018; Tariq et al. 2018; Hesham and Fahad 2020; Iqra et al. 2020; Akbar et al. 2020;

Mahar et al. 2020; Noor et al. 2020; Bayram et al. 2020; Amanullah and Fahad 2018a, 2018b; Amanullah and Fahad 2017; Amanullah et al. 2020; Amanullah et al. 2021; Rashid et al. 2020; Arif et al. 2020; Amir et al. 2020; Saman et al. 2020; Muhammad et al. 2020; Md Jakir and Allah 2020; Mahmood et al. 2021; Farah et al. 2020; Sadam et al. 2020; Unsar et al. 2020; Fazli et al. 2020; Md. Enamul et al. 2020; Gopakumar et al. 2020; Zia-Ur-Rehman 2020; Al-Wabel et al. 2020a, 2020b; Senol 2020; Amjad et al. 2020; Ibrar et al. 2020; Sajid et al. 2020; Muhammad et al. 2021; Sidra et al. 2021; Zahir et al. 2021; Sahrish et al. 2022).

There are various strategies for production improvement of wheat under heat stress after anthesis. Emphasis has been placed on exploiting prompt and inexpensive means of obtaining satisfactory productions from heat-stressed regions. Furthermore, one of the pragmatic approaches is the exogenous use of stress mitigating compounds either as seed or foliar application (Suryavanshi and Buttar 2016). They also stimulate photosynthesis machinery (Rasheed et al. 2011) and hence improve the yield and quality of crops. Wahid et al. (2007) reported that it has the potential to increase food production and quality much more quickly than plant breeding techniques.

Carotenoids (carotenes and xanthophylls), in addition to acting as accessory light-harvesting pigments, also show antioxidant action (de Pascale et al. 2001). Moreover, they protect the photosystems by (a) reacting with lipid peroxidation products to terminate chain reactions, (b) scavenging singlet oxygens and dissipating the energy as heat, (c) reacting with triplet-excited chlorophyll molecules to prevent formation of singlet oxygen, or (d) dissipating excess excitation energy through the xanthophyll cycle (Rmiki et al. 1999). A steady-state level of carotenoids was correlated with salt tolerance and forms important salinity-tolerance strategies of sugarcane (Wahid and Ghazanfar 2006), but comparable information is scanty for cases of heat stress.

Physiological characteristics and the possibility of mitigating the adversities of heat stress by soaking nodal buds in proline and glycine betaine solutions have improved the accumulation of soluble sugars and protected the developing tissues from heat stress effects (Rasheed et al. 2011). Seed priming with osmoprotectants such as proline and GB under optimal and suboptimal conditions is the promising technique for enhancing seed germination (Ashraf and Foolad 2017; Rasheed et al. 2011).

This improvement in wheat grains quality traits due to seed soaking and/or foliar spray with antioxidants may be due to the role of antioxidants in maintaining balanced plant growth and enhancing wheat grain quality as previously mentioned by Seadh and El-Metwally (2015). Salicylic acid (SA), a common phenolic compound, functions as a plant growth regulator and promotes photosynthesis under heat stress by influencing various physiological processes and biochemical reactions (Janda et al. 2014). Moreover, SA was reported to alleviate the adverse effects of heat stress on photosynthesis by the mechanism of induced proline accumulation, increased osmotic potential. This enhances higher water intake by plants, resulting in a positive influence on stomatal aperture and photosynthetic machinery; this leads to the higher efficiency of photosystem that cumulatively results in increased photosynthesis under heat stress.

Cotton can overcome many stresses if it has adequate soil and air temperatures for plant growth. The increase in temperature leads to the production of reactive oxygen species (ROS). Hydrogen peroxide has a long life, damaging the membranes of all organelles of the cells (Adachi et al. 2015). So, the enhancement of tolerance to the high temperature of plants is a viable approach to resolve the problem induced by global warming. There are known organic compounds viz. ascorbic acid, ascobine, and SA that have shown protective functions against heat stress. Ascorbate is an important metabolite in plants performing a key role in the protection of plants against various environmental stresses like high temperature (Vwioko et al. 2008). SA is known to influence different physiological and biochemical activities of plants and could play a key role in regulating their growth and yield (Hayat et al. 2010). In addition, foliar application of organic acids was reported to exhibit significant effects on fibre strength in flowering and boll development and maturation in cotton (Omar et al. 2018). The foliar application treatments had a significant effect on fibre elongation. All organic compounds significantly enhanced fibre elongation percentage compared to untreated control.

Lately, advances in nanotechnology have resulted in novel products for managing heat stress that allowed increasing productivity by reducing the losses due to stressful environments (Duhan et al. 2017). Nanosensors and field detection devices are tools that can provide useful information on the optimum times for sowing and crop harvesting and relevant information for the timely application of agrochemicals or some other specific products for cultivation (Chowdappa and Gowda 2013). In the near future, there will be nanostructured catalysts that will increase the efficacy of commercial products and allow reduction of the use of doses required for growing plants not only in favourable environments but also in stress conditions (Dimetry and Hussein 2016). Nanoparticles (NPs) containing essential metals are considered in fertilizer formulations to improve plant nutrition in soils with low bioavailability of metals. In several studies, researchers showed that some of the NPs have a beneficial effect to influence the germination of seeds and the growth of plants under stress conditions. For example, Zaimenko et al. reported that the application of NPs showed an increase in seed germination and growth of wheat plants, as result of an increase in photosynthetic pigments when plants were grown under unfavourable environmental conditions.

4.9 Conclusions

Abiotic stresses are important constraints limiting crop productivity universally. Plants show a varied range of responses to drought and heat stresses which are mostly depicted by a variety of alterations in the growth and morphology of plants. Climate changes are alarming the world by hampering cultivation and its products. Industrialization and poisonous gases cause global warming, which finally disturbs the world's environment. Climate change has devastating effects on plant growth and yield. The growth and yield hampering effects of heat stresses have been reported for several crops and emerging global warming is projected to make the situation even worse. Reproductive and grain-filling stages of crops are observed to be the most critical stages under stresses which adversely affect the crop growth, yield, and product quality of crops. To overcome the negative effects of heat stress, several strategies have been developed by researchers. However, in recent decades, the exogenous application of antioxidants has been observed to improve crop quality under heat stress. Exogenous application of osmoprotectants can ameliorate the adverse effects of heat stresses and improve seed quality of crops.

REFERENCES

Abdelaal AA, Hafez YM, EL Sabagh A (2017) Ameliorative effects of abscisic acid and yeast on morpho-physiological and yield characters of maize (*Zea mays* L.) plants under water deficit conditions. *Fres Environ Bul* 26(12):7372–7383.

Abdul S, Muhammad AA, Shabir H, Hesham A El E, Sajjad H, Niaz A, Abdul G, Sayyed RZ, Fahad S, Subhan D, Rahul D (2021a) Zinc nutrition and arbuscular mycorrhizal symbiosis effects on maize (*Zea mays* L.) growth and productivity. *J Saudi Society Agricultural Sci.* https://doi.org/10.1016/j.sjbs.2021.06.096

Abdul S, Muhammad AA, Subhan D, Niaz A, Fahad S, Rahul D, Mohammad JA, Omaima N, Muhammad HR, Bernard RG (2021b) Effect of arbuscular mycorrhizal fungi on the physiological functioning of maize under zinc-deficient soils. *Sci Rep* 11:18468.

Adachi H, Nakano T, Miyagawa N et al. (2015) WRKY transcription factors phosphorylated by MAPK regulate a plant immune NADPH oxidase in *Nicotiana benthamiana. Plant Cell* 27:2645–2663.

Adnan M, Fahad S, Muhammad Z, Shahen S, Ishaq AM, Subhan D, Zafar-ul-Hye M, Martin LB, Raja MMN, Beena S, Saud S, Imran A, Zhen Y, Martin B, Jiri H, Rahul D (2020) Coupling phosphate-solubilizing bacteria with phosphorus supplements improve maize phosphorus acquisition and growth under lime induced salinity stress. *Plants* 9(900). doi:10.3390/plants9070900

Adnan M, Fahad S, Khan IA, Saeed M, Ihsan MZ, Saud S, Riaz M, Wang D, Wu C (2019). Integration of poultry manure and phosphate solubilizing bacteria improved availability of Ca bound P in calcareous soils. *3 Biotech* 9(10):368.

Adnan M, Shah Z, Sharif M, Rahman H (2018b). Liming induces carbon dioxide (CO_2) emission in PSB inoculated alkaline soil supplemented with different phosphorus sources. *Environ Sci Poll Res* 25(10):9501–9509.

Adnan M, Zahir S, Fahad S, Arif M, Mukhtar A, Imtiaz AK, Ishaq AM, Abdul B, Hidayat U, Muhammad A, Inayat-Ur R, Saud S, Muhammad ZI, Yousaf J, Amanullah, Hafiz MH, Wajid N (2018a) Phosphate-solubilizing bacteria nullify the antagonistic effect of soil calcification on bioavailability of phosphorus in alkaline soils. *Sci Rep* 8:4339. https://doi.org/10.1038/s41598-018-22653-7

Ahmad S, Kamran M, Ding R, Meng X, Wang H, Ahmad I, Fahad S, Han Q (2019) Exogenous melatonin confers drought stress by promoting plant growth, photosynthetic capacity and antioxidant defense system of maize seedlings. *PeerJ* 7:e7793. http://doi.org/10.7717/peerj.7793

Ahmed N, Tetlow IJ, Nawaz S et al. (2015) Effect of high temperature on grain filling period, yield, amylose content and activity of starch biosynthesis enzymes in the endosperm of basmati rice. *J Sci Food Agric* 95:2237–2243.

Ainsworth EA, Ort DR (2010) How do we improve crop production in a warming world? *Plant Physiol* 154:526–530.

Akbar H, Timothy JK, Jagadish T, Golam M, Apurbo KC, Muhammad F, Rajan B, Fahad S, Hasanuzzaman M (2020) Agricultural land degradation: processes and problems undermining future food security. In: Fahad S, Hasanuzzaman M, Alam M, Ullah H, Saeed M, Khan AK, Adnan M (eds) *Environment, Climate, Plant and Vegetation Growth*. Springer Publications Ltd, Springer Nature Switzerland AG. Part of Springer Nature, pp. 17–62. https://doi.org/10.1007/978-3-030-49732-3

Akdeniz H, Hosaflioğlu I, Koç A et al. (2019) Evaluation of herbage yield and nutritive value of eight forage crop species. *Appl Eco Environ Res* 3:5571–5581.

Akram R, Turan V, Hammad HM, Ahmad S, Hussain S, Hasnain A, Maqbool MM, Rehmani MIA, Rasool A, Masood N, Mahmood F, Mubeen M, Sultana SR, Fahad S, Amanet K, Saleem M, Abbas Y, Akhtar HM, Waseem F, Murtaza R, Amin A, Zahoor SA, ul Din MS, Nasim W (2018a) Fate of organic and inorganic pollutants in paddy soils. In: Hashmi MZ, Varma A (eds) *Environmental Pollution of Paddy Soils, Soil Biology*. Springer International Publishing AG, Gewerbestrasse 11, Cham, Ch-6330, Switzerland, pp. 197–214.

Akram R, Turan V, Wahid A, Ijaz M, Shahid MA, Kaleem S, Hafeez A, Maqbool MM, Chaudhary HJ, Munis MFH, Mubeen M, Sadiq N, Murtaza R, Kazmi DH, Ali S, Khan N, Sultana SR, Fahad S, Amin A, Nasim W (2018b) Paddy land pollutants and their role in climate change. In: Hashmi MZ, Varma A (eds) *Environmental Pollution of Paddy Soils, Soil Biology*. Springer International Publishing AG, Gewerbestrasse 11, Cham, Ch-6330, Switzerland, pp. 113–124.

Ali S, Liu Y, Ishaq M et al. (2017) Climate change and its impact on the yield of major food crops. *Foods* 6(6):39. http://doi.org/10.3390/foods6060039

Al-Khatib K, Paulsen GM (1984) Mode of high temperature injury to wheat during grain development. *Physiol Plant* 61(3):363–368.

Altenbach SB, Tanaka CK, Hurkman WJ et al. (2011) Differential effects of a post-anthesis fertilizer regimen on the wheat flour proteome determined by quantitative 2-DE. *Proteome Sci* 9(1):46–49.

Amanullah, Muhammad I, Haider N, Shah K, Manzoor A, Asim M, Saif U, Izhar A, Fahad S, Adnan M et al. (2021) Integrated foliar nutrients application improve wheat (*Triticum Aestivum* L.) productivity under calcareous soils in drylands. *Commun Soil Sci Plant Anal.* https://doi.org/10.1080/00103624.2021.1956521

Amanullah, Fahad S (eds) (2017) Rice - technology and production. *Intech Open Croatia 2017*. http://dx.doi.org/10.5772/64480

Amanullah, Fahad S (eds) (2018a) Corn-production and human health in changing climate. *Intech Open United Kingdom* 2018. http://dx.doi.org/10.5772/intechopen.74074

Amanullah, Fahad S (eds) (2018b) Nitrogen in agriculture – updates. *Intech Open Croatia 2018*. http://dx.doi.org/10.5772/65846

Amanullah, Shah K, Imran, Hamdan AK, Muhammad A, Abdel RA, Muhammad A, Fahad S, Azizullah S, Brajendra P (2020) Effects of climate change on irrigation water quality. In: Fahad S, Hasanuzzaman M, Alam M, Ullah H, Saeed M, Khan AK, Adnan M (eds) *Environment, Climate, Plant and Vegetation Growth*. Springer Publications Ltd, Springer Nature Switzerland AG. Part of Springer Nature, pp. 123–132. https://doi.org/10.1007/978-3-030-49732-3

Amir M, Muhammad A, Allah B, Sevgi Ç, Haroon ZK, Muhammad A, Emre A (2020) Bio fortification under climate change: the fight between quality and quantity. In: Fahad S, Hasanuzzaman M, Alam M, Ullah H, Saeed M, Khan AK, Adnan M (eds) *Environment, Climate, Plant and Vegetation Growth*. Springer Publications Ltd, Springer Nature Switzerland AG. Part of Springer Nature, pp. 173–228. https://doi.org/10.1007/978-3-030-49732-3

Amjad I, Muhammad H, Farooq S, Anwar H (2020) Role of plant bioactives in sustainable agriculture. In: Fahad S, Hasanuzzaman M, Alam M, Ullah H, Saeed M, Khan AK, Adnan M (eds) *Environment, Climate, Plant and Vegetation Growth*. Springer Publications Ltd, Springer Nature Switzerland AG. Part of Springer Nature, pp. 591–606. https://doi.org/10.1007/978-3-030-49732-3

Amjad SF, Mansoora N, Din IU, Khalid IR, Jatoi GH, Murtaza G, Yaseen S, Naz M, Danish S, Fahad S et al. (2021) Application of zinc fertilizer and mycorrhizal inoculation on physio-biochemical parameters of wheat grown under water-stressed environment. *Sustainability* 13:11007. https://doi.org/10.3390/su131911007

Anis G, Hassan H, El-Sherif A et al. (2019) Evaluation of new promising rice hybrid and its parental lines for floral, agronomic traits and genetic purity assessment. *Pak J Agric Sci* 56(3):567–576.

Arif M, Talha J, Muhammad R, Fahad S, Muhammad A, Amanullah, Kawsar A, Ishaq AM, Bushra K, Fahd R (2020) Biochar; a remedy for climate change. In: Fahad S, Hasanuzzaman M, Alam M, Ullah H, Saeed M, Khan AK, Adnan M (eds) *Environment, Climate, Plant and Vegetation Growth*. Springer Publications Ltd, Springer Nature Switzerland AG. Part of Springer Nature, pp. 151–172. https://doi.org/10.1007/978-3-030-49732-3

Arshad M, Adnan M, Ali A, Khan AK, Khan F, Khan A, Kamal MA, Alam M, Ullah H, Saleem A, Hussain A, Shahwar D (2016) Integrated effect of phosphorous and zinc on wheat quality and soil properties. *Adv Environ Biol* 10(2):40–45.

Ashfaq AR, Uzma Y, Niaz A, Muhammad AA, Fahad S, Haider S, Tayebeh Z, Subhan D, Süleyman T, Hesham AE, Pramila T, Jamal MA, Sulaiman AA, Rahul D (2021) Toxicity of Cadmium and nickel in the context of applied activated carbon biochar for improvement in soil fertility. *Saudi Society Agricultural Sci.* https://doi.org/10.1016/j.sjbs.2021.09.035

Ashraf M, Foolad MR (2017) Roles of glycine betaine and proline in improving plant abiotic stress resistance. *Environ Exp Bot* 59:206–216.

Athar HR, Khan A, Ashraf M (2008). Exogenously applied ascorbic acid alleviates salt-induced oxidative stress in wheat. *Environ Exp Bot* 63:224–231.

Athar M, Masood IA, Sana S, Ahmed M, Xiukang W, Sajid F, Sher AK, Habib A, Faran M, Zafar H, Farhana G, Fahad S (2021) Bio-diesel production of sunflower through sulphur management in a semi-arid subtropical environment. *Environ Sci Pollution Res.* https://doi.org/10.1007/s11356-021-16688-z

Atif B, Hesham A, Fahad S (2021) Biochar coupling with phosphorus fertilization modifies antioxidant activity, osmolyte accumulation and reactive oxygen species synthesis in the leaves and xylem sap of rice cultivars under high-temperature stress. *Physiol Mol Biol Plants.* https://doi.org/10.1007/s12298-021-01062-7

Aziz K, Daniel KYT, Fazal M, Muhammad ZA, Farooq S, Fan W, Fahad S, Ruiyang Z (2017a) Nitrogen nutrition in cotton and control strategies for greenhouse gas emissions: a review. *Environ Sci Pollut Res* 24:23471–23487. https://doi.org/10.1007/s11356-017-0131-y

Aziz K, Daniel KYT, Muhammad ZA, Honghai L, Shahbaz AT, Mir A, Fahad S (2017b) Nitrogen fertility and abiotic stresses management in cotton crop: a review. *Environ Sci Pollut Res* 24:14551–14566. https://doi.org/10.1007/s11356-017-8920-x

Bachlava E, Cardinal AJ (2009) Correlation between temperature and oleic acid seed content in three segregating soybean populations. *Crop Sci* 49:1328–1335.

Balla K, Veisz O (2007) Changes in the quality of cereals in response to heat and drought stress. *Acta Agron Óvar* 49(2):451–455.

Barutcular C, EL Sabagh A, Koç M et al. (2017) Relationships between grain yield and physiological traits of durum wheat varieties under drought and high temperature stress in Mediterranean conditions. *Fresen Environ Bul* 26(4):4282–4291.

Barutcular C, Yildirim M, Koc M et al. (2016) Quality traits performance of bread wheat genotypes under drought and heat stress conditions. *Fresen Environ Bul* 25(12a):6159–6165.

Baseer M, Adnan M, Fazal M, Fahad S, Muhammad S, Fazli W, Muhammad A, Jr. Amanullah, Depeng W, Saud S, Muhammad N, Muhammad Z, Fazli S, Beena S, Mian AR, Ishaq AM (2019) Substituting urea by organic wastes for improving maize yield in alkaline soil. *J Plant Nutrition.* http://dx.doi.org/10.1080/01904167.2019.1659344

Basra AS, Malik CP (1984) Development of the cotton fiber. In *International review of cytology*, Vol. 89. Academic Press. pp. 65–113.

Bayram AY, Seher Ö, Nazlican A (2020) Climate change forecasting and modeling for the year of 2050. In: Fahad S, Hasanuzzaman M, Alam M, Ullah H, Saeed M, Khan AK, Adnan M (eds) *Environment, Climate, Plant and Vegetation Growth.* Springer Publications Ltd, Springer Nature Switzerland AG. Part of Springer Nature, pp. 109–122. https://doi.org/10.1007/978-3-030-49732-3

Bellaloui N, Mengistu A, Kassem MA (2013) Effects of genetics and environment on fatty acid stability in soybean seed. *Food Nut Sci* 4(9):165–173.
Ben-Asher J, Garcia AGY, Hoogenboom G (2008) Effect of high temperature on photosynthesis and transpiration of sweet corn (*Zea mays* L. var. rugosa). *Photosynthetica* 46:595–603.
Berry J, Bjorkman O (1980) Photosynthetic response and adaptation to temperature in higher plants. *An Rev Plant Physiol* 31:491–543.
Bishop J, Potts SG, Jones HE (2016) Susceptibility of faba bean (*Vicia faba* L.) to heat stress during floral development and anthesis. *J Agron Crop Sci* 202(6):508–517.
Byrla DR, Bouma TJ, Hartmond U, Eissenstat DM (2001) Influence of temperature and soil drying on respiration of individual roots in citrus, integrating green observations into a predictive model for the field. *Plant Cell Environ* 24:781–790.
Cairns JE, Crossa C, Zaidi PH et al. (2013) Identification of drought, heat and combined drought and heat tolerance donors in maize (*Zea mays* L.). *Crop Sci* 53:1335–1346.
Castro M, Peterson J, Rizza MD (2007) Influence of heat stress on wheat grain characteristics and protein molecular weight distribution. In: Buck HT (ed) *Wheat production in stressed environment*, pp. 365–371.
Ceccarelli S, Grando S, Maatougui M et al. (2010) Plant breeding and climate changes. *J Agric Sci* 148(6):627–637.
Chang W, Qiujuan J, Evgenios A, Haitao L, Gezi L, Jingjing Z, Fahad S, Ying J (2021) Hormetic effects of zinc on growth and antioxidant defense system of wheat plants. *Sci Total Environ.* https://doi.org/10.1016/j.scitotenv.2021.150992
Chen J, Liu L, Ma Y et al. (2014) Effect of late planting and shading on cellulose synthesis during cotton fiber secondary cell wall development. *PLoS One* 9:105088.
Chen Y, Guo Z, Dong L, Fu Z, Zheng Q, Zhang G, Qin L, Sun X, Shi Z, Fahad S, Xie F, Saud S (2021) Turf performance and physiological responses of native Poa species to summer stress in Northeast China. *PeerJ* 9:e12252 http://doi.org/10.7717/peerj.12252
Chowdappa P, Gowda S (2013) Nanotechnology in crop protection: status and scope. *Pest Manag Hort Eco* 19(2):131–151.
Connor DJ, Hall AJ (1997) Sunflower physiology. In: Schneiter AA (ed) *Sunflower technology and production.* American Society of Agronomy, Crop Science Society of America, and Soil Science Society of America, Madison, pp. 113–182.
Daniel C, Triboi E (2000) Effects of temperature and nitrogen nutrition on the grain composition of winter wheat: effects on gliadin content and composition. *J Cereal Sci 32*(1):45–56.
Darby HM, Lauer JG (2002) Planting date and hybrid influence on corn forage yield and quality *Agron J* 94:281–289.
De Oliveira, ED, Bramley H, Siddique S et al. (2013) Can elevated CO_2 combined with high temperature ameliorate the effect of terminal drought in wheat? *Fun Plant Biol* 40:160–171.
de Pascale S, Maggio A, Fogliano V et al. (2001) Irrigation with saline water improves carotenoids content and antioxidant activity of tomato. *J Hort Sci Biotechnol* 76:447–453.
Deepranjan S, Ardith S, Devika OS, Sonam S, Shikha, Manoj P, Amitava R, Sayyed RZ, Abdul G, Mohammad JA, Subhan D, Fahad S, Rahul D (2021) Optimizing nutrient use efficiency, productivity, energetics, and economics of red cabbage following mineral fertilization and biopriming with compatible rhizosphere microbes. *Sci Rep* 11:15680.
Demirevska-Kepova K, Holzer R, Simova-Stoilova L et al. (2005) Heat stress effects on ribulose-1,5-bisphosphate carboxylase/oxygenase, Rubisco binding protein and Rubisco activase in wheat leaves. *Biol Plant* 49:521–525.
Depeng W, Fahad S, Saud S, Muhammad K, Aziz K, Mohammad NK, Hafiz MH, Wajid N (2018) Morphological acclimation to agronomic manipulation in leaf dispersion and orientation to promote "Ideotype" breeding: evidence from 3D visual modeling of "super" rice (*Oryza sativa* L.). *Plant Physiol Biochem.* 135:499–510 https://doi.org/10.1016/j.plaphy.2018.11.010
Dimetry NZ, Hussein HM (2016) Role of nanotechnology in agriculture with special reference to pest control. *Int J Pharm Tech Res* 9:121–144.
Duhan JS, Kumar R, Kumar N et al. (2017) Nanotechnology: the new perspective in precision agriculture. *Biotechnol Rep* 15:11–23.
Dutta A, Mohanty S, Tripathy BC (2009) Role of temperature stress on chloroplast biogenesis and protein important in pea. *Plant Physiol* 150:1050–1061.

EL Sabagh A, Hossain A, Barutçular C (2019a) Sustainable soybean production and abiotic stress management in saline environments: a critical review. *Aust J Crop Sci* 13(2):228–236.

EL Sabagh A, Hossain A, Barutçular C et al. (2019b) Effects Of drought stress on the quality of major oilseed crops: implications and possible mitigation strategies – a review. *Appl Eco Environ Res* 17(2):4019–4043.

EL Sabagh A, Hossain A, Barutçular C et al. (2019c) Drought and salinity stress management for higher and sustainable canola (*Brassica napus* L.) production: A critical review. *Aust J Crop Sci* 13(1):88–97.

Emara MAA (2012) Response of cotton growth and productivity to application of potassium and zinc under normal and late sowing dates. *Mansoura J Plant Product* 3:509–524.

Emre B, Ömer SU, Martín LB, Andre D, Fahad S, Rahul D, Muhammad Z-ul-H, Ghulam SH, Subhan D (2021) Studying soil erosion by evaluating changes in physico-chemical properties of soils under different land-use types. *J Saudi Society Agric Sci* 20:190–197.

Fahad S, Bajwa AA, Nazir U, Anjum SA, Farooq A, Zohaib A, Sadia S, Nasim W, Adkins S, Saud S, Ihsan MZ, Alharby H, Wu C, Wang D, Huang J (2017) Crop production under drought and heat stress: plant responses and management options. *Front Plant Sci* 8:1147. https://doi.org/10.3389/fpls.2017.01147

Fahad S, Bano A (2012) Effect of salicylic acid on physiological and biochemical characterization of maize grown in saline area. *Pak J Bot* 44:1433–1438.

Fahad S, Chen Y, Saud S, Wang K, Xiong D, Chen C, Wu C, Shah F, Nie L, Huang J (2013) Ultraviolet radiation effect on photosynthetic pigments, biochemical attributes, antioxidant enzyme activity and hormonal contents of wheat. *J Food Agri Environ* 11(3&4):1635–1641.

Fahad S, Hussain S, Bano A, Saud S, Hassan S, Shan D, Khan FA, Khan F, Chen Y, Wu C, Tabassum MA, Chun MX, Afzal M, Jan A, Jan MT, Huang J (2014a) Potential role of phytohormones and plant growth-promoting rhizobacteria in abiotic stresses: consequences for changing environment. *Environ Sci Pollut Res* 22(7):4907–4921. https://doi.org/10.1007/s11356-014-3754-2

Fahad S, Hussain S, Matloob A, Khan FA, Khaliq A, Saud S, Hassan S, Shan D, Khan F, Ullah N, Faiq M, Khan MR, Tareen AK, Khan A, Ullah A, Ullah N, Huang J (2014b) Phytohormones and plant responses to salinity stress: a review. *Plant Growth Regul* 75(2):391–404. https://doi.org/10.1007/s10725-014-0013-y

Fahad S, Hussain S, Saud S, Hassan S, Chauhan BS, Khan F et al. (2016a) Responses of rapid viscoanalyzer profile and other rice grain qualities to exogenously applied plant growth regulators under high day and high night temperatures. *PLoS One* 11(7):e0159590. https://doi.org/10.1371/journal.pone.0159590

Fahad S, Hussain S, Saud S, Hassan S, Ihsan Z, Shah AN, Wu C, Yousaf M, Nasim W, Alharby H, Alghabari F, Huang J (2016c) Exogenously applied plant growth regulators enhance the morphophysiological growth and yield of rice under high temperature. *Front Plant Sci* 7:1250. https://doi.org/10.3389/fpls.2016.01250.

Fahad S, Hussain S, Saud S, Hassan S, Tanveer M, Ihsan MZ, Shah AN, Ullah A, Nasrullah KF, Ullah S, Alharby HNW, Wu C, Huang J (2016d) A combined application of biochar and phosphorus alleviates heat-induced adversities on physiological, agronomical and quality attributes of rice. *Plant Physiol Biochem* 103:191–198.

Fahad S, Hussain S, Saud S, Khan F, Hassan S, Nasim W, Arif M, Wang F, Huang J (2016b) Exogenously applied plant growth regulators affect heat-stressed rice pollens. *J Agron Crop Sci* 202:139–150.

Fahad S, Hussain S, Saud S, Tanveer M, Bajwa AA, Hassan S, Shah AN, Ullah A, Wu C, Khan FA, Shah F, Ullah S, Chen Y, Huang J (2015a) A biochar application protects rice pollen from high-temperature stress. *Plant Physiol Biochem* 96:281–287.

Fahad S, Muhammad ZI, Abdul K, Ihsanullah D, Saud S, Saleh A, Wajid N, Muhammad A, Imtiaz AK, Chao W, Depeng W, Jianliang H (2018) Consequences of high temperature under changing climate optima for rice pollen characteristics-concepts and perspectives. *Archives Agron Soil Sci* http://dx.doi.org/10.1080/03650340.2018.1443213

Fahad S, Nie L, Chen Y, Wu C, Xiong D, Saud S, Hongyan L, Cui K, Huang J (2015b) Crop plant hormones and environmental stress. *Sustain Agric Rev* 15:371–400.

Fahad S, Hasanuzzaman M, Alam M, Ullah H, Saeed M, Ali Khan I, Adnan M (eds) (2020) *Environment, climate, plant and vegetation growth.* Springer Nature Switzerland AG 2020. https://doi.org/10.1007/978-3-030-49732-3

Fahad S, Rehman A, Shahzad B, Tanveer M, Saud S, Kamran M, Ihtisham M, Khan SU, Turan V, Rahman MHU (2019). Rice responses and tolerance to metal/metalloid toxicity. In: Hasanuzzaman M, Fujita M, Nahar K, Biswas JK (eds) *Advances in rice research for abiotic stress tolerance.* Woodhead Publications Ltd, Abington, Cambridge, UK, pp. 299–312.

Fahad S, Sönmez O, Saud S, Wang D, Wu, C, Adnan M, Arif M, Amanullah (eds) (2021e) *Engineering tolerance in crop plants against abiotic stress, First edition. ed, Footprints of climate variability on plant diversity*. CRC Press, Boca Raton.

Fahad S, Sonmez O, Saud S, Wang D, Wu, C, Adnan M, Turan V (eds) (2021b) *Climate change and plants: biodiversity, growth and interactions, First edition. ed, Footprints of climate variability on plant diversity*. CRC Press, Boca Raton.

Fahad S, Sonmez O, Saud S, Wang D, Wu, C, Adnan M, Turan V (eds) (2021c) *Developing climate resilient crops: improving global food security and safety, First edition. ed, Footprints of climate variability on plant diversity*. CRC Press, Boca Raton.

Fahad S, Sönmez O, Saud S, Wang D, Wu, C, Adnan M, Turan V (eds) (2021a) *Plant growth regulators for climate-smart agriculture, First edition. ed, Footprints of climate variability on plant diversity*. CRC Press, Boca Raton, FL.

Fahad S, Sönmez O, Turan V, Adnan M, Saud S, Wu, C, Wang D (eds) (2021d) *Sustainable soil and land management and climate change, First edition. ed, Footprints of climate variability on plant diversity*. CRC Press, Boca Raton.

Farah R, Muhammad R, Muhammad SA, Tahira Y, Muhammad AA, Maryam A, Shafaqat A, Rashid M, Muhammad R, Qaiser H, Afia Z, Muhammad AA, Muhammad A, Fahad S (2020) Alternative and non-conventional soil and crop management strategies for increasing water use efficiency. In: Fahad S, Hasanuzzaman M, Alam M, Ullah H, Saeed M, Khan AK, Adnan M (eds) *Environment, Climate, Plant and Vegetation Growth*. Springer Publications Ltd, Switzerland, pp. 323–338. https://doi.org/10.1007/978-3-030-49732-3

Farhana G, Ishfaq A, Muhammad A, Dawood J, Fahad S, Xiuling L, Depeng W, Muhammad F, Muhammad F, Syed AS (2020) Use of crop growth model to simulate the impact of climate change on yield of various wheat cultivars under different agro-environmental conditions in Khyber Pakhtunkhwa, Pakistan. *Arabian J Geosci* 13:112. https://doi.org/10.1007/s12517-020-5118-1.

Farhat A, Hafiz MH, Wajid I, Aitazaz AF, Hafiz FB, Zahida Z, Fahad S, Wajid F, Artemi C (2020) A review of soil carbon dynamics resulting from agricultural practices. *J Environ Manage* 268:110319.

Fatima, Shah M, Usman A, Sohail K, Afzaal M, Shah B, Adnan M, Ahmed N, Junaid K, Shah SRA, Rahman LU (2016) Rearing and identification of Callosobruchusmaculatus (Bruchidae: Coleoptera) in Chickpea. *J Ento Zoo Studies* 4(2):264–266.

Fazli W, Muhmmad S, Amjad A, Fahad S, Muhammad A, Muhammad N, Ishaq AM, Imtiaz AK, Mukhtar A, Muhammad S, Muhammad I, Rafi U, Haroon I, Muhammad A (2020) Plant-microbes interactions and functions in changing climate. In: Fahad S, Hasanuzzaman M, Alam M, Ullah H, Saeed M, Khan AK, Adnan M (eds) *Environment, climate, plant and vegetation growth*. Springer Publications Ltd, Springer Nature Switzerland AG. Part of Springer Nature, pp. 397–420. https://doi.org/10.1007/978-3-030-49732-3.

Ghulam M, Muhammad AA, Donald LS, Sajid M, Muhammad FQ, Niaz A, Ateeq Ur R, Shakeel A, Sajjad H, Muhammad A, Summia M, Aqib HAK, Fahad S, Rahul D, Mazhar I, Timothy DS (2021) Formalin fumigation and steaming of various composts differentially influence the nutrient release, growth and yield of muskmelon (*Cucumis melo* L.). *Sci Rep* 11:21057.

Gifford RM (2003) Plant respiration in productivity models: conceptualization, representation and issues for global terrestrial carbon-cycle research. *Funct Plant Biol* 30:171–186.

Gopakumar L, Bernard NO, Donato V (2020) Soil microarthropods and nutrient cycling. In: Fahad S, Hasanuzzaman M, Alam M, Ullah H, Saeed M, Khan AK, Adnan M (eds) *Environment, Climate, Plant and Vegetation Growth*. Springer Publications Ltd, Springer Nature Switzerland AG. Part of Springer Nature, pp. 453–472. https://doi.org/10.1007/978-3-030-49732-3 .

Gormus O, EL Sabagh A, Harun R et al. (2017a) Enhancement of productivity and fiber quality by defining ideal defoliation and harvesting timing in cotton. *Rom Agric Res* 34:226–232.

Gormus O, Harun R, EL Sabagh A (2017b) Impact of defoliation timings and leaf pubescence on yield and fiber quality of cotton. *J Agric Sci Technol* 19(4):903–915.

Guofu L, Zhenjian B, Fahad S, Guowen C, Zhixin X, Hao G, Dandan L, Yulong L, Bing L, Guoxu J, Saud S (2021) Compositional and structural changes in soil microbial communities in response to straw mulching and plant revegetation in an abandoned artificial pasture in Northeast China. *Glob Ecol Conserv* 31:e01871.

Habib Ur R, Ashfaq A, Aftab W, Manzoor H, Fahd R, Wajid I, Aminul I, Vakhtang S, Muhammad A, Asmat U, Abdul W, Syeda RS, Shah S, Shahbaz K, Fahad S, Manzoor H, Saddam H, Wajid N (2017) Application of CSM-CROPGRO-Cotton model for cultivars and optimum planting dates: evaluation in changing semi-arid climate. *Field Crops Res*. http://dx.doi.org/10.1016/j.fcr.2017.07.007.

Hafeez M, Farman U, Muhammad MK, Xiaowei L, Zhijun Z, Sakhawat S, Muhammad I, Mohammed AA, Mandela F-G, · Nicolas D, Muzammal R, Fahad S, Yaobin L (2021) Metabolic-based insecticide resistance mechanism and ecofriendly approaches for controlling of beet armyworm Spodoptera exigua: a review. *Environ Sci Pollut Res*. https://doi.org/10.1007/s11356-021-16974-w.

Hafiz MH, Abdul K, Farhat A, Wajid F, Fahad S, Muhammad A, Ghulam MS, Wajid N, Muhammad M, Hafiz FB (2020b) Comparative effects of organic and inorganic fertilizers on soil organic carbon and wheat productivity under arid region. *Commun Soil Sci Plant Anal*. http://dx.doi.org/10.1080/00103624.2020.1763385

Hafiz MH, Farhat A, Ashfaq A, Hafiz FB, Wajid F, Carol Jo W, Fahad S, Gerrit H (2020a) Predicting kernel growth of maize under controlled water and nitrogen applications. *Int J Plant Prod*. https://doi.org/10.1007/s42106-020-00110-8.

Hafiz MH, Farhat A, Shafqat S, Fahad S, Artemi C, Wajid F, Chaves CB, Wajid N, Muhammad M, Hafiz FB (2018) Offsetting land degradation through nitrogen and water management during maize cultivation under arid conditions. *Land Degrad Dev* 1–10. http://dx.doi.org/10.1002/ldr.2933

Hafiz MH, Muhammad A, Farhat A, Hafiz FB, Saeed AQ, Muhammad M, Fahad S, Muhammad A (2019) Environmental factors affecting the frequency of road traffic accidents: a case study of sub-urban area of Pakistan. *Environ Sci Pollut Res*. https://doi.org/10.1007/s11356-019-04752-8.

Hafiz MH, Wajid F, Farhat A, Fahad S, Shafqat S, Wajid N, Hafiz FB (2016) Maize plant nitrogen uptake dynamics at limited irrigation water and nitrogen. *Environ Sci Pollut Res* 24(3):2549–2557. https://doi.org/10.1007/s11356-016-8031-0.

Haider SA, Lalarukh I, Amjad SF, Mansoora N, Naz M, Naeem M, Bukhari SA, Shahbaz M, Ali SA, Marfo TD, Subhan D, Rahul D, Fahad S (2021) Drought stress alleviation by potassium-nitrate-containing chitosan/montmorillonite microparticles confers changes in *Spinacia oleracea* L. *Sustain* 13:9903. https://doi.org/10.3390/su13179903.

Hake K, Silvertooth J (1990) High temperature effects on cotton. *Physiol Today* 1(10):110–124.

Hamza SM, Xiukang W, Sajjad A, Sadia Z, Muhammad N, Adnan M, Fahad S et al. (2021) Interactive effects of gibberellic acid and NPK on morpho-physio-biochemical traits and organic acid exudation pattern in coriander (*Coriandrum sativum* L.) grown in soil artificially spiked with boron. *Plant Physiol Biochem* 167:884–900.

Hancock RD, Morris WL, Ducreux LJ et al. (2014) Physiological, biochemical and molecular responses of the potato (*Solanum tuberosum* L.) plant to moderately elevated temperature. *Plant Cell Environ* 37(2): 439–450.

Hayat Q, Hayat S, Irfan M et al. (2010) Effect of exogenous salicylic acid under changing environment: a review. *Environ Exp Bot* 68:14–25.

Hesham FA and Fahad S (2020) Melatonin application enhances biochar efficiency for drought tolerance in maize varieties: modifications in physio-biochemical machinery. *Agron J* 112(4):1–22.

Hesketh JD, Low A (1968) Effect of temperature and fiber quality of cotton varieties of diverse origin. *Emp Cott Grow Rev* 45:243–257.

Hossain A, Sarker MAZ, Saifuzzaman M, et al. (2013) Evaluation of growth, yield, relative performance and heat susceptibility of eight wheat (*Triticum aestivum* L.) genotypes grown under heat stress. *Int J Plant Prod* 7:615–636.

Hossain MM, Hossain A, Alam MA et al. (2018) Evaluation of fifty spring wheat genotypes grown under heat stress condition in multiple environments of Bangladesh. *Fres Environ Bul* 27(9): 5993–6004.

Hrušková M, Švec I (2009) Wheat hardness in relation to other quality factors. *Czech J Food Sci* 27(4): 240–248.

Huang L-Y, Li X-X, Zhang Y-B, Fahad S, Wang F (2021) dep1 improves rice grain yield and nitrogen use efficiency simultaneously by enhancing nitrogen and dry matter translocation. *J Integrative Agri*. http://dx.doi.org/10.1016/S2095-3119(21)63795-4

Hurkman WJ, Vensel WH, Tanaka CK et al. (2009) Effect of high temperature on albumin and globulin accumulation in the endosperm proteome of the developing wheat grain. *J Cereal Sci* 49:12–23.

Hussain H, Raziq F, Khan I, Shah B, Altaf M, Attaullah, Ullah W, Naeem A, Adnan M, Junaid K, Shah SRA, Iqbal M (2016) Effect of bipolarismaydis (Y Nisik & C Miyake) shoemaker at various growth stages of different maize cultivars. *J Ento Zool Studies* 4(2):439–444.

Hussain MA, Fahad S, Rahat S, Muhammad FJ, Muhammad M, Qasid A, Ali A, Husain A, Nooral A, Babatope SA, Changbao S, Liya G, Ibrar A, Zhanmei J, Juncai H (2020) Multifunctional role of brassinosteroid and its analogues in plants. *Plant Growth Regul*. https://doi.org/10.1007/s10725-020-00647-8.

Hussain MW, Allen H, Bowes G (1999) Up-regulation of sucrose phosphate synthase in rice grown under elevated CO_2 and temperature. *Photosynth Res* 60:199–208.

Ibrar K, Aneela R, Khola Z, Urooba N, Sana B, Rabia S, Ishtiaq H, Mujaddad Ur R, Salvatore M (2020) Microbes and environment: global warming reverting the frozen zombies. In: Fahad S, Hasanuzzaman M, Alam M, Ullah H, Saeed M, Khan AK, Adnan M (eds) *Environment, climate, plant and vegetation growth*. Springer Publications Ltd, Springer Nature Switzerland AG. Part of Springer Nature, pp. 607–634. https://doi.org/10.1007/978-3-030-49732-3 .

Ikram U, Khadim D, Muhammad T, Muhammad S, Fahad S (2021) Gibberellic acid and urease inhibitor optimize nitrogen uptake and yield of maize at varying nitrogen levels under changing climate. *Environ Sci Pollution Res*. https://doi.org/10.1007/s11356-021-16049-w

Ilyas M, Mohammad N, Nadeem K, Ali H, Aamir HK, Kashif H, Fahad S, Aziz K, Abid U (2020) Drought tolerance strategies in plants: a mechanistic approach. *J Plant Growth Regul*. https://doi.org/10.1007/s00344-020-10174-5

Iqbal MA, Abdul H, Imtiaz H et al. (2019a) Competitive indices in cereal and legume mixtures in a South Asian environment. *Agron J* 111(1):242–249.

Iqbal MA, Hamid A, Ahmad T, Siddiqui MH, Hussain I, Ali S, Ali A, Ahmad Z (2019b) Forage sorghum-legumes intercropping: effect on growth, yields, nutritional quality and economic returns. *Bragantia* 78(1):82–95.

Iqbal MA, Asif I, Muzammil HS et al. (2018b) Bio-agronomic evaluation of forage sorghum-legumes binary crops on Haplic Yermosol soil of Pakistan. *Pak J Bot* 50(5):1991–1997.

Iqbal MA, Asif I, Rana NA (2018) Spatio-temporal reconciliation to lessen losses in yield and quality of forage soybean (*Glycine max* L.) in soybean-sorghum intercropping systems. *Bragantia* 77(2):283–291.

Iqbal MA, Asif I, Zahid M et al. (2018c) Revamping soil quality and correlation studies for yield and yield attributes in sorghum-legumes intercropping systems. *Biosci J* 34(3):1165–1176.

Iqbal MA, Muzammil HS, Sher A et al. (2018a) Forage productivity of cowpea [*Vigna unguiculata* (L.) Walp] cultivars improves by optimization of spatial arrangements. *Rev Mex De Cien Pec* 9(2): 203–219.

Iqra M, Amna B, Shakeel I, Fatima K, Sehrish L, Hamza A, Fahad S (2020) Carbon cycle in response to global warming. In: Fahad S, Hasanuzzaman M, Alam M, Ullah H, Saeed M, Khan AK, Adnan M (eds) *Environment, climate, plant and vegetation growth*. Springer Publications Ltd, Springer Nature Switzerland AG. Part of Springer Nature, pp. 1–16. https://doi.org/10.1007/978-3-030-49732-3.

Irfan M, Muhammad M, Muhammad JK, Khadim MD, Dost M, Ishaq AM, Waqas A, Fahad S, Saud S et al. (2021) Heavy metals immobilization and improvement in maize (*Zea mays* L.) growth amended with biochar and compost. *Sci Rep* 11:18416.

Islam MT (2011) Effect of temperature on photosynthesis, yield attributes and yield of tomato genotypes. *Int J Expt Agric* 2(1):8–11.

Izquierdo NG, Aguirrezábal LAN, Martínez E et al. (2013) Effect of growth temperature on the high stearic and high stearic-high oleic sunflower traits. *Crop Pasture Sci* 64:1825–1832.

Jabborova D, Sulaymanov K, Sayyed RZ, Alotaibi SH, Enakiev Y, Azimov A, Jabbarov Z, Ansari MJ, Fahad S, Danish S et al. (2021) Mineral fertilizers improves the quality of turmeric and soil. *Sustain* 13:9437. https://doi.org/10.3390/su13169437.

Jagadish SVK, Craufurd PQ, Wheeler TR (2007) High temperature stress and spikelet fertility in rice (*Oryza sativa* L.). *J Exp Bot* 58:1627–1635.

Jagadish SVK, Muthurajan R, Oane R et al. (2010) Physiological and proteomic approaches to address heat tolerance during anthesis in rice (*Oryza sativa* L.). *J Exp Bot* 61(1):143–156.

Jahan MAHS, Hossain A, Jaime A et al. (2019) Effect of naphthaleneacetic acid on root and plant growth and yield of ten irrigated wheat genotypes. *Pak J Bot* 51(2):451–459.

Jan M, Anwar-ul-Haq M, Adnan NS, Muhammad Y, Javaid I, Xiuling L, Depeng W, Fahad S (2019) Modulation in growth, gas exchange, and antioxidant activities of salt-stressed rice (*Oryza sativa* L.) genotypes by zinc fertilization. *Arabian J Geosci* 12:775. https://doi.org/10.1007/s12517-019-4939-2

Janda T, Gondor OK, Yordanova R et al. (2014) Salicylic acid and photosynthesis: signalling and effects. *Acta Physiol Plant* 36:2537 2546.

Jiang Y, Huang B (2001) Drought and heat stress injury to two cool season turfgrasses in relation to antioxidant metabolism and lipid peroxidation. *Crop Sci* 41:436–442.

Kakar NU, Oad FC, Tunio S et al. (2012) Response of sowing time to various cotton genotypes. *Sarhad J Agric* 28(3):379–385.

Kamarn M, Wenwen C, Irshad A, Xiangping M, Xudong Z, Wennan S, Junzhi C, Shakeel A, Fahad S, Qingfang H, Tiening L (2017) Effect of paclobutrazol, a potential growth regulator on stalk mechanical strength, lignin accumulation and its relation with lodging resistance of maize. *Plant Growth Regul* 84:317–332. https://doi.org/10.1007/s10725-017-0342-8.

Khadim D, Fahad S, Jahangir MMR, Iqbal M, Syed SA, Shah AK, Ishaq AM, Rahul D et al. (2021a) Biochar and urease inhibitor mitigate NH_3 and N_2O emissions and improve wheat yield in a urea fertilized alkaline soil. *Sci Rep* 11:17413.

Khadim D, Saif-Ur-R, Fahad S, Syed SA, Shah AK et al. (2021b) Influence of variable biochar concentration on yield-scaled nitrous oxide emissions, Wheat yield and nitrogen use efficiency. *Sci Rep* 11:16774.

Khan MMH, Niaz A, Umber G, Muqarrab A, Muhammad AA, Muhammad I, Shabir H, Shah F, Vibhor A, Abdel-Hafezh SH, Reham A, Syed MBA, Nadiyah MA, Ali TKZ, Subhan D, Rahul D (2021) *Synchronization of Boron application methods and rates is environmentally friendly approach to improve quality attributes of Mangifera indica L.* On sustainable basis. *Saudi J Bio Sci.* https://doi.org/10.1016/j.sjbs.2021.10.036.

Kobata T et al. (2004) "High temperatures during the grain-filling period do not reduce the potential grain dry matter increase of rice". *Agron J* 96:406–414.

Krieg DR (2002) Cotton yield and quality, genetic vs. environmental affectors. In: McRae J, Richter DA (eds) *Proceedings of Beltwide Cotton Conference*, Athens, GA. 8–13 January. Natl. Cotton Council of Am., Memphis, TN, pp. 8–12.

Labuschagne MT, Elago O, Koen E (2009) The influence of temperature extremes on some quality and starch characteristics in bread, biscuit and durum wheat. *J Cereal Sci* 49(2):184–189.

Lafta AM, Lorenzen JH (1995) Effect of high temperature on plant growth and carbohydrate metabolism in potato. *Plant Physiol* 109:637–643.

Lagravère T, Kleiber D, Surel O et al. (2004) Comparison of fatty acid metabolism of two oleic and one conventional sunflower hybrids: a new hypothesis. *J Agron Crop Sci* 190:223–229.

Lamaoui M, Jemo M, Datla R et al. (2018) Heat and drought stresses in crops and approaches for their mitigation. *Front Chem* 6:26–39.

Larkindale J, Knight MR (2002) Protection against heat stress induced oxidative damage in Arabidopsis involves calcium, abscisic acid, ethylene, and salicylic acid. *J Plant Physiol* 128:682–695.

Lea PJ, Leegood RC (1999) *Plant biochemistry and molecular biology* (eds) John Wiley and Sons, Chishter, England, p 384.

Leon AJ, Andrade FH, Lee M (2003) Genetic analysis of seed-oil concentration across generations and environments in sunflower. *Crop Sci* 43:135–140.

Li L, Haomeng Y, Peng L et al. (2018) Combined impact of heat stress and phosphate deficiency on growth and photochemical activity of sheepgrass (*Leymus chinensis*). *J Plant Physiol* 231:271–276.

Li SC, Bai P, Lu X et al. (2003) Ecological and sowing date effects on maize grain filling. *Acta Agron Sin* 29:775–778.

Liu Q, Wu X, Ma J et al. (2013) Effects of high air temperature on rice grain quality and yield under field condition. *Agron J* 105:446–454.

Lunde C, Drew PD, Jacobs AK, Tester M (2007) Exclusion of Na+ via sodium ATPase (PpENA1) ensures normal growth of Physcomitrella patens under moderate salt stress. *Plant Physiol* 144:1786–1796.

Machado S, Paulsen GM (2001) Combined effects of drought and high temperature on water relations of wheat and sorghum. *Plant Soil* 233:179–187.

Mahar A, Amjad A, Altaf HL, Fazli W, Ronghua L, Muhammad A, Fahad S, Muhammad A, Rafiullah, Imtiaz AK, Zengqiang Z (2020) Promising technologies for cd-contaminated soils: drawbacks and possibilities. In: Fahad S, Hasanuzzaman M, Alam M, Ullah H, Saeed M, Khan AK, Adnan M (eds) *Environment, climate, plant and vegetation growth.* Springer Publications Ltd, Springer Nature Switzerland AG. Part of Springer Nature, pp. 63–92. https://doi.org/10.1007/978-3-030-49732-3.

Mahmood S, Wahid A, Javed F, Basra SMA (2010) Heat stress efects on forage quality characteristics of maize (Zea mays) cultivars. *Int J Agri Biol* 12:701–706.

Mahmood Ul H, Tassaduq R, Chandni I, Adnan A, Muhammad A, Muhammad MA, Muhammad H-Ur-R, Mehmood AN, Alam S, Fahad S (2021) Linking plants functioning to adaptive responses under heat stress conditions: a mechanistic review. *J Plant Growth Regul.* https://doi.org/10.1007/s00344-021-10493-1.

Manzer HS, Saud A, Soumya M, Al-Amri AA, Qasi DA, Al-Munqedhi, BM, Hayssam MA, Hazem MK, Fahad S, Vishnu DR, Om PN (2021) Molybdenum and hydrogen sulfide synergistically mitigate arsenic toxicity by modulating defense system, nitrogen and cysteine assimilation in faba bean (*Vicia faba* L.) seedlings. *Environ Pollut* 290: 117953.

Martínez-Force E, Álvarez-Ortega R, Cantisán S, Garcés R. 1998. Fatty acid composition in developing high saturated sunflower (Helianthus annuus) seeds: maturation changes and temperature effect. *J Agric Food Chem* 46:3577–3582.

Mayer LI, Edreira JIR, Maddonni GA (2014) Oil yield components of maize crops exposed to heat stress during early and late grain-filling stages. *Crop Sci* 54:2236–2250.

Md Jakir H, Allah B (2020) Development and applications of transplastomic plants; A way towards eco-friendly agriculture. In: Fahad S, Hasanuzzaman M, Alam M, Ullah H, Saeed M, Khan AK, Adnan M (eds) *Environment, climate, plant and vegetation growth.* Springer Publications Ltd, Springer Nature Switzerland AG. Part of Springer Nature, pp. 285–322. https://doi.org/10.1007/978-3-030-49732-3

Md Enamul H, Shoeb AZM, Mallik AH, Fahad S, Kamruzzaman MM, Akib J, Nayyer S, Adnan KMM, Swati AS, Md Yeamin A, Most SS (2020) Measuring vulnerability to environmental hazards: qualitative to quantitative. In: Fahad S, Hasanuzzaman M, Alam M, Ullah H, Saeed M, Khan AK, Adnan M (eds) *Environment, climate, plant and vegetation growth.* Springer Publications Ltd, Springer Nature Switzerland AG. Part of Springer Nature, pp. 421–452. https://doi.org/10.1007/978-3-030-49732-3 .

Meredith WR (2005) *Influence of cotton breeding on yield and fi ber quality problems.* In Ch Chewing H (ed.) *Proceedings of the 18th Annual EFC Conf.* 6–8 June 2005. Memphis, TN.

Mitchell RB, Moser LE (2000) Developmental morphology and tiller dynamics of warm-season grass swards. In Moore KJ, Anderson BE (eds) *Native warm-season grasses: research trends and issues.* CSSA Spec. Pub. 30. Crop Science Society of America, Madison, WI, pp. 49–66.

Mittler R, Blumwald E (2010) Genetic engineering for modern agriculture: challenges and perspectives. *Ann Rev Plant Biol* 61:443–462.

Al-Wabel MI, Abdelazeem S, Munir A, Khalid E, Adel RAU (2020b) Extent of climate change in Saudi Arabia and its impacts on agriculture: a case study from Qassim region. In: Fahad S, Hasanuzzaman M, Alam M, Ullah H, Saeed M, Khan AK, Adnan M (eds) *Environment, climate, plant and vegetation growth.* Springer Publications Ltd, Springer Nature Switzerland AG. Part of Springer Nature, pp. 635–658. https://doi.org/10.1007/978-3-030-49732-3

Al-Wabel MI, Ahmad M, Usman A, Akanji M, Rafique MI (2020a) Advances in pyrolytic technologies with improved carbon capture and storage to combat climate change. In: Fahad S, Hasanuzzaman M, Alam M, Ullah H, Saeed M, Khan AK, Adnan M (eds) *Environment, climate, plant and vegetation growth.* Springer Publications Ltd, Springer Nature Switzerland AG. Part of Springer Nature, pp. 535–576. https://doi.org/10.1007/978-3-030-49732-3

Mohammadi V, Qannadha MR, Zali AA et al. (2006) Evaluation of screening techniques for heat tolerance in wheat. *Int J Agric Biol* 1:42–44.

Mohanty N, Mohanty P (1988) After-effect of elevated temperature on the development of photosynthetic characteristics in wheat seedlings. *J Plant Physiol* 133:235–239.

Molla MSH, Sutkhet N, Akkas MA et al. (2019) Influence of nitrogen application on dry biomass allocation and translocation in two maize varieties under short pre-anthesis and prolonged bracketing flowering periods of drought. *Arch Agron Soil Sci* 65(7):928–944.

Monjardino P, Machado J, Gil FS et al. (2007) Structural and ultrastructural characterization of maize coenocyte and endosperm cellularization. *Can J Bot* 85:216–223.

Motzo R, Fois S, Giunta F (2007) Protein content and gluten quality of durum wheat (*Triticum turgidum* subsp. durum) as affected by sowing date. *J Sci Food Agric* 87(8):1480–1488.

Mubeen M, Ashfaq A, Hafiz MH, Muhammad A, Hafiz UF, Mazhar S, Sami Ul Din M, Asad A, Amjed A, Fahad S, Wajid N (2020) Evaluating the climate change impact on water use efficiency of cotton-

wheat in semi-arid conditions using DSSAT model. *J Water Climate Change*. http://dx.doi.org/10.2166/wcc.2019.179/622035/jwc2019179.pdf
Muhammad N, Muqarrab A, Khurram S, Fiaz A, Fahim N, Muhammad A, Shazia A, Omaima N, Sulaiman AA, Fahad S, Subhan D, Rahul D (2021) Kaolin and Jasmonic acid improved cotton productivity under water stress conditions. *J Saudi Society Agric Sci.* 28:6606–6614. https://doi.org/10.1016/j.sjbs.2021.07.043.
Muhammad TQ, Amna F, Amna B, Barira Z, Xitong Z, Ling-Ling C (2020) Effectiveness of conventional crop improvement strategies vs. omics. In: Fahad S, Hasanuzzaman M, Alam M, Ullah H, Saeed M, Khan AK, Adnan M (eds) *Environment, climate, plant and vegetation growth*. Springer Publications Ltd, Springer Nature Switzerland AG. Part of Springer Nature, pp. 253–284. https://doi.org/10.1007/978-3-030-49732-3.
Muhammad Z, Abdul MK, Abdul MS, Kenneth BM, Muhammad S, Shahen S, Ibadullah J, Fahad S (2019) Performance of Aeluropus lagopoides (mangrove grass) ecotypes, a potential turfgrass, under high saline conditions. *Environ Sci Pollut Res*. https://doi.org/10.1007/s11356-019-04838-3.
Muzammal R, Fahad S, Guanghui D, Xia C, Yang Y, Kailei T, Lijun L, Fei-Hu L, Gang D (2021) Evaluation of hemp (*Cannabis sativa* L.) as an industrial crop: a review. *Environ Sci Pollution Res*. https://doi.org/10.1007/s11356-021-16264-5.
Nagai T, Makino A (2009) Differences betweenrice and wheat in temperature responses of photosynthesisand plant growth. *Plant Cell Physiol* 50, 744–755.
Noor M, Naveed Ur R, Ajmal J, Fahad S, Muhammad A, Fazli W, Saud S, Hassan S (2020) Climate change and costal plant lives. In: Fahad S, Hasanuzzaman M, Alam M, Ullah H, Saeed M, Khan AK, Adnan M (eds) *Environment, climate, plant and vegetation growth*. Springer Publications Ltd, Springer Nature Switzerland AG. Part of Springer Nature, pp. 93–108. https://doi.org/10.1007/978-3-030-49732-3.
Omar AM, El-Menshawy M, El-Okkiah SA, Sabbagh AE (2018) Foliar application of osmoprotectants stimulate cotton (*Gossypium barbadense* L.) to survive under late sown stress condition. *Open Agric* 3(1): 684–697.
Out H, Celiktas V, Duzenli S, Hossain A, El Sabagh A (2018) Germination and early seedling growth of five durum wheat cultivars (*Triticum durum* desf.) is affected by different levels of salinity. *Fresenius Environ Bullet* 27(11): 7746–7757.
Peng S, Huang J, Sheehy JE, Laza RC, Visperas RM et al. (2004) Rice yield decline with higher night temperature from global warming. In: Redona ED, Castro AP, Llanto GP (eds) *Rice integrated crop management: towards a rice check system in the philippines*, pp. 46–56.
Pipolo A, Sinclair TR, Camara GMS (2004) Effect of temperature on oil and protein concentration in soybean seeds cultured in vitro. *Ann Appl Biol* 144:70–76.
Pradhan GP, Prasad PV, Fritz AK et al. (2012) Effects of drought and high temperature stress on synthetic hexaploid wheat. *Funct Plant Biol* 39:190–198.
Prasad PVV, Boote KJ, Allen LH et al. (2006) Species, ecotype and cultivar differences in spikelet fertility and harvest index of rice in response to high temperature stress. *Field Crops Res* 95:398–411.
Prasad PVV, Staggenborg SA, Ristic Z (2008) Impacts of drought and/or heat stress on physiological, developmental, growth, and yield processes of crop plants. In: Ahuja LR, Reddy VR, Saseendran SA, Yu Q (eds) *Response of crops to limited water: understanding and modeling water stress effects on plant growth processes*. Adv. Agric. Syst. Model. 1. ASA, CSSA, SSSA, Madison, WI, pp. 301–355. https://doi.org/10.2134/advagricsystmodel1.c11.
Qamar-Uz Z, Zubair A, Muhammad Y, Muhammad ZI, Abdul K, Fahad S, Safder B, Ramzani PMA, Muhammad N (2017) Zinc biofortification in rice: leveraging agriculture to moderate hidden hunger in developing countries. *Arch Agron Soil Sci* 64:147–161. https://doi.org/10.1080/03650340.2017.1338343.
Rasheed R, Wahid A, Farooq M et al. (2011) Role of proline and glycinebetaine pretreatments in improving heat tolerance of sprouting sugarcane (*Saccharum* sp.) buds. *Plant Growth Regul* 65:35–45.
Rashid M, John GH, Phil R et al. (2018) Forage rape (*Brassica napus* L.) seed quality: impact of heat stress in the field during seed development. *Field Crops Res* 217:172–183.
Rashid M, Qaiser H, Khalid SK, Al-Wabel MI, Zhang A, Muhammad A, Shahzada SI, Rukhsanda A, Ghulam AS, Shahzada MM, Sarosh A, Muhammad FQ (2020) Prospects of biochar in alkaline soils to mitigate climate change. In: Fahad S, Hasanuzzaman M, Alam M, Ullah H, Saeed M, Khan AK, Adnan M (eds) *Environment, climate, plant and vegetation growth*. Springer Publications Ltd, Springer Nature Switzerland AG. Part of Springer Nature, pp. 133–150. https://doi.org/10.1007/978-3-030-49732-3
Rehman AU, Habib I, Ahmad H et al. (2009) Screening wheat germplasm for heat tolerance at terminal growth stage. *Plant Omics J* 2:9–19.

Rehman M, Fahad S, Saleem MH, Hafeez M, Rahman MH, Liu F, Deng G (2020) Red light optimized physiological traits and enhanced the growth of ramie (*Boehmeria nivea* L.). *Photosynthetica* 58(4):922–931.

Rharrabti Y, García del Moral LF, et al. (2003) Durum wheat quality in Mediterranean environments III. Stability and comparative methods in analysing G × E interaction. *Field Crops Res* 80:141–146.

Rizhsky L, Liang HJ, Shuman J et al. (2004) When defense pathways collide. The response of Arabidopsis to a combination of drought and heat stress. *Plant Physiol* 134:1683–1696.

Rmiki K-E, Lemoine Y, Schoeff B (1999) Carotenoids and stress in higher plants and algae. In: Pessarakli M (ed) *Handbook of plant and crop stress*, 2nd edn. Dekker, New York, pp. 465–482.

Roberts EB (1988) What we've learned: managing invention and innovation. *Res Technol Manage* 31:11–29.

Rondanini D, Savin R, Hall AJ (2003) Dynamics of fruit growth and oil quality of sunflower (*Helianthus annuus* L.) exposed to brief intervals of high temperature during grain filling. *Field Crops Res* 83(1):79–90.

Rose IA (1988) Effects of moisture stress on the oil and protein components of soybean seed. *Aust J Agric Res* 39:163–170.

Rykaczewska K (2015) The effect of high temperature occurring in subsequent stages of plant development on potato yield and tuber physiological defects. *Am J Potato Res* 92:339–349.

Sadam M, Qamar MT, Ghulam M, Muhammad SK, Faiz AJ (2020) Role of biotechnology in climate resilient agriculture. In: Fahad S, Hasanuzzaman M, Alam M, Ullah H, Saeed M, Khan AK, Adnan M (eds) *Environment, Climate, Plant and Vegetation Growth.* Springer Publications Ltd, Springer Nature Switzerland AG. Part of Springer Nature, pp. 339–366. https://doi.org/10.1007/978-3-030-49732-3 .

Sadras VO (2007) Evolutionary aspects of the trade-off between seed size and number in crops. *Field Crops Res* 100:125–138.

Safi UK, Ullah F, Mehmood S, Fahad S, Ahmad Rahi A, Althobaiti F et al. (2021) Antimicrobial, antioxidant and cytotoxic properties of *Chenopodium glaucum* L. *PLoS One* 16(10):e0255502. https://doi.org/10.1371/journal.pone.0255502.

Sahrish N, Shakeel A, Ghulam A, Zartash F, Sajjad H, Mukhtar A, Muhammad AK, Ahmad K, Fahad S, Wajid N, Sezai E, Carol Jo W, Gerrit H (2022) Modeling the impact of climate warming on potato phenology. *European J Agro N* 132:126404.

Sajid H, Jie H, Jing H, Shakeel A, Satyabrata N, Sumera A, Awais S, Chunquan Z, Lianfeng Z, Xiaochuang C, Qianyu J, Junhua Z (2020) Rice production under climate change: adaptations and mitigating strategies. In: Fahad S, Hasanuzzaman M, Alam M, Ullah H, Saeed M, Khan AK, Adnan M (eds) *Environment, climate, plant and vegetation growth.* Springer Publications Ltd, Springer Nature Switzerland AG. Part of Springer Nature, pp. 659–686. https://doi.org/10.1007/978-3-030-49732-3

Sajjad H, Muhammad M, Ashfaq A, Waseem A, Hafiz MH, Mazhar A, Nasir M, Asad A, Hafiz UF, Syeda RS, Fahad S, Depeng W, Wajid N (2019) Using GIS tools to detect the land use/land cover changes during forty years in Lodhran district of Pakistan. *Environ Sci Pollut Res.* https://doi.org/10.1007/s11356-019-06072-3.

Saleem MH, Fahad S, Adnan M, Mohsin A, Muhammad SR, Muhammad K, Qurban A, Inas AH, Parashuram B, Mubassir A, Reem MH (2020a) Foliar application of gibberellic acid endorsed phytoextraction of copper and alleviates oxidative stress in jute (*Corchorus capsularis* L.) plant grown in highly copper-contaminated soil of China. *Environ Sci Pollution Res.* https://doi.org/10.1007/s11356-020-09764-3.

Saleem MH, Fahad S, Shahid UK, Mairaj D, Abid U, Ayman ELS, Akbar H, Analía L, Lijun L (2020c) Copper-induced oxidative stress, initiation of antioxidants and phytoremediation potential of flax (*Linum usitatissimum* L.) seedlings grown under the mixing of two different soils of China. *Environ Sci Poll Res.* https://doi.org/10.1007/s11356-019-07264-7.

Saleem MH, Rehman M, Fahad S, Tung SA, Iqbal N, Hassan A, Ayub A, Wahid MA, Shaukat S, Liu L, Deng G (2020b) Leaf gas exchange, oxidative stress, and physiological attributes of rapeseed (*Brassica napus* L.) grown under different light-emitting diodes. *Photosynthetica* 58(3):836–845.

Salvucci ME, Crafts-Brandner SJ (2004) Inhibition of photosynthesis by heat stress: the activation state of Rubisco as a limiting factor in photosynthesis. *Physiol Plant* 120:179–186.

Saman S, Amna B, Bani A, ul Qamar MT, Rana MA, Muhammad SK (2020) QTL mapping for abiotic stresses in cereals. In: Fahad S, Hasanuzzaman M, Alam M, Ullah H, Saeed M, Khan AK, Adnan M (eds) *Environment, climate, plant and vegetation growth.* Springer Publications Ltd, Springer Nature Switzerland AG. Part of Springer Nature, pp. 229–252. https://doi.org/10.1007/978-3-030-49732-3.

Samol P, Umponstira C, Klomjek P et al. (2015) Responses of rice yield and grain quality to high temperature in the open-top chamber to predict the impact of future global warming in Thailand. *Aust J Crop Sci* 9:886–894.

Sarwar M, Saleem MF, Ullah N, Rizwan M, Ali S, Shahid MR, Alamri SA, Alyemeni MN, Ahmad P (2018) Exogenously applied growth regulators protect the cotton crop from heat induced injury by modulating plant defense mechanism. *Sci Rep* 8:17086.

Saud S, Chen Y, Fahad S, Hussain S, Na L, Xin L, Alhussien SA (2016) Silicate application increases the photosynthesis and its associated metabolic activities in Kentucky bluegrass under drought stress and post-drought recovery. *Environ Sci Pollut Res* 23(17):17647–17655. https://doi.org/10.1007/s11356-016-6957-x.

Saud S, Chen Y, Long B, Fahad S, Sadiq A (2013) The different impact on the growth of cool season turf grass under the various conditions on salinity and drought stress. *Int J Agric Sci Res* 3:77–84.

Saud S, Fahad S, Cui G, Chen Y, Anwar S (2020) Determining nitrogen isotopes discrimination under drought stress on enzymatic activities, nitrogen isotope abundance and water contents of Kentucky bluegrass. *Sci Rep* 10:6415. https://doi.org/10.1038/s41598-020-63548-w.

Saud S, Fahad S, Yajun C, Ihsan MZ, Hammad HM, Nasim W, Amanullah Jr, Arif M, Alharby H (2017) Effects of nitrogen supply on water stress and recovery mechanisms in Kentucky Bluegrass plants. *Front Plant Sci* 8:983. https://doi.org/10.3389/fpls.2017.00983.

Saud S, Li X, Chen Y, Zhang L, Fahad S, Hussain S, Sadiq A, Chen Y (2014) Silicon application increases drought tolerance of Kentucky bluegrass by improving plant water relations and morph physiological functions. *Sci World J* 2014:1–10. https://doi.org/10.1155/2014/368694.

Seadh SE, El-Metwally MA (2015) Influence of antioxidants on wheat productivity, quality and seedborne fungi management under NPK fertilization levels. *Asian J Crop Sci* 7:87–112.

Senol C (2020) The effects of climate change on human behaviors. In: Fahad S, Hasanuzzaman M, Alam M, Ullah H, Saeed M, Khan AK, Adnan M (eds) *Environment, climate, plant and vegetation growth.* Springer Publications Ltd, Springer Nature Switzerland AG. Part of Springer Nature, pp. 577–590. https://doi.org/10.1007/978-3-030-49732-3 .

Shafi MI, Adnan M, Fahad S, Fazli W, Ahsan K, Zhen Y, Subhan D, Zafar-ul-Hye M, Martin B, Rahul D (2020) Application of single superphosphate with humic acid improves the growth, yield and phosphorus uptake of wheat (*Triticum aestivum* L.) in calcareous soil. *Agron* 10(9):1224. https://doi.org/10.3390/agronomy10091224.

Shah F, Lixiao N, Kehui C, Tariq S, Wei W, Chang C, Liyang Z, Farhan A, Fahad S, Huang J (2013) Rice grain yield and component responses to near 2°C of warming. *Field Crop Res* 157:98–110.

Sidra K, Javed I, Subhan D, Allah B, Syed IUSB, Fatma B, Khaled DA, Fahad S, Omaima N, Ali TKZ, Rahul D (2021) Physio-chemical characterization of indigenous agricultural waste materials for the development of potting media. *J Saudi Soc Agric Sci.* https://doi.org/10.1016/j.sjbs.2021.08.058 .

Singh RP, Prasad PVV, Sunita K, Giri SN, Reddy KR (2007) Influence of high temperature and breeding for heat tolerance in cotton: a review. *Adv Agron* 93:313–385.

Subhan D, Zafar-ul-Hye M, Fahad S, Saud S, Martin B, Tereza H, Rahul D (2020) Drought stress alleviation by ACC deaminase producing *Achromobacter xylosoxidans* and *Enterobacter cloacae*, with and without Timber Waste Biochar in Maize. *Sustain* 12(6286). https://doi.org/10.3390/su12156286.

Suryavanshi P, Buttar GS (2016) Mitigating terminal heat stress in wheat. *Int J Bio-Resource Stress Manage* 7(1):142–150.

Taiz L, Zeiger E (2006) *Plant physiology*, 4th edn. Sinauer Associates Inc Publishers, Sunderland, MA, pp. 183–290.

Tao Z, Chen Y, Li C et al. (2016a) The causes and impacts for heat stress in spring maize during grain filling in the North China Plain — A review. *J Integ Agric* 15(12):2677–2687.

Tao ZQ, Chen YQ, Zou JX et al. (2016b) Spectral characteristics of different heat tolerance in spring maize varieties to high temperature. *Spectroscop Spectral Ana* 36:520–526.

Tao ZQ, Sui P, Chen YQ et al. (2013) Subsoiling and ridge tillage alleviate the high temperature stress in spring maize in the North China Plain. *J Integr Agric* 12:2179–2188.

Tariq M, Ahmad S, Fahad S, Abbas G, Hussain S, Fatima Z, Nasim W, Mubeen M, Ur Rehman MH, Khan MA, Adnan M (2018) The impact of climate warming and crop management on phenology of sunflower-based cropping systems in Punjab, Pakistan. *Agri Forest Met* 15(256):270–282.

Tashiro T, Wardlaw IF (1991) The effect of high-temperature on kernel dimensions and the type and occurrence of kerneldamage in rice. *Aust J Agric Res* 42:485–496.

Tewari AK, Tripathy BC (1998) Temperature stress-induced impairment of chlorophyll biosynthetic reactions in cucumber and wheat. *Plant Physiol* 117:851–858.

Tewolde H, Fernandez CJ, Erickson CA (2006) Wheat cultivars adapted to post-heading high temperature stress. *J Agric Crop Sci* 192:111–120.

Thitisaksakul M, Jimenez RC, Abias MC et al. (2012) Effects of environmental factors on cereal starch biosynthesis and composition. *J Cereal Sci* 56:67–80.

Turner LR, Donaghy DJ, Lane PA et al. (2006) Effect of defoliation management, based on leaf stage, on perennial ryegrass (*Lolium perenne* L.), prairie grass (*Bromus willdenowii* Kunth.) and cocksfoot (*Dactylis glomerata* L.) under dryland conditions. 1. Regrowth, tillering and water-soluble carbohydrate concentration. *Grass Forage Sci* 61:164–174.

Unsar N-U, Muhammad R, Syed HMB, Asad S, Mirza AQ, Naeem I, Muhammad HR, Fahad S, Shafqat S (2020) Insect Pests of Cotton Crop and Management Under Climate Change Scenarios. In: Fahad S, Hasanuzzaman M, Alam M, Ullah H, Saeed M, Khan AK, Adnan M (eds) *Environment, Climate, Plant and Vegetation Growth*. Springer Publications Ltd., Springer Nature Switzerland AG. Part of Springer Nature, pp. 367–396. https://doi.org/10.1007/978-3-030-49732-3

Vacca RA, De Pinto MC, Valenti D, et al. (2004) Production of reactive oxygen species, alteration of cytosolic ascorbate peroxidase, and impairment of mitochondrial metabolism are early events in heat shock-induced programmed cell death in tobacco bright yellow 2 cells. *J Plant Physiol* 134:1100–1112.

Van der Merwe JD, de Beer D, Joubert E et al. (2015) Short-term and sub-chronic dietary exposure to Aspalathin-Enriched Green Rooibos (*Aspalathus linearis*) extract affects rat liver function and antioxidant status. *Molecules* 20:22674–22690.

Vwioko ED, Osawaru ME, Eruogun OL (2008) Evaluation of okro (*Abelmoschus esculentus* L. Moench.) exposed to paint waste contaminated soil for growth, ascorbic acid and metal concentration. *Afr J Gen Agric* 4(1):39–48.

Wahid A, Gelani S, Ashraf M et al. (2007) Heat tolerance in plants: an overview. *Environ Exp Bot* 61: 199–223.

Wahid A, Ghazanfar A (2006) Possible involvement of some secondary metabolites in salt tolerance of sugarcane. *J Plant Physiol* 163:723–730.

Wahid F, Fahad S, Subhan D, Adnan M, Zhen Y, Saud S, Manzer HS, Martin B, Tereza H, Rahul D (2020) Sustainable management with mycorrhizae and phosphate solubilizing bacteria for enhanced phosphorus uptake in calcareous soils. *Agriculture* 10(334). https://doi.org/10.3390/agriculture10080334.

Wajid N, Ashfaq A, Asad A, Muhammad T, Muhammad A, Muhammad S, Khawar J, Ghulam MS, Syeda RS, Hafiz MH, Muhammad IAR, Muhammad ZH, Muhammad HR, Veysel T, Fahad S, Suad S, Aziz K, Shahzad A (2017) Radiation efficiency and nitrogen fertilizer impacts on sunflower crop in contrasting environments of Punjab. *Pakistan Environ Sci Pollut Res* 25:1822–1836. https://doi.org/10.1007/s11356-017-0592-z.

Wardlaw IF, Blumenthal C, Larroque O et al. (2002) Contrasting effects of heat stress and heat shock on kernel weight and flour quality in wheat. *Fun of Plant Biol* 29:25–34B94.

Wardlaw IF, Moncur L (1995) The response of wheat to high temperature following anthesis. I. The rate and duration of kernel filling. *Fun Plant Biol* 22(3):391–397.

Wei HU, Chen ML, Zhao WQ et al. (2017) The effects of sowing date on cottonseed properties at different fruiting-branch positions. *J Integr Agric* 16:1322–1330.

Weiss EA (2000) *Oilseed crops*, 2nd edn. Blackwell Science, Oxford, pp. 129–136.

Wilcox JR, Cavins JF (1992) Normal and low linolenic acid soybean strains: response to planting date. *Crop Sci* 32:1248–1251.

Wolf RB, Cavins JF, Kleiman R et al. (1982) Effect of temperature on soybean seed constituents: oil, protein, moisture, fatty acids, amino acids and sugars. *J Amer Oil Chem Soc* 59:230–232.

Wu C, Kehui C, She T, Ganghua L, Shaohua W, Fahad S, Lixiao N, Jianliang H, Shaobing P, Yanfeng D (2020) Intensified pollination and fertilization ameliorate heat injury in rice (*Oryza sativa* L.) during the flowering stage. *Field Crops Res* 252:107795.

Wu C, Tang S, Li G, Wang S, Fahad S, Ding Y (2019) Roles of phytohormone changes in the grain yield of rice plants exposed to heat: a review. *PeerJ* 7:e7792. https://doi.org/10.7717/peerj.7792.

Yakir D, DeNiro MJ (1990) Oxygen and hydrogen isotope fractionation during cellulose metabolism in *Lemna gibba* L. *Plant Physiol* 93:325–332.

Yamakawa H, Hirose T, Kuroda M et al. (2007) Comprehensive expression profiling of rice grain filling-related genes under high temperature using DNA microarray. *Plant Physiol* 144:258–277.

Yang H, Lu D, Shen X et al. (2015) Heat stress at different grain filling stages affects fresh waxy maize grain yield and quality. *Cereal Chem* 92:258–264. http://doi:10.1094/CCHEM-07-14-0146-R.

Yang Z, Zhang Z, Zhang T, Fahad S, Cui K, Nie L, Peng S, Huang J (2017) The effect of season-long temperature increases on rice cultivars grown in the central and southern regions of China. *Front Plant Sci* 8:1908. https://doi.org/10.3389/fpls.2017.01908.

Yassin M, Fara SA, Hossain A et al. (2019b) Assessment of salinity tolerance bread wheat genotypes: using stress tolerance indices. *Fresen Environ Bul* 28(5):4199–4217.

Yassin M, Mekawy AM, EL Sabagh A et al. (2019a) Physiological and biochemical responses of two bread wheat (*Triticum aestivum* L.) genotypes grown under salinity stress. *Appl Eco Environ Res* 17(2):5029–5041.

Yildirim M, Barutcular C, Koc M et al. (2018) Assessment of the grain quality of wheat genotypes grown under multiple environments using gge biplot analysis. *Fresen Environ Bul* 27(7):4830–4837.

Zafar-ul-Hye M, Muhammad TH, Muhammad A, Fahad S, Martin B, Tereza D, Rahul D, Subhan D (2020b) Potential role of compost mixed biochar with rhizobacteria in mitigating lead toxicity in spinach. *Scientific Rep* 10:12159. https://doi.org/10.1038/s41598-020-69183-9

Zafar-ul-Hye M, Muhammad N, Subhan D, Fahad S, Rahul D, Mazhar A, Ashfaq AR, Martin B, Ji ˇrí H, Zahid HT, Muhammad N (2020a) Alleviation of Cadmium adverse effects by improving nutrients uptake in bitter gourd through cadmium tolerant Rhizobacteria. *Environments* 7(54). https://doi.org/10.3390/environments7080054

Zafar-ul-Hye M, Akbar MN, Iftikhar Y, Abbas M, Zahid A, Fahad S, Datta R, Ali M, Elgorban AM, Ansari MJ et al. (2021) Rhizobacteria inoculation and caffeic acid alleviated drought stress in lentil plants. *Sustain* 13:9603. https://doi.org/10.3390/su13179603

Zahida Z, Hafiz FB, Zulfiqar AS, Ghulam MS, Fahad S, Muhammad RA, Hafiz MH, Wajid N, Muhammad S (2017) Effect of water management and silicon on germination, growth, phosphorus and arsenic uptake in rice. *Ecotoxicol Environ Saf* 144:11–18.

Zahir SM, Zheng HG, Ala Ud D, Amjad A, Ata UR, Kashif J, Shah F, Saud S, Adnan M, Fazli W, Saud A, Manzer HS, Shamsher A, Wajid N, Hafiz MH, Fahad S (2021) Synthesis of silver nanoparticles using *Plantago lanceolata* extract and assessing their antibacterial and antioxidant activities. *Sci Rep* 11:20754.

Zia-Ur-Rehman M (2020) Environment, climate change and biodiversity. In: Fahad S, Hasanuzzaman M, Alam M, Ullah H, Saeed M, Khan AK, Adnan M (eds) *Environment, climate, plant and vegetation growth.* Springer Publications Ltd, Springer Nature Switzerland AG. Part of Springer Nature, pp. 473–502. https://doi.org/10.1007/978-3-030-49732-3

5

Climate Change and Cotton Production

Asif Iqbal
Institute of Cotton Research of CAAS, Anyang, P.R. China

Mazhar Iqbal
Shaheed Benazir Bhutto University Sheringal Dir (U), Sheringal, Pakistan

Madeeha Alamzeb
The University of Agriculture Peshawar, Pakistan

Dong Qiang, Wang Xiangru, Gui Huiping, Zhang Hengheng, and Pang Nianchang
Institute of Cotton Research of CAAS, Anyang, P.R. China

Shah Fahad
Hainan University, Haikou, China; The University of Haripur, Haripur, Pakistan

Shazma Anwar, Asad Ali Khan, Amanullah, Muhammad Arif, Inamullah, Shaheen Shah, and Muhammad Saeed
The University of Agriculture Peshawar, Pakistan

Zhang Xiling and Meizhen Song
Institute of Cotton Research of CAAS, Anyang, P.R. China

CONTENTS

DOI: 10.1201/9781003286417-5

5.1 Introduction

5.1.1 Climate Change

Climate change and variability are the actual threats to crop production and food security (Rezaei et al. 2015; Fahad and Bano 2012; Fahad et al. 2013, 2014a, 2014b, 2015a, 2015b, 2016a, 2016b, 2016c, 2016d, 2017, 2018, 2019a, 2019b; Aziz et al. 2017b). According to the IPCC (Intergovernmental Panel on Climate Change), fourth report, 'Climate change refers to a change in the state of the climate that can be identified by changes in the mean and/or the variability of its properties, and that persists for an extended period, typically decades or longer.' This refers to climate change over time as a consequence of natural variability or by human activity. However, the UNFCCC (United Nation Framework Convention on Climate Change) described that 'climate change refers to a change of climate that is attributed directly or indirectly to human activities that alter the composition of the global atmosphere and that is in addition to natural climate variability observed over comparable time periods.' The report also stated that the global mean temperature raised by 0.74°C during the past century (Team et al. 2007). The previous studies reported that high temperature and uneven rainfall distribution have a negative impact on crop productivity over the entire world (Lobell and Field 2011). This negative impact of climate on crop productivity will affect the food security of the developing world, where the increasing population is already facing food insecurity (Lobell et al. 2011). The increasing global warming and climate change can result in extreme weather events such as an increase in temperature, uneven distribution of rainfall, etc. (Ahmed et al. 2018). These extreme events occur due to the shift in their distribution (Lewis and King 2017). It was reported that both daily maximum and minimum temperatures have shifted towards their higher values in almost all regions. The shift in temperature and rainfall distribution had greatly influenced the cropping patterns, phenology, development, antithesis, asynchrony between the flowering and pollinators harvest, and crop production (Ahmad et al. 2018; Adnan et al. 2018a, 2018b; Akram et al. 2018a, 2018b; Aziz et al. 2017a; Habib Ur et al. 2017; Hafiz et al. 2016, 2019; Kamarn et al. 2017; Muhammad et al. 2019; Sajjad et al. 2019; Saud et al. 2013, 2014, 2016, 2017; Shah et al. 2013; Qamar-Uz et al. 2017; Wajid et al. 2017; Yang et al. 2017; Zahida et al. 2017; Depeng et al. 2018). Along with this, it also increases evapotranspiration, reduces pollen germination, grain-filling duration, and ultimately lower crop yield (Ahmed et al. 2018). The combination of high temperature (above 35°C) along with high humidity and low wind lead to floret sterility and fruits (Lobell and Field 2011). Therefore, time-by-time assessment of climate change is needed to provide the basis for the development of adaptations to combat the negative effect of climate change. The current chapter was planned to know the impacts of climate change and adaptations for cotton production.

This effect varies among the countries. In this chapter, we highlight the climatic factors which affect cotton production of the main cotton producing countries and adaptive strategies to minimize the impact of climate change on cotton production.

5.1.2 Cotton Status and Production

By nature, cotton is a perennial crop; however, it has been grown as an annual crop. Four species of cotton are grown commercially, in which *Gossypium hirsutum*, or Upland cotton, is the major one in production worldwide. *Gossypium barbadense* comes in second having a high staple length. The other two species are *Gossypium arboreum* and *Gossypium herbaceum*. Cotton is mainly grown in the areas that lie on 37°N and 32°S and is extended to 45°N in China on longitudinal band (Rahman et al. 2016; Wakelyn and Chaudhry 2009). For good production, cotton needs favourable temperature, soil moisture, and plenty of sunshine. During the reproductive stage, it needs a dry season for boll opening and harvesting. After establishment, the cotton plant develops a good tap root system, which may help the plant during drought. Besides its help in water uptake from lower layers, the deep tap root system of cotton also uptakes more nutrients as compared to small grain crops. This characteristic makes cotton useful in crop rotation. However, this deep tap root system makes the cotton crop sensitive to waterlogging, floods, or heavy rains. Cotton crop needs adequate soil moisture for at least 105 to 125 days depending on the region. As in tropical region, it requires 2 to 4 mm of water daily at earlier growth; however, at the reproductive stage it

TABLE 5.1
Influence of Climate Change on Cotton Yield Reduction in Different Countries

Country	Yield Reduction (%)	References
China	−5.5	Chen et al. (2015)
USA	−17	Adhikari et al. (2016)
Africa	−7	Amouzou et al. (2018)
USA	−9	Reddy et al. (2002)
Pakistan	−8	Ur Rahman et al. (2018)
Burkina Faso	−13	Diarra et al. (2017)
Australia	−17	Williams et al. (2015)

needs 5 to 7 mm daily. Thus, a total of 500 to 700 mm water is required for successful cotton production. In rainfed areas, cotton can be grown, when the rainfall is 700 mm or more while keeping in mind the distribution and runoff (Ton 2012). Cotton has the capacity to withstand the changing environment. It has the capability of compensatory growth by which it loses vegetative and reproductive parts (Wakelyn and Chaudhry 2009).

Cotton is the leading fibre crop grown throughout the world. The leading cotton producing countries are shown in Table 5.1. In 2018–19, world cotton production was lower by 4.2% from the last year, as major producing countries showed a decrease in production except for Brazil (up) and China (no change). However, the US department of agriculture projected that world cotton production will exceed consumption, which raises the world stock by 1 million bales in 2019–20. Cotton production in the United States witnessed a decrease of 12.1% due to a severe storm and dry weather. Besides, this Australian production was reduced by drought. Production of Central Asia, India, Pakistan, and the USA was also affected by low precipitation. The production of cotton in China is estimated at 27.5 million bales, which remains the same as the previous year. In China, most of the cotton is produced in Xinjiang province, where the yield is almost double of the national average yield. In the current year, China harvested land is reduced by 1.5%; however, the yield is expected to increase by 1.5% (1,787 kg ha^{-1}) from the previous year. Cotton production in India is estimated at 27 million bales, which are 6.9% lower than the last year production. Therefore, the country is now second in production, after China. The reason behind this is a reduction in the cotton cultivable land, low monsoon rainfall, and infestation of pink bollworm during harvesting stage at the major cotton producing areas of Gujarat and Maharashtra. The estimated yield for the current year is about 480 kg ha^{-1}, which is 5.3% lower than the previous year. The United States Department of Agriculture (USDA) projected that world cotton production may exceed consumption in the coming year (2019–20) with a raise of 1 million bales in the world stock. The rise in world production is expected to be 6.8% in most of the countries with an increase in the cultivable land. It is also expected that China will increase imports in the following year. In the United States the production is expected to increase up to 22.5 million bales in the coming year, due to a slight increase in planted area and sharply lower abandonment, and therefore exports are expected to rise.

5.2 Impact of Climate Change on Cotton Production

Climate change has a great influence on agriculture (Adnan et al. 2018a; Akram et al. 2018a, 2018b; Aziz et al. 2017a; Habib Ur et al. 2017; Hafiz et al. 2016, 2019; Kamarn et al. 2017; Muhammad et al. 2019; Sajjad et al. 2019; Saud et al. 2013, 2014, 2016, 2017; Khan et al. 2016; Shah et al. 2013; Qamar-Uz et al. 2017; Wajid et al. 2017; Yang et al. 2017; Zahida et al. 2017; Depeng et al. 2018). Increase in global temperature due to climate change has a negative effect on crop yield while encouraging weed and pest infestation. Similarly, change in rainfall patterns may result in crop failure in the short run and reduction in crop production in the long run. Though some crops will get benefit in some part of the world, however, the overall impacts are expected to be negative (Change 2009).

5.2.1 Impact of Temperature on Cotton Production

Climate change is resulting in an increase in mean temperature, change in precipitation patterns, and some extreme weather events. In some regions, high temperature may lead to longer growing season and plenty of rainfall and in some regions, the growing season would become shorter with less rainfall (Basir et al. 2016; Fahad et al. 2014a, 2014b, 2015a, 2015b, 2018, 2019a, 2019b). For optimum growth and development, cotton needs warm days and cool nights. However, extreme weather may affect the plant growth at any time in the season, which varies from region to region. Low temperatures during sowing time may affect planting time in many countries. However, rising temperatures enhance planting time earlier than before. Moreover, the areas which are already facing high temperature will be affected more. Thus more shedding of flowers and buds will occur. The rising temperature may favour cotton production in those areas where crops face low temperature at flowering and maturity. Besides this temperature also affects crop phenology, leaf development, internodes elongation, biomass production, and assimilates partitioning (Reddy et al. 1999). It was reported that at the seedling stage, plants were insensitive to increase in temperature from 20°C/12°C to 40°C/32°C, and after two weeks they were temperature sensitive. A less biomass of 40% and 50% were observed at 20°C/10°C and 40°C/30°C, respectively, as compared to optimum temperature 30°C/20°C (Raja Reddy et al. 1992). Biomass of 13%, 15%, and 43% were partitioned to 'squares and bolls' at 20°C/10°C, 25°C/15°C, and 30°C/20°C, respectively, which shows slower development at a temperature lower than 30°C/20°C. The rise in temperature from 20°C/10°C to 30°C/20°C showed a 36% increase in plant biomass. Highest boll weight was recorded at 30°C/20°C and the lowest at both lower and higher temperatures. With respect to the vegetative stage, boll development stage was found to be more sensitive to temperature. Previous studies showed that 30°C/22°C was found optimum for the development of leaves, main stem, fruiting branches, squares, and bolls (Reddy et al. 1994).

The temperature has also a dramatic effect on boll set and fibre properties (Reddy et al. 1999). The suboptimal temperature during boll growth resulted in longer fibres. As the temperature increased, the fibre length and distribution were more uniform while fibre fineness and maturity increased linearly with increase in temperature up to 26°C; however, it decreased at 32°C. At higher temperature, short fibre content increases, while increasing drastically from 17°C to 26°C. Another report showed that most of the fibre quality parameters were little affected by the change in temperature (Pettigrew 2007). However, if the predicted global warming occurs, the temperature will become much higher and will have a negative effect on the current cultivars adapted to moderate temperature. Therefore, heat-tolerant cultivars will be needed to grow optimally. Furthermore, at high temperature plants also lost their reproductive capability due to sterility and boll retention problems. Retention of bolls is highly sensitive to rise in temperature than any other environmental factor, except nutrients which can be recovered easily. However, to avoid the impacts of high temperature is almost impossible, which is one of the major causes of bud shedding (Reddy and Rodrigues 1999). Previously, it was noted that boll growth decreases and fruits shed, when the temperature increases from 32°C, which showed that the upper limit for boll development and fruiting of cotton is 32°C. Cotton plants shed all their squares at 40°C daily temperature. Plants exposed to 40°C for 2 or 12 hr/day retained 64% and 0% bolls, respectively (Reddy et al. 1992). In another study, plants grown at 40°C /32°C abscise all flowers and buds and reduce the viability of the pollen (Reddy et al. 1994), which ultimately reduces boll size and seed numbers in the locks caused by poor seed set (McRae et al. 2007). Besides the effect on yield, the rising temperature also has a negative effect on fibre quality, as increasing temperature resulted in high micronaire values, which is not desirable in the areas already facing high temperature. However, increasing temperature would be beneficial for those areas which have low micronaire values and low maturity cotton.

5.2.2 Impact of Carbon Dioxide on Cotton Production

Among the four main greenhouse gases, CO_2 is the most important one with respect to cotton production. After the industrial revolution, CO_2 has dramatically increased by 80 ppm m^2 in the air at the end of the 18th century. In 2005, its value was recorded as 379 ppm and it continues to be increasing. This unstoppable increase in CO_2 concentration in the atmosphere has a great effect on global warming and leads to precipitation in different regions. Higher CO_2 increases net photosynthesis; biomass production; and ultimately fibre, oil, and seed yield. Besides these, higher inputs and their efficiency also needed to be improved especially in C_3 plants (cotton, rice, wheat) over C_4 plants (maize, sugarcane). Generally, cotton plants consumed about 11.34 tons CO_2 ha^{-1} to produce fibre (lint), oil (seed), protein (feed), and other plant parts. Out of the total, one-tenth is used for fibres, 0.5 tons for oil production, and about 7.9 tons of O_2 is released to the atmosphere. A total amount of 450 million tons of CO_2 were consumed by cotton plants during the season and more than 36.3 million tons of CO_2 were used to form cellulose in the fibre. It was reported that carbon sequestrated in cotton fibre is equal to removing 7.25 million vehicles from the highways (ICAC 2009).

Increase in CO_2 enhanced leaf size and expansion (Reddy et al. 1994). During early growth stages, plants exposed to higher CO_2 produced about 30% more biomass. This higher assimilation is due to increase in chlorophyll-a concentration in both wet (7.1%) treatment and dry (8.2%) treatment (Pinter Jr et al. 1994). The increment in whole plant leaf area under double CO_2 concentration was mainly due to an increase in leaf number and a slight increase in leaf size. In a CO_2 enriched environment (720 ppm), plants produced about 40% more squares and bolls than 360 ppm (Alam et al. 2016; Reddy et al. 1999). Thus plant growth and development in elevated CO_2 were more distinct due to higher leaf area and photosynthesis as compared to ambient CO_2. Cotton can fix 16% more CO_2 in elevated condition than at ambient conditions. Besides this, diurnal variation in net photosynthetic rate was also observed under free-air carbon dioxide enrichment (FACE). Leaf and canopy photosynthetic rates were 19%–41% higher in enriched condition than control during mid-day, however, stomatal conductance of leaves were 13%–44% more in control condition (Hileman et al. 1994). There was no significant effect of CO_2 enrichment on transpiration (Dugas et al. 1994) although CO_2 fluxes were significantly higher in FACE condition (550 ppm) as compared to ambient level (Nakayama et al. 1994). Roots and stem carbohydrates content were also increased under FACE condition. Similarly, starch and total soluble sugars in leaves were also high in FACE than control. Therefore, the effect of CO_2 enrichment on starch accumulating plants is the increase in nonstructural carbohydrates such as starch (Hendrix et al. 1994). In contrast, nitrogen and protein contents in the leaves, stems, and roots were significantly reduced in CO_2 enriched plants, while carbon to nitrogen ratio was found higher than control (Huluka et al. 1994), which has great implications in agriculture. CO_2 enriched condition (550 ppm) during day time decreased stomatal conductance and improved leaf water potential (Bhattacharya et al. 1994); it also improved water-use efficiency in both well-watered and watered condition (Kimball and Mauney 1993). Cotton root dry weight and density were also increased under CO_2 enrichment (Prior et al. 1994). This root development as a consequence of CO_2 enrichment may allow roots to store more carbohydrates for better growth to explore the soil profile and meet the nutrient and water need during peak demand periods (Prior et al. 1995). Seed cotton yield was also improved by 43% under 550 ppm of CO_2 throughout the growing season (Hunsaker et al. 1994; Mauney et al. 1994). This increase in yield might be attributed to the increase in leaf area, more number of flowers, squares, bolls, and a longer period of a good tap root system. Besides this, higher CO_2 has a pronounced effect on leaf number, leaf area, and the net photosynthetic rate, which ultimately increase cotton productivity. Another study concluded that higher CO_2 concentration increases the number of branches and fruiting sites and thus increases lint yield. The increase in photosynthesis will also increase the demands for inputs like water, nutrients, other cultural practices, etc. especially in those areas where water is not sufficiently available and therefore the results will be negative (Shah et al. 2017).

5.2.3 Impact of Water Availability on Cotton Production

Plants require sufficient water for normal growth and to maintain body temperature. Otherwise, they would suffer from heat stress. Therefore, water is supplied to maintain temperature in most of the growing areas. For cotton, both the amount and time of water through irrigation or rainfall is very important. However, oversupply of water through irrigation or rainfall has negative effect on cotton productivity (Karl et al. 2009). Irrigation is one of the most important factors for cotton (Chapagain et al. 2005), as the yield in irrigated areas is higher than in rainfed areas. This may be the reason that about 73% of cotton has been grown in irrigated areas. Irrigation is very important in top cotton-producing countries like China's Xinjiang province and Yellow River regions, the southwestern regions of the United States, Indus valley of Pakistan, Gujarat state of India, eastern Turkey, Israel, Egypt, Uzbekistan, Australia, and Peru. However, rainfed cotton is cultivated in Yangtze valley, China, southeastern part of USA, some states in India, Turkey, Brazil, Argentina, and Africa.

The availability of freshwater for irrigation is a limiting factor for cotton production in many countries. Out of the total, about 70% of the world's available water is used in the agriculture sector for irrigation. Therefore, any effect on freshwater from other sectors will directly affect the availability of water for agriculture. It is estimated that about 3% of the world agricultural water is used for cotton production (Hoekstra and Chapagain 2007). Cotton affects both the quality and quantity of fresh water through the use of pesticides and fertilizers and also plays a great role in soil degradation and salinization (Makhdum et al. 2011). It was reported that irrigated plants have high light utilization efficiency. However, plants grown without irrigation have low light-use efficiency. Both well-watered and moderate stressed plants have greater chlorophyll-a concentration in the free air carbon dioxide enrichment over the ambient CO_2 (Pinter Jr et al. 1994). However, assimilates partitioning to bolls was greatly affected by irrigation as compared to carbon dioxide (Mauney et al. 1994). It was reported that moisture-stressed plants reduced the yield by 42% and biomass by 36% in ambient condition, while the reduction under elevated carbon dioxide was 15% and 17% in yield and biomass, respectively. In semiarid climates, irrigation is the main cause of secondary salinization and elevated CO_2 may overcome this impact. However, the increase in temperature may increase secondary salinization. It was projected that water vapour, evaporation, and precipitation may increase (IPCC 2007). But overall, climate change will affect the pattern and distribution of precipitation, which had an adverse effect on agriculture in general and specifically on cotton production. Therefore, change in planting time, technology, and development of short-duration, heat- and drought-resistant cultivars are of prime importance to sustain cotton production for the global demand of fibres.

5.2.4 Impact of Pests, Diseases, and Weeds on Cotton Production

Increase in greenhouse gases in the atmosphere has a great impact on agriculture. Global warming and its consequence will affect pests and their population. Globally, insects are a well-known threat to cotton production. About 1,326 species of insects and mites attack cotton plants all over the world, and among them 24 have already attained the status of pests and 9 are considered as key pests (Dhawan 1998). As earlier mentioned, insects can adapt their body temperature to change in their environment. Therefore, the effect of global warming on cotton insects is low, and they can easily adjust to the rising temperature. Thus, the insects will continuously be affecting cotton even more than the current situation (Ton 2012) because it is predicted that rise in temperature will influence pest's metabolism and population. Increasing temperature will open new areas for insect's colonization and will spread to new areas. The most important insect of cotton – such as bollworms, the population will be increased due to the elimination of diapause during the winter season to avoid cold. These effects will be further increased if alternate host plants are available during the winter season (Ton 2012). Relative humidity during morning and evening has a positive and negative impact on the larval population of pink bollworm, respectively and was noted that weathers and related parameters have 90.9% contribution on larval population (Patil et al. 1992). In the case of climatic factors, the temperature is considered to be the most important factor affecting insects; however, little has been known about the direct effect of CO_2 or UV-B. Global warming will affect insects which show distinct life cycle strategies to exploit plants at different growth stages (Bale et al. 2002).

A previous study has shown that plants grown under CO_2-enriched environment depicted higher biomass and increase in carbon-nitrogen ratio, tannins, and other phenolics, and decrease in nitrogen concentration (Heagle 2003). It was also studied that plants grown under low nutrient conditions have higher concentrations of carbon-based allelochemicals (Fajer et al. 1992). In contrast, plants grown in elevated CO_2 (700 ppm) had similar or lower CO_2 conditions and similar or lower concentrations of carbon-based allelochemicals than ambient condition. Under elevated CO_2 and nitrogen fertilization, larvae consumed significantly more foliage than under ambient condition, while insects that fed on low nitrogen plants had higher mortality under elevated CO_2 (Coviella and Trumble 2000). The interaction of elevated CO_2 and nitrogen also affect the nitrogen-based toxin. It was reported that elevated CO_2 decreased N allocation to Bt and that's why in future the plant's defence mechanism may be altered. Elevated CO_2 and low nitrogen increase carbon-based defensive compounds, which adversely affect the survival of *Spodoptera exigua* in Bt cotton (Coviella et al. 2002). Similarly elevated CO_2 concentration increases the population of phloem feeders (Whittaker 1999), while densities of leaf minor species on plants were lower (Stiling and Cornelissen 2007). In contrast, the thrips population was not affected but laminar area feed by thrips was 90% higher in elevated CO_2 than ambient condition because of higher growth (Heagle 2003). Some of the insects will be benefitted due to climate change, which may increase the consumption of pesticides (Flores Araya 2008). Chewing insects have shown no change due to climate change. However, insects like leafhopper have reduced seed cotton yield by 390 kg ha^{-1} (Pandi 1997), and it was noted that even in complete protection, the leafhopper population increased with increase in temperature (Patel 2008).

Climate change regulates plants disease in three ways, which are the following: a direct effect on pathogens, creating a favourable environment, and by affecting host tissues. The rise in temperature will affect disease control by changing the time of pathogen emergence. Faster decomposition of chemicals by high temperature will make the chemical method less effective. Similarly, an increase in CO_2 concentration will enhance disease severity by inducing fungus growth and spore dissemination, which will ultimately destruct plant tissue (Chakraborty 2013). The effect of imidacloprid was high at a higher temperature (Gonias et al. 2008). It might be due to a reduction in glutathione reductase in the treated plants; hence, this defensive mechanism could be activated. Overall, the disease problem is very important and needs to be addressed (Ton 2012). The development of pest- and disease-resistant cultivars is the main focus of breeders, especially in cotton, to minimize the extent of future climate change.

5.3 Climate Change and Cotton Production of the Leading Countries

Climate change has a direct effect on cotton production as a result of a change in carbon dioxide concentration, rise in temperature, precipitation, soil moisture availability, evapotranspiration, pests, disease, and weeds infestation. Cotton is mostly grown in arid and semiarid regions in both irrigated and rainfed conditions in 70 different countries around the world. Globally, it was reported that nearly 53% of cotton and 73% of fibre growth needed full or partial irrigation. That's why climate change has negative impacts on cotton production due to a reduction in water availability for irrigation, especially in China (Xinjiang), Pakistan, western United States, and Australia. However, change in the rainfall due to climate change may have a positive effect on the Yellow river regions of China, India (Gujarat), southeastern United States, and Turkey (Anatolia). So far, the effect of rainfall on west and central Africa and Brazil is unpredicted. A slight increase in CO_2 concentration could increase cotton yield. The increase in temperature may favour cotton growth and increase the cotton growing season in some countries, while in Pakistan heat stress has already depressed the cotton yield. Cotton has some flexibility to high temperature and drought because of its long taproot system. Besides this, higher temperature will increase the area for cotton cultivation and is likely to increase the production and supplies. On the whole, the effect of climate change on cotton is unpredictable. Climate change has both positive and negative impacts on cotton production, varying among the countries. Therefore, climate change and its impact on cotton production of major cotton-producing countries are discussed as under.

5.4 Climate Change and Cotton Production of China

China is the world largest producer and consumer in the world. In 2017–2018, China's share was about 22% of the world cotton production (Figure 5.1). The estimated yield of cotton in 2018/2019 was 27.5 million bales, which makes China the top producer in the world as India's production was reduced to 27 million bales. The production of cotton has almost doubled in China since the last 20 years. This increase is highly attributed to the increase in the national average yield. There are three main regions of cotton production in China, which are Yangtze River valley, Yellow River valley, and Xinjiang (northwestern regions). The yield of the Xinjiang regions was 50% higher than Yangtze River valley and about 62% higher than Yellow River valley during 2006. The increase in the national average yield is attributed to the increase in irrigation. Hence the production was shifted from rainfed areas of Yangtze River and Yellow River areas

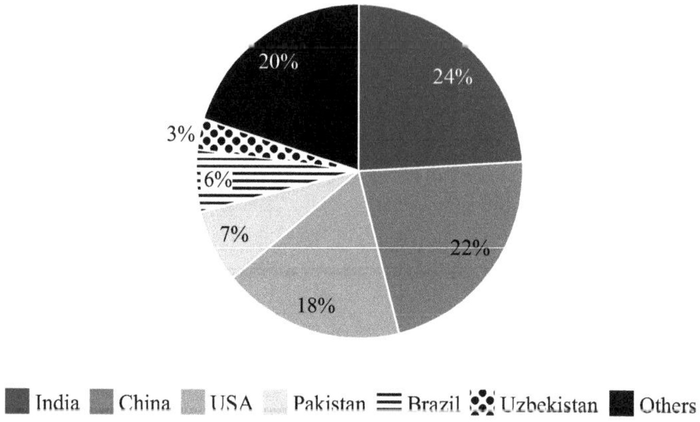

FIGURE 5.1 Global cotton share of different countries in 2017–2018.

to the fully irrigated areas of Xinjiang (Ton 2012). Xinjiang province contributed more to the national production, where the yield is near twice the national average yield. Besides, this cotton production in the eastern cotton areas also shows a stabilized yield. The regions have a huge difference in climate, soil, ecological condition, irrigation water availability, disease, pests, and weeds. The weather of the Yellow River region is mostly dry during the early growing season and irrigation is necessary to maintain higher cotton production. The Yangtze River valley is also one of the major producing regions, where the yield is relatively low and the competition of grain crops with cotton is high. The region also has adopted early maturing cultivars, which are double-cropped with wheat. Usually, excessive rainfall during late summer reduces cotton quality by increasing disease and pest infestation. Besides this, cotton production along Yellow River is also important as the growing season will become longer due to the rise in temperature. It was expected that the rainfall in China will increase, which may be beneficial for cotton production along with a slight increase in temperature and carbon dioxide. The northwest region includes Xinjiang and Gansu provinces. In China, Xinjiang is the largest cotton-producing province, which depends entirely on irrigation, and water availability is very crucial for Xinjiang cotton production. The use of both surface and groundwater has expanded the cotton cultivable land in Xinjiang from 1.5 to 4 m ha. Due to climate change, water availability is expected to decrease, which will limit cotton production due to high demand for water for agricultural and nonagricultural uses. The climate is arid in nature with rainfall less than 200 mm, which may limit the pests and diseases. The quality of Xinjiang cotton is very good and the long staple cotton is also grown in the area (Hsu and Gale 2001).

According to UNFCCC, climate change will result in global warming higher than now in central Asia, northeastern Asia, and Tibet (UNFCCC 2008). The rainfall is expected to increase in many parts of Asia; however, central Asia may face decrease in rainfall especially in summer. At the same time ice and snow will decrease in the Himalayan and Tibet Plateau, which will lead to water scarcity in central and East Asia due to a decrease in freshwater availability. However, that will lead to increase in cultivable land in the northern areas (Ayers and Dodman 2010). Chinese Yellow and Yangtze Rivers are not that much reliant like Indus and Brahmaputra. It was expected that the area along Yellow River will become humid due to the increase in precipitation in winter (Van Raaij 2010). In a previous report, it was expected that due to climate change most part of China will continue to warm up by 4.5°C by the 2080s. At the same time, temperature extremes and precipitation intensities will be increased (Piao et al. 2010). Overall, rising temperature has a positive effect on cotton production as the increase in 1°C of annual temperature will lead to 10 frost-free days. Therefore, the cotton growing period would be increased by 10 days. Similarly, a 5%–10% increase in bolls opening will occur before frost, which will improve the strength and maturity of fibre (Ton 2012). Along with this natural disaster related to climate change such as drought, floods, extreme temperature events will also increase. If the production system varieties were not changed, the cotton production would drop by 10% in the period 2030–2050. As mentioned earlier, climate change has decreased cotton production (–5.5%) as shown in Table 5.1. Thus, climate change will not directly affect China's capacity; rather it will increase the demand for management techniques to sustain higher productivity (Ton 2012).

5.5 Climate Change and Cotton Production of India

Worldwide, India is the second largest cotton producer. The country's share was about 24% of the global cotton production in the year 2017–2018 (Figure 5.1). Like many regions in the world, temperature will also increase in India in the future. Besides this, the increasing intensity of monsoon rains will become a problem in many areas. The areas facing high temperature will be affected more due to rise in temperature, which will hinder cotton production. The areas, which are totally dependent on rain water will suffer more from severe drought or flood as a result of climate change. Besides this, irrigated lands will also suffer due to low water availability from the Himalayas and Tibet as a result of snowfall decline.

The total production in 2017/2018 was 28.5 million bales, which was 5.6% up from the previous years. However, the yield was decreased by 7% from the previous year's (504 kg ha^{-1}) because of pest problems. However, cotton production in India in 2018/2019 is estimated at 27 million bales, which is 6.9% down from 2017/2018. As a result, India falls down to the second largest cotton producer after China. This reduction in cotton production is associated with low monsoon rainfall and pink bollworm attack in the major cotton producing areas of Gujarat and Maharashtra during harvesting. In India, cotton is cultivated in three main zones, i.e. the central zone comprised of 65%, south zone (20%), and the north zone, which comprised of 14% of the total cotton cultivated land. In the central zone, the climate is hot and semiarid with the driest land cotton. The monsoon rain is poorly distributed; the soil is shallow black in nature with poor water retaining capacity. In the south zone, both irrigated and rainfed cotton is cultivated. The climate is suitable for cotton production. In the northern zone, all the cotton grown is irrigated. The crop faces high temperature at sowing time and the season is comparatively short (six months). The seed cotton yield was severely damaged by adverse climate and pests (Dendage 2017). The availability of water for irrigation is a big problem, the soil is saline with a lower germination rate due to high soil temperature (CICR 2009).

Climate change will result in warming in South Asia above the global mean (UNFCCC 2008). The frequency and intensity of rainfall will increase in some part of South Asia. A significant decline in snowfall in the northern areas will reduce water availability for irrigation (UNFCCC 2008). World average temperature has been raised by 0.4°C in the past century. A significant variation in monsoon rainfall was recorded in the past century and a slight increase is also projected up to 2050. However, the duration of rainfall will decrease in most parts of the country. The western and central parts will face a decrease of 15 days, while in northern part (Himalayan foothills) a 5–10 days' increase is expected. Similarly, the rainfall intensity is expected to increase by 1–4 mm per day; however, in the northwest India a 1 mm per day increase may occur.

5.6 Climate Change and Cotton Production of USA

Worldwide, USA is the third largest and top cotton exporter country. The United States of America shared about 18% of the world cotton production (Figure 5.1). The yield of cotton is expected to increase with a slight increase in temperature and carbon dioxide. However, its impact on rainfall is diverse depending on the regions, and most of the rainfed areas will face increase in the rainfall along with extreme weather events. However, production in the southwest mostly relies on irrigation. These areas will face water shortage because of the depletion in groundwater status and irregular melting of snow in the summer. In the USA, cotton is grown in four geographic areas: the southeast, mid-south, southwest, and west zone. The southwest and mid-south areas comprise 70% of the total cotton cultivable land. The south comprises of 35% of the total cotton growing land. The western regions comprising of Arizona, California, and New Mexico account for 9% of the total. In these regions, the climate is dry and mostly warm, which is favourable for cotton production. In the United States, cotton production is highly mechanized, which consumes a lot of fuel. The crop management system is very intensive and is carried out according to field variation such as soil fertility, crop growth, etc. due to the introduction of genetically modified varieties in 2009, which almost covered 95% of the cotton growing areas. About 36% acreage of the total were irrigated during 2007 (Cotton Incorporated 2009).

Global climate change is already observed in the United States also. Climate change leads to increase in temperature, reduced frost days, increased rainfall intensity, rise in sea levels, increase in growing seasons, water vapour in the atmosphere, reduced snow cover, glaciers, permafrost, and sea ice. During the past 30 years, rise in temperature about 4°C was noted in winter season in the Midwest and northern Great Plains. These changes are expected to continue with the new one developing with the passage of time. In the future, changes in the United States are associated with an increase in wind, storms, rain, and dryness in the southwest Caribbean. As a result, it will affect water availability, agricultural production, and other socioeconomic aspects (Karl et al. 2009). If the farmers adopt new agricultural practices against climate change, cotton yields will increase with an increase in carbon dioxide concentration. The increasing CO_2 concentration is likely to increase photosynthesis (National Center for Atmospheric Research, NCAR 2001). In the southeastern United States, a modal was developed, which predicted a 15% increase in cotton yield. The second modal, which is for adaptation of farming system for elevated CO_2 and to take full benefit of longer growing season, accounts for 26% and 36% increase in cotton yield (NCAR 2001). However, the impacts vary according to the regions. As the Great Plains and the Mississippi Delta showed an increase in yield, the Midwest corn belt suffered, and the southwest lost at least 20% of the yield (NCAR 2004). Previous reports stated that a reduction (9%–17%) was found in US cotton yield (Table 5.1). It is projected that climate change may enhance the number of days facing temperature of 37.8°C or above. A dramatic increase in this context will be observed in Texas, where the current 10 to 20 days of 37.8°C would increase up to 100 days per year (Karl et al. 2009). In the Great Plains of the United States, water availability is one of the most important factors limiting agricultural production. The use of water is not sustainable as the water consumption is more than the recharge. The increasing population and irrigation systems increased the annual withdrawal over the natural recharge. Subsequently, in the irrigated areas of Texas, Oklahoma, and Kansas the ground water level has fallen from 30 m to 75 m. Besides this increasing temperature, evaporation and drought due to climate change will further decrease water resources (Karl et al. 2009). Many areas have already faced water deficiency and the shortage will increase in the future. Pumping is reducing ground water table and rising temperature is reducing river flows like Colorado River. Similarly, reduction in rainfall and snow will further reduce water supplies to the reservoirs. In western part of the United States, climate change will bring more drought and high risk of floods (Karl et al. 2009). A warming in all the seasons is projected for the southeast due to climate change. The mean temperature is expected to increase by 2.5°C–5°C by 2080, which will increase the number of hot days (Karl et al. 2009).

5.7 Climate Change and Cotton Production of Brazil

Brazil is the fourth largest cotton producer, which has surpassed Pakistan this year. In the previous year, the country shared 6% to the world cotton production (Figure 5.1). In Brazil cotton production is rapidly increasing since past decade due to increase in cultivation land. As mentioned before, climate change will lead to rise in global temperature. The impacts of rainfall are unpredictable in Brazil. A reduction of 20%–40% and increase of 5%–10% is predicted in the Central and tropical America by the year 2080. It was predicated that in the future, cotton production in Brazil will mostly depend on deforestation, competition with soybean cultivation, and soil fertility of the newly cultivated lands. Currently, Brazil produces 6% and consumes 4% of the world cotton fibre. Cotton production is mainly associated in the centre-west regions. The soil of the centre-west regions is very poor, but the use of effective fertilizers and good weather condition for cotton improved the productivity, which made Brazil a cotton exporter from being an importer. This efficient cotton production is associated with the most technically advanced farmers of Brazil (Graham 2009).

Climate change will affect Latin America due to decline in the Andean glaciers before the end of this century. However, Brazil cotton production mostly relies on rainfall instead of Andean glaciers.

The rainfall in Brazil is uncertain, and it is predicted that the arid and semiarid areas will receive less rain, which will lead to land degradation and disturb the food security. Throughout Latin America, yields are expected to be decreased, except some areas, where increasing CO_2 concentration may overcome the negative effects of climate change (UNFCCC 2008). The expected increase in temperature is 0.4°C–1.8°C for 2020 and 1.0°C–7.5°C, and central and tropical south America will face reduction in rainfall in 2080 (Kundzewicz et al. 2008).

5.8 Climate Change and Cotton Production of Pakistan

Pakistan is the world's fifth largest cotton producer with about 7% of the global cotton production during 2018 (Figure 5.1). Cotton is mainly produced in Punjab (79%) and Sindh provinces (20%) of Pakistan. The soils of the growing areas vary from sandy loam, clay, to clay loam in the south (Gillham et al. 1995). The country faces high temperature, especially in May and June 40–45°C, and in some parts at May will reach up to 50°C. However, in winter the temperatures fall below freezing but the lower Sindh part is frost-free. These areas have mainly two cropping system, summer (Kharif) and winter (Rabi) cropping. Cotton is the main commercial crop of the country (Gillham et al. 1995). The yield varies among the areas, as southern Sindh parts have the higher yield (850 kg/ha) as compared to Punjab (692 kg/ha); however, the yield of irrigated cotton is low in both regions. The previous study reported that this is due to the high temperature in Pakistan as compared to other countries (Raza 2009). The boll weight in Pakistan is lower (2–3 g per boll), than that of United States, Australia, Egypt (each 5–6 g per boll), and Turkey (4–5 g per boll). Due to limited rainfall (150–750 mm), the production is entirely dependent on irrigation (WWF, 2005). Irrigation water is supplied from the dams and reservoirs on a weekly basis according to the crop requirement (Gillham et al. 1995).

Pakistan is one of the most vulnerable countries to climate change and cotton production is suffering more from it. The country has already faced a yield reduction of 8% as shown in Table 5.1. The reason behind this is the country is totally dependent on the Indus River coming from the Himalayan and Tibetan glaciers, which is affected due to the decline in the snowfields (Van Raaij 2010), which will ultimately affect the Indus River. If the irrigation water availability is continuing to decrease, the farmers will switch to crops that have less demand for water (Pakistan 2003). The maximum temperature in the central and south parts of the country exceeds 40°C in the summer. Therefore, an increase in temperature and reduction in rainfall in the future would have a negative impact on cotton production by increasing the demands for irrigation water, particularly for cotton. Due to high temperature, cotton production is already in suboptimal condition. However, a further rise in temperature will reduce the yield. Previously it was reported that increase in temperature by 0.3°C per decade will affect the production of many major crops including cotton due to heat stress. However, the prediction (10°C) is more than the optimum, which will negatively affect cotton growth and productivity. Moreover, the need for irrigation water will increase to compensate the heat stress and maintain optimum cotton production. Besides this, the use of poor-quality groundwater in irrigation will increase secondary salinization (Pakistan 2003).

5.9 Adaptation Strategies for Cotton to Address Climate Change

Climate change is affecting cotton production; as a result different adaptive strategies need to be applied to mitigate the negative impact of climate change and sustain cotton production. In the case of cotton, the crop has a lower capacity to respond to changes in climatic conditions (ICAC 2007). The amazing response of cotton to stress is that it drops the overloaded flowers, buds, and bolls as a result of compensatory growth. The vertical tap root system of the crop helps to withstand drought condition, however, it is highly vulnerable to waterlogging condition. The crop is highly dependent on irrigation water availability. Half the cotton of the world is produced in irrigated areas.

Adaptation is the action to climate change to reduce vulnerabilities in the social and biological system. Agriculture is highly vulnerable to changing climate. Climatic factors like rising temperature, CO_2 concentration, precipitation, etc. affect crop growth and yields. Increase in temperature at crop's critical

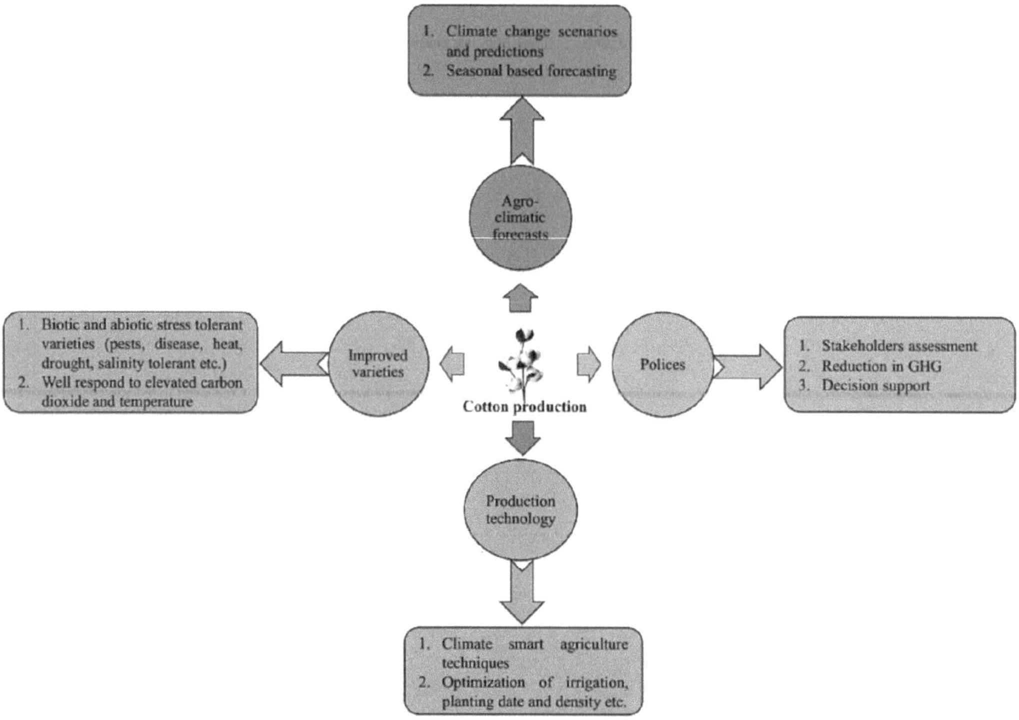

FIGURE 5.2 Adaptation strategies for cotton to address climate change.

stages leads to floret sterility, reduces grain-filling duration, and finally crop yield (Ahmed et al. 2018). Therefore, different strategies should be adapted to avoid the negative impact of climate change on crop production (Figure 5.2). Adaptation provides the opportunities to address climate change and sustain global crop production (Fischer et al. 2002).

Recently farmers have adapted many, ways such as the farmers of Pakistan and Brazil who have adjusted planting time and optimized plant population as a consequence of climate change (Ahmad et al. 2018). It is very important to explore the potential of the crop. High temperatures during reproductive stage decrease the yield. Therefore, adjusting planting date based on rain and heat stress onset will reduce the chances of yield losses. Moreover, it will also help prevent the pest attack and utilize the resources fully in the growing season. Similarly, the number of plants per unit area is also important to potentially use the available resources. It was noted that optimum plant population compensates yield losses. Another option of climate change adaptation is to develop site-specific varieties such as early maturing, drought, heat, pests, disease, and weeds that are tolerant to sustain the productivity in the changing climate. Both conventional and biotechnology approach should be applied for the development of cultivars with improved water and nutrient-use efficiency, as well as resistant to different biotic and abiotic stresses. Identification and screening of genotypes can fit the new cropping system with high resource-use efficiency and respond to the increasing temperature and carbon dioxide concentration. It is necessary to develop varieties which emit low greenhouse gases and build soil organic matter. These newly developed cultivars can increase the production under moisture stress and high temperature (Deressa et al. 2009). Moreover, advanced research is needed to enhance photosynthetic efficiency of leaves by keeping the green leaves for a long time to consume more atmospheric carbon dioxide. Improved understanding is needed to know the interactive effects of climate change and management on carbon source–sink relations in cotton ecosystems. Similarly, the role of integrated management as a means for providing better management options for cotton ecosystems needs to be explored. Precision management of nutrients can increase the resilience in the crops by increasing the efficiency of fertilizers. Precision management of fertilizers is important to enhance productivity while reducing the greenhouse gas emission (Srinivasan 2006). Stop or reduce nutrients loss from the farming system, prevent soil erosion, avoid burning cotton residues. Encourage crop rotation to favour soil fertility by including legumes, cover crops, etc. Properly use natural fertilizer to increase soil carbon, such as organic manure, compost, microbes, and residue management should be encouraged to reduce the impact of climate change. The reduction in water availability due to climate change will impact irrigated cotton production. Therefore, improvement in off-farm and on-farm water-use efficiencies is needed. Optimization of water-use efficiency can be done by providing irrigation in the peak time of crop water requirement and avoid the overuse of irrigation. At the same time in situ soil moisture conservation techniques are used to conserve soil moisture, along with this promotion of diploid cotton in low and high erratic rainfall and saline environment. Weeds will also become more important to control in the future. Therefore, integrated weed management techniques should be followed to improve the yield and reduce synthetic weedicides, which will subsequently reduce greenhouse gas emissions (Pedde et al. 2013).

5.10 Conclusion

The current chapter aims to shed light on climate change and cotton production of major cotton-producing countries and to address the changing climate by various adaptive strategies. Climate change has both positive and negative impacts on cotton production, which varies from region to region. For most of the regions the effect of climate change is positive. The overall impact of climate change on cotton production is positive. As the simulation study showed an acceleration in the phenological development due to increase in temperature, the yield will increase up to 2050 with an increase in carbon dioxide. It is predicted that due to climate change, the temperature will rise by 1.8°C–4°C at the end of this century. This rise in temperature will reduce the crop growing period by enhancing crop growth and development. For this reason, planting date will be delayed for 5–10 days and careful management of weeds and nitrogen will be required between crop emergence and the first rain. Another way is to modify crop phenology is by delaying flowering and extending boll development stage, which will delay crop maturity. This delay in flowering will enable the plants to get more biomass, which can be later converted into reproductive organs. The projections for rainfall and evapotranspiration are uncertain for most of the cotton-growing regions. However, climate change will also increase the intensity of extreme weather events such as heat stress, erratic rainfall, drought, salinity, etc. which have a negative impact on crop growth and yield. The intensity of monsoon rainfall will damage cotton production in Asia. Therefore, to minimize these impacts and sustain cotton production, certain adaptive strategies will be followed. Moreover, adaptation strategies should be quantified based on modelling approaches.

REFERENCES

Adhikari P, Ale S, Bordovsky JP, et al. (2016) Simulating future climate change impacts on seed cotton yield in the Texas High Plains using the CSM-CROPGRO-Cotton model. *Agric Water Manag* 164:317–330.

Adnan M, Shah Z, Sharif M, Rahman H. (2018a) Liming induces carbon dioxide (CO_2) emission in PSB inoculated alkaline soil supplemented with different phosphorus sources. *Environ Sci Pol Res* 25:9501–9509.

Adnan M, Zahir S, Fahad S, Arif M, Mukhtar A, Imtiaz AK, Ishaq AM, Abdul B, Hidayat U, Muhammad A, Inayat-Ur R, Saud S, Muhammad ZI, Yousaf J, Amanullah Hafiz MH, Wajid N (2018b) Phosphate-solubilizing bacteria nullify the antagonistic effect of soil calcification on bioavailability of phosphorus in alkaline soils. *Sci Rep* 8:4339. https://doi.org/10.1038/s41598-018-22653-7

Ahmad I, Wajid SA, Ahmad A, et al. (2018) Assessing the impact of thermo-temporal changes on the productivity of spring maize under semi-arid environment. *Int J Agric Biol* 20:2203–2210.

Ahmed I, Ur Rahman MH, Ahmed S, et al. (2018) Assessing the impact of climate variability on maize using simulation modeling under semi-arid environment of Punjab, Pakistan. *Environ Sci Pollut Res* 25:28413–28430.

Akram R, Turan V, Hammad HM, Ahmad S, Hussain S, Hasnain A, Maqbool MM, Rehmani MIA, Rasool A, Masood N, Mahmood F, Mubeen M, Sultana SR, Fahad S, Amanet K, Saleem M, Abbas Y, Akhtar HM, Waseem F, Murtaza R, Amin A, Zahoor SA, ul Din MS, Nasim W (2018a) Fate of organic and inorganic

pollutants in paddy soils. In: Hashmi MZ, Varma A (ed) *Environmental pollution of paddy soils, soil biology*. Springer International Publishing Ag, Cham, Switzerland, pp. 197–214.
Akram R, Turan V, Wahid A, Ijaz M, Shahid MA, Kaleem S, Hafeez A, Maqbool MM, Chaudhary HJ, Munis, MFH, Mubeen M, Sadiq N, Murtaza R, Kazmi DH, Ali S, Khan N, Sultana SR, Fahad S, Amin A, Nasim W (2018b) Paddy land pollutants and their role in climate change. In: Hashmi MZ, Varma A (ed) *Environmental pollution of paddy soils, soil biology*. Springer International Publishing Ag, Cham, Switzerland, pp. 113–124.
Alam A, Baig J, Din SU, Arshad M, Adnan M, Khan A (2016) Relative abundance of benthic Macro-invertebrates in relation to abiotic environment in Hussainbadnallah, Hunza, Gilgit Baltistan, Pakistan. *Int J Biosci* 9(3):185–193.
Amouzou KA, Naab JB, Lamers JPA, et al. (2018) CROPGRO-Cotton model for determining climate change impacts on yield, water-and N-use efficiencies of cotton in the Dry Savanna of West Africa. *Agric Syst* 165:85–96.
Ayers J, Dodman D (2010) Climate change adaptation and development I: the state of the debate. *Prog Dev Stud* 10:161–168.
Aziz K, Daniel KYT, Fazal M, Muhammad ZA, Farooq S, Fan W, Fahad S, Ruiyang Z (2017a) Nitrogen nutrition in cotton and control strategies for greenhouse gas emissions: a review. *Environ Sci Pollut Res* 24:23471–23487. https://doi.org/10.1007/s11356-017-0131-y
Aziz K, Daniel KYT, Muhammad ZA, Honghai L, Shahbaz AT, Mir A, Fahad S (2017b) Nitrogen fertility and abiotic stresses management in cotton crop: a review. *Environ Sci Pollut Res* 24:14551–14566. https://doi.org/10.1007/s11356-017-8920-x
Bale JS, Masters GJ, Hodkinson ID, et al. (2002) Herbivory in global climate change research: direct effects of rising temperature on insect herbivores. *Glob Chang Biol* 8:1–16.
Basir A, Jan MT, Alam M, Shah AS, Afridi K, Adnan M, Ali K, Mian IA (2016) Impacts of tillage, stubble management and nitrogen on wheat production and soil properties. *Can J Soil Sci* 97(2):133–140.
Bhattacharya NC, Radin JW, Kimball BA, et al. (1994) Leaf water relations of cotton in a free-air CO_2-enriched environment. *Agric For Meteorol* 70:171–182.
Chakraborty S (2013) Migrate or evolve: options for plant pathogens under climate change. *Glob Chang Biol* 19:1985–2000.
Change C (2009) Impact on agriculture and costs of adaptation. Rep by Int Food Policy Res Institute, October.
Chapagain A, Hoekstra AY, Savenije HHG, Gautam R (2005) The water footprint of cotton consumption.
Chen C, Pang Y, Pan X, Zhang L (2015) Impacts of climate change on cotton yield in China from 1961 to 2010 based on provincial data. *J Meteorol Res* 29:515–524.
CICR (2009) Constraints analysis of cotton in India. Central Institute for Cotton Research (CICR) India (website).
Cotton Incorporated (2009) Summary of Life Cycle Inventory Data for Cotton (Field to Bale – version 1.1 –2 July 2009). Cotton Incorporated, United States of America, 31 p.
Coviella CE, Stipanovic RD, Trumble JT (2002) Plant allocation to defensive compounds: interactions between elevated CO2 and nitrogen in transgenic cotton plants. *J Exp Bot* 53:323–331.
Coviella CE, Trumble JT (2000) Effect of elevated atmospheric carbon dioxide on the use of foliar application of Bacillus thuringiensis. *Biocontrol* 45:325–336.
Dendage VR (2017) Agroclimatic characterization of cotton crop under parbhani district.
Deressa TT, Hassan RM, Ringler C, et al. (2009) Determinants of farmers' choice of adaptation methods to climate change in the Nile Basin of Ethiopia. *Glob Environ Chang* 19:248–255.
Dhawan AK (1998) Integrated pest management in cotton: status of different management options and future perspectives. *Pestology* 22:21–33.
Diarra A, Barbier B, Zongo B, Yacouba H (2017) Impact of climate change on cotton production in Burkina Faso. *African J Agric Res* 12:494–501.
Depeng W, Fahad S, Saud S, Muhammad K, Aziz K, Mohammad NK, Hafiz MH, Wajid N (2018) Morphological acclimation to agronomic manipulation in leaf dispersion and orientation to promote "Ideotype" breeding: evidence from 3D visual modeling of "super" rice (*Oryza sativa* L.). *Plant Physiol Biochem.* https://doi.org/10.1016/j.plaphy.2018.11.010
Dugas WA, Heuer ML, Hunsaker D, et al. (1994) Sap flow measurements of transpiration from cotton grown under ambient and enriched CO_2 concentrations. *Agric For Meteorol* 70:231–245.

Fahad S, Bajwa AA, Nazir U, Anjum SA, Farooq A, Zohaib A, Sadia S, Nasim W, Adkins S, Saud S, Ihsan MZ, Alharby H, Wu C, Wang D, Huang J (2017) Crop production under drought and heat stress: plant responses and management options. *Front Plant Sci* 8:1147. https://doi.org/10.3389/fpls.2017.01147

Fahad S, Bano A (2012) Effect of salicylic acid on physiological and biochemical characterization of maize grown in saline area. *Pak J Bot* 44:1433–1438.

Fahad S, Chen Y, Saud S, Wang K, Xiong D, Chen C, Wu C, Shah F, Nie L, Huang J (2013) Ultraviolet radiation effect on photosynthetic pigments, biochemical attributes, antioxidant enzyme activity and hormonal contents of wheat. *J Food Agri Environ* 11(3&4):1635–1641.

Fahad S, Hussain S, Bano A, Saud S, Hassan S, Shan D, Khan FA, Khan F, Chen Y, Wu C, Tabassum MA, Chun MX, Afzal M, Jan A, Jan MT, Huang J (2014a) Potential role of phytohormones and plant growth-promoting rhizobacteria in abiotic stresses: consequences for changing environment. *Environ Sci Pollut Res* 22(7):4907–4921. https://doi.org/10.1007/s11356-014-3754-2

Fahad S, Hussain S, Matloob A, Khan FA, Khaliq A, Saud S, Hassan S, Shan D, Khan F, Ullah N, Faiq M, Khan MR, Tareen AK, Khan A, Ullah A, Ullah N, Huang J (2014b) Phytohormones and plant responses to salinity stress: a review. *Plant Growth Regul* 75(2):391–404. https://doi.org/10.1007/s10725-014-0013-y

Fahad S, Hussain S, Saud S, Hassan S, Chauhan BS, Khan F, et al. (2016a) Responses of rapid viscoanalyzer profile and other rice grain qualities to exogenously applied plant growth regulators under high day and high night temperatures. *PLoS One* 11(7):e0159590. https://doi.org/10.1371/journal.pone.0159590

Fahad S, Hussain S, Saud S, Hassan S, Ihsan Z, Shah AN, Wu C, Yousaf M, Nasim W, Alharby H, Alghabari F, Huang J (2016c) Exogenously applied plant growth regulators enhance the morphophysiological growth and yield of rice under high temperature. *Front Plant Sci* 7:1250. https://doi.org/10.3389/fpls.2016.01250

Fahad S, Hussain S, Saud S, Hassan S, Tanveer M, Ihsan MZ, Shah AN, Ullah A, Nasrullah KF, Ullah S, Alharby HNW, Wu C, Huang J (2016d) A combined application of biochar and phosphorus alleviates heat-induced adversities on physiological, agronomical and quality attributes of rice. *Plant Physiol Biochem* 103:191–198.

Fahad S, Hussain S, Saud S, Khan F, Hassan S, Jr., Nasim W, Arif M, Wang F, Huang J (2016b) Exogenously applied plant growth regulators affect heat-stressed rice pollens. *J Agron Crop Sci* 202:139–150.

Fahad S, Hussain S, Saud S, Tanveer M, Bajwa AA, Hassan S, Shah AN, Ullah A, Wu C, Khan FA, Shah F, Ullah S, Chen Y, Huang J (2015a) A biochar application protects rice pollen from high-temperature stress. *Plant Physiol Biochem* 96:281–287.

Fahad S, Muhammad ZI, Abdul K, Ihsanullah D, Saud S, Saleh A, Wajid N, Muhammad A, Imtiaz AK, Chao W, Depeng W, Jianliang H (2018) Consequences of high temperature under changing climate optima for rice pollen characteristics-concepts and perspectives. *Archives Agron Soil Sci.* https://doi.org/10.1080/03650340.2018.1443213

Fahad S, Nie L, Chen Y, Wu C, Xiong D, Saud S, Hongyan L, Cui K, Huang J (2015b) Crop plant hormones and environmental stress. *Sustain Agric Rev* 15:371–400.

Fahad S, Adnan M, Hassan S, Saud S, Hussain S, Wu C, Wang D, Hakeem KR, Alharby HF, Turan V, Khan MA, Huang J (2019b) Rice responses and tolerance to high temperature. In: Hasanuzzaman M, Fujita M, Nahar K, Biswas JK (ed) *Advances in rice research for abiotic stress tolerance.* Woodhead Publ Ltd, Abington, Cambridge, UK, pp. 201–224.

Fahad S, Rehman A, Shahzad B, Tanveer M, Saud S, Kamran M, Ihtisham M, Khan SU, Turan V, Rahman MHU (2019a) Rice responses and tolerance to metal/metalloid toxicity. In: Hasanuzzaman M, Fujita M, Nahar K, Biswas JK (ed) *Advances in rice research for abiotic stress tolerance.* Woodhead Publ Ltd, Abington, Cambridge, UK, pp. 299–312.

Fajer ED, Bowers MD, Bazzaz FA (1992) The effect of nutrients and enriched CO_2 environments on production of carbon-based allelochemicals in Plantago: a test of the carbon/nutrient balance hypothesis. *Am Nat* 140:707–723.

Fischer G, Shah MM, Van Velthuizen HT (2002) Climate change and agricultural vulnerability.

Flores Araya J (2008) The effects on cotton production due to climate change: an assessment on water availability and pesticide use in two different cotton growing regions in India.

Gillham FEM, Bell TM, Arin T, Matthews GA, Le Rumeur C, Hearn AB (1995) Cotton production prospects in the next decade. World Bank, United States of America, 277 p.

Gonias ED, Oosterhuis DM, Bibi AC (2008) Physiologic response of cotton to the insecticide imidacloprid under high-temperature stress. *J Plant Growth Regul* 27:77–82.

Graham, P (2009) Brazil. In: Cotton Outlook (2009) Cotton trading relationships with China, pp. 22–26 Cotton Outlook, Special Feature June 2009
Habib Ur R, Ashfaq A, Aftab W, Manzoor H, Fahd R, Wajid I, Md. Aminul I, Vakhtang S, Muhammad A, Asmat U, Abdul W, Syeda RS, Shah S, Shahbaz K, Fahad S, Manzoor H, Saddam H, Wajid N (2017) Application of CSM-CROPGRO-Cotton model for cultivars and optimum planting dates: Evaluation in changing semi-arid climate. *Field Crops Res.* http://dx.doi.org/10.1016/j.fcr.2017.07.007
Hafiz MH, Muhammad A, Farhat A, Hafiz FB, Saeed AQ, Muhammad M, Fahad S, Muhammad A (2019) Environmental factors affecting the frequency of road traffic accidents: a case study of sub-urban area of Pakistan. *Environ Sci Pollut Res.* https://doi.org/10.1007/s11356-019-04752-8
Hafiz MH, Wajid F, Farhat A, Fahad S, Shafqat S, Wajid N, Hafiz FB (2016) Maize plant nitrogen uptake dynamics at limited irrigation water and nitrogen. *Environ Sci Pollut Res* 24(3):2549–2557. https://doi.org/10.1007/s11356-016-8031-0
Heagle AS (2003) Influence of elevated carbon dioxide on interactions between *Frankliniella occidentalis* and *Trifolium repens*. *Environ Entomol* 32:421–424.
Hendrix DL, Mauney JR, Kimball BA, et al. (1994) Influence of elevated CO_2 and mild water stress on nonstructural carbohydrates in field-grown cotton tissues. *Agric For Meteorol* 70:153–162.
Hileman DR, Huluka G, Kenjige PK, et al. (1994) Canopy photosynthesis and transpiration of field-grown cotton exposed to free-air CO_2 enrichment (FACE) and differential irrigation. *Agric For Meteorol* 70:189–207.
Hoekstra AY, Chapagain AK (2007) The water footprints of Morocco and the Netherlands: global water use as a result of domestic consumption of agricultural commodities. *Ecol Econ* 64:143–151.
Hsu H, Gale F (2001) Regional shifts in China's cotton production and use. *Cott wool Situat Outlook* 11:19–25.
Huluka G, Hileman DR, Biswas PK, et al. (1994) Effects of elevated CO_2 and water stress on mineral concentration of cotton. *Agric For Meteorol* 70:141–152.
Hunsaker DJ, Hendrey GR, Kimball BA, et al. (1994) Cotton evapotranspiration under field conditions with CO_2 enrichment and variable soil moisture regimes. *Agric For Meteorol* 70:247–258.
ICAC (2007) Global warming and cotton production – Part 1. In: ICAC Recorder, Vol. 25, No. 4 (December 2007). pp. 12–16. International Cotton Advisory Committee (ICAC). United States of America.
ICAC (2009) Global warming and cotton production – Part 2. In: ICAC Recorder, Vol. 27, No. 1 (March 2009). pp. 9–13. International Cotton Advisory Committee (ICAC). United States of America.
IPCC (2007) Climate Change 2007: Synthesis Report. Intergovernmental Panel on Climate Change (IPCC). Germany, 52 p.
Kamarn M, Wenwen C, Irshad A, Xiangping M, Xudong Z, Wennan S, Junzhi C, Shakeel A, Fahad S, Qingfang H, Tiening L (2017) Effect of paclobutrazol, a potential growth regulator on stalk mechanical strength, lignin accumulation and its relation with lodging resistance of maize. *Plant Growth Regul* 84:317–332. https://doi.org/10.1007/s10725-017-0342-8
Karl TR, Melillo JM, Peterson TC, Hassol SJ (2009) *Global climate change impacts in the United States.* Cambridge University Press.
Khan N, Shah Z, Adnan M, Ali M, Khan B, Mian IA, Ali A, Zahoor M, Roman M, Ullah L, Khaliq A, Khan WA, Alam A (2016) Evaluation of soil for important properties and chromium concentration in the basin of chromite hills in lower Malakand. *Adv Environ Biol* 10(7):141–147.
Kimball BA, Mauney JR (1993) Response of cotton to varying CO_2, irrigation, and nitrogen: yield and growth. *Agron J* 85:706–712.
Kundzewicz ZW, Mata LJ, Arnell NW, et al. (2008) The implications of projected climate change for freshwater resources and their management. *Hydrol Sci J* 53:3–10.
Lewis SC, King AD (2017) Evolution of mean, variance and extremes in 21st century temperatures. *Weather Clim Extrem* 15:1–10.
Lobell DB, Field CB (2011) California perennial crops in a changing climate. *Clim Change* 109:317–333.
Lobell DB, Schlenker W, Costa-Roberts J (2011) Climate trends and global crop production since 1980. *Science* (80-) 333:616–620.
Makhdum AH, Khan HN, Ahmad S (2011) *Reducing cotton footprints through implementation of better management practices in cotton production; a step towards better cotton initiative.* In: *Proceedings of the Fifth Meeting of the Asian Cotton Research and Development Network*, Lahore, Pakistan, pp. 23–26.
Mauney JR, Kimball BA, Pinter Jr PJ, et al. (1994) Growth and yield of cotton in response to a free-air carbon dioxide enrichment (FACE) environment. *Agric For Meteorol* 70:49–67.

McRae D, Roth G, Bange M (2007) Climate change in cotton catchment communities–a scoping study. Cott Catchment Communities CRC.

Muhammad Z, Abdul MK, Abdul MS, Kenneth BM, Muhammad S, Shahen S, Ibadullah J, Fahad S (2019) Performance of Aeluropus lagopoides (mangrove grass) ecotypes, a potential turfgrass, under high saline conditions. *Environ Sci Pollut Res.* https://doi.org/10.1007/s11356-019-04838-3

Nakayama FS, Huluka G, Kimball BA, et al. (1994) Soil carbon dioxide fluxes in natural and CO_2-enriched systems. *Agric For Meteorol* 70:131–140.

NCAR (2001) The Effect of the Spatial Resolution of Climate Scenarios on Cotton Production in the Southeastern United States. National Center for Atmospheric Research, United States of America.

NCAR (2004) Climate change impacts. Brochure. National Center for Atmospheric Research, United States of America.

Pakistan (2003) Pakistan's Initial National Communication on Climate Change. Ministry of Environment, Islamabad, Pakistan, 92 p.

Pandi R (1997) Ecology and Management of Leafhopper, Amrasca devastans (Distant) in cotton. M Sc(Ag) Thesis Tamil Nadu Agrl University, Kill.

Patel JJ (2008) Population Dynamics And Eco-Friendly Management Of Serpentine Leaf Miner, Liriomyza Trifolii (Burgess)(Diptera: Agromyzidae) In Cucumber, Cucumis Sativus Linnaeus.

Patil B V, Nandihalli BS, Hugar PS 1992. Influence of weather parameters on pheromone trap catches of cotton bollworms. *Karnataka J Agric Sci* 5:346–350.

Pedde KC, Gore AK, Chavan AS (2013) Integrated weed management in chickpea. *Indian J Weed Sci* 45:299.

Pettigrew TW (2007) *High temperature effects on cotton yield, yield components, and fibre quality.* In: *Back to The ASA-CSSASSSA International Annual Meetings*, November. pp. 4–8.

Piao S, Ciais P, Huang Y, et al. (2010) The impacts of climate change on water resources and agriculture in China. *Nature* 467:43.

Pinter Jr PJ, Idso SB, Hendrix DL, et al. (1994) Effect of free-air CO_2 enrichment on the chlorophyll content of cotton leaves. *Agric For Meteorol* 70:163–169.

Prior SA, Rogers HH, Runion GB, et al. (1995) Free-air carbon dioxide enrichment of cotton: Root morphological characteristics. *J Environ Qual* 24:678–683.

Prior SA, Rogers HH, Runion GB, Hendrey GR (1994) Free-air CO_2 enrichment of cotton: vertical and lateral root distribution patterns. In: *Belowground responses to rising atmospheric CO2: implications for plants, soil biota, and ecosystem processes.* Springer, pp. 33–44.

Qamar-Uz Z, Zubair A, Muhammad Y, Muhammad ZI, Abdul K, Fahad S, Safder B, Ramzani PMA, Muhammad N (2017) Zinc biofortification in rice: leveraging agriculture to moderate hidden hunger in developing countries. *Arch Agron Soil Sci* 64:147–161. https://doi.org/10.1080/03650340.2017.1338343

Rahman I, Ali S, Rahman I, Adnan M, Ullah H, Basir A, Malik FA, Shah AS, Ibrahim M, Arshad M (2016) Effect of pre-storage seed priming on biochemical changes in okra seed. *Pure Appl Biol* 5(1):165–171.

Raja Reddy K, Hodges HF, Reddy VR. (1992) Temperature effects on cotton fruit retention. *Agron J* 84: 26–30.

Raza, SH (2009) *Cotton production in Pakistan. A grower's view.* In: *Presentation (.ppt) at the 68th ICAC Plenary Meeting.* International Cotton Advisory Committee (ICAC). United States of America.

Reddy GV, Rodrigues V (1999) A glial cell arises from an additional division within the mechanosensory lineage during development of the microchaete on the *Drosophila notum. Development* 126:4617–4622.

Reddy KR, Davidonis GH, Johnson AS, Vinyard BT (1999) Temperature regime and carbon dioxide enrichment alter cotton boll development and fiber properties. *Agron J* 91:851–858.

Reddy KR, Doma PR, Mearns LO, et al. (2002) Simulating the impacts of climate change on cotton production in the Mississippi delta. *Clim Res* 22:271–281.

Reddy KR, Hodges HF, Reddy VR (1992) Temperature effects on cotton fruit retention. *Agron J* 84:26–30.

Reddy VR, Reddy KR, Acock B (1994) Carbon dioxide and temperature effects on cotton leaf initiation and development. *Biotronics* 23:59–74.

Rezaei EE, Webber H, Gaiser T, et al. (2015) Heat stress in cereals: mechanisms and modelling. *Eur J Agron* 64:98–113.

Sajjad H, Muhammad M, Ashfaq A, Waseem A, Hafiz MH, Mazhar A, Nasir M, Asad A, Hafiz UF, Syeda RS, Fahad S, Depeng W, Wajid N (2019) Using GIS tools to detect the land use/land cover changes during forty years in Lodhran district of Pakistan. *Environ Sci Pollut Res.* https://doi.org/10.1007/s11356-019-06072-3

Saud S, Chen Y, Fahad S, Hussain S, Na L, Xin L, Alhussien SA (2016) Silicate application increases the photosynthesis and its associated metabolic activities in Kentucky bluegrass under drought stress and post-drought recovery. *Environ Sci Pollut Res* 23(17):17647–17655. https://doi.org/10.1007/s11356-016-6957-x

Saud S, Chen Y, Long B, Fahad S, Sadiq A (2013) The different impact on the growth of cool season turf grass under the various conditions on salinity and drought stress. *Int J Agric Sci Res* 3:77–84.

Saud S, Fahad S, Yajun C, Ihsan MZ, Hammad HM, Nasim W, Amanullah Jr, Arif M, Alharby H (2017) Effects of nitrogen supply on water stress and recovery mechanisms in Kentucky Bluegrass Plants. *Front Plant Sci* 8:983. https://doi.org/10.3389/fpls.2017.00983

Saud S, Li X, Chen Y, Zhang L, Fahad S, Hussain S, Sadiq A, Chen Y (2014) Silicon application increases drought tolerance of Kentucky bluegrass by improving plant water relations and morph physiological functions. *SciWorld J* 2014:1–10. https://doi.org/10.1155/2014/368694

Shah F, Lixiao N, Kehui C, Tariq S, Wei W, Chang C, Liyang Z, Farhan A, Fahad S, Huang J (2013) Rice grain yield and component responses to near 2°C of warming. *Field Crop Res* 157:98–110.

Shah N, Tanzeela, Khan A, Khisroon M, Adnan M, Jawad SM (2017) Seroprevalence and risk factors of toxoplasmosis in pregnant women of district Swabi. *Pure Appl Biol* 6(4): 1306–1313.

Srinivasan A (2006) *Handbook of precision agriculture: principles and applications.* CRC Press.

Stiling P, Cornelissen T (2007) How does elevated carbon dioxide (CO_2) affect plant–herbivore interactions? A field experiment and meta-analysis of CO_2-mediated changes on plant chemistry and herbivore performance. *Glob Chang Biol* 13:1823–1842.

Team CW, Pachauri RK, Reisinger A (2007) *Climate change 2007: synthesis report.* Geneva, Switz IPCC 104.

Ton P (2012) *Cotton and climate change: impacts and options to mitigate and adapt.* In: *EGU General Assembly Conference Abstracts.* p. 421.

Ur Rahman MH, Ahmad A, Wang X, et al. (2018) Multi-model projections of future climate and climate change impacts uncertainty assessment for cotton production in Pakistan. *Agric For Meteorol* 253:94–113.

Van Raaij B (2010) Circa 60 miljoen mensen bedreigd door afname smeltwater Himalaya. De Volkskrant 11:

UNFCCC (2008) Climate change: Impacts, vulnerabilities and adaptation in developing countries. United Nations Framework Convention on Climate Change (UNFCCC). 68 p.

Wakelyn PJ, Chaudhry MR (2009) Organic cotton: production practices and post-harvest considerations. In: *Sustainable textiles.* Elsevier, pp. 231–301.

Wajid N, Ashfaq A, Asad A, Muhammad T, Muhammad A, Muhammad S, Khawar J, Ghulam MS, Syeda RS, Hafiz MH, Muhammad IAR, Muhammad ZH, Ur Rahman HR, Veysel T, Fahad S, Suad S, Aziz K, Shahzad A (2017) Radiation efficiency and nitrogen fertilizer impacts on sunflower crop in contrasting environments of Punjab. *Pakistan Environ Sci Pollut Res* 25:1822–1836. https://doi.org/10.1007/s11356-017-0592-z

Whittaker JB (1999) Impacts and responses at population level of herbivorous insects to elevated CO_2. *Eur J Entomol* 96:149–156.

Williams A, White N, Mushtaq S, et al. (2015) Quantifying the response of cotton production in eastern Australia to climate change. *Clim Change* 129:183–196.

WWF (2005) Pakistan Sustainable Cotton Initiative (PSCI) World Wildlife Fund (website).

Yang Z, Zhang Z, Zhang T, Fahad S, Cui K, Nie L, Peng S, Huang J (2017) The effect of season-long temperature increases on rice cultivars grown in the central and southern regions of China. *Front Plant Sci* 8:1908. https://doi.org/10.3389/fpls.2017.01908

Zahida Z, Hafiz FB, Zulfiqar AS, Ghulam MS, Fahad S, Muhammad RA, Hafiz MH, Wajid N, Muhammad S (2017) Effect of water management and silicon on germination, growth, phosphorus and arsenic uptake in rice. *Ecotoxicol Environ Saf* 144:11–18.

6

Effect of Climate Change on Cereal Crops

Amanullah, Shah Khalid, Imran, Hamdan Ali Khan, and Muhammad Arif
The University of Agriculture Peshawar, Peshawar, Pakistan

Abdel Rahman M. Al-Tawaha
Al-Hussein bin Talal University, Maan, Jordan

Muhammad Adnan
The University of Swabi, Swabi, Pakistan

Shah Fahad
Hainan University, Haikou, China; The University of Haripur, Haripur, Pakistan

Azizullah Shah
The University of Haripur, Haripur, Pakistan

Sahar Mumtaz
University of Education, Lahore, Pakistan

Muhammad Zamin
The University of Swabi, Swabi, Pakistan

Shah Saud
Linyi University, Linyi, China

Brajendra Parmar
ICAR-Indian Institute of Rice Research, Soil Science, Hyderabad, India

CONTENTS

DOI: 10.1201/9781003286417-6

6.1 Introduction

Cereals are those crops which are grown for their edible grains. Cereal crops are the largest consumable crops in the world. Among cereal crops, wheat, rice, maize, sorghum, and millet are the most popular crops in the world. These cereals are the staple food for the humans and provide feed for their animals in the form of fodder and straw. World population is increasing day by day, and to provide enough food it is immensely necessary to increase the cereal crop production and improve it grain quality. But due to the ongoing climate change, cereals production along with its quality is significantly affected in terms of decreasing essential micronutrient concentration in their grains. Micronutrients are very essential for healthy life and wellbeing of humans, animals, as well as plants (Ali et al. 2018; Nakandalage and Seneweera 2018; Wang et al. 2011). Deficiency of micronutrients generates global health problem for the mankind with estimated two billion people being affected (Anandan et al. 2011; Bouis 2003). Intake of low-level micronutrient cereal-based diets is considered the major cause of micronutrients deficiency (Bouis 2003; Imtiaz et al. 2010; Nakandalage and Seneweera 2018). Cereal grain crops are the chief source of energy, minerals, and proteins for global production (Cakmak 2008); however, most of the cereals like rice are genetically low in micronutrient concentration (Cakmak 2009). Additional cereal grains have antinutrients, for example, trypsin inhibitor, phytate, and polyphenols, which decrease micronutrient availability and thus reduce the quality of the grains (Graham and Welch 1996; Nadeem et al. 2010). Additionally, climate change aggravated the intrinsic low concentration of micronutrients in cereal crops (Nakandalage and Seneweera 2018; Sinclair 1992). Increases in temperature, change in precipitation patterns and soil water have reflective consequence on the nutritive quality of the cereals' grains. The planet Earth CO_2 [e (CO_2)] concentration is increasing rapidly and the currents level of 411 ppm is expected to reach 550 ppm by 2050. Among greenhouse gases, CO_2 is the most important one, which is responsible for the climate change by changing rainfall patterns, increasing temperature, and creating extreme weather events like flooding and drought (Adnan et al. 2018; Akram et al. 2018a, 2018b; Aziz et al. 2017a; Habib Ur et al. 2017; Hafiz et al. 2016, 2019; Kamarn et al. 2017; Muhammad et al. 2019; Sajjad et al. 2019; Saud et al. 2013, 2014, 2016, 2017; Arif et al. 2015; Shah et al. 2013; Qamar-Uz et al. 2017; Wajid et al. 2017; Yang et al. 2017; Zahida et al. 2017; Depeng et al. 2018; Fahad and Bano 2012; Fahad et al. 2013, 2014a, 2014b, 2015a, 2015b, 2016a, 2016b, 2016c, 2016d, 2017, 2018, 2019a, 2019b; Ilahi et al. 2020). According to Drake et al. (1997), Earth CO_2 has a reflective impact on the yield, quality of grain, plant growth for staples food. Elevated CO_2 can enhance photosynthesis and increase water-use efficiency under sufficient nutrients availability (Erice et al. 2019; Hasegawa et al. 2013; Tausz-Posch et al. 2012; Thilakarathne et al. 2015). However, positive response of plant to e (CO_2) has been overshadowed by the negative effects of climate change on grain quality, particularly in wheat and rice (Högy et al. 2009; Seneweera et al. 1996). Additionally, Earth CO_2 changes nutrient cycle of the cropping system by influencing the activity of soil microorganisms. Furthermore, Earth CO_2 coupled with increasing temperature may cause a serious threat to the global nutrient and food security, by affecting nutrient quality of the major cereal crops like wheat, rice, barley, maize, and legumes (Seneweera et al. 1996). Additionally, rise in temperature along with drought considerably affects grains, mineral contents particularly Fe and Zn concentration in wheat. The variation in the mineral content can change the process of plant growth, grain quality, grain yield, and thereby affect the human nutrition. Therefore, an inclusive understanding of the behaviour of plant to climate change, predominant increase in temperature, drought, and e (CO_2) is necessary. Adaptation approaches will play an essential role in explaining the cereal crops issue in a changing climate change.

6.2 Grains Quality of Crops under Climate Change

To cope with environmental stress, biotic and abiotic, balanced nutrition in terms of quality and quantity of the cereal crops is the key factor (Fahad and Bano 2012; Fahad et al. 2013, 2014a, 2014b, 2015a, 2015b, 2016a, 2016b, 2016c, 2016d, 2017, 2018, 2019a, 2019b; Aziz et al. 2017b). For example, e (CO_2) will considerably be increased biomass, yield, and photosynthesis in C_3 plant, whereas a little increase will be overserved for C_4 crops. In spite of the good understanding regarding impacts of changing climate on physiological process of plants, evidence interrelated to the effect of climate change on essential micronutrients like iron, zinc, selenium, manganese, and copper is scarce. Each of these mineral elements is interrelated with different climate variables in a complex way. However, there is very limited information about their complex interaction, which produces a fruitful result contentiously.

6.2.1 Wheat

Earth CO_2 decrease both macro and micronutrient concentration like Mg, S, Ca, Fe, Zn in wheat crops regardless of crop variety and environment, a slight increase in K concentration and concentration of Cu and P was inconsistent. Additionally, crop genotypes have also a significant effect on mineral content of the grains in wheat.

6.2.2 Rice

Decrease in concentration of both macro and micronutrients like Mg, S, Ca, Fe, Zn was observed in rice. In addition, large variation in under Earth CO_2 has been observed in various genotypes of rice. According to under e (CO_2) a reduction in Zn concentration was observed under P defect condition, suggesting it might be due to low root growth due to low P concentration in soil. In contrast there is an increase in the concentration of Mg, Ca, Zn, and Mn under e (CO_2) in rice shoot. Ozone is a strong oxidizing agent, which breaks down into reactive oxygen in the apoplast. Thus, it is very harmful to cereal crops like rice. Under e (CO_2) condition milling and grain quality of rice decreased considerably (Bilqees et al. 2019). Similarly like e (CO_2), elevated ozone also has negative affect on brown rice percentage and milling and grain quality. The increase of temperature increases high temperature stress and as a result there is shorter growth period of plants, which adversely affects crop growth. Increase in temperature during grain-filling stage decreases rice grain quality significantly (Fahad et al. 2014a, 2014b, 2015a, 2015b, 2016a, 2016b, 2016c, 2016d, 2017, 2018, 2019a, 2019b). Combination effect of high temperature and CO_2 and/or ozone has deteriorative effect on rice grain and milling quality.

6.2.3 Maize

Increase in temperature decreased maize yield by 1.8% to 3% for each 1% shortening of growth period. The effect of increase in temperatures on maize yield might be a result of lessening of kernel set. Lobell et al. testified that increase in temperature above 30°C could decrease by about 1%–1.7% maize kernel set per degree day. Exposure of maize to above optimal temperature could reduce maize yield under future changing climate. Increase in temperature increases senescence of the maize leaves and as a result decreases photosynthesis period. Rate of photosynthesis and water-use efficiency was decreased with increase in temperature, which leads towards yield reduction.

6.3 Impact of Different Climatic Variables on Cereals Crops

6.3.1 The Impact of Elevated CO_2 on Grain Quality

In only a few researches the effect of e (CO_2) on grain mineral content of cereal crops has been studied. Under e (CO_2) plant change nutrient metabolism at both cellular and plant levels. Generally, e (CO_2) decreases most of the macronutrients and all micronutrients in various crop species. It was observed that CO_2 decreases the concentration of macronutrients including N, P, Ca, Mg, and micronutrients including Zn, and Cu in both grain and straw. As cereals like wheat, rice, and maize are globally very important staple crops, the main focus of this chapter will be on these crops.

6.3.2 Effect of Temperature and Drought on Quality of Grain

Water supply and temperature determine the availability of the nutrients in the soil to the plants. Among various stresses, plants mostly experience temperature and water stress, which are most common in different cropping systems. Plant water-use efficiency is considerably increased under e (CO_2) and need more for the crop growth and development and on other side transpiration rate increases with higher temperature and thus adjustment of transpiration at e (CO_2) determines the crop response to e (CO_2). Despite these adjustments, the potential grain yield of C_3 crops like wheat and rice improve at e (CO_2). However, cereal crop response mostly depends upon the interaction of water supply and temperature; under water-deficit and high temperature conditions, plant growth response to e (CO_2) is reduced or even diminished. Concentration of micronutrients like Zn, Fe, Ca, and S is improved in cereal grain under high temperature and water stress condition when grown under FACE condition, suggesting that drought and higher temperature reduced the negative effect of e (CO_2) on grain quality (Table 6.1).

TABLE 6.1

Different Environmental Stresses and Their Effects on Cereal Crops

Crop	Stress	Treatment	Effect
Wheat	Temperature	30°C/25°C (day/night)	Reductions in grain size, yield, grains per spike
	Flooding	Waterlogged	Decreased leaf nitrogen content
	Drought	90, 60, and 30 of field capacity	Decreased carotenoids, Ca, K, Mg, and increased organic soluble sugars and proline
	CO_2	700 ppm	Increased test weight, TSS, and starch in seeds, and decreased protein and amino acids
	Ozone	Ambient 110 and 120 ppb	Damage to vegetative and reproductive parts and reduced yield
	SO_2	0.06 ppm	Significantly reduced plant height, leaf area, relative growth rate, biomass, and yield
Rice	Temperature	32°C (night temperature)	Reductions in yield, grain size, width, and weight, and increased spikelet sterility
	Flooding	Stagnant flooding (50 cm above soil surface)	Increased aerenchyma gas space, and declined root oxidase activity
	Drought	Mild to severe	Significant decline in yield along with spikelet fertility, reduced plant height and number of tillers
	CO_2	Ambient 1,200 μmol mol^{21}	Declined iron, zinc, vitamins, and proteins
	Ozone	Ambient 110 and120 ppb	Induction of catalase, peroxidase, superoxide dismutase ascorbic acid, glutathione reductase and phenolics
	SO_2	0.0010.008 ppm	Reduced plant height, tillers and leaves, leaf area, chlorophyll content, biomass, and yield
Maize	Temperature	35°C/27°C (day/night), 14 days	Reduced hemicelluloses and cellulose synthesis
	Flooding	Waterlogged (2 cm above soil surface)	Reduced biomass, shoot growth, increased adventitious roots
	Drought	Moderate to severe	Reduced test weight, harvest index, yield, and biomass
	CO_2	550 ppm	Decreased N and P concentrations in grain but increased K content
	Ozone	Ambient 115 and 130 ppb	Induction of ROS, phenol, flavonoids anthocyanin, antioxidative enzymes Reduced yield
	SO_2	110, 70, and 45 nL SO_2 L^{21}	Decrease in leaf protein, aspartic acid, glycine, glutamine, and arginine concentration, while glutamic acid, asparagine, and alanine contents were increased

6.4 Approaches to Enhance Micronutrient Status of Grains

In cereal grain micronutrient contents can be enhanced by adjusting a multidisciplinary method, joining both agronomic and genetic management. This approach contains investigation of adaptive traits and novel genetic resources. However, an inadequate understanding of nutrient acquisitions mechanism and its adjustment to change in climate are documented as major tasks to advance the cereal grain micronutrient status.

6.4.1 Germplasm Screening

Among the nutrient concentration in cereals, micronutrients Zn and Fe in particular are the most susceptible to climate change. In cereal crops and their uncultivated relatives a huge variation exists in terms of micronutrient concentration; at CIMMYT about 3,000 accession were studied for the Fe and Zn concentration in their grains, and several of them were found to have high micronutrient concentration, particularly Zn and Fe including einkorn, emmer, and landraces of wheat in their grains. But inappropriately, there is small genetic divergence in terms of Fe and Zn in enhanced wheat varieties. Therefore, currently, investigation concentration needs to deeply evaluate all cereals landraces (Cakmak 2008; Monasterio and Graham 2000). For wheat, it has been studied that Zn concentrations ranged between 30 and 98 mg kg^{-1} (Cakmak 2008). In rice, considerable genetic variation of Zn concentration was observed ranging from 13.5 mg to 58.4 mg kg^{-1} in brown rice. On the other side, like in wheat and rice, under e (CO_2) cereals showed wide genetic variation in term of grain micronutrients (Dietterich et al. 2015). Therefore in future, research is needed to find out promising genotypes, which are compatible with changing climate (DaMatta et al. 2010; Naher et al. 2019).

6.4.2 Management of Crops

A number of management strategies have been used to improve cereal grain quality, which includes cultural and fertilizer management to pull micronutrients into the cereal grains. Manipulation of soil pH and carbon content, and peripheral application of Fe/Zn to improve the minerals and their accessibility in soil are other management strategies (Cakmak 2009). Fertilizer and irrigation management have been testified for intensification of Zn in grains of wheat and rice (Hao et al. 2007). Judicious use of Zn fertilizer increases crop yield along with Zn concentration in plant organs including grains. Arai-Sanoh et al. (2010) and; Khan et al. (2002) anticipated the importance of appropriate agronomic practices and selecting rice genotypes, to deal with climate change. Practices like the management of soil water by mulching and conservation tillage would be more effective in ensuring crop nutrient availability. In addition, choice of appropriate planting depth, fertilizer application policies, planting dates, and new cultivars are important to sustain and improve micronutrient content in grains under changing climate in future (Cakmak 2009; Donatelli et al. 2015). Therefore, to improve cereal grain quality micronutrient application should be managed according to the current status of the soil.

6.4.3 Knowledge Gaps

Plant breeding for high concentration of nutrients in cereal crops needs a systematic knowledge of the molecular and physiological processes manipulating grain quality. Various strategies have been adopted by the crop plants to combat environmental stress (Bhattacharya 2019; Lynch and Clair 2004), but the mechanism of climate change which affects cereal grain quality has not been well understood. Thus, a proper understanding of cereal grain quality control mechanism will vastly help to articulate plant breeding approaches to improve grain quality under climate change (Graham and Welch 1996; McGrath and Lobell 2013). Collaborating effects of environmental variables on crops act in different ways from their individual effects. Up to now t only a very few research studies have been directed to estimate the effect of environmental and genotype interaction on cereal grain quality. Therefore, it is useful to study the combined interactive effect of climate variables on cereal grain quality. Additionally, molecular markers helped plant breeding to combat various environmental stresses and improve cereal grain quality. There is very less knowledge regarding identification of both molecular and physiological qualities that can be exploited to improve crop quality under climate change (Dietterich et al. 2014).

REFERENCES

Adnan M, Zahir S, Fahad S, Arif M, Mukhtar A, Imtiaz AK, Ishaq AM, Abdul B, Hidayat U, Muhammad A, Inayat-Ur R, Saud S, Muhammad ZI, Yousaf J, Amanullah Hafiz MH, Wajid N (2018) Phosphate-solubilizing bacteria nullify the antagonistic effect of soil calcification on bioavailability of phosphorus in alkaline soils. *Sci Rep* 8:4339. https://doi.org/10.1038/s41598-018-22653-7

Akram R, Turan V, Hammad HM, Ahmad S, Hussain S, Hasnain A, Maqbool MM, Rehmani MIA, Rasool A, Masood N, Mahmood F, Mubeen M, Sultana SR, Fahad S, Amanet K, Saleem M, Abbas Y, Akhtar HM, Waseem F, Murtaza R, Amin A, Zahoor SA, ul Din MS, Nasim W (2018a) Fate of organic and inorganic pollutants in paddy soils. In: Hashmi MZ, Varma A (ed) *Environmental pollution of paddy soils, soil biology*. Springer International Publishing Ag, Cham, Switzerland, pp. 197–214.

Akram R, Turan V, Wahid A, Ijaz M, Shahid MA, Kaleem S, Hafeez A, Maqbool MM, Chaudhary HJ, Munis, MFH, Mubeen M, Sadiq N, Murtaza R, Kazmi DH, Ali S, Khan N, Sultana SR, Fahad S, Amin A, Nasim W (2018b) Paddy land pollutants and their role in climate change. In: Hashmi MZ, Varma A (ed) *Environmental pollution of paddy soils, soil biology*. Springer International Publishing Ag, Cham, Switzerland, pp. 113–124.

Ali S, Xua S, Ahmad I, Jia Q, Ma X, Ullah H, Alam M, Adnan M, Daur I, Ren X, Cai T, Zhang J, Jia Z (2018) Tillage and deficit irrigation strategies to improve winter wheat production through regulating root development under simulated rainfall conditions. *Agric Water Manag* 209:44–54

Anandan AR, Eswaran G, Prakash R (2011) Genotypic variation and relationships between quality traits and trace elements in traditional and improved rice (*Oryza sativa* L.) genotypes. *J Food Sci* 76(4):122–130.

Arai-Sanoh YI, Tsutomu O, Akihiro K, et al. (2010) Effects of soil temperature on growth and root function in rice. *J Plant Prod Sci* 13(3):235–242.

Arif M, Shah T, Ilyas M, Ahmad W, Mian AA, Jadoon MA, Adnan M (2015) Effect of organic manures and their levels on weeds density and maize yield. *Pak J Weed Sci Res* 21(4):517–522.

Aziz K, Daniel KYT, Fazal M, Muhammad ZA, Farooq S, Fan W, Fahad S, Ruiyang Z (2017a) Nitrogen nutrition in cotton and control strategies for greenhouse gas emissions: a review. *Environ Sci Pollut Res* 24:23471–23487. https://doi.org/10.1007/s11356-017-0131-y

Aziz K, Daniel KYT, Muhammad ZA, Honghai L, Shahbaz AT, Mir A, Fahad S (2017b) Nitrogen fertility and abiotic stresses management in cotton crop: a review. *Environ Sci Pollut Res* 24:14551–14566. https://doi.org/10.1007/s11356-017-8920-x

Bhattacharya A (2019) Chapter 1 - Global climate change and its impact on agriculture. In: Bhattacharya A (ed) *Changing climate and resource use efficiency in plants*. Academic Press, pp. 1–50. https://doi.org/10.1016/B978-0-12-816209-5.00001-5

Bilqees R, Romman M, Subhan M, Jan S, Parvez R, Adnan M, Ikram M, Iqbal S (2019) Cardiac function profiling and antidiabetic effect of *Vigna radiata* in Alloxan monohydrate induced diabetic rabbits. *Pure Appl Biol* 9(1):390–395.

Bouis HE (2003) Micronutrient fortification of plants through plant breeding: can it improve nutrition in man at low cost? *Proc Nutr Soc* 62(2):403–411.

Cakmak I (2008) Enrichment of cereal grains with zinc: agronomic or genetic biofortification? *Plant Soil* 302(12):1–17.

Cakmak I (2009) Enrichment of fertilizers with zinc: an excellent investment for humanity and crop production in India. *J Trace Ele Med Bio* 23(4):281–289.

DaMatta FMG, Adriana A, Bruna C et al. (2010) Impacts of climate changes on crop physiology and food quality. *J Food Res Intern* 43(7):1814–1823.

Depeng W, Fahad S, Saud S, Muhammad K, Aziz K, Mohammad NK, Hafiz MH, Wajid N (2018) Morphological acclimation to agronomic manipulation in leaf dispersion and orientation to promote "Ideotype" breeding: Evidence from 3D visual modeling of "super" rice (*Oryza sativa* L.). *Plant Physiol Biochem*. https://doi.org/10.1016/j.plaphy.2018.11.010

Dietterich LH, Zanobetti A, Kloog I, et al. (2015) Impacts of elevated atmospheric CO_2 on nutrient content of important food crops. *J Sci Data* 2:1–6.

Dietterich LHZ, Antonella K, Itai H, et al. (2014) Increasing CO_2 threatens human nutrition. *J Sci Data* 2:1–8.

Donatelli MS, Amit KD, Gregory N, et al. (2015) Climate change impact and potential adaptation strategies under alternate realizations of climate scenarios for three major crops in Europe. *J Envir Res Let* 10(7):50–55.

Drake BG, Gonzàlez-Meler MA, Long SP (1997) More efficient plants: a consequence of rising atmospheric CO_2? *J Ann Rev Plant Bio* 48(1):609–639.

Erice G, Sanz-Sáez Á, González-Torralba J, et al. (2019) Impact of elevated CO_2 and drought on yield and quality traits of a historical (Blanqueta) and a modern (Sula) durum wheat. *J Cer Sci* 87:194–201.

Fahad S, Bajwa AA, Nazir U, Anjum SA, Farooq A, Zohaib A, Sadia S, Nasim W, Adkins S, Saud S, Ihsan MZ, Alharby H, Wu C, Wang D, Huang J (2017) Crop production under drought and heat stress: plant responses and management options. *Front Plant Sci* 8:1147. https://doi.org/10.3389/fpls.2017.01147

Fahad S, Bano A (2012) Effect of salicylic acid on physiological and biochemical characterization of maize grown in saline area. *Pak J Bot* 44:1433–1438.

Fahad S, Chen Y, Saud S, Wang K, Xiong D, Chen C, Wu C, Shah F, Nie L, Huang J (2013) Ultraviolet radiation effect on photosynthetic pigments, biochemical attributes, antioxidant enzyme activity and hormonal contents of wheat. *J Food Agri Environ* 11(3&4):1635–1641

Fahad S, Hussain S, Bano A, Saud S, Hassan S, Shan D, Khan FA, Khan F, Chen Y, Wu C, Tabassum MA, Chun MX, Afzal M, Jan A, Jan MT, Huang J (2014a) Potential role of phytohormones and plant growth-promoting rhizobacteria in abiotic stresses: consequences for changing environment. *Environ Sci Pollut Res* 22(7):4907–4921. https://doi.org/10.1007/s11356-014-3754-2

Fahad S, Hussain S, Matloob A, Khan FA, Khaliq A, Saud S, Hassan S, Shan D, Khan F, Ullah N, Faiq M, Khan MR, Tareen AK, Khan A, Ullah A, Ullah N, Huang J (2014b) Phytohormones and plant responses to salinity stress: A review. *Plant Growth Regul* 75(2):391–404. https://doi.org/10.1007/s10725-014-0013-y

Fahad S, Hussain S, Saud S, Hassan S, Chauhan BS, Khan F, et al. (2016a) Responses of rapid viscoanalyzer profile and other rice grain qualities to exogenously applied plant growth regulators under high day and high night temperatures. *PLoS One* 11(7):e0159590. https://doi.org/10.1371/journal.pone.0159590

Fahad S, Hussain S, Saud S, Hassan S, Ihsan Z, Shah AN, Wu C, Yousaf M, Nasim W, Alharby H, Alghabari F, Huang J (2016c) Exogenously applied plant growth regulators enhance the morphophysiological growth and yield of rice under high temperature. *Front Plant Sci* 7:1250. https://doi.org/10.3389/fpls.2016.01250

Fahad S, Hussain S, Saud S, Hassan S, Tanveer M, Ihsan MZ, Shah AN, Ullah A, Nasrullah KF, Ullah S, Alharby HNW, Wu C, Huang J (2016d) A combined application of biochar and phosphorus alleviates heat-induced adversities on physiological, agronomical and quality attributes of rice. *Plant Physiol Biochem* 103:191–198.

Fahad S, Hussain S, Saud S, Khan F, Hassan S, Jr, Nasim W, Arif M, Wang F, Huang J (2016b) Exogenously applied plant growth regulators affect heat-stressed rice pollens. *J Agron Crop Sci* 202:139–150.

Fahad S, Hussain S, Saud S, Tanveer M, Bajwa AA, Hassan S, Shah AN, Ullah A, Wu C, Khan FA, Shah F, Ullah S, Chen Y, Huang J (2015a) A biochar application protects rice pollen from high-temperature stress. *Plant Physiol Biochem* 96:281–287.

Fahad S, Muhammad ZI, Abdul K, Ihsanullah D, Saud S, Saleh A, Wajid N, Muhammad A, Imtiaz AK, Chao W, Depeng W, Jianliang H (2018) Consequences of high temperature under changing climate optima for rice pollen characteristics-concepts and perspectives. *Archiv Agron Soil Sci.* https://doi.org/10.1080/03650340.2018.1443213

Fahad S, Nie L, Chen Y, Wu C, Xiong D, Saud S, Hongyan L, Cui K, Huang J (2015b) Crop plant hormones and environmental stress. *Sustain Agric Rev* 15:371–400.

Fahad S, Adnan M, Hassan S, Saud S, Hussain S, Wu C, Wang D, Hakeem KR, Alharby HF, Turan V, Khan MA, Huang J (2019b) Rice responses and tolerance to high temperature. In: Hasanuzzaman M, Fujita M, Nahar K, Biswas JK (ed) *Advances in rice research for abiotic stress tolerance*. Woodhead Publ Ltd, Abington, Cambridge, UK, pp. 201–224.

Fahad S, Rehman A, Shahzad B, Tanveer M, Saud S, Kamran M, Ihtisham M, Khan SU, Turan V, Rahman MHU (2019a) Rice responses and tolerance to metal/metalloid toxicity. In: Hasanuzzaman M, Fujita M, Nahar K, Biswas JK (ed) *Advances in rice research for abiotic stress tolerance*. Woodhead Publ Ltd, Abington, Cambridge, UK, pp. 299–312.

Graham RD, Welch RM (1996) *Breeding for staple food crops with high micronutrient density*, Vol. 3. Intl Food Policy Res Inst.

Habib Ur R, Ashfaq A, Aftab W, Manzoor H, Fahd R, Wajid I, Md. Aminul I, Vakhtang S, Muhammad A, Asmat U, Abdul W, Syeda RS, Shah S, Shahbaz K, Fahad S, Manzoor H, Saddam H, Wajid N (2017) Application of CSM-CROPGRO-Cotton model for cultivars and optimum planting dates: Evaluation in changing semi-arid climate. *Field Crops Res.* http://dx.doi.org/10.1016/j.fcr.2017.07.007

Hafiz MH, Muhammad A, Farhat A, Hafiz FB, Saeed AQ, Muhammad M, Fahad S, Muhammad A (2019) Environmental factors affecting the frequency of road traffic accidents: a case study of sub-urban area of Pakistan. *Environ Sci Pollut Res.* https://doi.org/10.1007/s11356-019-04752-8

Hafiz MH, Wajid F, Farhat A, Fahad S, Shafqat S, Wajid N, Hafiz FB (2016) Maize plant nitrogen uptake dynamics at limited irrigation water and nitrogen. *Environ Sci Pollut Res* 24(3):2549–2557. https://doi.org/10.1007/s11356-016-8031-0

Hao HL, Wei YZ, Yang XE, et al. (2007) Effects of different nitrogen fertilizer levels on Fe, Mn, Cu and Zn concentrations in shoot and grain quality in rice (*Oryza sativa*). *J Rice Sci* 14(4):289–294.

Hasegawa T, Sakai H, Tokida T, et al. (2013) Rice cultivar responses to elevated CO_2 at two free-air CO_2 enrichment (FACE) sites in Japan. *J Funct Plant Bio* 40(2):148–159.

Högy P, Wieser H, Köhler P, et al. (2009) Does elevated atmospheric CO_2 allow for sufficient wheat grain quality in the future? *J App Bot Food Qua* 82(2):114–121.

Ilahi H, Wahid F, Ullah R, et al. (2020) Evaluation of bacterial and fertility status of Rawalpindi cultivated soil with special emphasis on fluoride content. *Int J Agric Environ Res* 6(3):177–184.

Imtiaz M, Rashid A, Khan P, et al. (2010) The role of micronutrients in crop production and human health. *Pak J Bot* 42(4):2565–2578.

Kamarn M, Wenwen C, Irshad A, Xiangping M, Xudong Z, Wennan S, Junzhi C, Shakeel A, Fahad S, Qingfang H, Tiening L (2017) Effect of paclobutrazol, a potential growth regulator on stalk mechanical strength, lignin accumulation and its relation with lodging resistance of maize. *Plant Growth Regul* 84:317–332. https://doi.org/10.1007/s10725-017-0342-8

Khan MU, Qasim M, Jamil M (2002) Response of rice to zinc fertilizer in calcareous soils of DI Khan. *Asian J Plant Sci* 1:1–2.

Lynch JP, Clair SBS (2004) Mineral stress: the missing link in understanding how global climate change will affect plants in real world soils. *Field Crops Res* 90(1):101–115.

McGrath JM, Lobell DB (2013) Reduction of transpiration and altered nutrient allocation contribute to nutrient decline of crops grown in elevated CO_2 concentrations. *Plant Cell Envi Behav* 36(3):697–705.

Monasterio I, Graham RD (2000) Breeding for trace minerals in wheat. *Food Nut Bul* 21(4):392–396.

Muhammad Z, Abdul MK, Abdul MS, Kenneth BM, Muhammad S, Shahen S, Ibadullah J, Fahad S (2019) Performance of *Aeluropus lagopoides* (mangrove grass) ecotypes, a potential turfgrass, under high saline conditions. *Environ Sci Pollut Res.* https://doi.org/10.1007/s11356-019-04838-3

Nadeem M, Anjum FM, Amir RM, et al. (2010) An overview of anti-nutritional factors in cereal grains with special reference to wheat-A review. *Pak J Food Sci* 20(1–4):54–61.

Naher UA, Ahmed M, Sarkar MIU, et al. (2019) Fertilizer management strategies for sustainable rice production. In: *Organic farming*. Elsevier, pp. 251–267.

Nakandalage N, Seneweera S (2018) Chapter 12 - Micronutrients use efficiency of crop-plants under changing climate. In: Hossain MA, Kamiya T, Burritt DJ, Phan Tran L-S, Fujiwara T (eds) *Plant micronutrient use efficiency*. Academic Press, pp. 209–224. https://doi.org/10.1016/B978-0-12-812104-7.00015-0

Qamar-Uz Z, Zubair A, Muhammad Y, Muhammad ZI, Abdul K, Fahad S, Safder B, Ramzani PMA, Muhammad N (2017) Zinc biofortification in rice: leveraging agriculture to moderate hidden hunger in developing countries. *Arch Agron Soil Sci* 64:147–161. https://doi.org/10.1080/03650340.2017.1338343

Sajjad H, Muhammad M, Ashfaq A, Waseem A, Hafiz MH, Mazhar A, Nasir M, Asad A, Hafiz UF, Syeda RS, Fahad S, Depeng W, Wajid N (2019) Using GIS tools to detect the land use/land cover changes during forty years in Lodhran district of Pakistan. *Environ Sci Pollut Res.* https://doi.org/10.1007/s11356-019-06072-3

Saud S, Chen Y, Fahad S, Hussain S, Na L, Xin L, Alhussien SA (2016) Silicate application increases the photosynthesis and its associated metabolic activities in Kentucky bluegrass under drought stress and post-drought recovery. *Environ Sci Pollut Res* 23(17):17647–17655. https://doi.org/10.1007/s11356-016-6957-x

Saud S, Chen Y, Long B, Fahad S, Sadiq A (2013) The different impact on the growth of cool season turf grass under the various conditions on salinity and drought stress. *Int J Agric Sci Res* 3:77–84.

Saud S, Fahad S, Yajun C, Ihsan MZ, Hammad HM, Nasim W, Amanullah Jr, Arif M and Alharby H (2017) Effects of nitrogen supply on water stress and recovery mechanisms in kentucky bluegrass plants. *Front Plant Sci* 8:983. https://doi.org/10.3389/fpls.2017.00983

Saud S, Li X, Chen Y, Zhang L, Fahad S, Hussain S, Sadiq A, Chen Y (2014) Silicon application increases drought tolerance of Kentucky bluegrass by improving plant water relations and morph physiological functions. *SciWorld J* 2014:1–10. https://doi.org/10.1155/2014/368694

Seneweera S, Blakeney A, Milham P (1996) Influence of rising atmospheric CO_2 and phosphorus nutrition on the grain yield and quality of rice (*Oryza sativa* cv. Jarrah). *Cereal Chem* 73(2):239–243.

Shah F, Lixiao N, Kehui C, Tariq S, Wei W, Chang C, Liyang Z, Farhan A, Fahad S, Huang J (2013) Rice grain yield and component responses to near 2°C of warming. *Field Crop Res* 157:98–110.

Sinclair T (1992) Mineral nutrition and plant growth response to climate change. *J Exp Bot* 43(8):1141–1146.

Tausz-Posch S, Seneweera S, Norton RM, et al. (2012) Can a wheat cultivar with high transpiration efficiency maintain its yield advantage over a near-isogenic cultivar under elevated CO_2? *Field Crops Res* 133:160–166.

Thilakarathne CL, Tausz-Posch S, Cane K, et al. (2015) Intraspecific variation in leaf growth of wheat (*Triticum aestivum*) under Australian Grain Free Air CO_2 Enrichment (AGFACE): is it regulated through carbon and/or nitrogen supply? *Funct Plant Biol* 42(3):299–308.

Wajid N, Ashfaq A, Asad A, Muhammad T, Muhammad A, Muhammad S, Khawar J, Ghulam MS, Syeda RS, Hafiz MH, Muhammad IAR, Muhammad ZH, Rahman MHR, Veysel T, Fahad S, Suad S, Aziz K, Shahzad A (2017) Radiation efficiency and nitrogen fertilizer impacts on sunflower crop in contrasting environments of Punjab. *Pakistan Environ Sci Pollut Res* 25:1822–1836. https://doi.org/10.1007/s11356-017-0592-z

Wang KM, Wu JG, Li G, et al. (2011) Distribution of phytic acid and mineral elements in three indica rice (*Oryza sativa* L.) cultivars. *J Cereal Sci* 54(1):116–121.

Yang Z, Zhang Z, Zhang T, Fahad S, Cui K, Nie L, Peng S, Huang J (2017) The effect of season-long temperature increases on rice cultivars grown in the central and southern regions of China. *Front Plant Sci* 8:1908. https://doi.org/10.3389/fpls.2017.01908

Zahida Z, Hafiz FB, Zulfiqar AS, Ghulam MS, Fahad S, Muhammad RA, Hafiz MH, Wajid N, Muhammad S (2017) Effect of water management and silicon on germination, growth, phosphorus and arsenic uptake in rice. *Ecotoxicol Environ Saf* 144:11–18.

7

Abiotic Stress in Plants and Metabolic Responses

Muhammad Hussaan, Muhammad Tariq Javed, and Muhammad Sohail Akram
Government College University, Faisalabad, Pakistan

Muhammad Hamzah Saleem
Hauzhong Agricultural University, Wuhan, China

Hassan Javed Chaudhary
Quaid-i-Azam University, Islamabad, Pakistan

CONTENTS

7.1 Introduction

In recent years, abiotic stresses have become a global constraint, which affect plant growth and crop productivity worldwide. These are expected to increase with the increasing world population, industrial revolution, global warming, and climatic changes (Kumari et al. 2019). Abiotic stresses include

DOI: 10.1201/9781003286417-7

heavy metals, heat, cold, salinity, drought, waterlogging, nutrient imbalance, ultraviolet radiation, etc., which are proven major constraints for global crop production. According to estimates, almost 50% of crop yield reduction is the outcome of abiotic stress. Abiotic stresses are negatively affecting the plant development, growth, and yield by altering the plant physiology and biochemical processes (Mahajan et al. 2020; Chaki et al. 2020; Hasanuzzaman et al. 2020; Deepranjan et al. 2021; Haider et al. 2021; Huang Li et al. 2021; Ikram et al. 2021; Jabborova et al. 2021; Khadim et al. 2021a, 2021b; Manzer et al. 2021; Muzammal et al. 2021; Abdul et al. 2021a, 2021b; Ashfaq et al. 2021; Amjad et al. 2021; Atif et al. 2021; Athar et al. 2021; Adnan et al. 2018a, 2018b, 2019; Akram et al. 2018a, 2018b; Aziz et al. 2017a, 2017b; Chang et al. 2021; Chen et al. 2021; Emre et al. 2021; Habib et al. 2017; Hafiz et al. 2016, 2019; Ghulam et al. 2021; Guofu et al. 2021; Hafeez et al. 2021; Khan et al. 2021; Kamarn et al. 2017; Muhammad et al. 2019; Safi et al. 2021; Sajjad et al. 2019). Abiotic stress cause disruption in water uptake, gas exchange, protein synthesis, nutrients acquisition (Saud et al. 2013, 2014, 2016, 2017, 2020; Shah et al. 2013; Qamar et al. 2017; Hamza et al. 2021; Irfan et al. 2021; Wajid et al. 2017; Yang et al. 2017; Zahida et al. 2017; Depeng et al. 2018; Hussain et al. 2020; Hafiz et al. 2020a, 2020b; Shafi et al. 2020; Wahid et al. 2020; Subhan et al. 2020; Zafar-ul-Hye et al. 2020a, 2020b; Zafar-ul-Hye et al. 2021; Adnan et al. 2020; Ilyas et al. 2020; Saleem et al. 2020a, 2020b, 2020c; Rehman et al. 2020; Farhat et al. 2020; Wu et al. 2020; Mubeen et al. 2020; Farhana et al. 2020; Jan et al. 2019; Wu et al. 2019; Ahmad et al. 2019; Baseer et al. 2019; Hafiz et al. 2018; Tariq et al. 2018). These stresses also decreased the photosynthetic efficiency probably due to negative impact on the plant pigments, chlorophyll synthesis, electron transport mechanism, photosystem performance, and gas exchange parameters (Sharma et al. 2020). Furthermore, these stresses led to increase in the production of reactive oxygen species (singlet oxygen, hydrogen peroxide, hydroxyl radicals, superoxide radicals) which damage a number of important macromolecules such as protein, lipid, nucleic acid and finally cause cell death (Chaki et al. 2020). Abiotic stresses affect the plant yields and fertility of soil along with accumulation of various harmful heavy metals and elements in food chains which are crucial for human survival. These are affecting the biotic organization, as well as leading to changing the chemistry of ecological environment (Cai et al. 2019; Nadarajah 2020).

In response to abiotic stresses, plants protect themselves by activating special mechanisms against these stresses (Fahad and Bano 2012; Fahad et al. 2013, 2014a, 2014b, 2015a, 2015b, 2016a, 2016b, 2016c, 2016d, 2017, 2018a, 2018b, 2019a, 2019b, 2020, 2021a, 2021b, 2021c, 2021d, 2021e, 2021f; Hesham and Fahad 2020). They can enhance biosynthesis and accumulation of various organic solutes, alkaloids, terpenoids, and activate antioxidant defence mechanism. Accumulation of antioxidant molecules and organic solutes protect plants from abiotic stresses by maintaining the osmotic strength in vacuole and environment. Plants protect themselves from reactive oxygen species (ROS)–mediated oxidative damage by scavenging through activating enzymatic and nonenzymatic antioxidants. Furthermore, plants also activate the phytohormone response, which acts as signalling molecules as well as synthesis of organic solutes (Mansour and Salama 2020; Mahajan et al. 2020; Khare et al. 2020; Iqra et al. 2020; Akbar et al. 2020; Mahar et al. 2020; Noor et al. 2020; Bayram et al. 2020; Amanullah and Fahad 2017, 2018a, 2018b; Amanullah et al. 2020, 2021; Rashid et al. 2020; Arif et al. 2020a, 2020b; Amir et al. 2020; Saman et al. 2020; Muhammad et al. 2020; Md Jakir and Allah 2020; Mahmood et al. 2021; Farah et al. 2020; Sadam et al. 2020; Unsar et al. 2020; Fazli et al. 2020; Md. Enamul et al. 2020; Gopakumar et al. 2020; Zia-Ur-Rehman 2020; Senol 2020; Amjad et al. 2020; Ibrar et al. 2020; Sajid et al. 2020; Muhammad et al. 2021; Sidra et al. 2021; Zahir et al. 2021; Sahrish et al. 2022).

7.2 Types of Abiotic Stresses

Plants face a wide range of abiotic stresses such as salt, drought, heavy metals, nutrient deficiency, UV light, cold, high temperature, waterlogging, ozone, etc., which severely affect the plant growth and development. This book chapter summarizes the stresses that affect the crops significantly (Hasanuzzaman et al. 2012; Fahad and Bano 2012; Fahad et al. 2013, 2014a, 2014b, 2015a, 2015b, 2016a, 2016b, 2016c, 2016d, 2017, 2018a, 2018b, 2019a, 2019b, 2020, 2021a, 2021b, 2021c, 2021d, 2021e, 2021f; Hesham and Fahad 2020).

7.2.1 Heavy Metal Stress

Soil contamination by a variety of heavy metals including Cd, Fe, Cu, Zn, Co, Ni, Hg, and Mn has become a worldwide serious environmental hazard. Sources of heavy metals are industrial effluents, agriculture practices, mining activities; some natural processes also contribute to heavy metals pollution such as soil erosion, forest fires, and volcanic eruptions (Ghori et al. 2019; Javaid et al. 2020). Among the heavy metals, Cu, Zn, and Mn are essential micronutrients for the growth of plants. These are required in very small concentrations for the growth of plants, but concentration of these metals above the threshold capacity is toxic for the plant's metabolic processes (Ghori et al. 2019; Choudhary et al. 2020; Saleem et al. 2020). Other heavy metals like cadmium, arsenic, chromium, lead, nickel, etc. are nonbiodegradable, nonessential, and very toxic for plants' growth and soil microorganisms. These affect the plant sugar, photosynthetic pigments, protein and nonprotein molecules leading to plant cell death (Tang et al. 2019; Li et al. 2019; Mahajan et al. 2020).

7.2.2 Salinity Stress

Salt stress is one of the major limiting factors for agricultural production on worldwide scale (Riyazuddin et al. 2020). According to the previous studies, salinity stress affects almost 930 million ha of arable land, which represents 6% of the total world land area (Munns and Tester 2008). Salt stress adversely affects the growth and yield of plants by disturbing the nutrients and water balance through higher accumulation of Na^+ and Cl^- ions. Moreover, salt stress also reduced the photosynthesis process by decreasing the stomatal conductance, transpiration rate, and leaf area (Majeed and Siyyar 2020). Salt stress also effects the plant growth and physiological parameters by increasing the production of higher concentration of reactive oxygen species (ROS) in plant tissues. ROS-induced oxidative stress impairs the cell membrane and affects the proteins of cell organelles, mostly chloroplast, mitochondria, peroxisome (Mushtaq et al. 2020).

7.2.3 Drought Stress

Drought stress is a major contributor of abiotic stress, which hampers global crop production. At global scale, 45% of land is occupied by drought stress (Cairns et al. 2012). Drought stress stops the growth and development process by altering the plant's biological and metabolic function (Mahmood et al. 2020). Drought stress exerts significant effect on the photosynthetic process, and it reduces the photosynthesis by decreasing the CO_2 assimilation. Prolonged drought stress condition lowers the CO_2 concentration in chloroplast, which as a result deactivates the Rubisco activities and reduces the nitrate reductase, sucrose phosphate synthesis, and regeneration of ribulose biphosphate (Reddy et al. 2004). Lower levels of CO_2 in plants due to drought stress also stimulate the processes of photorespiration, which enhance the production of reactive oxygen species (ROS) in the form of H_2O_2 (Noctor et al. 2014).

7.2.4 Heat Stress

Climatic changes that lead to global warming and increase CO_2 concentration in the atmosphere are major causes of heat stress. When temperature rises from 10°C to 15°C from ambient temperature, it is called *heat stress* and *heat shock* (Wahid et al. 2007). Heat stress negatively affects the plants' growth, especially seed germination, photosynthetic process, cell turgidity, grain-filling stage, pollen development, water-use efficiency, and alters the enzyme activities and yield of plant (Akter and Islam 2017; Ihsan et al. 2019). Under heat stress, photosystem-II (P-II) is the most sensitive part because thermal stress also replaces the 2 Mn from the P-II which has 4 Mn atoms per photosystem-II for proper functioning. P-II reaction centre, light capturing centre, and oxygen evolving complex are the primary sites of the plants that are highly damaged by heat stress (Tyystjarvi 2008; Chen et al. 2012; Mathur et al. 2014). Heat stress also changes the structure of the thylakoid membrane, more specifically destacking the thylakoid membrane (Mathur et al. 2014). Heat stress also affects the carbon assimilation process, and elevated temperature also reduced the C_3 cycle by decreasing the enzyme activity (Weis 1981; Sharkey 2005).

Elevated temperature initiates the oxygenase reaction processes of Rubisco and stimulates the photorespiratory pathway that produced H_2O_2 as a byproduct (Sharkey 2005).

7.2.5 Cold Stress

Cold stress or chilling stress occurs in plants when subjected to low temperature from 0°C to 15°C. Cold stress has negative impact on plant morphologies, which affects growth repression and reduced plant yield (Pradhan et al. 2017). Cold stress changes the plant–water relation, protein structure, enzyme activities, nutrients uptake, destruction of plant pigments, leading to cell death. Cold stress also enhances the oxidative stress by the production of reactive oxygen species (hydrogen peroxide, superoxide, malondialdehyde) which cause visible toxicity, stunted growth, decreased chlorophyll content, water content, and antioxidants (Sohag et al. 2020). Elkelish et al. (2020) describe that chilling stress induced phenotypic deformation, shrinkage of leaf blade, wilting, and leaf curling in tomato plant. Thakur et al. (2020) stated that cold stress also caused disruption in the reproductive component, floral abortion, and development process. Cold stress also damages the cell membrane of photosystem-II, pollen germination, chlorophyll, stomatal conductance, hydration status, photo assimilation capacity, and stomatal conductance.

7.3 Metabolic Responses

It is well known that plants have a variety of metabolites which have diverse roles in plants. They are responses to abiotic stresses, which comprise of morphological, physiological, and biochemical changes through regulation of antioxidants, osmolytes, alkaloids, terpenoids, and phytohormones.

7.3.1 Reactive Oxygen Species Production

In response to abiotic stress, plants start the production of ROS (reactive oxygen species) in mitochondria, chloroplasts, and peroxisomes of plant parts, which is considered a primary response to abiotic stresses. ROS perform dual role in plants where less amount of ROS production during abiotic stresses function as signal for initiating the responses and adaptation against stresses, while elevated concentration of ROS production enhances the lipid peroxidation; protein oxidation; enzyme denaturing; oxidative damage of various biomolecules such as DNA, RNA, protein, nucleic acid, and ultimately cause cell death. In order to overcome oxidative injury, plant activates both enzymatic and nonenzymatic antioxidants (Farooq et al. 2019; Ejaz et al. 2020; Nadarajah 2020).

7.3.2 Antioxidant Defence System

Plants possess a variety of antioxidants as presented in Figure 7.1 including enzymatic (superoxide dismutase, peroxidase, catalase, monodehydroascorbate reductase, glutathione reductase, dehydroascorbate reductase, ascorbate peroxidase, glutathione reductase, guaiacol peroxidase) and nonenzymatic antioxidants (ascorbic acid, phenolics, carotenoids, α-tocopherol, flavonoids, proline, glutathione), which are upregulated to cope with hazardous effect of ROS production (Table 7.1) during abiotic stresses (Ejaz et al. 2020). All these enzymes work collectively in coordination to scavenge free radicals. Of the various ROS radicals, O_2^- and OH^- cannot be properly scavenged by the enzymatic antioxidants and their scavenging and detoxification were mostly dependent upon the nonenzymatic antioxidants. However, production of ROS under various abiotic stresses is universal and their production depends upon the type of stress and based on plants species (Uzilday et al. 2015).

The Ascorbate-Glutathione pathway, also called as *Asada–Halliwell pathway*, includes two nonenzymatic antioxidants, ascorbate (AsA) and glutathione (GSH), and four enzymatic antioxidants ascorbate peroxidase (APX), monodehydroascorbate reductase (MDHAR), dehydroascorbate reductase (DHAR), and glutathione reductase (GR), which play dynamic roles in plants against abiotic stresses by ROS detoxification and also work with defence system to save the plants. Numerous plant studies depicted that

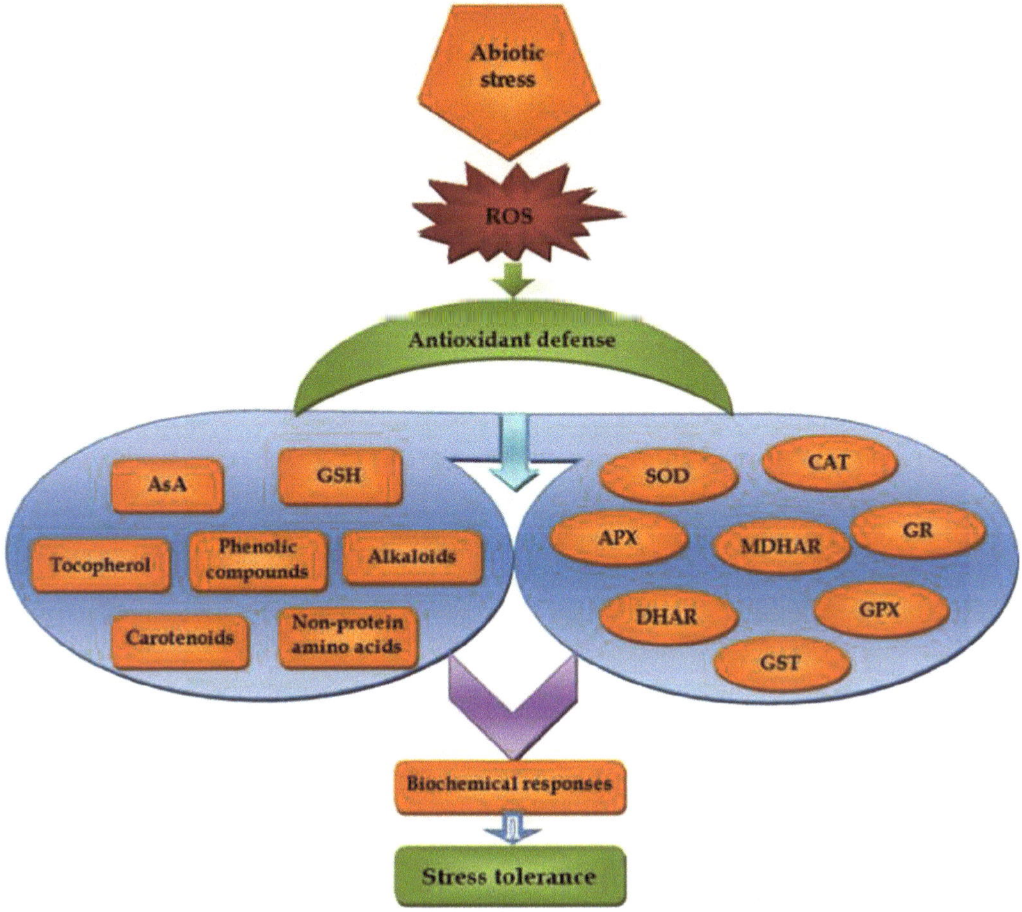

FIGURE 7.1 Abiotic stresses and major antioxidants defence system in plants.

upregulation of AsA-GSH pathway induced better tolerance against different stresses (Hasanuzzaman et al. 2019). AsA-GSH pathway is also considered as redox regulating centre of the cell. In AsA-GSH cycle, APX scavenge the hydrogen peroxide that is produced by detoxification of superoxide radical by the action of SOD and then produce H_2O molecule. During this process AsA acts as a reducing agent and is oxidized to MDHA. MDHA is usually recycled back to AsA by the action of enzyme MDHAR. MDAH produces DHA which also gets recycled and changes into AsA by the action of DHAR. For conversion of DHA into AsA, DHAR uses the GSH which is oxidized into GSSG. Glutathione reductase (GSH) is converted into GSSG by the catalytic action of GR via the process of oxidation of NADPH to NADP. The AsA-GSH cycle regulates the process of GSH:GSSG concentration, which scavenges reactive oxygen species and minimizes oxidative hazards (Banerjee and Roychoudhury 2018).

7.3.2.1 Enzymatic Antioxidants

Enzymatic antioxidants, located in various cellular parts of plants, work together for detoxification of ROS. The major enzymatic antioxidants are SOD, CAT, APX, GST, and AsA-GSH regulating enzyme. The AsA-GSH cycle is regulated by different enzymatic (APX, DHAR, MDHAR, and GR) and non-enzymatic antioxidants for detoxification of H_2O_2 and regeneration of GSH and AsA (Hasanuzzaman et al. 2012).

TABLE 7.1

Modulation of osmolytes and antioxidative defence mechanisms in various plant species in response to different abiotic stresses

Stress	Stress Conc.	Plant Name	Antioxidant Activity	References
Heavy metals	250, 500, 1000 μM Cd	Wheat	Proline, APX, GR, POD, SOD	Çatav et al. (2020)
	50, 100, 200 Cd μmol L^{-1}	Wheat	CAT, SOD, POD, proline	Rahman et al. (2020)
	0.5, 5, and 50 Cd μM	Wheat	SOD, POD, CAT, ASA, GSH	Qin et al. (2018)
	0.05, 0.2mM Boron	Wheat	SOD, CAT, POD, LOX, sugar, protein, phenol	Kaya et al. (2019)
	0.4, 10 PbmM	Wheat	SOD, APX, CAT, POX	Murtaza et al. (2019)
	150, 250, 350, 450, 600, 1000 CrμM	Castor bean	Anthocyanin, flavonoids, phenolics, ASA, protein, amino acid, proline	Qureshi et al. (2020)
	0.2, 1, 5, 8, 10 Cu mg L^{-1}	*Salvinia natans* (L)	SOD, POD, APX, CAT	Liu et al. (2019)
	1000 mM As	*Isatis cappadocica* desv	SOD, APX, GR, GST, GSH, Proline	Souri et al. (2020)
Salt	20, 40 mg/L	Maize	SOD, CAT, POD, protein, phenolics, flavonoids	Ashraf et al. (2018)
	0.1, and 0.2mM	Wheat	CAT, POD, SOD, phenolics, ASA, Proline	—
	6, 12 dms^{-1} NaCl	Bean	SOD, POD, CAT, APX, soluble sugar, proline, glycinebetaine	Farhangi-Abriz and Torabian (2017)
	100 mM	Okra	Phenolics, flavonoids, anthocyanin, protein, proline, amino acid, POD, CAT, ASA	Ashraf et al. (2019)
	0, 150 mM	Wheat	Proline, amino acid, glycinebetaine, Total soluble sugar	Shahnaz et al. (2020)
	200 mM NaCl	*Artemisia annua* L	Proline, SOD, APX, CAT, GR, carotenoids, phenolics, flavonoids	Arora et al. (2020)
Drought		Rice	SOD, POD, CAT, soluble sugar, protein	Wang et al. (2021)
		Sunflower	Carotenoids, phenolics, flavonoids, ascorbic acid, SOD, POD, CAT	Rasheed et al. (2020)
		Helianthus annuus L	Protein, ASA, phenolics, glycinebetaine, CAT, SOD, POD	Kosar et al. (2020)
		Olive	SOD, APX, POD	Denaxa et al. (2020)
		Tobacco	Sugar, amino acid, proline	—
		Soybean	APX, CAT, glycine betaine, proline	Rezayian et al. (2020)
		Maize	POD, CAT, APX, ASA, proline, phenolics, soluble sugar	Ibrahim et al. (2020)
Heat	25, 30, 35°C	Barley	SOD, POD, CAT, anthocyanin	Hussain et al. (2019)
	35,41°C	Maize	SOD, APX, POD, proline	Hussain et al. (2016)
	35°C	*Mentha piperita*	Sugar, proline, glycinebetaine, mannitol, inositol	Alhaithloul et al. (2020)
	45°C	Tomato	SOD, APX, GR, ASA, DHA, GSH	—

(*Continued*)

TABLE 7.1 (CONTINUED)

Stress	Stress Conc.	Plant Name	Antioxidant Activity	References
Cold	5.0, 1.0°C 8 h day	Rice	Proline, total soluble sugar, soluble protein	Sohag et al. (2020)
	4°C	*Calendula officinalis*	SOD, APX, GR, proline, soluble sugar, trehalose, glycinebetaine	Jan et al. (2018)
	5°C	Chickpea	Proline, glycinebetaine, trehalose	Thakur et al. (2020)
	10°C	*Capsella bursa pastoris*	SOD, CAT, POD, APX, GR, proline, amino acid, glycinebetaine, trehalose, total sugar	Wani et al. (2018)
Waterlogging		*Sesamum indicum*	GSH, APX, MDAR, GR	Anee et al. (2019)
		Cajanus cajan L.	SOD, APX, CAT, GR	Kumutha et al. (2009)
		Citrus	SOD, CAT, APX, ASA	Hossain et al. (2009)
		Cajanus cajan	POX, CAT, APX, SOD,	Bansal and Srivastava (2012)

7.3.2.1.1 Superoxide Dismutase

From enzymatic antioxidants, superoxide dismutase (SOD) is the only enzyme that has a central role in scavenging of O_2^- by converting it to H_2O. Among ROS, O_2^- radical was mostly produced during respiration and photosynthesis process; therefore, SOD is the main defence mechanism against ROS production in respiration and photosynthetic process. SOD is considered as a first line of defence because it initially acts against O_2^- production (Uzilday et al. 2015). It is well known that O_2^- does not easily pass through phospholipid membrane due to having negative charges. So, scavenging of O_2^- needs to destroy in the same organelles where they were produced. For this purpose, SOD found in various compartments of the cell performs scavenging of O_2^- in a specific cellular part. Another basic function of the SOD is to prevent the formation of OH^- radical by the Haber-Weiss reaction. SOD has been usually involved in catabolizing the ROS through dismutation process by approaching the two molecules of O_2^-. Of these two molecules, one molecule is reduced to H_2O_2 and another one is oxidized to O_2. SOD exists in three types CU/Zn-SOD (copper/zinc superoxide dismutase), Fe-SOD (iron superoxide dismutase), Mn-SOD (manganese superoxide dismutase) due to the presence of various cofactors at their active sites (Fe, Zn, Mn, and Cu) (Mittler 2002; Das and Roychoudhury 2014).

7.3.2.1.2 Catalase

CAT (Catalase) is a heme protein containing enzymatic antioxidant, which has a special role in defence mechanism against the ROS, because CAT degrades the H_2O_2 into oxygen and water. They have high affinity for H_2O_2, and therefore have an important role in H_2O_2 elimination. Mostly these enzymes are found in the peroxisome; however, these are also activated in various compartments of the cell under abiotic stresses, and these are constantly increased under various abiotic stresses (Das and Roychoudhury 2014; Tyagi et al. 2020). Tyagi et al. (2020) also observed that TACAT gene family has three classes (I, II, and III) which are modulated under various abiotic stresses.

7.3.2.1.3 Ascorbate Peroxidase

Ascorbate peroxidase (APX) is an important part of the AsA-GSH cycle. APX is involved in the detoxification of H_2O_2 in chloroplast and cytosol by changing into the substrates DHA and H_2O by using ascorbic acid as reducing agent. APX have various isoenzymes found in various cellular parts such as stroma, mitochondria, cytosol, peroxisomes, and thylakoid (Das and Roychoudhury 2014; Pandey et al. 2017). In response to various environmental stresses, APX also increases together with other enzymatic antioxidants. Higher production of APX also stimulates the activity of SOD to overcome the biosynthesis of ROS (Sarowar et al. 2005).

7.3.2.1.4 Glutathione Peroxidase

Glutathione is a member of the family that reduces GSH into H_2O_2, and then into organic and lipid hydroxide (Noctor et al. 2002). So, among enzymes, Glutathione Peroxidase (GPX) has a principal role in plant protection against oxidative stress by repairing the membrane lipid peroxidation (Kühn and Borchert 2002). Overall, GPX has a role in H_2O_2 detoxification and also acts as an oxidative signal transducer (Miao et al. 2006).

7.3.2.1.5 Glutathione Reductase

Glutathione reductase (GR) is a part of enzymatic antioxidants, which maintains the reduced position of GSH via AsA-GSH pathway and plays an efficient role in the stabilization of sulfhydryl (–SH) group and works as a substrate for glutathione-*S*-transferases. These are mostly found in the chloroplast; however, they are found in small concentrations in the cytosol, mitochondria, and peroxisomes. Glutathione play a vital role in the response against oxidative stress by stabilizing the intracellular glutathione level mostly in reduced state. Another important function is to work as antioxidant which scavenges reactive oxygen species (Yousuf et al. 2012). It is also known as glutathione disulphide reductase and acts as a flavoprotein and belongs to the family of NADPH-dependent oxidoreductases. Glutathione reductase catalyses the reduction of GSSG (glutathione disulphide) to GSH (glutathione), which has a central role in defence mechanism against ROS (Liu et al. 2020a).

7.3.2.1.6 Dehydroascorbate Reductase and Monodehydroascorbate Reductase

Dehydroascorbate reductase (DHAR) and monodehydroascorbate reductase (MDHAR) are both significant in regulating the ascorbic acid level and its redox state during abiotic stresses (Eltayeb et al. 2007). DHAR has an important role in AsA recycling system which has function to regenerate the ascorbic acid from the DHA and regulates the ascorbic acid redox state. It is maintaining the plant against abiotic stresses by the production of signalling molecules. MDHAR has a role in regeneration of ascorbic acid (ASA), which is a very important step to regulate antioxidative activity against abiotic stresses. Enhanced activities of DHAR and the MDHAR were reported under various abiotic stresses (Mittova et al. 2000; Hasanuzzaman et al. 2011).

7.3.2.2 Nonenzymatic Antioxidants

7.3.2.2.1 Ascorbic Acid

Ascorbic acid is a water-soluble antioxidant and works as cofactor for many enzymes, redox buffer, and signalling molecules. It also has a role in the growth of plants by regulation of cell division, seed germination, photosynthesis, respiration, and biosynthesis of antioxidants. It is the most powerful antioxidant and has the capability to denote electron in many nonenzymatic and enzymatic reactions. Ascorbic acid under abiotic stress neutralizes the ROS such as $^1O_2$, O_2^-, and OH^-. It also upregulates the APX activity by working as a cofactor of APX. Since, ascorbic acid is present in the chloroplast, peroxisomes, and cytosol, it plays a promising role in scavenging mechanism of H_2O_2 through interaction with APX, which is mainly recycled by AsA-GSH cycle. Plants maintain higher level of ascorbic acid contents under environmental stress to cope with stress-mediated ROS production (Hameed et al. 2016; Akram et al. 2017; Ali et al. 2019; Acosta-Gamboa et al. 2020). Ascorbic acid is the coenzyme of gibberellin 2-oxidases, aminocyclopropane carboxylate, 9-*cis*-epoxycarotenoid dioxygenase which are involved in the biosynthesis of gibberellic acid, ethylene, and ABA hormones (Pastori et al. 2003). In addition, ascorbic acid also acts as cofactor of salicylic acid-3 hydrolase which function in salicylic acid catabolism (Kawai et al. 2014). Hence, ascorbic acid participates in regulation of growth and defence mechanisms against various abiotic and biotic stresses through its positive interaction with phytohormones (Veljović-Jovanović et al. 2017).

7.3.2.2.2 Glutathione

Glutathione (GSH) is a thiol compound having L-glutamic acid, glycine, and L-cysteine. GSH is found in plants, microorganisms, and animals that are considered to possess more functional components against radical scavenger. GSH is involved in various plant cellular processes and provides additional protection against the toxic hazards of ROS (Monneveux et al. 2006; Petropoulos et al. 2008). GSH also participates in different biological processes, i.e. protein activities, DNA biosynthesis, and stabilization of cellular membrane. During oxidative stress, main function of GSH is to control the H_2O_2 production by the oxidation and reduction mechanism of glutathione and GSHG. GSH also has role in redox signalling (Pastori and Foyer 2002). GSH is the tripeptide which is mostly found in the chloroplast, cytosol, mitochondria, vacuole, and ER (Sankar et al. 2007). GSH are nonprotein thiols which are nucleophilic in nature and most important for the formation of mercaptide bonds with different metals for reacting with various selective electrophiles. It is considered an important nonenzymatic antioxidant due to its high-water solubility and stability (Samarah 2005).

7.3.2.2.3 Phenolic Compounds

Phenolic compounds are the most imperative class of plant metabolites which have powerful antioxidant properties and help to counteract the harmful effect of abiotic stresses. Synthesis and accumulation of phenolic compounds such as flavonoids, anthocyanin, phenolic acid, and flavanols under abiotic stress are enhanced under stressed condition compared to nonstressed one to protect plant cell from hostile conditions. Plant under various abiotic stresses stimulate the synthesis of phenolic compounds by overexpression of the different genes (PAL, CHI, CH4, CHS, 4CL, F3H, DFR, FNS, FLS, IFS, ANS, IFR, and UFGT, which are involved in biosynthesis of phenolic compounds) (Sharma et al. 2019a). Some phenolic compounds such as caftaric acid, caffeic acid, gallic compounds, cinnamylmalic acid, vanillic acid, and ferulic acid are synthesized under saline condition in response to ROS (Dkhil and Denden 2012). To combat chilling stress, plants accumulate phenolic compounds such as lignin and suberin which ultimately increased resistance against chilling stress (Griffith and Yaish 2004). Zhou et al. (2018) described that enhanced biosynthesis of phenolic compounds under chilling stress may be due to increased overexpression of CAD (cinnamylalcohol dehydrogenase), PAL (phenylalanine ammonia lyase), and HCT (hydroxycinnamoyl transferase). Under heavy metal stress, phenolics enhance the uptake of nutrient contents by the chelation of metallic ions and accelerate the absorption of nutrient elements such as Ca, K, Mg, Fe, Mn, Zn from soil (Seneviratne and Jayasinghearachchi 2003).

7.3.2.2.3.1 Flavonoids

Flavonoids are the important participants of secondary metabolites and perform different roles in plants including UV protection mechanism, pollination, and auxin transport. In response to biotic and abiotic stresses, flavonoids biosynthesis genes that are activated to enhance the flavonoids accumulation in plants, especially during heavy metal stress, nutrient deficiency, and drought stress (Baskar et al. 2018). Among plant antioxidants which work against oxidative stress, flavonoids are the main antioxidants for savaging of ROS and inhibiting the ROS production enzymes such as xanthine oxidase, monooxygenase, lipoxygenase, and cyclooxygenase. Flavonoids also have a role in chelation of heavy metal ions in order to recycle the various antioxidants. Antioxidative function carried out by flavonoids due to possession of functional groups also assist the scavenging of ROS and metal chelation (Kumar and Pandey 2013; Mierziak et al. 2014). Some studies revealed that higher accumulation of flavonoids in leaf mesophyll cells, particularly in chloroplast and vacuole, indicated their important role as ROS quencher. Their production in chloroplast has an important role in stabilization of out membrane of chloroplast by quenching the singlet oxygen (Agati et al. 2012). Bilger et al. (2007) reported that flavonoids accumulation also increased in the epidermal cell in response to cold stress. Similarly, in *A. thaliana,* positive link was observed between the concentration of flavonoids and cold stress responses (Korn et al. 2008).

7.3.2.2.3.2 Anthocyanin

Anthocyanin, found in chloroplast, is a derivative of flavonoids originated from phenylpropanoid pathway and also possess antioxidant activity (Kähkönen and Heinonen 2003; Kovinich et al. 2015). Anthocyanins are low molecular weight antioxidant molecules that provide protection to plant against abiotic stresses including heavy metal–mediated oxidative stress (Stambulska et al. 2018). Higher synthesis of anthocyanin contents in response to abiotic stress is an adaptive mechanism in plants (Mahmood et al. 2016). Previously it was well documented that higher synthesis of anthocyanin occurs in response to various abiotic stresses such as heavy metal (Qureshi et al. 2020), salt (Ali et al. 2013; Ashraf et al. 2019), heat (Hussain et al. 2019). Anthocyanin as an antioxidant also has a role in scavenging of abiotic stress through mediation of ROS production in plants (Nakabayashi et al. 2014).

7.3.2.2.4 Carotenoids

Carotenoids are lipid-soluble antioxidants that have diverse function in plants. Under abiotic stress conditions they protect the plant from oxidative damage. Another important function is that they protect the photosynthetic apparatus of plants by scavenging of ROS, which are produced in chloroplast during photosynthetic processes (Collins 2001). Carotenoids are also important for the PS1 and promote the integrity of light harvesting complex. They also perform structural role by protecting the thylakoid membrane in plants under stressed conditions (Niyogi et al. 2001). Carotenoids also protect plants by scavenging the $^1O_2$ (singlet oxygen) and lipid radicals. Carotenoids comprise xanthophyll (zeaxanthin, lutein) and carotenes (β-carotene) (Pospíšil and Prasad 2014). Rai et al. (2005) reported that carotenoid content increased in plant to counteract Cd stress. It also protects the plant from the high temperature stress and oxidative stress (Loreto et al. 2004).

7.3.3 Tocopherol

Tocopherols are lipid-soluble powerful antioxidants, which have higher substantial role in the degradation of ROS together with glutathione and ascorbate (Ma et al. 2020). Tocopherols are abundantly present in the membranes of plants and accomplish both nonenzymatic and enzymatic functions. Tocopherols are localized in chloroplast thylakoid membrane. At present, plants have four various kinds of tocopherols such as (α, β, γ, and δ). Under abiotic stresses, tocopherols have a role in maintaining of redox state of plant and to detoxify the ROS especially singlet oxygen (Čamagajevac et al. 2018). Among the various tocopherols, α-tocopherol has greater ability to scavenge oxidative radicals with combination of different antioxidants. Under stressed condition, it has role in protection of PS11 structure (Kamal-Eldin and Appelqvist 1996; Massacci et al. 2008). Important function of α-tocopherol is to alleviate and stabilize the plant during oxidative stress. Previously it is well documented that plant generates and enhances the level of α-tocopherol under various environmental stresses. Ali et al. (2020) also reported that application of α-tocopherol in response to water stress reduces the stress by increasing the enzymatic (SOD, POD, CAT), and nonenzymatic antioxidants and nutrient uptake (Ca, K, P, and N).

7.3.4 Osmolytes

Abiotic stresses like cold, salt, heat, and drought stress dramatically elevate the intracellular loss of water and as a result trigger the dehydration process. To cope with these type of stresses, plants synthesize and accumulate different ranges of osmoprotective compounds such as amino acid (proline, glycine betaine, lysine, arginine), polyamines (spermine, spermidine, and putrescine), and various types of carbohydrates (glucose, sucrose, fructose, trehalose, raffinose), sugar alcohols (glycerol, inositol, sorbitol, mannitol), and some organic acids. These act as osmoprotectants that protect cellular structure to minimize the harmful effect of ROS and stabilize the enzymatic activities (Salama and Mansour 2015; Rios et al. 2017; Sharma et al. 2019b). Moreover, they reduce water loss and promote osmotic adjustment (El-Bassiouny and Sadak 2015).

7.3.4.1 Amino Acid Response

Amino acids are considered as precursor of various nitrogenous compounds and building block of protein. Under stressed conditions, amino acids play an inclusive role as nonenzymatic antioxidants in plants and most probably function as signalling and regulatory molecules, as well as regulate metabolic activities (Rai 2002; Hammad et al. 2015; Shekari and Javanmardi 2017). Amino acids such as proline, arginine, and glycine also provide protection to plants by osmotic adjustments, ROS detoxification, membrane integrity, and protein-enzyme stabilization (Ashraf and Foolad 2007). Amino acid also acts as chelating agent by forming metal complexes with metal ions at root and soil interface. Theses complexes are mainly formed through amine and carboxylate groups, which significantly enhance the bioavailability of metal ions for plants (Nowack et al. 2006).

7.3.4.1.1 Proline

Accumulation of proline in plant in response to abiotic stresses makes us consider proline to be a good stress indicator. Proline works as an osmolyte that plays an important role in osmotic adjustment, maintaining cytoplasmic pH, and function as ROS scavenger. Proline also works as molecular chaperone, which prevents cell membrane damage by stabilizing protein and buffering the cellular redox potential. During stress, proline also functions as nitrogen/carbon sources and recovers the plant from stress phase (Mansour and Salama 2020). In some plants like olive and tobacco, proline boosts up the enzymes which are involved in photosynthesis and antioxidant defence system (Gupta and Huang 2014). Under stress conditions, proline stabilizes protein complex and DNA, scavenge different free radicals and act as sources of carbon and energy for plant growth (Rejeb et al. 2015; Liu et al. 2020b). Proline also works as antioxidant defence molecule, signalling molecule and during heavy metal stress proline works as metal chelator by inducing the synthesis of phytochelatins (Hayat et al. 2012). Noreen et al. (2018) reported that application of proline enhanced the Cd tolerance in wheat plant by increasing the SOD activity and remarkably reduced the ROS generation and lipid peroxidation. Proline accumulation enhances electron transport and PS-II activity which regulates the enzymatic activities. Sharma et al. (1998) reported that proline application protects the nitrate reductase and glucose-6-phosphate activity of plants from the Cd and Zn stress.

7.3.4.1.2 Arginine

Amino acid arginine has an important role in nitrogen storage and their transport in plants. During senescence and germination processes, arginine is degraded by the enzyme arginase, which induces the nitrogen mobilization from its sources (Hildebrandt et al. 2015). Decarboxylation of ornithine and arginine induces the synthesis of polyamines such as putrescine, spermidine, and spermine, which are involved in stress response mechanisms (Minocha et al. 2014).

7.3.4.1.3 Glycinebetaine

Glycinebetaine is an amino acid–derived quaternary compound that is synthesized in plant chloroplast from choline, ethanolamine, and betaine aldehyde (Krasensky and Jonak 2012). In response to various types of stresses, plants accumulate and synthesize higher concentration of glycinebetaine. Glycinebetaine improves the tolerance in plants against various abiotic stresses by stabilizing the membrane and PS-II, translation, transcription processes and boosting up the antioxidant activities, particularly catalase activity (Park et al. 2007; Huang et al. 2020). Activities of other antioxidants such as SOD, CAT, APX, DHAR, GR, GPX were also increased by the application of glycinebetaine under water stress in various crops (Hisyam et al. 2017). Exogenous application lowers the oxidative stress by decreasing the H_2O_2 concentration, which depicts that glycinebetaine has a functional role in scavenging H_2O_2. Another important function is the expression of catalase gene which is also linked to glycinebetaine indicating that glycinebetaine has an important role in H_2O_2 quenching (Park et al. 2007). Furthermore, evidence revealed that glycinebetaine has a role in ROS detoxification and enhances the glutathione content which in return maintains the glutathione redox state (Hossain and Fujita 2010).

7.3.4.2 Polyamines

Polyamines (PAs) are small, low molecular weight organic molecules including spermine, spermidine, and putrescine, derived from decarboxylation of ornithine and arginine, which are ubiquitously found in all cellular compartments (Minocha et al. 2014; Handa et al. 2018; Wang et al. 2021). The synthesis and catabolism of PAs maintain the homeostasis of cell under abiotic stresses. Higher tolerance in plant against abiotic stresses is also linked with increased enzymatic activity along with overexpression of gene which is associated with biosynthesis of polyamines. Polyamines play the role of compatible solutes, and are involved in scavenging of ROS by modulating the antioxidant activities (Shi and Chan 2014; Liu et al. 2015). Antagonistic effects of PAs against abiotic stresses are reported (Bano et al. 2020). Exogenous application of spermidine and spermine was reported to increase the Cd tolerance in wheat crop through improving water uptake, antioxidative machinery, and membrane integrity. Moreover, enhanced concentration of protein, starch, ascorbic acid, and glutathione was recorded by the application of spermidine and spermine (Rady and Hemida 2015).

7.3.4.3 Carbohydrate

Sugars, as carbohydrates, have an important role in signalling, structure, storage, and transport. Sugars as osmolytes including fructose, sucrose, raffinose, and hexoses help plants to face the various environmental stresses by maintaining its metabolic activities (Peshev et al. 2013). Sugar also acts as an antioxidant and plays an antagonistic function against the reactive oxygen species. Furthermore, sugars also have a role in the generation of NADPH for glutathione synthesis, which contributes towards ROS scavenging (Bolouri-Moghaddam et al. 2010). Sugars are also involved in many metabolic processes such as generation of metabolites, energy production, regulation of source-sink activity, and biosynthesis of biopolymers. Under stressed conditions, sugars also work as osmoprotectants and stabilize the membrane. Sugars are also having role in growth, flowering, senescence, and development under abiotic stress conditions (Sami et al. 2016).

Disaccharides trehalose and sucrose are water-soluble carbohydrates that maintain the osmotic adjustment and perform various regulatory functions in plants. Sucrose is an important sugar, which is the product of photosynthesis and participates in various growth and stress acclimation processes in plants (Keunen et al. 2013). Sucrose works as transport molecule, excellent osmotic solute, fuel for anabolic reaction, and as energy and carbon storage compound (Patzke et al. 2019). Metabolism of sugars under different abiotic stresses also promotes the upregulation of sucrose-synthesizing enzymes that activate the enzymatic antioxidants and decrease the partitioning of sucrose-starch by lowering the activity of these enzymes (Rosa et al. 2009). However, trehalose, being a nonreducing agent, helps the stabilization of lipid membrane, protein, dehydrated enzyme, and osmotic balance (Alam et al. 2014; Kosar et al. 2019). Overexpression of trehalose biosynthesis gene (OsTPP3) is also linked to tolerance of drought in rice (Jiang et al. 2019).

7.3.4.4 Sugar Alcohols

Mannitol and pinitol are significant sugar alcohols that facilitate osmotic adjustments and also maintain redox control. Mannitol has a multifunctional role in plants and protects them from abiotic stresses by working as osmoprotectant, membrane stabilizer, and radical scavenger (Seckin et al. 2009). Moreover, Seckin et al. (2009) reported that salt-sensitive wheat cultivars with mannitol have reduced the abiotic stresses by modulating the antioxidant activities such as SOD, APX, POD, CAT, and glutathione reductase. However, pinitol is obtained by the methylation of myoinositol through inositol-O-methyltransferase and after subsequent epimerization of ononitol by the deed of ononitol epimerase (Sureshan et al. 2009).

7.3.5 Terpenoids

Terpenoids are secondary metabolites having greater than 40,000 various compounds (Aharoni et al. 2005). Terpenoids are organic compounds obtained from isoprene units that also have antioxidation function in plants (Grassmann et al. 2002). Terpenoids are classified into various types including mono, di-, tri-, and tetraterpenes due to varied composition (Rabi and Bishayee 2009). Among all terpenoids, tetraterpenes showed strong antioxidant property both in in vitro and in vivo studies (Kasote et al. 2015).

7.3.6 Alkaloids

Alkaloids are nitrogen containing secondary metabolites constituting about 10%–15% of the plant's tissue (Schardl et al. 2006). These have antioxidant properties due to presence of negatively charged nitrogen compounds. Alkaloids are mostly accumulated in plant in response to environmental stresses (Ali and

Alqurainy 2006). Some alkaloids such as indole alkaloids (brucine, strychnine) function as most operative inhibitor for singlet oxygen ($^1O_2$) radicals. Srivastava and Srivastava (2010) reported the enhanced contents of alkaloids in *Catharanthus roseus* root in response to Ni and Cd stress.

7.3.7 Phytohormone Responses

Phytohormones are organic substances produced endogenously and play a vital role in promoting growth, development, and metabolism of the plant, as well as a protector against various environmental stresses (Arif et al. 2020a, 2020b; Khan et al. 2020). Hormonal transduction cascades that operate as a crosstalk are most essential for plant growth. Hormones also work as second messenger through phosphorylation cascade. Transduction cascade regulates the synthesis of various hormones and plant growth processes in combination with different stimuli (Khan et al. 2020). Phytohormones also act as central mechanism in signalling and sensing of various environmental stresses (Lymperopoulos et al. 2018). There are almost nine major classes of phytohormones including gibberellins, auxins, abscisic acid, cytokinins, jasmonic acid, brassinosteroids, strigolactones, and salicylic acid, which are involved in many various responses against stress in plants (Ali and Baek 2020).

7.3.7.1 Abscisic Acid

Abscisic acid (ABA) plays an important role in plant against abiotic stresses by stimulating the various signalling pathways (Bucker-Neto et al. 2017). Under abiotic stresses, plant hormone ABA level rises, ABA in guard cell initiates the signalling cascades which enhance the efflux of K^+ ions from the guard cell and as a result lower the turgor pressure and consequently close the stomata (Salazar et al. 2015; Lim et al. 2015).

7.3.7.2 Salicylic Acid

Salicylic acid (SA) is a phenolic acid compound that works as growth regulator in plants under different types of environmental stresses. In response to abiotic stresses, salicylic acid interacts with abscisic acid, auxin, and gibberellin hormones and stimulates the antioxidant systems and enzymes, which alter the different stresses including heavy metal stress (Sharma et al. 2020). SA signalling initiates the synthesis and modulation of different protein structures. It also has a role in photosynthesis, seed germination, growth, flowering, and development process. It also enhances the activity of both enzymatic and nonenzymatic antioxidants for ROS scavenging. Under normal condition, SA interacts with polyamines and phytohormones, modulates plant responses, and provides immunity against stressed conditions (Arif et al. 2020a, 2020b).

7.3.7.3 Ethylene

Ethylene is the most diverse signalling hormone that has a role in acclimation of abiotic stresses. Its role in petiole and stem elongation, adventitious root formation, and closing of stomata is well established (Husain et al. 2020). In response to different stresses, ethylene is produced in plants by enzymes such as aminocyclopropane-1-carboxylic acid (ACC) synthase and ACC oxidases (ACOs) (Bleecker and Kende 2000). Ethylene combines with salicylic acid and jasmonic acid to modulate the defence mechanism against abiotic stresses (Kazan 2015).

7.3.7.4 Jasmonic Acid

Jasmonates are polyunsaturated fatty acid derivative phytohormones that play an important role in plants under abiotic stresses. Jasmonates participate in different signalling transduction pathways including regulatory protein, gene network, and work as signalling intermediates for enzymes, proteins, and molecules that provide protection to plants from stresses. Moreover, jasmonic acid enhances tolerance in plants by activating the plant defence systems, which mostly include antioxidant defence and others defence-related compounds (Ali and Baek 2020). Jasmonic acid also plays a role in the regulation of various physiological processes of plants, i.e. root growth, stamen growth, and senescence, and is also involved in regulation of different metabolites such as terpenoids and phytoalexins (Gomi 2020).

7.4 Future Prospects

According to the report of FAO (Food and Agriculture Organization), human population is constantly increasing and may reach up to 10 billion in 2050. Consequently, humans need 70% more food in future. On the other hand, food production is suffering from various abiotic stresses (Nooghabi et al. 2018; Newton 2019). Thus, it is important to obtain such crops which exhibit better metabolic tolerance to diverse environmental stresses to face the future challenges related to food security (Mabhaudhi et al. 2019). To meet this type of challenges, metabolites, genes, proteins, and transcriptional changes should be explored in crops that control the plant metabolism and enhance the stress resistance in response to varied range of environments stresses.

REFERENCES

Abdul S, Muhammad AA, Shabir H, Hesham A, Sajjad H, Niaz A, Abdul G, RZ Sayyed, Fahad S, Subhan D, Rahul D (2021a) Zinc nutrition and arbuscular mycorrhizal symbiosis effects on maize (*Zea mays* L.) growth and productivity. *J Saudi Soc Agric Sci*. https://doi.org/10.1016/j.sjbs.2021.06.096

Abdul S, Muhammad AA, Subhan D, Niaz A, Fahad S, Rahul D, Mohammad JA, Omaima N, Ur Habib MR, Bernard RG (2021b) Effect of arbuscular mycorrhizal fungi on the physiological functioning of maizeunder zinc-deficient soils. *Sci Rep* 11:18468

Acosta-Gamboa LM, Suxing L, Raquel T, Argelia L (2020) Characterization of the response to abiotic stresses of high ascorbate Arabidopsis lines using phenomic approaches. *Plant Physiol Biochem* 151:500–515.

Adnan M, Fahad S, Muhammad Z, Shahen S, Ishaq AM, Subhan D, Zafar-ul-Hye M, Martin LB, Raja MMN, Beena S, Saud S, Imran A, Zhen Y, Martin B, Jiri H, Rahul D (2020) Coupling phosphate-solubilizing bacteria with phosphorus supplements improve maize phosphorus acquisition and growth under lime induced salinity stress. *Plants* 9(900). https://doi.org/10.3390/plants9070900

Adnan M, Fahad S, Khan IA, Saeed M, Ihsan MZ, Saud S, Riaz M, Wang D, Wu C (2019). Integration of poultry manure and phosphate solubilizing bacteria improved availability of Ca bound P in calcareous soils. *3 Biotech. 1* 9(10):368.

Adnan M, Shah Z, Sharif M, Rahman H (2018b) Liming induces carbon dioxide (CO_2) emission in PSB inoculated alkaline soil supplemented with different phosphorus sources. *Environ Sci Poll Res* 25(10):9501–9509.

Adnan M, Zahir S, Fahad S, Arif M, Mukhtar A, Imtiaz AK, Ishaq AM, Abdul B, Hidayat U, Muhammad A, Inayat-Ur R, Saud S, Muhammad ZI, Yousaf J, Amanullah, Hafiz MH, Wajid N (2018a) Phosphate-solubilizing bacteria nullify the antagonistic effect of soil calcification on bioavailability of phosphorus in alkaline soils. *Sci Rep* 8:4339. https://doi.org/10.1038/s41598-018-22653-7

Agati G, Azzarello E, Pollastri S, Tattini M (2012) Flavonoids as antioxidants in plants: location and functional significance. *Plant Sci* 196:67–76.

Aharoni A, Jongsma MA, Bouwmeester HJ (2005) Volatile science? Metabolic engineering of terpenoids in plants. *Trends Plant Sci* 10:594–602.

Ahmad S, Kamran M, Ding R, Meng X, Wang H, Ahmad I, Fahad S, Han Q (2019) Exogenous melatonin confers drought stress by promoting plant growth, photosynthetic capacity and antioxidant defense system of maize seedlings. *PeerJ* 7:e7793. http://doi.org/10.7717/peerj.7793

Akbar H, Timothy JK, Jagadish T, Golam M, Apurbo KC, Muhammad F, Rajan B, Fahad S, Hasanuzzaman M (2020) Agricultural land degradation: processes and problems undermining future food security. In: Fahad S, Hasanuzzaman M, Alam M, Ullah H, Saeed M, Khan AK, Adnan M (eds) *Environment, climate, plant and vegetation growth*. Springer Publ Ltd, Springer Nature Switzerland AG. Part of Springer Nature, pp. 17–62. https://doi.org/10.1007/978-3-030-49732-3

Akram NA, Shafiq F, Ashraf M (2017) Ascorbic acid-a potential oxidant scavenger and its role in plant development and abiotic stress tolerance. *Front Plant Sci* 8:613. https://doi.org/10.3389/fpls.2017.00613.

Akram R, Turan V, Hammad HM, Ahmad S, Hussain S, Hasnain A, Maqbool MM, Rehmani MIA, Rasool A, Masood N, Mahmood F, Mubeen M, Sultana SR, Fahad S, Amanet K, Saleem M, Abbas Y, Akhtar HM, Waseem F, Murtaza R, Amin A, Zahoor SA, ul Din MS, Nasim W (2018a) Fate of organic and inorganic pollutants in paddy soils. In: Hashmi MZ, Varma A (eds) *Environmental pollution of paddy soils, soil biology*. Springer International Publishing AG, Gewerbestrasse 11, Cham, CH-6330, Switzerland, pp. 197–214

Akram R, Turan V, Wahid A, Ijaz M, Shahid MA, Kaleem S, Hafeez A, Maqbool MM, Chaudhary HJ, Munis MFH, Mubeen M, Sadiq N, Murtaza R, Kazmi DH, Ali S, Khan N, Sultana SR, Fahad S, Amin A, Nasim W (2018b) Paddy land pollutants and their role in climate change. In Hashmi MZ, Varma A (eds) *Environmental pollution of paddy soils, soil biology*. Springer International Publishing AG, Gewerbestrasse 11, Cham, CH-6330, Switzerland, pp. 113–124.

Akter N, Islam MR (2017) Heat stress effects and management in wheat: a review. *Agron Sustain Dev* 37(5):37

Alam MM, Nahar K, Hasanuzzaman M, Fujita M (2014) Trehalose-induced drought stress tolerance: a comparative study among different Brassica species. *Plant Omics* 7(4):271–283.

Alhaithloul HA, Soliman MH, Ameta KL, El-Esawi MA, Elkelish A (2020) Changes in ecophysiology, osmolytes, and secondary metabolites of the medicinal plants of *Mentha piperita* and *Catharanthus roseus* subjected to drought and heat stress. *Biomolecules* 10(1):43. https://doi.org/10.3390/biom10010043.

Ali AA, Alqurainy F (2006) Activities of antioxidants in plants under environmental stress. The lutein-prevention and treatment for diseases. In: Motohashi N (ed) *The lutein prevention and treatment for disease*. Transworld Research Network, Kerala, pp. 187–256.

Ali K, Arif M, Khan Z, Tariq M, Waqas M, Gul B, Bibi S, Zia-ud-Din, Ali M, Shafi B, Adnan M (2013) Effect of cutting on productivity and associated weeds of canola. *Pak J Weed Sci Res* 19(4):393–401.

Ali M, Baek KH (2020) Jasmonic acid signaling pathway in response to abiotic stresses in plants. *Int J Mol Sci* 21(2):621.

Ali Q, Javed MT, Haider MZ, Habib N, Rizwan M, Perveen R, Ali S, Alyemeni MN, El-Serehy HA, Al-Misned FA (2020) α-Tocopherol foliar spray and translocation mediates growth, photosynthetic pigments, nutrient uptake, and oxidative defense in Maize (*Zea mays* L.) under drought Stress. *Agronomy 10*:1235.

Ali S, Nawaz A, Hussain S, Khan SM, Ejaz S, Ahmad S (2019) Abiotic stress tolerance in plants by priming and pretreatments with ascorbic acid. In: Hasanuzzaman M, Fotopoulos V (eds) *Priming and pretreatment of seeds and seedlings*. Springer, Singapore, pp. 459–493.

Amanullah, Muhammad I, Haider N, Shah K, Manzoor A, Asim M, Saif U, Izhar A, Fahad S, Adnan M et al. (2021) Integrated foliar nutrients application improve wheat (*Triticum Aestivum* L.) productivity under calcareous soils in drylands. *Commun Soil Sci Plant Anal*. https://doi.org/10.1080/00103624.2021.1956521

Amanullah, Fahad S (Eds.) (2017) Rice - technology and production. *IntechOpen Croatia 2017*. http://dx.doi.org/10.5772/64480

Amanullah, Fahad S (Eds.) (2018a) Corn-Production and human Health in changing climate. *IntechOpen United Kingdom 2018*. http://dx.doi.org/10.5772/intechopen.74074

Amanullah, Fahad S (Eds.) (2018b) Nitrogen in agriculture – updates. *IntechOpen Croatia 2018*. http://dx.doi.org/10.5772/65846

Amanullah, Shah K, Imran, Hamdan AK, Muhammad A, Abdel RA, Muhammad A, Fahad S, Azizullah S, Brajendra P (2020) Effects of climate change on irrigation water quality. In: Fahad S, Hasanuzzaman M, Alam M, Ullah H, Saeed M, Khan AK, Adnan M (eds.) *Environment, climate, plant and vegetation growth*. Springer Publ Ltd, Springer Nature Switzerland AG. Part of Springer Nature, pp. 123–132. https://doi.org/10.1007/978-3-030-49732-3

Amir M, Muhammad A, Allah B, Sevgi Ç, Haroon ZK, Muhammad A, Emre A (2020) Bio fortification under climate change: the fight between quality and quantity. In: Fahad S, Hasanuzzaman M, Alam M, Ullah H, Saeed M, Khan AK, Adnan M (eds.) *Environment, climate, plant and vegetation growth*. Springer Publ Ltd, Springer Nature Switzerland AG. Part of Springer Nature, pp. 173–228. https://doi.org/10.1007/978-3-030-49732-3

Amjad I, Muhammad H, Farooq S, Anwar H (2020) Role of plant bioactives in sustainable agriculture. In: Fahad S, Hasanuzzaman M, Alam M, Ullah H, Saeed M, Khan AK, Adnan M (eds.) *Environment, climate, plant and vegetation growth*. Springer Publ Ltd, Springer Nature Switzerland AG. Part of Springer Nature, pp. 591–606. https://doi.org/10.1007/978-3-030-49732-3

Amjad SF, Mansoora N, Din IU, Khalid IR, Jatoi GH, Murtaza G, Yaseen S, Naz M, Danish S, Fahad S et al. (2021) Application of Zinc fertilizer and mycorrhizal inoculation on physio- biochemical parameters of wheat grown under water-stressed environment. *Sustainability* 13:11007. https://doi.org/10.3390/su131911007

Anee TI, Nahar K, Rahman A, Mahmud JA, Bhuiyan TF, Alam MU, Hasanuzzaman M (2019) Oxidative damage and antioxidant defense in *Sesamum indicum* after different waterlogging durations. *Plants* 8(7):196.

Arif M, Talha J, Muhammad R, Fahad S, Muhammad A, Amanullah, Kawsar A, Ishaq AM, Bushra K, Fahd R (2020a) Biochar; a remedy for climate change. In: Fahad S, Hasanuzzaman M, Alam M, Ullah H, Saeed M, Khan AK, Adnan M (eds) *Environment, climate, plant and vegetation growth*. Springer Publ Ltd, Springer Nature Switzerland AG. Part of Springer Nature, pp. 151–172. https://doi.org/10.1007/978-3-030-49732-3

Arif Y, Sami F, Siddiqui H, Bajguz A, Hayat S (2020b) Salicylic acid in relation to other phytohormones in plant: a study towards physiology and signal transduction under challenging environment. *Environ Exp Bot* 175:104040.

Arora M, Saxena P, Abdin MZ, Varma A (2020) interaction between *Piriformospora indica* and azotobacter chroococcum diminish the effect of salt stress in *Artemisia annua* L. by enhancing enzymatic and non-enzymatic antioxidants. *Symbiosis* 80(1):61–73.

Ashfaq AR, Uzma Y, Niaz A, Muhammad AA, Fahad S, Haider S, Tayebeh Z, Subhan D, Süleyman T, Hesham AE, Pramila T, Jamal MA, Sulaiman AA, Rahul D (2021) Toxicity of cadmium and nickel in the context of applied activated carbon biochar for improvement in soil fertility. *Saudi Soc Agric Sci*. https://doi.org/10.1016/j.sjbs.2021.09.035

Ashraf M, Foolad MR (2007) Roles of glycine betaine and proline in improving plant abiotic stress resistance. *Environ Exp Bot* 59(2):206–216.

Ashraf MA, Akbar A, Parveen A, Rasheed R, Hussain I, Iqbal M (2018) Phenological application of selenium differentially improves growth, oxidative defense and ion homeostasis in maize under salinity stress. *Plant Physiol Biochem* 123:268–280.

Ashraf MA, Asma HF, Iqbal M (2019) Exogenous menadione sodium bisulfite mitigates specific ion toxicity and oxidative damage in salinity-stressed Okra (*Abelmoschus esculentus* Moench). *Acta Physiol Plant* 41(12):187.

Athar M, Masood IA, Sana S, Ahmed M, Xiukang W, Sajid F, Sher AK, Habib A, Faran M, Zafar H, Farhana G, Fahad S (2021) Bio-diesel production of sunflower through sulphur management in a semi-arid subtropical environment. *Environ Sci Pollution Res*. https://doi.org/10.1007/s11356-021-16688-z

Atif B, Hesham A, Fahad S (2021) Biochar coupling with phosphorus fertilization modifies antioxidant activity, osmolyte accumulation and reactive oxygen species synthesis in the leaves and xylem sap of rice cultivars under high-temperature stress. *Physiol Mol Biol Plants*. https://doi.org/10.1007/s12298-021-01062-7

Aziz K, Daniel KYT, Fazal M, Muhammad ZA, Farooq S, Fan W, Fahad S, Ruiyang Z (2017a) Nitrogen nutrition in cotton and control strategies for greenhouse gas emissions: a review. *Environ Sci Pollut Res* 24:23471–23487. https://doi.org/10.1007/s11356-017-0131-y

Aziz K, Daniel KYT, Muhammad ZA, Honghai L, Shahbaz AT, Mir A, Fahad S (2017b) Nitrogen fertility and abiotic stresses management in cotton crop: a review. *Environ Sci Pollut Res* 24:14551–14566. https://doi.org/10.1007/s11356-017-8920-x

Banerjee A, Roychoudhury A (2018) Abiotic stress, generation of reactive oxygen species, and their consequences: an overview. In: Singh VP, Singh S, Tripathi D, Mohan Prasad S, Chauhan DK (eds) *Revisiting the role of reactive oxygen species (ROS) in plants: ROS Boon or bane for plants?* Wiley, New York, pp. 23–50.

Bano C, Amist N, Singh NB (2020) Role of polyamines in plants abiotic stress tolerance: advances and future prospects. In: Tripathi DK, Singh VP, Ramavat N (eds) *Plant life under changing environment*. Academic Press, pp 481–496. https://doi.org/10.1016/B978-0-12-818204-8.00021-7.

Bansal R, Srivastava JP (2012) Antioxidative defense system in pigeonpea roots under waterlogging stress. *Acta Physiol Plant* 34(2):515–522.

Baseer M, Adnan M, Fazal M, Fahad S, Muhammad S, Fazli W, Muhammad A, Jr. Amanullah, Depeng W, Saud S, Muhammad N, Muhammad Z, Fazli S, Beena S, Mian AR, Ishaq AM (2019) Substituting urea by organic wastes for improving maize yield in alkaline soil. *J Plant Nutrition*. https://doi.org/10.1080/01904167.2019.1659344

Baskar V, Venkatesh R, Ramalingam S (2018) Flavonoids (antioxidants systems) in higher plants and their response to stresses. In: Gupta DK, Palma JM, Corpas JM (eds) *Antioxidants and antioxidant enzymes in higher plants*. Springer, Cham, pp 253–268.

Bayram AY, Seher Ö, Nazlican A (2020) Climate change forecasting and modeling for the year of 2050. In: Fahad S, Hasanuzzaman M, Alam M, Ullah H, Saeed M, Khan AK, Adnan M (eds) *Environment, climate, plant and vegetation growth*. Springer Publ Ltd, Springer Nature Switzerland AG. Part of Springer Nature, pp. 109–122. https://doi.org/10.1007/978-3-030-49732-3

Bilger W, Rolland M, Nybakken L (2007) UV screening in higher plants induced by low temperature in the absence of UV-B radiation. *Photochem Photobiol Sci* 6:190–195.

Bleecker AB, Kende H (2000) Ethylene: a gaseous signal molecule in plants. *Annu Rev Cell Dev Biol* 16:1–18.

Bolouri-Moghaddam MR, Le Roy K, Xiang L, Rolland F, Van den Ende W (2010) Sugar signaling and antioxidant network connections in plant cells. *FEBS J* 277(9):2022–2037.

Bucker-Neto L, Paiva ALS, Machado RD, Arenhart RA, Margis-Pinheiro M (2017) Interactions between plant hormones and heavy metals responses. *Genet Mol Biol* 40:373–386.

Cai LM, Wang QS, Luo J, Chen LG, Zhu RL, Wang S, Tang CH (2019) Heavy metal contamination and health risk assessment for children near a large Cu-smelter in central China. *Sci Total Environ* 650:725–733.

Cairns JE, Sanchez C, Vargas M, Ordoñez R, Araus JL (2012) Dissecting maize productivity: ideotypes associated with grain yield under drought stress and well-watered conditions. *J Integr Plant Biol* 54:1007–1020.

Camagajevac ŠI, Pfeiffer ŽT, Maronić ŠD (2018) Abiotic stress response in plants: the relevance of tocopherols. In: Gupta D, Palma J, Corpas F (eds) *Antioxidants and antioxidant enzymes in higher plants*. Springer, Cham, Berlin, pp 233–251.

Çatav ŞS, Genç TO, Oktay MK, Küçükakyüz K (2020) Cadmium toxicity in wheat: impacts on element contents, antioxidant enzyme activities, oxidative stress, and genotoxicity. *Bull Environ Contam Toxicol* 104(1):71–77.

Chaki M, Begara-Morales JC, Barroso JB (2020) Oxidative stress in plants. *Antioxidants 9*:481.

Chang W, Qiujuan J, Evgenios A, Haitao L, Gezi L, Jingjing Z, Fahad S, Ying J (2021) Hormetic effects of zinc on growth and antioxidant defense system of wheat plants. *Sci Total Environ*. https://doi.org/10.1016/j.scitotenv.2021.150992

Chen WR, Zheng JS, Li YQ, Guo WD (2012) Effects of high temperature on photosynthesis, chlorophyll fluorescence, chloroplast ultrastructure, and antioxidant activities in fingered citron. *Russ J Plant Physiol* 59:732–740.

Chen Y, Guo Z, Dong L, Fu Z, Zheng Q, Zhang G, Qin L, Sun X, Shi Z, Fahad S, Xie F, Saud S. (2021) Turf performance and physiological responses of native Poa species to summer stress in Northeast China. *PeerJ* 9:e12252 http://doi.org/10.7717/peerj.12252

Choudhary S, Zehra A, Wani KI, Naeem M, Hakeem KR, Aftab T (2020) The role of micronutrients in growth and development: transport and signalling pathways from crosstalk perspective. In: Aftab T, Hakem KR (eds) *Plant micronutrients*. Springer, Cham, pp 73–81.

Collins AR (2001) Carotenoids and genomic stability. *Mutat Res* 475:21–28.

Das K, Roychoudhury A (2014) Reactive oxygen species (ROS) and response of antioxidants as ROS-scavengers during environmental stress in plants. *Front Environ Sci* 2:53.

Deepranjan S, Ardith S, Siva D, Sonam S, Shikha, Manoj P, Amitava R, Sayyed RZ, Abdul G, Mohammad JA, Subhan D, Fahad S, Rahul D (2021) Optimizing nutrient use efficiency, productivity, energetics, and economics of red cabbage following mineral fertilization and biopriming with compatible rhizosphere microbes. *Sci Rep* 11:15680.

Denaxa NK, Damvakaris T, Roussos PA (2020) Antioxidant defense system in young olive plants against drought stress and mitigation of adverse effects through external application of alleviating products. *Sci Hortic* 259:108812.

Depeng W, Fahad S, Saud S, Muhammad K, Aziz K, Mohammad NK, Hafiz MH, Wajid N (2018) Morphological acclimation to agronomic manipulation in leaf dispersion and orientation to promote "Ideotype" breeding: Evidence from 3D visual modeling of "super" rice (*Oryza sativa* L.). *Plant Physiol Biochem.* 135:499-510 https://doi.org/10.1016/j.plaphy.2018.11.010

Dkhil B, Denden M (2012) Effect of salt stress on growth, anthocyanins, membrane permeability and chlorophyll fluorescence of okra (*Abelmoschus esculentus* L.) seedlings. *Am J Plant Physiol* 7(4):174–183.

Ejaz S, Fahad S, Anjum MA, Nawaz A, Naz S, Hussain S, Ahmad S (2020) Role of osmolytes in the mechanisms of antioxidant defense of plants. In: Aftab T, Hakeem KR (eds) *Sustainable agriculture reviews 39*. Springer, Cham, pp 95–117.

El-Bassiouny HMS, Sadak MS (2015) Impact of foliar application of ascorbic acid and α-tocopherol on antioxidant activity and some biochemical aspects of flax cultivars under salinity stress. *Acta Biol Colomb* 20(2):209–222.

Elkelish A, Qari SH, Mazrou YS, Abdelaal KA, Hafez YM, Abu-Elsaoud AM, El Nahhas N (2020) Exogenous ascorbic acid induced chilling tolerance in tomato plants through modulating metabolism, osmolytes, antioxidants, and transcriptional regulation of catalase and heat shock proteins. *Plants* 9(4):431.

Eltayeb AE, Kawano N, Badawi GH, Kaminaka H, Sanekata T, Shibahara T, Inanaga S, Tanaka K (2007) Overexpression of monodehydroascorbate reductase in transgenic tobacco confers enhanced tolerance to ozone, salt and polyethylene glycol stresses. *Planta* 225:1255–1264.

Emre B, Ömer SU, Martín LB, Andre D, Fahad S, Rahul D, Muhammad Z-ul-H, Ghulam SH, Subhan D (2021) Studying soil erosion by evaluating changes in physico-chemical properties of soils under different land-use types. *J Saudi Society Agricultural Sci* 20:190–197.

Fahad S, Abdul B, Adnan M. (Eds.), (2018b) Global wheat production. *IntechOpen United Kingdom 2018.* http://dx.doi.org/10.5772/intechopen.72559

Fahad S, Bajwa AA, Nazir U, Anjum SA, Farooq A, Zohaib A, Sadia S, NasimW, Adkins S, Saud S, Ihsan MZ, Alharby H, Wu C, Wang D, Huang J (2017) Crop production under drought and heat stress: plant responses and management options. *Front Plant Sci* 8:1147. https://doi.org/10.3389/fpls.2017.01147

Fahad S, Bano A (2012) Effect of salicylic acid on physiological and biochemical characterization of maize grown in saline area. *Pak J Bot* 44:1433–1438.

Fahad S, Chen Y, Saud S, Wang K, Xiong D, Chen C, Wu C, Shah F, Nie L, Huang J (2013) Ultraviolet radiation effect on photosynthetic pigments, biochemical attributes, antioxidant enzyme activity and hormonal contents of wheat. *J Food Agri Environ* 11(3&4):1635–1641.

Fahad S, Hussain S, Bano A, Saud S, Hassan S, Shan D, Khan FA, Khan F, Chen Y, Wu C, Tabassum MA, Chun MX, Afzal M, Jan A, Jan MT, Huang J (2014a) Potential role of phytohormones and plant growth-promoting rhizobacteria in abiotic stresses: consequences for changing environment. *Environ Sci Pollut Res* 22(7):4907– 4921. https://doi.org/10.1007/s11356-014-3754-2

Fahad S, Hussain S, Matloob A, Khan FA, Khaliq A, Saud S, Hassan S, Shan D, Khan F, Ullah N, Faiq M, Khan MR, Tareen AK, Khan A, Ullah A, Ullah N, Huang J (2014b) Phytohormones and plant responses to salinity stress: A review. *Plant Growth Regul* 75(2):391– 404. https://doi.org/10.1007/s10725-014-0013-y

Fahad S, Hussain S, Saud S, Hassan S, Chauhan BS, Khan F et al (2016a) Responses of rapid viscoanalyzer profile and other rice grain qualities to exogenously applied plant growth regulators under high day and high night temperatures. *PLoS One* 11(7):e0159590. https://doi.org/10.1371/journal.pone.0159590

Fahad S, Hussain S, Saud S, Hassan S, Ihsan Z, Shah AN,Wu C, Yousaf M, Nasim W, Alharby H, Alghabari F, Huang J (2016c) Exogenously applied plant growth regulators enhance the morphophysiological growth and yield of rice under high temperature. *Front Plant Sci* 7:1250. https://doi.org/10.3389/fpls.2016.01250

Fahad S, Hussain S, Saud S, Hassan S, Tanveer M, Ihsan MZ, Shah AN, Ullah A, Nasrullah KF, Ullah S, Alharby HNW, Wu C, Huang J (2016d) A combined application of biochar and phosphorus alleviates heat-induced adversities on physiological, agronomical and quality attributes of rice. *Plant Physiol Biochem* 103:191–198.

Fahad S, Hussain S, Saud S, Khan F, Hassan S, Nasim W, Arif M, Wang F, Huang J (2016b) Exogenously applied plant growth regulators affect heat-stressed rice pollens. *J Agron Crop Sci* 202:139–150.

Fahad S, Hussain S, Saud S, Tanveer M, Bajwa AA, Hassan S, Shah AN, Ullah A, Wu C, Khan FA, Shah F, Ullah S, Chen Y, Huang J (2015a) A biochar application protects rice pollen from high-temperature stress. *Plant Physiol Biochem* 96:281–287.

Fahad S, Muhammad ZI, Abdul K, Ihsanullah D, Saud S, Saleh A, Wajid N, Muhammad A, Imtiaz AK, Chao W, Depeng W, Jianliang H (2018a): Consequences of high temperature under changing climate optima for rice pollen characteristics-concepts and perspectives, *Archives Agron Soil Sci.* https://doi.org/10.1080/03650340.2018.1443213

Fahad S, Nie L, Chen Y, Wu C, Xiong D, Saud S, Hongyan L, Cui K, Huang J (2015b) Crop plant hormones and environmental stress. *Sustain Agric Rev* 15:371–400.

Fahad S, Saud S, Yajun C, Chao W, Depeng W (Eds.), (2021f) Abiotic stress in plants. *IntechOpen United Kingdom 2021.* http://dx.doi.org/10.5772/intechopen.91549

Fahad S, Hasanuzzaman M, Alam M, Ullah H, Saeed M, Ali Khan I, Adnan M. (Eds.) (2020) *Environment, climate, plant and vegetation growth.* Springer Nature Switzerland AG 2020. doi: https://doi.org/10.1007/978-3-030-49732-3

Fahad S, Adnan M, Hassan S, Saud S, Hussain S, Wu, C, Wang D, Hakeem KR, Alharby HF, Turan V, Khan MA, Huang J (2019b). *Rice responses and tolerance to high temperature.* In: Hasanuzzaman M and Fujita M and Nahar K and Biswas JK (eds) *Advances in rice research for abiotic stress tolerance.* Woodhead Publ Ltd, Abington Hall, Cambridge, UK, pp. 201–224.

Fahad S, Rehman A, Shahzad B, Tanveer M, Saud S, Kamran M, Ihtisham M, Khan SU, Turan V, Rahman MHU (2019a) Rice responses and tolerance to metal/metalloid toxicity. In: Hasanuzzaman M and Fujita M and Nahar K and Biswas JK (eds) *Advances in rice research for abiotic stress tolerance*. Woodhead Publ Ltd, Abington Hall Abington, Cambridge CB1 6AH, CAMBS, England, pp. 299–312.

Fahad S, Sonmez O, Saud S, Wang D, Wu C, Adnan M, Turan V (Eds.), (2021a) *Plant growth regulators for climate-smart agriculture, First edition. ed, Footprints of climate variability on plant diversity*. CRC Press, Boca Raton, FL.

Fahad S, Sonmez O, Saud S, Wang D, Wu C, Adnan M, Turan V (Eds.), (2021b) *Climate change and plants: biodiversity, growth and interactions, First edition. ed, Footprints of climate variability on plant diversity*. CRC Press, Boca Raton.

Fahad S, Sonmez O, Saud S, Wang D, Wu C, Adnan M, Turan V (Eds.), (2021c) *Developing climate resilient crops: improving global food security and safety, First edition. ed, Footprints of climate variability on plant diversity*. CRC Press, Boca Raton.

Fahad S, Sonmez O, Turan V, Adnan M, Saud S, Wu C, Wang D (Eds.), (2021d) *Sustainable soil and land management and climate change, First edition. ed, Footprints of climate variability on plant diversity*. CRC Press, Boca Raton.

Fahad S, Sönmez O, Saud S, Wang D, Wu C, Adnan M, Arif M, Amanullah. (eds.), (2021e) *Engineering Tolerance in Crop Plants Against Abiotic Stress, First edition. ed, Footprints of climate variability on plant diversity*. CRC Press, Boca Raton.

Farah R, Muhammad R, Muhammad SA, Tahira Y, Muhammad AA, Maryam A, Shafaqat A, Rashid M, Muhammad R, Qaiser H, Afia Z, Muhammad AA, Muhammad A, Fahad S (2020) Alternative and non-conventional soil and crop management strategies for increasing water use efficiency. In: Fahad S, Hasanuzzaman M, Alam M, Ullah H, Saeed M, Khan AK, Adnan M (eds) *Environment, climate, plant and vegetation growth*. Springer Publ Ltd, Springer Nature Switzerland AG. Part of Springer Nature, pp. 323–338. https://doi.org/10.1007/978-3-030-49732-3

Farhana G, Ishfaq A, Muhammad A, Dawood J, Fahad S, Xiuling L, Depeng W, Muhammad F, Muhammad F, Syed AS (2020) Use of crop growth model to simulate the impact of climate change on yield of various wheat cultivars under different agro-environmental conditions in Khyber Pakhtunkhwa, Pakistan. *Arabian J Geosci* 13:112 https://doi.org/10.1007/s12517-020-5118-1

Farhangi-Abriz S, Torabian S (2017) Antioxidant enzyme and osmotic adjustment changes in bean seedlings as affected by biochar under salt stress. *Ecotoxicol Environ Saf* 137:64–70.

Farhat A, Hafiz MH, Wajid I, Aitazaz AF, Hafiz FB, Zahida Z, Fahad S, Wajid F, Artemi C (2020) A review of soil carbon dynamics resulting from agricultural practices. *J Environ Manage* 268 (2020) 110319.

Farooq MA, Niazi AK, Akhtar J, Farooq M, Souri Z, Karimi N, Rengel Z (2019) Acquiring control: The evolution of ROS-Induced oxidative stress and redox signaling pathways in plant stress responses. *Plant Physiol Biochem* 141:353–369.

Fazli W, Muhmmad S, Amjad A, Fahad S, Muhammad A, Muhammad N, Ishaq AM, Imtiaz AK, Mukhtar A, Muhammad S, Muhammad I, Rafi U, Haroon I, Muhammad A (2020) Plant-Microbes Interactions and Functions in Changing Climate. In: Fahad S, Hasanuzzaman M, Alam M, Ullah H, Saeed M, Khan AK, Adnan M (Ed.), *Environment, Climate, Plant and Vegetation Growth*. Springer Publ Ltd, Springer Nature Switzerland AG. Part of Springer Nature, pp. 397–420. https://doi.org/10.1007/978-3-030-49732-3

Ghori NH, Ghori T, Hayat MQ, Imadi SR, Gul A, Altay V, Ozturk M (2019) Heavy metal stress and responses in plants. *Int J Environ Sci Technol* 16(3):1807–1828.

Ghulam M, Muhammad AA, Donald LS, Sajid M, Muhammad FQ, Niaz A, Ateeq R, Shakeel A, Sajjad H, Muhammad A, Summia M, Aqib HAK, Fahad S, Rahul D, Mazhar I, Timothy DS (2021) Formalin fumigation and steaming of various composts differentially influence the nutrient release, growth and yield of muskmelon (*Cucumis melo* L.). *Sci Rep* 11:21057.

Gomi K (2020) Jasmonic acid: an essential plant hormone. *Int J Mol Sci* 21(4):1261.

Gopakumar L, Bernard NO, Donato V (2020) Soil microarthropods and nutrient cycling. In: Fahad S, Hasanuzzaman M, Alam M, Ullah H, Saeed M, Khan AK, Adnan M (eds) *Environment, climate, plant and vegetation growth*. Springer Publ Ltd, Springer Nature Switzerland AG. Part of Springer Nature, pp. 453–472. https://doi.org/10.1007/978-3-030-49732-3

Grassmann J, Hippeli S, Elstner EF (2002) Plant's defence and its benefts for animals and medicine: role of phenolics and terpenoids in avoiding oxygen stress. *Plant Physiol Biochem* 40:471–478.

Griffith M, Yaish MW (2004) Antifreeze proteins in overwintering plants: a tale of two activities. *Trends Plant Sci* 9:399–405.

Guofu L, Zhenjian B, Fahad S, Guowen C, Zhixin X, Hao G, Dandan L, Yulong L, Bing L, Guoxu J, Saud S (2021) Compositional and structural changes in soil microbial communities in response to straw mulching and plant revegetation in an abandoned artificial pasture in Northeast China. *Global Ecology and Conservation* 31 (2021) e01871.

Gupta B, Huang B (2014) Mechanism of salinity tolerance in plants: physiological, biochemical, and molecular characterization. *Int J Genom.* https://doi.org/10.1155/2014/701596

Habib Ur R, Ashfaq A, Aftab W, Manzoor H, Fahd R, Wajid I, Md. Aminul I, Vakhtang S, Muhammad A, Asmat U, Abdul W, Syeda RS, Shah S, Shahbaz K, Fahad S, Manzoor H, Saddam H, Wajid N (2017) Application of CSM-CROPGRO-Cotton model for cultivars and optimum planting dates: evaluation in changing semi-arid climate. *Field Crops Res.* http://dx.doi.org/10.1016/j.fcr.2017.07.007

Hafeez M, Farman U, Muhammad MK, Xiaowei L, Zhijun Z, Sakhawat S, Muhammad I, Mohammed AA, Mandela F-G, Nicolas D, Muzammal R, Fahad S, Yaobin L (2021) Metabolic-based insecticide resistance mechanism and ecofriendly approaches for controlling of beet armyworm Spodoptera exigua:a review. *Environ Sci Pollution Res.* https://doi.org/10.1007/s11356-021-16974-w

Hafiz MH, Abdul K, Farhat A, Wajid F, Fahad S , Muhammad A, Ghulam MS, Wajid N, Muhammad M, Hafiz FB (2020b): Comparative effects of organic and inorganic fertilizers on soil organic carbon and wheat productivity under arid region, *Commun Soil Sci Plant Anal.* https://doi.org/10.1080/00103624.2020.1763385

Hafiz MH, Farhat A, Ashfaq A, Hafiz FB, Wajid F, Carol Jo W, Fahad S, Gerrit H (2020a) Predicting kernel growth of maize under controlled water and nitrogen applications. *Int J Plant Prod.* https://doi.org/10.1007/s42106-020-00110-8

Hafiz MH, Farhat A, Shafqat S, Fahad S, Artemi C, Wajid F, Chaves CB, Wajid N, Muhammad M, Hafiz FB (2018) Offsetting land degradation through nitrogen and water management during maize cultivation under arid conditions. *Land Degrad Dev* 1–10. https://doi.org/10.1002/ldr.2933

Hafiz MH, Muhammad A, Farhat A, Hafiz FB, Saeed AQ, Muhammad M, Fahad S, Muhammad A (2019) Environmental factors affecting the frequency of road traffic accidents: a case study of sub-urban area of Pakistan. *Environ Sci Pollut Res.* https://doi.org/10.1007/s11356-019-04752-8

Hafiz MH, Wajid F, Farhat A, Fahad S, Shafqat S, Wajid N, Hafiz FB (2016) Maize plant nitrogen uptake dynamics at limited irrigation water and nitrogen. *Environ Sci Pollut Res* 24(3):2549–2557. https://doi.org/10.1007/s11356-016-8031-0

Haider SA, Lalarukh I, Amjad SF, Mansoora N, Naz M, Naeem M, Bukhari SA, Shahbaz M, Ali SA, Marfo TD, Subhan D, Rahul D, Fahad S (2021) Drought stress alleviation by potassium-nitrate-containing Chitosan/Montmorillonite microparticles confers changes in *Spinacia oleracea* L. *Sustain* 13: 9903. https://doi.org/10.3390/su13179903

Hameed A, Rasool S, Azooz MM, Hossain MA, Ahanger MA, Ahmad P (2016) Heavy metal stress: plant responses and signaling. In: Ahmad P (ed) *Plant metal interaction: emerging remediation techniques*, Elsevier Academic Press, pp. 557–584.

Hammad M, Ahmed N, Khan IA, Shah B, Khan A, Ali M, Rasheed MT, Junaid K, Adnan M, Zaki AB, Ahmed S (2015) Studies on the efficacy of selected insecticides against anopheles mosquitoes of village Goth. Bhoorji (Sindh) Pakistan. *J Ento Zool Stud* 3(6):169–173.

Hamza SM, Xiukang W, Sajjad A, Sadia Z, Muhammad N, Adnan M, Fahad S et al. (2021) Interactive effects of gibberellic acid and NPK on morpho-physio-biochemical traits and organic acid exudation pattern in coriander (*Coriandrum sativum* L.) grown in soil artificially spiked with boron. *Plant Physiol Biochem* 167(2021):884–900

Handa AK, Fatima T, Mattoo AK (2018) Polyamines: bio-molecules with diverse functions in plant and human health and disease. *Front Chem* 6:10. https://doi.org/10.3389/fchem.2018.00010

Hasanuzzaman M, Bhuyan MHM, Anee TI, Parvin K, Nahar K, Mahmud JA, Fujita M (2019) Regulation of ascorbate-glutathione pathway in mitigating oxidative damage in plants under abiotic stress. *Antioxidants* 8(9):384.

Hasanuzzaman M, Bhuyan MHM, Zulfiqar F, Raza A, Mohsin SM, Mahmud JA, Fotopoulos V (2020) Reactive oxygen species and antioxidant defense in plants under abiotic stress: revisiting the crucial role of a universal defense regulator. *Antioxidants* 9(8):681.

Hasanuzzaman M, Hossain MA, da Silva JAT, Fujita M (2012) Plant response and tolerance to abiotic oxidative stress: antioxidant defense is a key factor. In: Venkateshwarulu B, Shanker AK, Shanker C, Mandapaka M (eds) *Crop stress and its management: perspectives and strategies*. Springer, Berlin, pp. 261–315.

Hasanuzzaman M, Hossain MA, Fujita M (2011) Selenium-induced up-regulation of the antioxidant defense and methylglyoxal detoxification system reduces salinity-induced damage in rapeseed seedlings. *Biol Trace Elem Res*. https://doi.org/10.1007/s12011-011-8958-4

Hayat S, Hayat Q, Alyemeni MN, Wani AS, Pichtel J, Ahmad A (2012) Role of proline under changing environments: a review. *Plant Signal Behav* 7(11):1456–1466.

Hesham FA, Fahad S (2020) Melatonin application enhances biochar efficiency for drought tolerance in maize varieties: Modifications in physio-biochemical machinery. *Agron J* 112(4):1–22.

Hildebrandt TM, Adriano NN, Wagner LA, Hans-Peter B (2015) Amino acid catabolism in plants. *Mol Plant* 8(11):1563–1579.

Hisyam B, Alam MA, Naimah N, Jahan MS (2017) Roles of glycinebetaine on antioxidants and gene function in rice plants under water stress. *Asian J Plant Sci* 16:132–140.

Hossain MA, Fujita M (2010) Evidence for a role of exogenous glycinebetaine and proline in antioxidant defense and methylglyoxal detoxification systems in mung bean seedlings under salt stress. *Physiol Mol Biol Plants* 16(1):19–29.

Hossain Z, López-Climent MF, Arbona V, Pérez-Clemente RM, Gómez-Cadenas A (2009) Modulation of the antioxidant system in citrus under waterlogging and subsequent drainage. *J Plant Physiol* 166(13):1391–1404.

Huang Li Y, Li X-X, Zhang Y-B, Fahad S, Wang F (2021) dep1 improves rice grain yield and nitrogen use efficiency simultaneously by enhancing nitrogen and dry matter translocation. *J Integrative Agri*. https://doi.org/10.1016/S2095-3119(21)63795-4

Huang S, Zuo T, Ni W (2020) Important roles of glycinebetaine in stabilizing the structure and function of the photosystem II complex under abiotic stresses. *Planta* 251(2):1–14.

Husain T, Fatima A, Suhel M, Singh S, Sharma A, Prasad SM, Singh VP (2020) A brief appraisal of ethylene signaling under abiotic stress in plants. *Plant Signal Behav* 15(9):1782051.

Hussain I, Ashraf MA, Rasheed R, Iqbal M, Ibrahim M, Ashraf S (2016) Heat shock increases oxidative stress to modulate growth and physico-chemical attributes in diverse maize cultivars. *Int Agrophys* 30(4):519–531.

Hussain I, Parveen A, Rasheed R, Ashraf MA, Ibrahim M, Raza S, Afzaal Z, Iqbal M (2019) Exogenous silicon modulates growth, physio-chemicals and antioxidants in barley (*Hordeum vulgare* L.) exposed to different temperature regimes. *Silicon* 11:2753–2762.

Hussain MA, Fahad S, Rahat S, Muhammad FJ, Muhammad M, Qasid A, Ali A, Husain A, Nooral A, Babatope SA, Changbao S, Liya G, Ibrar A, Zhanmei J, Juncai H (2020) Multifunctional role of brassinosteroid and its analogues in plants. *Plant Growth Regul*. https://doi.org/10.1007/s10725-020-00647-8

Ibrahim MF, El-Samad A, Ashour H, El-Sawy AM, Hikal M, Elkelish A, Farag R (2020) Regulation of agronomic traits, nutrient uptake, osmolytes and antioxidants of maize as influenced by exogenous potassium silicate under deficit irrigation and semiarid conditions. *Agronomy* 10(8):1212.

Ibrar K, Aneela R, Khola Z, Urooba N, Sana B, Rabia S, Ishtiaq H, Mujaddad UR, Salvatore M (2020) Microbes and environment: global warming reverting the frozen zombies. In: Fahad S, Hasanuzzaman M, Alam M, Ullah H, Saeed M, Khan AK, Adnan M (eds) *Environment, climate, plant and vegetation growth*. Springer Publ Ltd, Springer Nature Switzerland AG. Part of Springer Nature, pp. 607–634. https://doi.org/10.1007/978-3-030-49732-3

Ihsan MZ, Daur I, Alghabari F, Alzamanan S, Rizwan S, Ahmad M, Shafqat W (2019) Heat stress and plant development: role of sulphur metabolites and management strategies. *Acta Agric Scand Sec B Soil Plant Sci* 69(4):332–342.

Ikram U, Khadim D, Muhammad T, Muhammad S, Fahad S (2021) Gibberellic acid and urease inhibitor optimize nitrogen uptake and yield of maize at varying nitrogen levels under changing climate. *Environ Sci Pollution Res*. https://doi.org/10.1007/s11356-021-16049-w

Ilyas M, Mohammad N, Nadeem K, Ali H, Aamir HK, Kashif H, Fahad S, Aziz K, Abid U (2020) Drought tolerance strategies in plants: a mechanistic approach. *J Plant Growth Regul*. https://doi.org/10.1007/s00344-020-10174-5

Iqra M, Amna B, Shakeel I, Fatima K, Sehrish L, Hamza A, Fahad S (2020) Carbon cycle in response to global warming. In: Fahad S, Hasanuzzaman M, Alam M, Ullah H, Saeed M, Khan AK, Adnan M (eds) *Environment, climate, plant and vegetation growth*. Springer Publ Ltd, Springer Nature Switzerland AG. Part of Springer Nature, pp. 1–16. https://doi.org/10.1007/978-3-030-49732-3

Irfan M, Muhammad M, Muhammad JK, Khadim MD, Dost M, Ishaq AM, Waqas A, Fahad S, Saud S et al. (2021) Heavy metals immobilization and improvement in maize (*Zea mays* L.) growth amended with biochar and compost. *Sci Rep* 11:18416.

Jabborova D, Sulaymanov K, Sayyed RZ, Alotaibi SH, Enakiev Y, Azimov A, Jabbarov Z, Ansari MJ, Fahad S, Danish S et al. (2021) Mineral fertilizers improves the quality of turmeric and soil. *Sustain* 13: 9437. https://doi.org/10.3390/su13169437

Jan M, Anwar-ul-Haq M, Adnan NS, Muhammad Y, Javaid I, Xiuling L, Depeng W, Fahad S, (2019) Modulation in growth, gas exchange, and antioxidant activities of salt-stressed rice (*Oryza sativa* L.) genotypes by zinc fertilization. *Arabian J Geosci* 12:775. https://doi.org/10.1007/s12517-019-4939-2

Jan N, Umer M, Khurshid IA, Riffat J (2018) Cold stress modulates osmolytes and antioxidant system in Calendula officinalis. *Acta Physiol Plant* 40:73. https://doi.org/10.1007/s11738-018-2649-0

Javaid S, Zaman Q, Sultan K, Riaz U, Aslam A, Sharif SNE, Ibraheem S (2020) 5. Heavy metals stress, mechanism and remediation techniques in rice (*Oryza sativa* L.): a review. *Pure Appl Biol* 9(1):403–426.

Jiang D, Chen W, Gao J, Yang F, Zhuang C (2019) Overexpression of the trehalose-6-phosphate phosphatase OsTPP3 increases drought tolerance in rice. *Plant Biotechnol Rep* 13(3): 285–292.

Kähkönen MP, Heinonen M (2003) Antioxidant activity of anthocyanins and their aglycons. *J Agric Food Chem* 51:628–633.

Kamal-Eldin A, Appelqvist LA (1996) The chemistry and antioxidant properties of tocopherols and tocotrienols. *Lipids* 31(7): 671–701.

Kamarn M, Wenwen C, Irshad A, Xiangping M, Xudong Z, Wennan S, Junzhi C, Shakeel A, Fahad S, Qingfang H, Tiening L (2017) Effect of paclobutrazol, a potential growth regulator on stalk mechanical strength, lignin accumulation and its relation with lodging resistance of maize. *Plant Growth Regul* 84:317–332. https://doi.org/10.1007/s10725-017-0342-8

Kasote DM, Katyare SS, Hegde MV, Bae H (2015) Signifcance of antioxidant potential of plants and its relevance to therapeutic applications. *Int J Biol Sci* 11:982–991.

Kawai Y, Ono E, Mizutani M (2014) Evolution and diversity of the 2–oxoglutarate-dependent dioxygenase superfamily in plants. *Plant J* 78(2):328–343.

Kaya C, Sarioğlu A, Akram NA, Ashraf M (2019) Thiourea-mediated nitric oxide production enhances tolerance to boron toxicity by reducing oxidative stress in bread wheat (*Triticum aestivum* L.) and durum wheat (*Triticum durum* Desf.) plants. *J Plant Growth Regul* 38(3):1094–1109.

Kazan K (2015) Diverse roles of jasmonates and ethylene in abiotic stress tolerance. *Trends Plant Sci* 20: 219–229.

Keunen LS, Peshev D, Vangronsveld J, Van Den Ende IM, Cuypers NN (2013) Plant sugars are crucial players in the oxidative challenge during abiotic stress: extending the traditional concept. *Plant Cell Environ* 36(7):1242–1255.

Khadim D, Fahad S, Jahangir MMR, Iqbal M, Syed SA, Shah AK, Ishaq AM, Rahul D et al. (2021a) Biochar and urease inhibitor mitigate NH_3 and N_2O emissions and improve wheat yield in a urea fertilized alkaline soil. *Sci Rep* 11:17413.

Khadim D, Saif R, Fahad S, Syed SA, Shah AK et al. (2021b) Influence of variable biochar concentration on yield-scaled nitrous oxide emissions, Wheat yield and nitrogen use efficiency. *Sci Rep* 11:16774.

Khan MMH, Niaz A, Umber G, Muqarrab A, Muhammad AA, Muhammad I, Shabir H, Shah F, Vibhor A, Shams A-H, Reham A, Syed MBA, Nadiyah MA, Ali TKZ, Subhan D, Rahul D (2021) *Synchronization of Boron application methods and rates is environmentally friendly approach to improve quality attributes of Mangifera indica L. On sustainable basis. Saudi J Bio Sci.* https://doi.org/10.1016/j.sjbs.2021.10.036

Khan N, Bano A, Ali S, Babar MA (2020) Crosstalk amongst phytohormones from planta and PGPR under biotic and abiotic stresses. *Plant Growth Regul* 90:189–203.

Khare S, Singh NB, Singh A, Hussain I, Niharika K, Yadav V, Amist N (2020) Plant secondary metabolites synthesis and their regulations under biotic and abiotic constraints. *J Plant Biol* 63(3):203–216.

Korn M, Peterek S, Mock, HP, Heyer AG, Hincha DK (2008) Heterosis in the freezing tolerance, and sugar and flavonoid contents of crosses between Arabidopsis thaliana accessions of widely varying freezing tolerance. *Plant Cell Environ* 31(6):813–827.

Kosar F, Akram NA, Ashraf M, Ahmad A, Alyemeni MN, Ahmad P (2020) Impact of exogenously applied trehalose on leaf biochemistry, achene yield and oil composition of sunflower under drought stress. *Physiol Plant.* https://doi.org/10.1111/ppl.13155

Kosar F, Akram NA, Sadiq M, Al-Qurainy F, Ashraf M (2019) Trehalose: a key organic osmolyte effectively involved in plant abiotic stress tolerance. *J Plant Growth Regul* 38(2):606–618.

Kovinich N, Kayanja G, Chanoca A, Otegui MS, Grotewold E (2015) Abiotic stresses induce different localizations of anthocyanins in Arabidopsis. *Plant Signal Behav* 10:1027850.

Krasensky J, Jonak C (2012) Drought, salt, and temperature stress-induced metabolic rearrange ments and regulatory networks. *J Exp Bot* 63:1593–1608.

Kühn H, Borchert A (2002) Regulation of enzymatic lipid peroxidation: the interplay of peroxidizing and peroxide reducing enzymes. *Free Radic Biol Med* 33:154–172.

Kumar S, Pandey AK (2013) Chemistry and biological activities of flavonoids: an overview. *Sci World J Article* 2013:1–16. https://doi.org/10.1155/2013/162750

Kumari P, Kumari R, Sharma B, Prasad SK, Singh RK (2019) Abiotic Stress Response of Field Crops: Recent Approach. *Int J Curr Microbiol App Sci* 8(4):1761–1769.

Kumutha D, Ezhilmathi K, Sairam RK, Srivastava GC, Deshmukh PS, Meena RC (2009) Waterlogging induced oxidative stress and antioxidant activity in pigeonpea genotypes. *Biol Plant* 53(1): 75–84.

Li C, Zhou K, Qin W, Tian C, Qi M, Yan X, Han W (2019) A review on heavy metals contamination in soil: effects, sources, and remediation techniques. *Soil Sediment Contam* 28(4): 380–394.

Lim CW, Baek W, Jung J, Kim JH, Lee SC (2015) Function of ABA in stomatal defense against biotic and drought stresses. *Int J Mol Sci* 16(7):15251–15270.

Liu HJ, Wang X, Yang ZL, Ren LL, Qian TT (2020a) Identification and biochemical characterization of the glutathione reductase family from *Populus trichocarpa. Plant Sci* 294:110459.

Liu HJ, Wang X, Yang ZL, Ren LL, Qian TT (2020b) Identification and biochemical characterization of the glutathione reductase family from *Populus trichocarpa. Plant Sci* 294:110459.

Liu JH, Wang W, Wu H, Gong X, Moriguchi T (2015) Polyamines function in stress tolerance: from synthesis to regulation. *Front Plant Sci* 6:827.

Liu N, Zhong G, Zhou J, Liu Y, Pang Y, Cai H, Wu Z (2019) Separate and combined effects of glyphosate and copper on growth and antioxidative enzymes in *Salvinia natans* (L.) All. *Sci Total Environ* 655:1448–1456.

Loreto F, Pinelli P, Manes F, Kollist H (2004) Impact of ozone on monoterpene emissions and evidence for an isoprene-like antioxidant action of monoterpenes emitted by *Quercus ilex* leaves. *Tree Physiol* 24(4):361–367.

Lymperopoulos P, Msanne J, Rabara R (2018) Phytochrome and phytohormones: working in tandem for plant growth and development. *Front Plant Sci* 9:1–14.

Ma J, Qiu D, Pang Y, Gao H, Wang X, Qin Y (2020). Diverse roles of tocopherols in response to abiotic and biotic stresses and strategies for genetic biofortification in plants. *Mol Breed* 40(2):1–15.

Mabhaudhi T, Chimonyo VGP, Hlahla S, Massawe F, Mayes S, Nhamo L, Modi AT (2019) Prospects of orphan crops in climate change. *Planta* 250:695–708.

Mahajan M, Kuiry R, Pal PK (2020) Understanding the consequence of environmental stress for accumulation of secondary metabolites in medicinal and aromatic plants. *J Appl Res Med Aromat Plants* 18:100255.

Mahar A, Amjad A, Altaf HL, Fazli W, Ronghua L, Muhammad A, Fahad S, Muhammad A, Rafiullah, Imtiaz AK, Zengqiang Z (2020) Promising technologies for Cd-Contaminated soils: drawbacks and possibilities. In: Fahad S, Hasanuzzaman M, Alam M, Ullah H,Saeed M, Khan AK, Adnan M (eds) *Environment, climate, plant and vegetation growth.* Springer Publ Ltd, Springer Nature Switzerland AG. Part of Springer Nature, pp. 63–92. https://doi.org/10.1007/978-3-030-49732-3

Mahmood K, Xu Z, El-Kereamy A, Casaretto JA, Rothstein SJ (2016) The Arabidopsis transcription factor ANAC032 represses anthocyanin biosynthesis in response to high sucrose and oxidative and abiotic stresses. *Front Plant Sci* 7:1548.

Mahmood T, Khalid S, Abdullah M, Ahmed Z, Shah KN, Ghafoor A, Du X (2020) Insights into drought stress signaling in plants and the molecular genetic basis of cotton drought tolerance. *Cells* 9(1):105.

Mahmood UH, Tassaduq R, Chandni I, Adnan A, Muhammad A, Muhammad MA, Muhammad HR, Mehmood AN, Alam S, Fahad S (2021) Linking plants functioning to adaptive responses under heat stress conditions: a mechanistic review. *J Plant Growth Regul.* https://doi.org/10.1007/s00344-021-10493-1

Majeed A, Siyyar S (2020) Salinity stress management in field crops: an overview of the agronomic approaches. In: Hasanuzzaman M (ed) *Plant ecophysiology and adaptation under climate change: mechanisms and perspectives II.* Springer, Singapore, pp 1–16.

Mansour MMF, Salama KHA (2020) Proline and abiotic stresses*: responses and adaptation.* In: Hasanuzzaman M (ed) *Plant ecophysiology and adaptation under climate change: mechanisms and perspectives II.* Springer, Singapore, pp. 357–397.

Manzer HS, Saud A, Soumya M, Abdullah A, Qasi DA, Bander MA, Hayssam MA, Hazem MK, Fahad S, Vishnu DR, Om PN (2021) Molybdenum and hydrogen sulfide synergistically mitigate arsenic toxicity by modulating defense system, nitrogen and cysteine assimilation in faba bean (*Vicia faba* L.) seedlings. *Environ Pollut* 290:117953.

Massacci A, Nabiev SM, Pietrosanti L, Nematov SK, Chernikova TN, Thor K, Leipner J (2008) Response of the photosynthetic apparatus of cotton (*Gossypium hirsutum*) to the onset of drought stress under field conditions studied by gas-exchange analysis and chlorophyll fluorescence imaging. *Plant Physiol Biochem* 46:189–195.

Mathur S, Agrawal D, Jajoo A (2014) Photosynthesis: response to high temperature stress. *J Photochem Photobiol B Biol* 137:116–126.

Md Jakir H, Allah B (2020) Development and applications of transplastomic plants; A way towards eco-friendly agriculture. In: Fahad S, Hasanuzzaman M, Alam M, Ullah H, Saeed M, Khan AK, Adnan M (Ed.), *Environment, climate, plant and vegetation growth.* Springer Publ Ltd, Springer Nature Switzerland AG. Part of Springer Nature, pp. 285–322. https://doi.org/10.1007/978-3-030-49732-3

Md. Enamul H, Shoeb AZM, Mallik AH, Fahad S, Kamruzzaman MM, Akib J, Nayyer S, Mehedi A, Swati AS, Md Yeamin A, Most SS (2020) Measuring vulnerability to environmental hazards: qualitative to quantitative. In: Fahad S, Hasanuzzaman M, Alam M, Ullah H, Saeed M, Khan AK, Adnan M (eds) *Environment, climate, plant and vegetation growth.* Springer Publ Ltd, Springer Nature Switzerland AG. Part of Springer Nature, pp. 421–452. https://doi.org/10.1007/978-3-030-49732-3

Miao Y, Lv D, Wang P, Wang XC, Chen J, Miao C, Song CP (2006) An *Arabidopsis* glutathione peroxidase functions as both a redox transducer and a scavenger in abscisic acid and drought stress responses. *Plant Cell* 18:2749–2766.

Mierziak J, Kostyn K, Kulma A (2014) Flavonoids as important molecules of plant interactions with the environment. *Molecules* 19:16240–16265.

Minocha R, Majumdar R, Minocha SC (2014) Polyamines and abiotic stress in plants: a complex relationship 1. *Front Plant Sci* 5:175.

Mittler R (2002) Oxidative stress, antioxidants and stress tolerance. *Trends Plant Sci* 7:405–410.

Mittova V, Volokita M, Guy M, Tal M (2000) Activities of SOD and the ascorbate-glutathione cycle enzymes in subcellular compartments in leaves and roots of the cultivated tomato and its wild salt-tolerant relative *Lycopersicon pennellii. Physiol Plant* 110:42–51.

Monneveux P, Sánchez C, Beck D, Edmeades GO (2006) Drought tolerance improvement in tropical maize source populations: evidence of progress. *Crop Sci* 46:180–191.

Mubeen M, Ashfaq A, Hafiz MH, Muhammad A, Hafiz UF, Mazhar S, Muhammad SD, Asad A, Amjed A, Fahad S, Wajid N (2020) Evaluating the climate change impact on water use efficiency of cotton-wheat in semi-arid conditions using DSSAT model. *J Water Climate Change.* https://doi.org/10.2166/wcc.2019.179/622035/jwc2019179.pdf

Muhammad N, Muqarrab A, Khurram S, Fiaz A, Fahim N, Muhammad A, Shazia A, Omaima N, Sulaiman AA, Fahad S, Subhan D, Rahul D (2021) Kaolin and Jasmonic acid improved cotton productivity under water stress conditions. *J Saudi Society Agricultural Sci* 28 (2021) 6606–6614. https://doi.org/10.1016/j.sjbs.2021.07.043

Muhammad TQ, Amna F, Amna B, Barira Z, Xitong Z, Ling-Ling C (2020) Effectiveness of conventional crop improvement strategies vs. omics. In: Fahad S, Hasanuzzaman M, Alam M, Ullah H, Saeed M, Khan AK, Adnan M (Ed.), *Environment, climate, plant and vegetation growth.* Springer Publ Ltd, Springer Nature Switzerland AG. Part of Springer Nature, pp. 253–284. https://doi.org/10.1007/978-3-030-49732-3

Muhammad Z, Abdul MK, Abdul MS, Kenneth BM, Muhammad S, Shahen S, Ibadullah J, Fahad S (2019) Performance of Aeluropus lagopoides (mangrove grass) ecotypes, a potential turfgrass, under high saline conditions. *Environ Sci Pollut Res.* https://doi.org/10.1007/s11356-019-04838-3

Munns R, Tester M (2008) Mechanisms of salinity tolerance. *Annu Rev Plant Biol* 59: 651–681.

Murtaza B, Naeem F, Shahid M, Abbas G, Shah NS, Amjad M, Murtaza G (2019) A multivariate analysis of physiological and antioxidant responses and health hazards of wheat under cadmium and lead stress. *Environ Sci Pollut Res* 26(1):362–370.

Mushtaq Z, Faizan S, Gulzar B (2020) Salt stress, its impacts on plants and the strategies plants are employing against it: A review. *J Appl Biol Biotech* 8(3):81–91.

Muzammal R, Fahad S, Guanghui D, Xia C, Yang Y, Kailei T, Lijun L, Fei-Hu L, Gang D (2021) Evaluation of hemp (*Cannabis sativa* L.) as an industrial crop: a review. *Environ Sci Pollution Res.* https://doi.org/10.1007/s11356-021-16264-5

Nadarajah KK (2020) ROS homeostasis in abiotic stress tolerance in plants. *Int J Mol Sci* 21(15):5208.

Nakabayashi R, Yonekura-Sakakibara K, Urano K, Suzuki M, Yamada Y, Nishizawa T, Michael AJ (2014) Enhancement of oxidative and drought tolerance in Arabidopsis by overaccumulation of antioxidant flavonoids. *Plant J* 7: 367–379.

Newton DE (2019) *Spotlight on current events: essays on contemporary world issues.* ABC-CLIO, Santa Barbara, CA. ISBN 978-1-4408-7063-7.
Niyogi KK, Shih C, Soon CW, Pogson BJ, Dellapenna D, Björkman O (2001) Photoprotection in a zeaxanthin- and lutein-defcient double mutant of Arabidopsis. *Photosynthesis Res* 67(1–2):139–145.
Noctor G, Gomez L, Vanacker H, Foyer CH (2002) Interactions between biosynthesis, compartmentation and transport in the control of glutathione homeostasis and signalling. *J Exp Bot* 53:1283–1304.
Noctor G, Mhamdi A, Foyer CH (2014) The roles of reactive oxygen metabolism in drought: not so cut and dried. *Plant Physiol* 164:1636–1648.
Nooghabi SN, Burkart S, Mahmoudi H, Taheri F, Damghani AM, Yazdanpanah M, Hosseininia G, Zadi H (2018) More food or better distribution? Reviewing food policy options in developing countries. *Food Rev Int* 34:566–580.
Noor M, Naveed R, Ajmal J, Fahad S, Muhammad A, Fazli W, Saud S, Hassan S (2020) Climate change and costal plant lives. In: Fahad S, Hasanuzzaman M, Alam M, Ullah H, Saeed M, Khan AK, Adnan M (eds) *Environment, climate, plant and vegetation growth.* Springer Publ Ltd, Springer Nature Switzerland AG. Part of Springer Nature, pp. 93–108. https://doi.org/10.1007/978-3-030-49732-3
Noreen S, Akhter MS, Yaamin T, Arfan M (2018) The ameliorative effects of exogenously applied proline on physiological and biochemical parameters of wheat (*Triticum aestivum* L.) crop under copper stress condition. *J Plant Interact* 13:221–230.
Nowack B, Schulin R, Robinson BH (2006) Critical assessment of chelant-enhanced metal phytoextraction. *Environ Sci Technol* 40:5225–5232.
Pandey S, Fartyal D, Agarwal A, Shukla T, James D, Kaul T, Reddy MK (2017) Abiotic stress tolerance in plants: myriad roles of ascorbate peroxidase. *Front Plant Sci* 8:581.
Park EJ, Jeknic Z, Pino MT, Murata N, Chen TH (2007) Glycinebetaine accumulation is more effective in chloroplasts than in the cytosol for protecting transgenic tomato plants against abiotic stress. *Plant Cell Environ* 30(8):994–1005.
Pastori GM, Foyer CH (2002) Common components, networks, and pathways of cross- tolerance to stress. The central role of "redox" and abscisic acid-mediated controls. *Plant Physiol* 129:460–468.
Pastori GM, Kiddle G, Antoniw J, Bernard S, Veljovic-Jovanovic S, Verrier PJ, Foyer CH (2003) Leaf vitamin C contents modulate plant defense transcripts and regulate genes that control development through hormone signaling. *Plant Cell* 15(4): 939–951.
Patzke K, Prananingrum P, Klemens PA, Trentmann O, Rodrigues CM, Keller I, Schmitz-Esser S (2019) The plastidic sugar transporter pSuT influences flowering and affects cold responses. *Plant Physiol* 179(2):569–587.
Peshev D, Vergauwen R, Moglia A, Hideg É, Van den Ende W (2013) Towards understanding vacuolar antioxidant mechanisms: a role for fructans? *J Exp Bot* 64(4):1025–1038.
Petropoulos SA, Daferera D, Polissiou MG, Passam HC (2008) The effect of water deficit stress on the growth, yield and composition of essential oils of parsley. *Sci Hortic* 115:393–397.
Pospíšil P, Prasad A (2014) Formation of singlet oxygen and protection against its oxidative damage in Photosystem II under abiotic stress. *J Photochem Photobiol B Biol* 137:39–48.
Pradhan J, Sahoo SK, Laloltra S, Sarma RS (2017) Positive impact of abiotic stress on medicinal and aromatic plants. *Int J Plant Sci* 12(2):309–313.
Qamar-Uz Z, Zubair A, Muhammad Y, Muhammad ZI, Abdul K, Fahad S, Safder B, Ramzani PMA, Muhammad N (2017) Zinc biofortification in rice: leveraging agriculture to moderate hidden hunger in developing countries. *Arch Agron Soil Sci* 64:147–161. https://doi.org/10.1080/03650340.2017.1338343
Qin S, Liu H, Nie Z, Gao W, Li C, Lin Y, Zhao P (2018) AsA–GSH cycle and antioxidant enzymes play important roles in cd tolerance of wheat. *Bull Environ Contam Toxicol* 101(5):684–690.
Qureshi FF, Ashraf MA, Rasheed R, Ali S, Hussain I, Ahmed A, Iqbal M (2020) Organic chelates decrease phytotoxic effects and enhance chromium uptake by regulating chromium-speciation in castor bean (*Ricinus communis* L.). *Sci Total Environ* 716:137061.
Rabi T, Bishayee A (2009) Terpenoids and breast cancer chemoprevention. *Breast Cancer Res Treat* 115: 223–239.
Rady MM, Hemida KA (2015) Modulation of cadmium toxicity and enhancing cadmium-tolerance in wheat seedlings by exogenous application of polyamines. *Ecotoxicol Environ Saf* 119:178–185.
Rahman S, Xuebin Q, Yatao X, Ahmad MI, Shehzad M, Zain M (2020) silicon and its application methods improve physiological traits and antioxidants in *Triticum aestivum* (L.) under cadmium stress. *J Soil Sci Plant Nutr* 1–12. https://doi.org/10.1007/s42729-020-00197-y

Rai V, Khatoon S, Bisht SS, Mehrotra S (2005) Effect of cadmium on growth, ultramorphology of leaf and secondary metabolites of *Phyllanthus amarus* Schum. And Thonn. *Chemosphere* 61(11):1644–1650.
Rai VK (2002) Role of amino acids in plant responses to stresses. *Biol Plant* 45(4):481–487.
Rasheed R, Yasmeen H, Hussain I, Iqbal M, Ashraf MA, Parveen A (2020) Exogenously applied 5-aminolevulinic acid modulates growth, secondary metabolism and oxidative defense in sunflower under water deficit stress. *Physiol Mol Biol Plants* 26(3):489–499.
Rashid M, Qaiser H, Khalid SK, Mohammad IAW, Zhang A, Muhammad A, Shahzada SI, Rukhsanda A, Ghulam AS, Shahzada MM, Sarosh A, Muhammad FQ (2020) Prospects of biochar in alkaline soils to mitigate climate change. In: Fahad S, Hasanuzzaman M, Alam M, Ullah H, Saeed M, Khan AK, Adnan M (Ed.), *Environment, climate, plant and vegetation growth.* Springer Publ Ltd, Springer Nature Switzerland AG. Part of Springer Nature, pp. 133–150. https://doi.org/10.1007/978-3-030-49732-3
Reddy AR, Chaitanya KV, Vivekanandan M (2004) Drought-induced responses of photosynthesis and antioxidant metabolism in higher plants. *J Plant Physiol* 161:1189–1202.
Rehman M, Fahad S, Saleem MH, Hafeez M, Muhammad HR, Liu F, Deng G (2020) Red light optimized physiological traits and enhanced the growth of ramie (*Boehmeria nivea* L.). *Photosynthetica* 58 (4): 922–931
Rejeb KB, Lefebvre-De Vos D, Le Disquet I, Leprince AS, Bordenave M, Maldiney R, Jdey A, Abdelly C, Savoure A (2015) Hydrogen peroxide produced by NADPH oxidases increases proline accumulation during salt or mannitol stress in Arabidopsis thaliana. *New Phytol* 208:1138–1148.
Rezayian M, Ebrahimzadeh H, Niknam V (2020) Nitric oxide stimulates antioxidant system and osmotic adjustment in soybean under drought stress. *J Soil Sci Plant Nutr* 1–11. https://doi.org/10.1007/s42729-020-00198-x
Rios JJ, Martínez-Ballesta MC, Ruiz JM, Blasco B, Carvajal M (2017) Silicon mediated improvement in plant salinity tolerance: the role of aquaporins. *Front Plant Sci* 8:948.
Riyazuddin R, Verma R, Singh K, Nisha N, Keisham M, Bhati KK, Gupta R (2020) Ethylene: a master regulator of salinity stress tolerance in plants. *Biomolecules* 10(6):959.
Rosa M, Hilal M, González, JA, Prado FE (2009) Low-temperature effect on enzyme activities involved in sucrose–starch partitioning in salt-stressed and salt-acclimated cotyledons of quinoa (*Chenopodium quinoa* Willd.) seedlings. *Plant Physiol Biochem* 47(4):300–307.
Sadam M, Muhammad TQ, Ghulam M, Muhammad SK, Faiz AJ (2020) Role of biotechnology in climate resilient agriculture. In: Fahad S, Hasanuzzaman M, Alam M, Ullah H, Saeed M, Khan AK, Adnan M (Ed.), *Environment, climate, plant and vegetation growth.* Springer Publ Ltd, Springer Nature Switzerland AG. Part of Springer Nature, pp. 339–366. https://doi.org/10.1007/978-3-030-49732-3
Safi UK, Ullah F, Mehmood S, Fahad S, Ahmad Rahi A, Althobaiti F, et al. (2021) Antimicrobial, antioxidant and cytotoxic properties of *Chenopodium glaucum* L. *PLoS ONE* 16(10): e0255502. https://doi.org/10.1371/journal.pone.0255502
Sahrish N, Shakeel A, Ghulam A, Zartash F, Sajjad H, Mukhtar A, Muhammad AK, Ahmad K, Fahad S, Wajid N, Sezai E, Carol Jo W, Gerrit H (2022) Modeling the impact of climate warming on potato phenology. *European J AgroN* 132:126404.
Sajid H, Jie H, Jing H, Shakeel A, Satyabrata N, Sumera A, Awais S, Chunquan Z, Lianfeng Z, Xiaochuang C, Qianyu J, Junhua Z (2020) Rice production under climate change: adaptations and mitigating strategies. In: Fahad S, Hasanuzzaman M, Alam M, Ullah H, Saeed M, Khan AK, Adnan M (eds) *Environment, climate, plant and vegetation growth.* Springer Publ Ltd, Springer Nature Switzerland AG. Part of Springer Nature, pp. 659–686. https://doi.org/10.1007/978-3-030-49732-3
Sajjad H, Muhammad M, Ashfaq A, Waseem A, Hafiz MH, Mazhar A, Nasir M, Asad A, Hafiz UF, Syeda RS, Fahad S, Depeng W, Wajid N (2019) Using GIS tools to detect the land use/land cover changes during forty years in Lodhran district of Pakistan. *Environ Sci Pollut Res.* https://doi.org/10.1007/s11356-019-06072-3
Salama KH, Mansour MMF (2015) Choline priming-induced plasma membrane lipid alterations contributed to improved wheat salt tolerance. *Acta Physiol Plant* 37:170.
Salazar C, Hernandez C, Pino MT (2015) Plant water stress: associations between ethylene and abscisic acid response. *Chil J Agric Res* 75:71–79.
Saleem MH, Ali S, Irshad S, Hussaan M, Rizwan M, Rana MS, Ahmad P (2020) Copper uptake and accumulation, ultra-structural alteration, and bast fibre yield and quality of fibrous jute (*Corchorus capsularis* L.) plants grown under two different soils of China. *Plants* 9(3):404.
Saleem MH, Fahad S, Adnan M, Mohsin A, Muhammad SR, Muhammad K, Qurban A, Inas AH, Parashuram B, Mubassir A, Reem MH (2020a) Foliar application of gibberellic acid endorsed phytoextraction of

copper and alleviates oxidative stress in jute (*Corchorus capsularis* L.) plant grown in highly copper-contaminated soil of China. *Environ Sci Pollution Res*. https://doi.org/10.1007/s11356-020-09764-3

Saleem MH, Fahad S, Shahid UK, Mairaj D, Abid U, Ayman ELS, Akbar H, Analía L, Lijun L (2020c) Copper-induced oxidative stress, initiation of antioxidants and phytoremediation potential of flax (*Linum usitatissimum* L.) seedlings grown under the mixing of two different soils of China. *Environ Sci Poll Res*. https://doi.org/10.1007/s11356-019-07264-7

Saleem MH, Rehman M, Fahad S, Tung SA, Iqbal N, Hassan A, Ayub A, Wahid MA, Shaukat S, Liu L, Deng G (2020b) Leaf gas exchange, oxidative stress, and physiological attributes of rapeseed (*Brassica napus* L.) grown under different light-emitting diodes. *Photosynthetica* 58 (3): 836–845.

Saman S, Amna B, Bani A, Muhammad TQ, Rana MA, Muhammad SK (2020) QTL mapping for abiotic stresses in cereals. In: Fahad S, Hasanuzzaman M, Alam M, Ullah H, Saeed M, Khan AK, Adnan M (Ed.), *Environment, climate, plant and vegetation growth*. Springer Publ Ltd, Springer Nature Switzerland AG. Part of Springer Nature, pp. 229–252. https://doi.org/10.1007/978-3-030-49732-3

Samarah NH (2005) Effects of drought stress on growth and yield of barley. *Agron Sustain Dev* 25:145–149.

Sami F, Yusuf M, Faizan M, Faraz A, Hayat S (2016) Role of sugars under abiotic stress. *Plant Physiol Biochem* 109:54–61.

Sankar B, Jaleel CA, Manivannan P, Kishorekumar A, Somasundaram R, Panneerselvam R (2007) Effect of paclobutrazol on water stress amelioration through antioxidants and free radical scavenging enzymes in *Arachis hypogaea* L. *Colloids Surf B Biointerfaces* 60:229–235.

Sarowar S, Kim EN, Kim YJ, Ok SH, Kim KD, Hwang BK, Shin JS (2005) Overexpression of a pepper ascorbate peroxidase-like 1 gene in tobacco plants enhances tolerance to oxidative stress and pathogens. *Plant Sci* 169(1):55–63.

Saud S, Chen Y, Fahad S, Hussain S, Na L, Xin L, Alhussien SA (2016) Silicate application increases the photosynthesis and its associated metabolic activities in Kentucky bluegrass under drought stress and post-drought recovery. *Environ Sci Pollut Res* 23(17):17647– 17655. https://doi.org/10.1007/s11356-016-6957-x

Saud S, Chen Y, Long B, Fahad S, Sadiq A (2013) The different impact on the growth of cool season turf grass under the various conditions on salinity and drought stress. *Int J Agric Sci Res* 3:77–84

Saud S, Fahad S, Cui G, Chen Y, Anwar S (2020) Determining nitrogen isotopes discrimination under drought stress on enzymatic activities, nitrogen isotope abundance and water contents of Kentucky bluegrass. *Sci Rep* 10:6415. https://doi.org/10.1038/s41598-020-63548-w

Saud S, Fahad S, Yajun C, Ihsan MZ, Hammad HM, Nasim W, Amanullah Jr, Arif M and Alharby H (2017) Effects of nitrogen supply on water stress and recovery mechanisms in Kentucky bluegrass plants. *Front Plant Sci* 8:983. https://doi.org/10.3389/fpls.2017.00983

Saud S, Li X, Chen Y, Zhang L, Fahad S, Hussain S, Sadiq A, Chen Y (2014) Silicon application increases drought tolerance of Kentucky bluegrass by improving plant water relations and morph physiological functions. *SciWorld J* 2014:1–10. https://doi.org/10.1155/2014/ 368694

Schardl CL, Panaccione DG, Tudzynski P (2006) Ergot alkaloids–biology and molecular biology. *Alkaloids Chem Boil* 63:45–86.

Seckin B, Sekmen AH, Türkan I (2009) An enhancing effect of exogenous mannitol on the antioxidant enzyme activities in roots of wheat under salt stress. *J Plant Growth Regul* 28(1):12–20.

Seneviratne G, Jayasinghearachchi HS (2003) Mycelial colonization by bradyrhizobia and azorhizobia. *J Biosci* 28(2):243–247.

Senol C (2020) The effects of climate change on human behaviors. In: Fahad S, Hasanuzzaman M, Alam M, Ullah H, Saeed M, Khan AK, Adnan M (eds) *Environment, climate, plant and vegetation growth*. Springer Publ Ltd, Springer Nature Switzerland AG. Part of Springer Nature, pp. 577–590. https://doi.org/10.1007/978-3-030-49732-3

Shafi MI, Adnan M, Fahad S, Fazli W, Ahsan K, Zhen Y, Subhan D, Zafar-ul-Hye M, Martin B, Rahul D (2020) Application of single superphosphate with humic acid improves the growth, yield and phosphorus uptake of wheat (*Triticum aestivum* L.) in calcareous soil. *Agron* 10:1224. https://doi.org/10.3390/agronomy10091224

Shah F, Lixiao N, Kehui C, Tariq S, Wei W, Chang C, Liyang Z, Farhan A, Fahad S, Huang J (2013) Rice grain yield and component responses to near 2°C of warming. *Field Crop Res* 157:98–110.

Shahnaz MM, Iqbal M, Akhter N (2020) Tissue-specific modulation of metabolism and nutrients acquisition through seed priming with sodium selenate confers salt tolerance in wheat. *Arch Agron Soil Sci*. https://doi.org/10.1080/03650340.2020.1797999

Sharkey TD (2005) Effects of moderate heat stress on photosynthesis: importance of thylakoid reactions, rubisco deactivation, reactive oxygen species, and thermotolerance provided by isoprene. *Plant Cell Environ* 28:269–277.

Sharma A, Kumar V, Shahzad B, Ramakrishnan M, Sidhu GP, Ali AS, Bakshi P (2020). Photosynthetic response of plants under different abiotic stresses: a review. *J Plant Growth Regul* 39:509–531.

Sharma A, Shahzad B, Kumar V, Kohli SK, Sidhu GP, Bali AS, Zheng B (2019b) Phytohormones regulate accumulation of osmolytes under abiotic stress. *Biomolecules* 9(7):285.

Sharma A, Shahzad B, Rehman A, Bhardwaj R, Landi M, Zheng B (2019a) Response of phenylpropanoid pathway and the role of polyphenols in plants under abiotic stress. *Molecules* 24(13):2452.

Sharma SS, Schat H, Vooijs R (1998) In vitro alleviation of heavy metal-induced enzyme inhibition by proline. *Phytochemistry* 49(6):1531–1535.

Shekari G, Javanmardi J (2017) Effects of foliar application pure amino acid and amino acid containing fertilizer on broccoli (*Brassica oleracea* L. var. italica) transplants. *Adv Crop Sci Tech* 5(3):1–4.

Shi H, Chan Z (2014) Improvement of plant abiotic stress tolerance through modulation of the polyamine pathway. *J Int Plant Biol* 56:114–121.

Sidra K, Javed I, Subhan D, Allah B, Syed IUSB, Fatma B, Khaled DA, Fahad S, Omaima N, Ali TKZ, Rahul D (2021) Physio-chemical characterization of indigenous agricultural waste materials for the development of potting media. *J Saudi Society Agricultural Sci.* https://doi.org/10.1016/j.sjbs.2021.08.058

Sohag AAM, Tahjib-Ul-Arif M, Afrin S, Khan MK, Hannan MA, Skalicky M, Murata Y (2020) Insights into nitric oxide-mediated water balance, antioxidant defence and mineral homeostasis in rice (*Oryza sativa* L.) under chilling stress. *Nitric Oxide* 100:7–16.

Souri Z, Karimi N, Farooq MA, Sandalio LM (2020) Nitric oxide improves tolerance to arsenic stress in Isatis cappadocica desv. *Shoots by enhancing antioxidant defenses. Chemosphere* 239:124523.

Srivastava NK, Srivastava AK (2010) Influence of some heavy metals on growth, alkaloid content and composition in *Catharanthus roseus* L. *Indian J Pharm Sci* 72(6):775–778.

Stambulska UY, Bayliak MM, Lushchak VI (2018) Chromium (VI) toxicity in legume plants: modulation effects of rhizobial symbiosis. *BioMed Res Int.* https://doi.org/10.1155/2018/8031213

Subhan D, Zafar-ul-Hye M, Fahad S, Saud S, Martin B, Tereza H, Rahul D (2020) Drought Stress Alleviation by ACC Deaminase Producing *Achromobacter xylosoxidans* and *Enterobacter cloacae*, with and without Timber Waste Biochar in Maize. *Sustain* 12(6286). https://doi.org/10.3390/su12156286

Sureshan KM, Murakami T, Watanabe Y (2009) Total syntheses of cyclitol based natural products from myo-inositol: brahol and pinpollitol. *Tetrahedron* 65(20):3998–4006.

Tang L, Deng S, Tan D, Long J, Lei M (2019) Heavy metal distribution, translocation, and human health risk assessment in the soil-rice system around Dongting Lake area, *China. Environ Sci Pollut Res* 26(17):17655–17665.

Tariq M, Ahmad S, Fahad S, Abbas G, Hussain S, Fatima Z, Nasim W, Mubeen M, Rehman MH, Khan MA, Adnan M (2018). The impact of climate warming and crop management on phenology of sunflower-based cropping systems in Punjab, Pakistan. *Agric Forest Met.* 15(256):270–282.

Thakur A, Sharma KD, Siddique KH, Nayyar H (2020) Cold priming the chickpea seeds imparts reproductive cold tolerance by reprogramming the turnover of carbohydrates, osmo-protectants and redox components in leaves. *Sci Hortic* 261:108929.

Tyagi S, Singh K, Upadhyay SK (2020) Molecular characterization revealed the role of catalases under abiotic and arsenic stress in bread wheat (*Triticum aestivum* L.). *J Hazard Mater* 403:123585.

Tyystjarvi E (2008) Photoinhibition of photosystem II and photodamage of the oxygen evolving manganese cluster. *Coord Chem Rev* 252:361–376.

Unsar N-U, Muhammad R, Syed HMB, Asad S, Mirza AQ, Naeem I, Muhammad HR, Fahad S, Shafqat S (2020) Insect pests of cotton crop and management under climate change scenarios. In: Fahad S, Hasanuzzaman M, Alam M, Ullah H, Saeed M, Khan AK, Adnan M (Ed.), *Environment, climate, plant and vegetation growth.* Springer Publ Ltd, Springer Nature Switzerland AG. Part of Springer Nature, pp. 367–396. https://doi.org/10.1007/978-3-030-49732-3

Uzilday B, Ozgur R, Sekmen AH, Turkan I (2015) Redox regulation and antioxidant defence during abiotic stress: what have we learned from *Arabidopsis* and its relatives?. In: Gupta DK, Palma JM, Corpas FJ (eds) *Reactive oxygen species and oxidative damage in plants under stress.* Springer, Cham, pp. 83–113.

Veljović-Jovanović S, Vidović M, Morina F (2017) Ascorbate as a key player in plant abiotic stress response and tolerance. In: Hossain MA, Munné-Bosch S, Burritt DJ, Diaz-Vivancos P (eds) *Ascorbic acid in plant growth, development and stress tolerance*. Springer, Cham, pp. 47–109.

Wahid A, Gelani S, Ashraf M, Foolad MR (2007) *Heat tolerance in plants: an overview. Environ Exp Bot* 61:199–223.

Wahid F, Fahad S, Subhan D, Adnan M,, Zhen Y, Saud S, Manzer HS, Martin B, Tereza H, Rahul D (2020) Sustainable management with mycorrhizae and phosphate solubilizing bacteria for enhanced phosphorus uptake in calcareous soils. *Agri* 10 (334). https://doi.org/10.3390/agriculture10080334

Wajid N, Ashfaq A, Asad A, Muhammad T, Muhammad A,Muhammad S, Khawar J, Ghulam MS, Syeda RS, Hafiz MH, Muhammad IAR, Muhammad ZH, Muhammad HR, Veysel T, Fahad S, Suad S, Aziz K, Shahzad A (2017) Radiation efficiency and nitrogen fertilizer impacts on sunflower crop in contrasting environments of Punjab. *Pakistan Environ Sci Pollut Res* 25:1822–1836. https://doi.org/10.1007/s11356-017-0592-z

Wang, X., Saleem, M. H., Parveen, A., Mumtaz, S., Hassan, A., Adnan, M., Yasin, G. (2021). Proximate composition and nutritive value of some leafy vegetables from Faisalabad, Pakistan. *Sustainability* 13(15), 8444.

Wani MA, Jan N, Qazi HA, Andrabi KI, John R (2018) Cold stress induces biochemical changes, fatty acid profile, antioxidant system and gene expression in *Capsella bursa pastoris* L. *Acta Physiol Plant* 40(9):167.

Weis E (1981) Reversible heat-inactivation of the calvin cycle: a possible mechanism of the temperature regulation of photosynthesis. *Planta* 151:33–39.

Wu C, Kehui C, She T, Ganghua L, Shaohua W, Fahad S, Lixiao N,Jianliang H, Shaobing P, Yanfeng D (2020) Intensified pollination and fertilization ameliorate heat injury in rice (*Oryza sativa* L.) during the flowering stage. *Field Crops Res* 252:107795.

Wu C, Tang S, Li G, Wang S, Fahad S, Ding Y (2019) Roles of phytohormone changes in the grain yield of rice plants exposed to heat: a review. *PeerJ* 7:e7792. https://doi.org/10.7717/peerj.7792

Yang Z, Zhang Z, Zhang T, Fahad S, Cui K, Nie L, Peng S, Huang J (2017) The effect of season-long temperature increases on rice cultivars grown in the central and southern regions of China. *Front Plant Sci* 8:1908. https://doi.org/10.3389/fpls.2017.01908

Yousuf PY, Hakeem KUR, Chandna R, Ahmad P (2012) Role of glutathione reductase in plant abiotic stress. In: Ahmad P, Prasad MNV (eds) *Abiotic stress responses in plants*. Springer, New York, NY, pp. 149–158.

Zafar-ul-Hye M, Muhammad TH, Muhammad A, Fahad S, Martin B, Tereza D, Rahul D, Subhan D (2020b) Potential role of compost mixed biochar with rhizobacteria in mitigating lead toxicity in spinach. *Scientific Rep* 10:12159. https://doi.org/10.1038/s41598-020-69183-9

Zafar-ul-Hye M, Muhammad N, Subhan D, Fahad S, Rahul D, Mazhar A, Ashfaq AR, Martin B, Ji ˇrí H, Zahid HT, Muhammad N (2020a) Alleviation of cadmium adverse effects by improving nutrients uptake in bitter gourd through cadmium tolerant rhizobacteria. *Environ* 7 (54). https://doi.org/10.3390/environments7080054

Zafar-ul-Hye M, Akbar MN, Iftikhar Y, Abbas M, Zahid A, Fahad S, Datta R, Ali M, Elgorban AM, Ansari MJ et al. (2021) Rhizobacteria inoculation and Caffeic acid alleviated drought stress in lentil plants. *Sustain* 13: 9603. https://doi.org/10.3390/su13179603

Zahida Z, Hafiz FB, Zulfiqar AS, Ghulam MS, Fahad S, Muhammad RA, Hafiz MH, Wajid N, Muhammad S (2017) Effect of water management and silicon on germination, growth, phosphorus and arsenic uptake in rice. *Ecotoxicol Environ Saf* 144:11–18.

Zahir SM, Zheng-HG, Ala UD, Amjad A, Ata UR, Kashif J, Shah F, Saud S, Adnan M, Fazli W, Saud A, Manzer HS, Shamsher A, Wajid N, Hafiz MH, Fahad S (2021) Synthesis of silver nanoparticles using Plantago lanceolata extract and assessing their antibacterial and antioxidant activities. *Sci Rep* 11:20754.

Zhou P, Li Q, Liu G, Xu N, Yang Y, Zeng W, Chen A, Wang S (2018) Integrated analysis of transcriptomic and metabolomic data reveals critical metabolic pathways involved in polyphenol biosynthesis in *Nicotiana tabacum* under chilling stress. *Funct Plant Biol* 46:30–43.

Zia-Ur-Rehman M (2020) Environment, Climate Change and Biodiversity. In: Fahad S, Hasanuzzaman M, Alam M, Ullah H, Saeed M, Khan AK, Adnan M (Ed.), *Environment, climate, plant and vegetation growth*. Springer Publ Ltd, Springer Nature Switzerland AG. Part of Springer Nature, pp. 473–502. https://doi.org/10.1007/978-3-030-49732-3

8

Reactive Oxygen Species and Antioxidants Defence in Plants under Hostile Environment

Kashif Tanwir
Government College University, Faisalabad, Pakistan

Muhammad Tariq Javed and Muhammad Shahid
Government College University, Faisalabad, Pakistan

Nabeel Khan Niazi
University of Agriculture Faisalabad, Faisalabad, Pakistan

Sabaz Ali Khan
Institute of Information Technology, Abbottabad, Pakistan

Naeem Iqbal
Government College University, Faisalabad, Pakistan

CONTENTS

DOI: 10.1201/9781003286417-8

8.1 Introduction

The hostile environments of various kinds faced by plants are generally characterized by reduced rhizospheric microbial community, which ultimately decreased fertility of soil and plant growth, with low biomass accumulation. Hostile environments also imbalance the nutrients availability to plants, which result in stunted plant metabolic activities. Plants are fixed entities; so they cannot avoid exposure to such environmental fluctuations. Plants are encountering several kinds of hostile environments like scarcity of water, water logging, temperature fluctuations (chilling, frost, hot), nutrients deficiency, higher concentrations of organic pollutants, salinity, heavy metals and metalloids toxicity. All these climate changes generate biotic and abiotic stresses in plants frequently and with more intensity. These climate changes produced either naturally or generated by human activities have now become a worldwide danger for sustainable agricultural growth (Ahmad et al. 2017; Tomar et al. 2019; Deepranjan et al. 2021; Haider et al. 2021; Huang Li et al. 2021; Ikram et al. 2021; Jabborova et al. 2021; Khadim et al. 2021a, 2021b; Manzer et al. 2021; Muzammal 2021; Abdul et al. 2021a, 2021b; Ashfaq et al. 2021; Amjad et al. 2021; Atif et al. 2021; Athar et al. 2021; Adnan et al. 2018a, 2018b, 2019; Akram et al. 2018a, 2018b; Aziz et al. 2017a, 2017b; Chang et al. 2021; Chen et al. 2021; Emre et al. 2021; Habib et al. 2017; Hafiz et al. 2016, 2019; Ghulam et al. 2021; Guofu et al. 2021; Hafeez et al. 2021; Khan et al. 2021; Kamarn et al. 2017; Muhammad et al. 2019; Safi et al. 2021; Sajjad et al. 2019; Saud et al. 2013, 2014, 2016, 2017, 2020; Shah et al. 2013; Qamar-Uz et al. 2017; Hamza et al. 2021; Irfan et al. 2021; Wajid et al. 2017; Yang et al. 2017; Zahida et al. 2017; Depeng et al. 2018; Hussain et al. 2020; Hafiz et al. 2020a, 2020b; Shafi et al. 2020; Wahid et al. 2020; Subhan et al. 2020; Zafar-ul-Hye et al. 2020a, 2020b; Zafar-ul-Hye et al. 2021; Adnan et al. 2020; Ilyas et al. 2020; Saleem et al. 2020a, 2020b, 2020c; Rehman et al. 2020; Farhat et al. 2020; Wu et al. 2020; Mubeen et al. 2020; Farhana et al. 2020; Jan et al. 2019; Wu et al. 2019; Ahmad et al. 2019; Baseer et al. 2019; Hafiz et al. 2018; Tariq et al. 2018).

These biotic and abiotic stresses stunted the plant growth and yield (Ramegowda and Senthil-Kumar 2015; Fahad and Bano 2012; Fahad et al. 2013, 2014a, 2014b, 2015a, 2015b, 2016a, 2016b, 2016c, 2016d, 2017a, 2017b, 2018a, 2018b, 2019a, 2019b, 2020, 2021a, 2021b, 2021c, 2021d, 2021e, 2021f; Hesham and Fahad 2020. Iqra et al. 2020; Akbar et al. 2020; Mahar et al. 2020; Noor et al. 2020; Bayram et al. 2020; Amanullah and Fahad 2017, 2018a, 2018b; Amanullah et al. 2020, 2021; Rashid et al. 2020; Arif et al. 2020; Amir et al. 2020; Saman et al. 2020; Muhammad et al. 2020; Md Jakir and Allah 2020; Mahmood et al. 2021; Farah et al. 2020; Sadam et al. 2020; Unsar et al. 2020; Fazli et al. 2020; Md. Enamul et al. 2020; Gopakumar et al. 2020; Zia-Ur-Rehman 2020; Ayman et al. 2020; Mohammad et al. 2020; Senol 2020; Amjad et al. 2020; Ibrar et al. 2020; Sajid et al. 2020; Muhammad et al. 2021; Sidra et al. 2021; Zahir et al. 2021; Sahrish et al. 2022). Soil salinization, deficiency of water, increased environmental temperature, cold and frost stresses are harmful for normal plant health. Plants are unable to avoid these environments; instead they have developed various complex mechanisms that confer protection and hence support survival during unfavourable times. Hence, farmers should give importance to these stress ameliorating strategies of crop plants for obtaining sustainable yield and growth (Fracasso et al. 2016). Plants have a system to sense these stress factors and in response trigger an appropriate mechanism to cope. Environmental stresses trigger the production of reactive oxygen species (ROS) as a stress indicator in some vital cellular biochemical processes, which work in different cellular organelles such as chloroplast, cell membrane, peroxisomes, cell wall, mitochondria (Ahanger et al. 2017; Arafat et al. 2016). ROS are essential and common factor in almost all kinds of environmental stresses that generate downstream responses in stress signalling (Singh et al. 2019). Accumulation of ROS in plants leads to the generation of oxidative stress. The ROS are of two types including free radicals like superoxide anion ($O_2^{\bullet-}$), peroxyl radical ($RO_2^{\bullet}$), hydroperoxyl radical ($HO_2^{\bullet}$), hydroxyl radical ($OH^{\bullet}$), alkoxy radical ($RO^{\bullet}$), carbonate ($CO_3^{\bullet-}$), semiquinone ($SQ^{\bullet-}$) and nonradicals hydrogen peroxide (H_2O_2), singlet oxygen ($^1O_2$), hydro peroxide (HOOH), hypobromous acid (HOBr), ozone (O_3), hydroperoxide (ROOH), hypochlorous acid (HOCl) (Mehla et al. 2017; Hasanuzzaman et al. 2019). The atmospheric oxygen was partially reduced and converted into activated oxygen radical by electron transfer or high-energy based initiation reactions (Choudhury et al. 2017). These ROS generating reactions take place in cellular compartments such as mitochondria, chloroplast, cellular membrane, peroxisomes, apoplast, and cell wall (Singh et al. 2019). The production of ROS inside plant cell is a component of cellular metabolism; however oxidative stress base overproduction relentlessly breakdown the essential cellular elements like lipids, DNA, carbohydrates, and proteins due to its higher reactivity (Raja et al. 2017; Hasanuzzaman et al. 2020).

The higher accumulation of ROS leads to the following damages at molecular levels in cell.

(i) It disrupts the nucleic acids structure by initiating various reactions like nucleotides consumption, deoxyribose oxidation, breakdown nucleic acid chains, cross-linked DNA-proteins and changes in arrangements of nitrogen bases. (ii) Lipid peroxidation takes place due to the breakdown of the peptide bonds, which enhanced the lipids viscosity and cross-membrane diffusion. (iii) The denaturation of proteins takes place via oxidation through various reactions like breakage of peptide linkages, accretion of cross-linked reaction outcomes, and finally the alteration of electric charge. The damage caused by overproduction of ROS results in apoptosis of cell, which eventually decreased the plant growth and yield (Mittler 2017).

Although plants have developed various mechanisms to cope with ROS-based stress damages at cellular level. Primarily plants produced various kinds of antioxidants to deal with overproduction of ROS. This endogenous self-protective machinery comprising of various enzymatic (superoxide dismutase (SOD), peroxiredoxins (PRXs), catalase (CAT), oxidase-like alternative oxidase (AOX), ascorbate peroxidase (APX), glutaredoxin (GRX), glutathione reductase (GR), monodehydroascorbate reductase (MDHAR), dehydroascorbate reductase (DHAR), glutathione peroxidase (GPX), thioredoxins (TRXs), guaiacol peroxidase (GOPX), glutathione S-transferase (GST)) and nonenzymatic antioxidants (tocopherol, ascorbic acid (AsA), phenolic acids, glutathione (GSH), alkaloids, flavonoids, nonprotein amino acids and carotenoids) (Kaur et al. 2019; Al Mahmud et al. 2019). In plants, the antioxidants and ROS are present in equilibrium to uphold the normal cellular metabolism (Hasanuzzaman et al. 2020). Sustaining this ROS and antioxidants balance in the cell assists suitable redox biological reactions and the regulation of different cellular processes critical for normal growth and development of plants (Mittler 2017). The ROS scavengers are playing a critical role in maintaining this intermediate level of ROS production (Hasanuzzaman et al. 2019). However, unfavourable environment generates oxidative stress conditions which start over-accumulation of ROS in plants, which ultimately imbalances the equilibrium and starts damaging cellular and molecular processes, leading to apoptosis and declining plant yield (Raja et al. 2017).

Additionally ROS also have some beneficial roles in plants' normal growth like it acts as signalling molecule or secondary messengers, which are responsible for carrying signals through redox reactions using MAPK pathway to the nucleus in a number of cellular reactions to enhanced biotic and abiotic stress tolerance (Singh et al. 2019). The ROS also act as major molecules in the acclimation process of plants under environmental stimuli. They primarily act as signal transducing agent, which control various pathways throughout the acclimation of the plant under stress conditions (Antoniou et al. 2016). Numerous studies exhibited that ROS are necessary for the success of many important natural metabolic reactions, comprising cellular differentiation and proliferation (Mittler, 2017). Additionally, H_2O_2 also have an important constituent of oxidative stress response regulation in various plants such as wheat (Habib et al. 2020), rice (Sohag et al. 2020), mung bean (Fariduddin et al. 2014), strawberry (Christou et al. 2014), maize (Tanwir et al. 2015), soybean (Guler and Pehlivan 2016), basil (Gohari et al. 2020), cucumber (Sun et al. 2016), rapeseed (Naseer et al. 2016; Hasanuzzaman et al. 2017) and sour orange (Tanou et al. 2012). The main objective of this chapter is to explain the role of ROS in generating oxidative stress during different types of hostile environments and their mitigation by enzymatic and nonenzymatic antioxidants.

8.2 Hostile Environments

Any sort of environment – intrinsic or extrinsic, anthropogenic or natural – which is unfavourable for the normal metabolic processes, growth, development, and yield of plants is hostile in nature. Plants being sessile have to face various kinds of such environments during their life cycle (Figure 8.1).

8.2.1 Heat Stress

Temperature is highly important among the various components of plant environment; continuously increasing ambient temperature is considered as one of the most harmful stresses. Worldwide 0.2°C rise in temperature was recorded after every ten years, which will result in 1.8–4.0°C increase in global temperature at the end of 21st century (IPCC 2007). This rise in temperature is creating unrest amongst the scientists, because heat stress is highly dangerous for living organisms. The heat stress has dual effects on organisms; it directly affects plant metabolisms or indirectly causes fluctuations in surrounding environments, which ultimately affects plant. Continuous exposure to high temperature stress leads to stunted growth and cell death (Lobell and Field 2007; Hasanuzzaman et al. 2013). The heat stress in plants generates oxidative stress via the production of excess ROS, which leads to oxidative injury

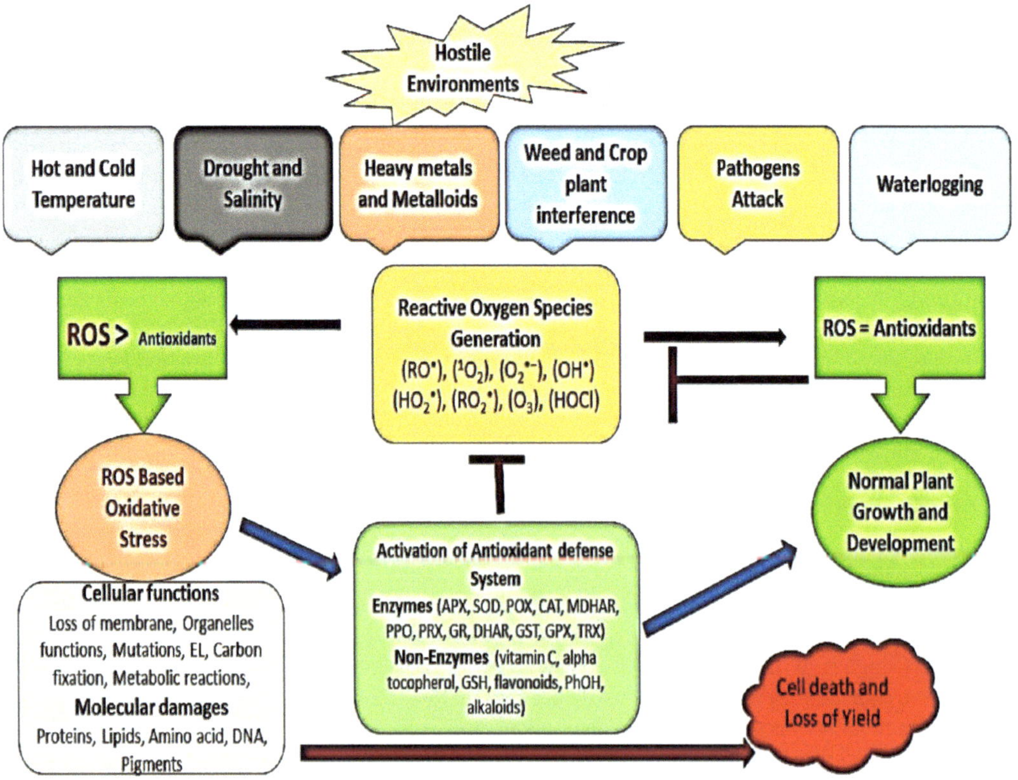

FIGURE 8.1 Hostile environments and their effects on the generation of reactive oxygen species (ROS) and antioxidants-based mitigation of ROS.

(Hasanuzzaman et al. 2013). Heat stress in plants increased the production of ROS, which resulted in alteration of cellular and molecular metabolism, enhanced lipid peroxidation, cell membrane injury, disabled oxygen-evolving compound, and nicking of plant DNA, all of which ultimately drive plants towards cell death (Hasanuzzaman et al. 2013). Plants which are facing higher temperature than 35°C /32°C day/night reduced photosynthesis by blocking electron flow from reaction center of PS-II, reducing quantum efficiency of photosystems (Fv/Fm), and damaging PS-II structures (Fahad et al. 2016). Cucumber plant increased its $O_2^{•-}$ content up to 80% when growing heat stress under 35°C (Ding et al. 2016). Plants unremittingly fight for their existence under different environmental stresses including hot temperature. Plants have two main strategies to encounter unfavourable heat: (i) physical adjustments within the plant body and (ii) by producing regular signalling cascade for changing internal metabolic processes (Rahman et al. 2016; Liu et al. 2019). They alter their metabolism by producing attuned solutes, which act as organizer of proteins, cellular structures, and also sustain cell turgor pressure by osmotic alterations and amend the antioxidant system to readjust the homeostasis and cellular redox balance (Janska et al. 2010).

8.2.2 Metals and Metalloids-based Toxicity

Some essential metal elements are required for the normal growth of the plants, like iron (Fe), copper (Cu), zinc (Zn), manganese (Mn), nickel (Ni), and boron (B) up to a specific optimum limit (Javed et al. 2017). These elements are involved in a variety of biochemical and physiological reactions in plants for the formation and activation of various metal-dependent proteins and enzymes. Some heavy metals like lead (Pb), cadmium (Cd), aluminium (Al), chromium (Cr), mercury (Hg) and also some metalloids like arsenic (As) are not required for plant metabolism and all these are potentially toxic for normal plant metabolic processes (Tanwir et al. 2015; Javed et al. 2018, 2019). These toxic metals and metalloids are highly mobile, reactive, and generate toxicity for plants' cellular metabolism reliant on their nature, concentration, and types. After accumulation in plant body, they rapidly generate ROS leading towards oxidative stress (Mahmud et al. 2018). They produce ROS in plants by disturbing the mitochondrial and chloroplast membrane electron transformation activities together with peroxisome-based oxidative metabolism (Parvez et al. 2020). Many studies reported that under heavy metal toxicity the plants encountered ROS-based oxidative stress. A notable rise in lipid peroxidation and production of H_2O_2 was recorded in *O. sativa* plant seedlings when cultivated under various concentrations of nickel (Hasanuzzaman et al. 2019). Lower concentrations of Ni up to 100mM also increased remarkable ROS generation in *Pisum sativum* (El-Amier et al. 2019). Similarly, under Cd stress increase in H_2O_2, MDA, and $O_2^{•-}$ levels in diverse crop plants (Al Mahmud et al. 2018) were seen. For example, cadmium toxicity (100µM $CdCl_2$) caused higher accumulation of $O_2^{•-}$, H_2O_2, and MDA in *Z. mays* (Javed et al. 2017), in *A. thaliana* (Gupta et al. 2017), and *C. sativum* at seedling stage (Kabala et al. 2019).

8.2.3 Organic Pollutants Stress

In recent years organic pollutants utilization in household and agriculture commodities was immensely increased, which is also creating hostile environment for the normal growth and development of agricultural plants. These organic pollutants include polychlorinated biphenyls (PCBs), herbicides, PCDD/Fs, bisphenol-A, and antibiotics that are frequently present in agricultural environments. Organic pollutants themselves and in soil are transformed into various products that are toxic for plants. Being hydrophobic in nature they easily enter into the plants, animals, and humans bodies. These organic pollutants are entering into the soil-plant system by various anthropogenic and natural activities. They are highly mobile, toxic, and abundant everywhere in soil, water, and air. These organic pollutants inside plant body generate oxidative stress, ROS which alter numerous physiobiochemical reactions inside plants at cellular and molecular levels. They are also able to change the kinetic energies of metabolic processes inside plants, which are linked with development and growth (Tomar et al. 2019). Many researchers reported the increased production of MDA, H_2O_2, and ROS in plants growing under different organic pollutants (Yildiztekin et al. 2015; Kreslavski et al. 2017; Jajoo et al. 2014; Jajoo, 2017; Sharma et al. 2018; Tomar et al. 2019).

8.2.4 Crop and Weed Interference

Weeds are important competitors of the crop plants, and they produce an important biotic stress which severely reduced crop yield (Radosevich et al. 2007). This competition depends upon various factors including crop type, weed species, potential allelopathy, time period of coexistence, edaphoclimatic conditions, plant density, and spatial distribution (Barroso et al. 2010; Caverzan et al. 2019).

Competition amongst the weeds and crops is the communal type of interference for the same resources (Radosevich et al. 2007). This competition can be both interspecific and intraspecific, such as high crop density. The specie with more competency and ability adapt themselves according to the environment and get more resources, grow faster, and produce higher yield (Barroso et al. 2010). Weed interference stress also showed variations in ROS production in crops together with changes in gene expression and enzyme concentration (Afifi and Swanton 2012; Gal et al. 2015; Agostinetto et al. 2017; Piasecki et al. 2018).

According to Afifi, and Swanton (2012), the interference between maize and ryegrass seedlings amplified the H_2O_2 production and the expression of antioxidant gene like ZmSOD2, ZmGST1, ZmCAT3, and ZmAPX2. The production of H_2O_2 leads the plant towards oxidative stress by disturbing physiological and biochemical processes, which ultimately result in yield loss. Gal et al. (2015) also studied a similar relationship between ryegrass and soybean crop, which showed reduction in flavonoids production and increase of LPO and H_2O_2 in soybean. Soybean also enhanced the transcription of antioxidant genes related to antioxidant enzymes like CAT, APX, GPX, and SOD (Gal et al. 2015). To overcome the effect of weed interference, crop plants enhanced the production of enzymatic and nonenzymatic antioxidants for countering the effect of ROS and restoring of cellular metabolic processes (Agostinetto et al. 2017; Piasecki et al. 2018).

8.2.5 Salinization of Soil

Saline soil is abundantly present in the arid and semiarid areas of the world and affects the plants' growth along with high temperature, low carbon storage in soil, and low precipitation (Lashari et al. 2013). Soil with higher concentration of salts is saline soil, and it becomes more harmful for plant yield and growth under water-deficit conditions (Negrao et al. 2017). About 20% agricultural soils and 50% croplands of the world have now become saline in nature and unfavourable for plants growth (Qadir et al. 2014). It is also estimated that if the sources of salinity in soil are not controlled, in future 50% of the total cultivated soil of world will become saline by 2050. Higher concentration of Na^+ and Cl^- salts in soil alters the soil properties like soil porosity, soil aeration, water percolation, availability of water and nutrients for roots (Hmaeid et al. 2018). It was estimated that in Pakistan 6.63 million ha lands are saline in nature and still increasing (Akram et al. 2010). Saline stress has affected around one-third of agriculture lands of the world, which in response is highly deleterious for world food production, demand, and supply.

Crop plants rarely survive under saline conditions, and their stress combating mechanism is also poor against salinity because it destroys the osmotic potential of water present in soil, which ultimately reduced the accumulation of water and minerals in plant tissue (Panuccio et al. 2014). Salinity also produces nutrient deficiency, physiological drought, and Na ions toxicity in crop plants (Taiz and Zeiger 2006). Alteration in metabolism of cell produced osmotic stress, which ultimately reduced cell division, cell expansion, and closure of stomata due to change in turgor pressure altered cell homeostasis and dropping of cell turgor (Miller et al. 2010). Plants growing in saline soil also produce reactive oxygen species (ROS) like singlet oxygen ($1O_2$), hydrogen peroxide (H_2O_2), superoxide ($O_2^{\bullet -}$), and hydroxyl radicals ($OH^{\bullet}$) (Yamasaki et al. 2019).

8.2.6 Water-deficit Conditions and Oxidative Stress

Drought stress negatively affects the closure of stomata, abridged CO_2 entry, rate of photosynthesis, light harvesting in photosystems, and alteration of chloroplasts photochemistry; however, the production of ROS was increased in plants (Hasanuzzaman et al. 2020). Furthermore, the overproduction of ROS also denatures membrane and protein structures through photorespiration, production of TCA cycle enzymes, and reduced carboxylation efficiency (Fahad et al. 2017a, 2017b; Rady et al. 2020). Moreover, a minor production of $NADP^+$ provides higher reduction of electron transport chain and electrolyte leakage; ultimately it triggers higher production of ROS-based oxidative stress (Hasanuzzaman et al. 2018). Abideen et al. (2020) planted *Phragmites karka* for 35 days under drought stress and 40% water holding capacity and reported 22% increase in MDA production. However, in *Coffea arabica* L. the MDA production was increased after 20 days of drought stress (Campos et al. 2019). In rice plant increase in 1.8 fold $^1O_2$, 2.1 fold H_2O_2 and 1.66 fold MDA was recorded after eight days of drought (Saha et al. 2018). In another study, finger millet cultivated in 75% water shortage condition significantly amplified the concentration of H_2O_2 and electrolyte leakage (Satish et al. 2018). Tomato plants also enhanced 39% MDA production under 6 days of water deficiency (Malhotra et al. 2017).

8.2.7 Waterlogging or Flooding

Availability of too much water for a xerophytic plant is termed as *waterlogging* or *flooding*, which produced anoxic or hypoxic environment and various noxious compounds that damage plant metabolism. The abridged plant metabolism is due to the ROS-induced oxidative stress in response to waterlogging (Loreti et al. 2016). A remarkable increase in MDA content was reported in two waterlogged varieties of *Sorghum bicolor* cultivated in 6, 9, 12 days waterlogging conditions. In another study, Anee et al. (2019) reported that in *Sesamum indicum* L. the MDA and H_2O_2 content was increased depending upon the duration of waterlogged conditions. Similarly, the overproduction of reactive oxygen species was also reported in *S. lycopersicum* (Rasheed et al. 2018) and *Deschampsia antarctica* (Antarctic plant) when cultivated in waterlogged soil for 7 days (Park and Lee 2019).

8.2.8 Oxidative Stress under Low Temperature

The plants facing low or chilling temperature for a long time also enhanced the production of ROS by degrading fluidity of plasma membrane, lowering the rate of photosynthesis by inhibiting photosystems activity, and inactivating of ROS detoxification mechanism, which enhanced the electrolyte leakage and lipid peroxidation (Repetto et al. 2012). When rice plant seedlings were cultivated for 6 days at 12°C temperature, they exhibited 49% increase in electrolyte leakage and 180 higher accumulation of MDA (Han et al. 2017). Another study also reported the higher accumulation of MDA and H_2O_2 in *S. lycopersicum* (cold sensitive) at low temperature treatment of 15°C/8°C day and night for 48 hours (Liu et al. 2020). Xue et al. (2019) also checked the effect of time-based low temperature treatment on transgenic and wildtype cultivars of *Ammopiptanthus mongolicus*. The plants were kept at 4°C for first 24 hours, then at 0°C for 12 hours, and followed by 6°C for another 12 hours. The low temperature stress produced higher ROS and oxidative stress in wild type as compared to transgenic plants.

8.3 Generation of ROS under Oxidative Stress

First molecular oxygen in this world was produced around 2.7 billion years ago. It was originated by the reduction of atmosphere in first oxygen producing photosynthetic organism. This reaction gives us a hint about ROS that all oxygen entities are the byproducts of a biological reaction (Halliwell 2006). All reactive oxygen species (ROS) are emitted as a result of aerobic metabolism in organisms. Most commonly they are found in plant cell organelles like peroxisomes, chloroplasts, plasma membrane, mitochondria, and apoplast (Roychoudhury et al. 2012). Only 1%–2% of total O_2 of the plant is consumed for the formation of ROS. Under normal or controlled conditions plants also generate minute quantity of ROS, which is useful for the normal growth and development of the plant. However, these ROS acts as natural vaccines of plant and facilitate plant to activate its antioxidant defence system without imparting any critical effect on metabolic processes (Foyer and Noctor 2011). Homeostasis of ROS production and the concentration of scavengers of ROS were kept in balance to ensure the rapid growth and development of plants. The oxidative stress is a condition in which plants balance between ROS, and its scavengers abruptly change in favour of ROS due to external factors. The antioxidants which act as scavengers of

ROS cannot detect that sudden shift in homeostasis in sensitive crop plants and the plants metabolism collapses. However, the tolerant crop plants have a higher system of genes expression for ROS scavengers and can survive oxidative stress induced by abrupt ROS accumulation (Banerjee and Roychoudhury 2018). Hence, to survive in ROS-induced oxidative stress, crop plants have to modulate their physiobiochemical metabolism and also quickly acclimatize themselves according to the changing energy equation (Miller et al. 2010).

8.3.1 Types of ROS and Damages

The ROS are of two major types: first one is radical ROS and second one is nonradical ROS (Figure 8.2). Oxidative stress produces a number of ROS including hydrogen peroxide (H_2O_2), singlet oxygen ($^1O_2$), superoxide radical (O_2^-), and hydroxyl radical ($^\bullet OH$). They all are toxic for plant metabolism and caused lethal changes in the structures of nucleic acid, lipids, and proteins (Foyer and Noctor, 2011). Superoxide radical ($O_2^{\bullet-}$) is necessary for the propagation of various other species, which are related to the generation of oxidative stress. The superoxide radical produced inside the chloroplast by the reduction of oxygen through electrons moving inside electron transport chain is a process called as Mehler reaction. The breakdown of $O_2^{\bullet-}$, into H_2O_2 takes place inside chloroplast by the scavenging activity of CuZn (SOD) (Takagi et al. 2016). Singlet oxygen ($1O_2$) is a highly reactive specie and is an inevitable byproduct of oxygen-based photosynthesis (Telfer 2014). In photosystem-II, the triplet state of chlorophyll reacts with oxygen to produce singlet oxygen during the process of photosynthesis. The generation of singlet oxygen during photosynthesis imparts toxic effects on both photosystems (PS-I and PS-II) (Wang and Apel 2016). Hydroxyl radical ($^\bullet OH$) exists in a neutral form with a high reactivity particularly with $3O_2$. The $^\bullet OH$ production is harmful for cellular metabolic processes because it is capable of destroying the conformation of proteins, lipids, and membranes (Bailly et al. 2008). The $^\bullet OH$ is generated in the cytoplasm, and its generation is linked with the discharge of H_2O_2 or $O_2^{\bullet-}$ from mitochondria and chloroplast of the cell (Richards et al. 2015). The reactivity of hydrogen peroxide (H_2O_2) is lower than the other ROS due to the unavailability of unpaired electrons. But it is highly toxic due to its mobility far away from its site of production (Bienert et al. 2006). The H_2O_2 is produced by various cell organelles like nucleus, chloroplast, endoplasmic reticulum, plasma membranes, mitochondria, and peroxisomes, particularly during photosynthetic carbon oxidation cycle (Miller et al. 2010). Overproduction of hydrogen peroxide in a

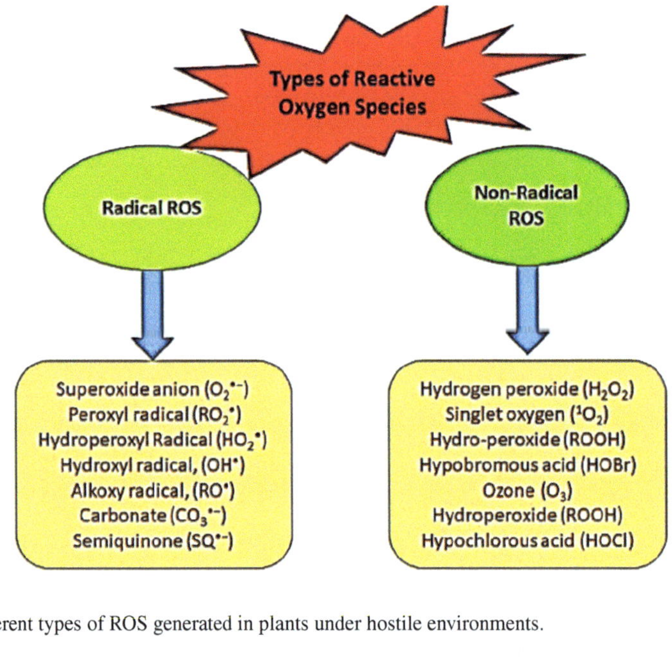

FIGURE 8.2 Different types of ROS generated in plants under hostile environments.

plant negatively affects the various physiological activities like stomatal opening and closing, germination of seed, aging, and programmed cell death (Gechev and Hille 2005).

8.3.2 Production Sites of ROS

In plants, ROS are produced generally inside the cell organelles such as chloroplasts, peroxisomes, endoplasmic reticulum, cell wall, mitochondria, and apoplast.

8.3.2.1 Chloroplast

The chloroplast is the reaction centre of photosynthesis in plants, and during this process ROS are produced as byproducts of PS-I and PS-II. The formation of ROS is influenced by various factors and production is significantly enhanced when photosynthetic flux is higher than the requirements of CO_2 assimilation (Kim and Dogra 2019). The generation of ROS in chloroplast is 30–100 times higher than the mitochondria. The singlet oxygen is produced due to the reaction between $^3Chl^*$ and $^3O_2$ (Singh et al. 2019). The chlorophyll in light harvesting complex of PS-II on absorption of light converted into a high energy $^1Chl^*$. Some light energy is transferred from P-II to P-680 by a process of photochemical quenching (pQ) to run electron transport chain. However, if the energy absorption is higher than pQ capacity, the surplus energy is released as fluorescence or as heat or through intersystem by producing $^3Chl^*$. The lutein and zeaxanthin carotenoids present in light harvesting complex quench $^3Chl^*$ to prevent further transfer of energy. If this quenching of $^3Chl^*$ is not proper, it reacts with $^3O_2$ that is produced from water during its splitting in OEC complex and produced $^1O_2$ as a byproduct (Dietz et al. 2016). Additionally, plants under abiotic stresses closed stomata due to which carbon dioxide concentration decreased in chloroplast start reduction of ETC and increase the chances of charge recombination between plastoquinone and $^1P680^*$ in PS-II, also generates $^1O_2$. However, in PS- I the $O_2^{\bullet -}$ is produced in Mahler reaction instead of $^1O_2$ which converts into H_2O_2 by the action of superoxide dismutase. After some time, H_2O_2 and $O_2^{\bullet -}$ both are changed into $^*OH$, which is highly stable with the help of metal ions (Bose et al. 2014).

8.3.2.2 Plant Cell Wall

Cell wall in plants also acts as an active organelle for the production of ROS (Sharma et al. 2012). The H_2O_2 production in cell wall by cell wall peroxidase enzymes was reported in cotton plant during oversensitive response produced against the infection of *Xanthomonas campestris* and also in Arabidopsis growing in potassium-deficit conditions (Kim et al. 2010). The cell wall contains lipoxygenase enzymes which catalyze the hydroperoxidation of PUFA and release ROS during the process (Das and Roychoudhury 2014). The cells also have another activated enzyme diamine oxidase, which is also releasing $^*O_2$ during the polyamines or diamine-based reduction of quinone that further self-oxidizes to produce peroxides. The lignin precursors undergo a high density cross-linking controlled by H_2O_2-mediated pathways for the lignification of cell wall during pathogens attack (Higuchi 2006).

8.3.2.3 Cellular Membranes

Plasma membrane also contains electron-transporting oxidoreductases for the production of ROS under unfavourable environments (Sharma et al. 2012). The NADPH-dependent oxidases catalyze transformation of electrons releasing from cytoplasmic NADPH and shifted to O_2 for the formation of $O_2^{\bullet -}$. This $O_2^{\bullet -}$ was directly transformed to H_2O_2 or by the action of SOD (Sharma et al. 2012; Das and Roychoudhury 2014). The NADPH oxidase plays a significant role in survival of plants through ROS-mediated activation of antioxidants during biotic and abiotic stresses.

8.3.2.4 Endoplasmic Reticulum

The ROS ($O2^{\bullet -}$) generation in endoplasmic reticulum takes place through NADPH-dependent electron transport and cytochrome P450 (Mittler 2017). A flavoprotein was formed in response to the interaction of R-H present in organic compounds with Cyt. P450, the further reduction of this flavoprotein generates a free radical Cyt. P450 R^-. This P450 R^- is an unstable stage and it quickly reacts with $^3O_2$ to form oxygenated product (Cyt P450-ROO^-). This oxygenated radical also undergoes self-decomposition or Cyt-b-mediated reduction into P450-Rh forming $O_2^{\bullet-}$ (Sharma et al. 2012; Das and Roychoudhury 2014).

8.3.2.5 Peroxisomes

Peroxisome is a single-membrane-bound cytoplasmic organelle that acts as a centre for various metabolic activities (Sandalio and Romero-Puertas 2015). Peroxisomes generate ROS through fatty acid β-oxidation and photorespiration pathways (Corpas 2015). The peroxisomes also produce $O_2^{\bullet-}$ via a short electron transport chain mediated by various enzymes located in peroxisomal membrane such as NADH/NADPH and peroxisome matrix like xanthine oxidoreductase (XOD/XDH) and uricase (Wang et al. 2015).

The key enzyme which is involved in ROS generation in peroxisome is glycolate oxidase (GOX). Furthermore, xanthine oxidase (XOD) has a main role in the production of $O_2^{\bullet-}$ together with uric acid in peroxisomal matrix, which is converted to H_2O_2 by the action of urate oxidase (UO) and SOD enzymes. Moreover, the activity of flavin oxidase, $O_2^{\bullet-}$ disproportionation, and oxidation of fatty acids also promote the synthesis of H_2O_2 in peroxisomes (Corpas, 2015). In addition, copper amine oxidase, polyamine oxidase, sarcosine oxidase, and sulphite oxidase enzyme action are able to synthesize H_2O_2 in peroxisome, while, the MDHAR enzymes have a role in scavenging of H_2O_2 via the AsA-GSH cycle and reviving ascorbic acid (Das and Roychoudhury 2014).

8.3.2.6 Mitochondria

In mitochondria the ROS production is lower as compared to chloroplast, and the ETC Complex I and III are the major ROS producing sites. The ROS ($O_2^{\bullet-}$) is synthesized by the intermediate ubisemiquinone formation at complex I and III by donating electrons to molecular oxygen and followed by reduction to form H_2O_2 and $^{\bullet}OH$ with the help of CuZn-SOD and Mn-SOD (Huang et al. 2016). If the complex I is restricted with the restriction of NAD^+-linked substrates, the transport of electron also reverses from complex II to I. In this situation, the reaction needs ATP hydrolysis for its completion (Das et al. 2015).

8.3.2.7 Apoplast

An apoplast is an empty space between adjacent cells which is meant for the intracellular transport of material. The reduction of low molecular weight antioxidants (glutathione and ascorbate) and cysteine takes place inside apoplast due to its lower pH and antioxidant capacity as compared to cytoplasm (Qi et al. 2017). Both in normal and stressed plants apoplasts produce ROS. The synthesis of ROS in apoplast is executed by amino oxidases, quinine reductases, lipoxygenases, NADPH oxidases, and peroxidases belonging to class III (Liu and He 2016).

8.4 Plant Antioxidant Defence System

Antioxidants are the substances that are meant for the direct or indirect scavenging of ROS at cellular level in plants (Carocho and Ferreira, 2013). The two major types of antioxidant defence systems (enzymatic and nonenzymatic) coexist in plants and are produced against oxidative stress. The nonenzymatic antioxidants like vitamin C, nonprotein amino acids alpha tocopherol, GSH, flavonoids, phenolic compound (PhOH), and alkaloids and antioxidant enzymes like APX, SOD, POX, CAT, MDHAR, polyphenol oxidase (PPO), PRX, GR, DHAR, GST, GPX and TRX both can exist together and work in coordination with each other to restrict ROS (Nath et al. 2018; Laxa et al. 2019). The two major antioxidant pathways (Asada–Halliwell cycle and AsA-GSH cycle) are working at cellular level for the detoxification of H_2O_2 and these comprise of two nonenzymatic (ascorbic acid, GSH) and four enzymatic (MDHAR, APX, GR, DHAR) antioxidants (Figure 8.1). Furthermore, GST and GPX are also necessary for xenobiotics and

H_2O_2 detoxification (Hasanuzzaman et al. 2018). Moreover, beta-carotene also breaks down the $O2^{\bullet-}$, $^{*}OH$, and ROO^{-} radicals at cellular level (Kapoor et al. 2019).

8.5 Enzymes as Antioxidants

8.5.1 Superoxide Dismutase (SOD)

SOD is a metalloprotein with EC (1.15.1.1) and acts as a first-line defence against reactive oxygen species (Del Rio et al. 2018). The disproportionation reaction is carried out by SOD in which $O_2^{\bullet-}$ is first converted into H_2O_2 and further into O_2 (Wang et al. 2009). It exists in all subcellular aerobic components that are vulnerable to oxidative stress. SOD contains three types on the basis of their cofactor and localization at cellular level such as Fe-SOD found in chloroplast, Mn-SOD present in mitochondria, and Cu/Zn-SOD exists in peroxisome, chloroplast, and cytosol (Hasanuzzaman et al. 2018).

8.5.2 Catalase

Catalase (CAT) is a tetrameric enzyme with heme group having oxidation number EC (1.11.1.6) and also named H_2O_2 oxidoreductase. It is located in all aerobic eukaryotes, it works for the depletion of hydrogen peroxide produced in peroxisome during oxidative stress, β-oxidation of fatty acids, purine catabolism, and photorespiration (Garg and Manchanda 2009). Catalase controls peroxisomal H_2O_2 without limiting its production. Catalase is highly reactive against ROS, and it can convert 6,000,000 units of H_2O_2 only in one minute to O_2 and water. Catalase exists in three isomeric forms including CAT_1, CAT_2 (peroxisomes), and CAT_3 (mitochondria) each having three classes I, II, and III. It also does not require any cofactor and energy power for ROS transformation (Fotopoulos et al. 2010).

8.5.3 Peroxidase

Peroxidase (POD) belongs to the oxidoreductase group of enzymes, and is found in aerobic organisms having systematic number EC (1.11.1.7). It utilizes H_2O_2 for the oxidation of other substrates. It is substrate-specific in case of GSH for glutathione peroxidase but mostly it has a broad specificity for substrates (Nath et al. 2018). Major PODs are specific in nature such as NADH peroxidase, cytochrome C peroxidase, ascorbate peroxidase, and some are nonspecific peroxidases like horseradish peroxidase and guaiacol peroxidase (Sharma 2016). It also excites oxidation of many hydrogen donor entities like aromatic amines, phenol compounds, hydroquinone amines, hydroquinone, and gasoline derivatives with the utilization of hydrogen peroxide as a hydrogen receptor (Kawano 2003).

8.5.4 Ascorbate Peroxidase

Ascorbate peroxidase (APX) is present in plants, euglena, algae, and other aerobic organisms at cellular level in cytosol, vacuole, and chloroplast (Sade et al. 2011). It reduced the H_2O_2 found in cytoplasm and chloroplast while CAT enzyme can reduce hydrogen peroxide in peroxisome. The ascorbate is utilized as the source of hydrogen for the APX-mediated reduction of H_2O_2 into monodehydroascorbate (MDHA) and water (Hasanuzzaman et al. 2020). The APX naturally exists in four varying isomeric forms such as thylakoid (tAPX), chloroplast stromal soluble (sAPX), cytosolic (cAPX), and glyoxisome membrane (gmAPX) (Gill and Tuteja 2010).

8.5.5 Guaiacol Peroxidase

The heme group is an essential component of these guaiacol peroxidase (GPoX) enzymes, which are composed of 40–50 kDa protein monomers (Hasanuzzaman et al. 2018). These GPoX are involved in the oxidation of certain substrates at the expenditure of H_2O_2 and clear the cell from peroxide evolved during metabolic processes under normal and stress environment. It is also involved in indole-3-acetic acid break down and formation of lignin (Nath et al. 2018). GPoX together with other peroxidases fixes the cell wall of plants and carries out the oxidation of phenols into phenoxyl radicals, which are the polymers of lignin.

8.5.6 Glutathione Reductase

Glutathione reductase (GR) is a flavoprotein containing oxidoreductase that exists in almost all prokaryotic and eukaryotic organisms (Romero-Puertas et al. 2006). It mainly exists in the chloroplasts of cells but also in minute concentrations in cytosol and mitochondria. It is an essential component of Halliwell-Asada pathway in which breakdown of ROS takes place. It plays a key role in production of redox state of GSH. Further, it also takes part in NADPH-dependent synthesis of a disulphide linkage in GSSG pool. The redox potential of a cell under stress conditions was also maintained by GR (Hasanuzzaman et al. 2020).

8.5.7 Monodehydroascorbate Reductase

Monodehydroascorbate reductase (MDHAR) is an FAD enzyme, which stimulates ascorbate reduction and occurs in chloroplast plant cytosol (Mittler, 2017). MDHAR along with APX found in mitochondria and peroxisome scavenges H_2O_2; it acts as an electron acceptor from NADH instead of NADPH. It is reported that MDHAR, GR, and DHAR dehydrogenase reductase were found enhanced under drought stress in rice plants (Hasanuzzaman et al. 2018). The regeneration mechanism facilitates FAD reduction to establish charge transfer complex, thus giving an electron to MDHA creating two ascorbate molecules through semiquinone form.

8.5.8 Dehydroascorbate Reductase

Dehydroascorbate reductase (DHAR) restores ascorbate which acts as ascorbic acid regulator recycler (Paciolla et al. 2001). The ascorbic acid univalent oxidation triggers MDHA formation which later results in divalent oxidation product (DHA) through rapid disproportionation or additional oxidation. Moreover DHA later reduced to ascorbic acid by DHAR via GSH required reaction (Laxa et al. 2019).

8.5.9 Glutathione-S-Transferase

Glutathione-s-trasferase (GST) is a distinct group of enzymes that facilitate electrophilic xenobiotic conjugation with tripeptide GSH, which can degrade peroxidase with GST and trigger genotoxic and cytotoxic compound scavenging (Gill and Tuteja 2010). Plant gene families are distinct and diverse with 25 numbers recorded in soybean plant, 54 in Arabidopsis, and 42 in maize plants. GSTs detoxify herbicides in plants causing hormone homeostasis, anthocyanin vacuolar sequestration, H_2O_2 detoxification, and tyrosine metabolism in plants under various biotic and abiotic stresses (Laxa et al. 2019).

8.5.10 Glutathione Peroxide

Glutathione peroxide (GPX) belongs to isozymes family, which uses GSH to reduce H_2O_2, organic and lipid hydroperoxides produced under oxidative stress. Seven associated proteins of GPX family are identified in chloroplast, cytosol, and mitochondria termed as AtGPX1–AtGPX7 in Arabidopsis plants (Fahad et al. 2017a, 2017b).

8.6 Nonenzymatic Antioxidants

8.6.1 Ascorbic Acid

Ascorbic acid is an essential antioxidant found in most plants, which has the ability to minimize the dangerous impact of ROS in plants and ranges from 20mM cell cytosol and 20–300mM in chloroplast stroma. Ascorbic acid is involved in different, essential cell functions and can contribute electrons to different enzymatic, nonenzymatic reactions involved in protecting membrane by reducing oxygen and hydroxyl radicals associated with vitamin E generation from tocopheroxyl radical (Lopez-Mungia et al. 2011). Ascorbate transport requires facilitated diffusion and different mechanisms are involved in ascorbate transport between apoplast and cytosol. Ascorbic acid acts as violaxanthin de-epoxidase cofactor in chloroplast and performs excess energy dissipation. It also regulates cell division, cell enlargement, and progression of cell cycle from the G1 to S phase (Lopez-Mungia et al. 2011).

8.6.2 Tocopherols

Tocopherols are present in all phototropic organisms except cyanobacteria. Tocopherol is a lipid-soluble antioxidant, which is an important ROS scavenger and considered as essential antioxidant. Four isomers α, β, ϒ, and δ, which comprise of chroman head, give vitamin E amphipathic character. It is reported that α tocopherol having three different methyl substituents possesses the highest antioxidative ability (Nishio et al. 2013). They perform membrane stability by ROS scavenging. Tocopherols shield lipids and protect PS-II structure and membrane components by reaction with O_2 in cell chloroplast. One α tocopherol molecule can quench up to 120 RO-, ROO^- by resonance energy transfer (Hasanuzzaman et al. 2020).

8.6.3 Carotenoids

Carotenoids are considered as natural soluble lipophilic compounds found in plastids of plants; they are also known as terpenoid or isoprenoid compounds (Rezayian et al. 2020). They serve as essential photoprotective by dissipating additional excitation energy in the form of heat, or they act as ROS scavenging and lipid peroxidation suppressors in photosynthetic organisms. Over six hundred carotenoids occur in nature and serve as antenna molecules in photosynthesis, as they absorb light from 450 to 570 nm. They also facilitate photosynthetic pigments by (Chl3) and ROS inhibition and help photosynthesis in plants (AL-Aloosy et al. 2019).

8.6.4 Proline

Proline is an amino acid, which facilitates protein synthesis, and its quantity is proportional to stress duration when plants face different biotic and abiotic stress (Abbas et al. 2018). Proline plays essential role in membrane stability by reducing free radicals and regulating of oxidation-reduction attempts of cell. Proline is considered as antioxidant by its ability to reduce the (OH•) and ($1O_2$) associated with lipid oxidation inhibition (Ashraf and Foolad 2007; Trovato et al. 2008).

8.6.5 Phenolic Compounds

Phenolics such as flavonoids, hydroxycinnamate esters, and tannins are secondary metabolites in plant tissues. Flavonoids are found in floral parts and leaves are present widely in plants and facilitate various functions such as fruit formation, flowering seed pigmentation, pollen germination, protection against UV light and pathogens in plants (Mehla et al. 2017). They modify lipid packing order to reduce membrane fluidity and alter the peroxidation kinetics. They are considered to have high reactivity and ability to delocalize unpaired electron associated with chelate transition metal ions (Liu et al. 2019).

8.7 Conclusion

The key factor in the better growth and yield of a plant is stress-free environment. The soil is the growth medium of the plant, and it provides nutrients and water in required quantity to plant. Plants are nonmotile and stay in the same environment for whole life. The imbalance of any single or multiple environmental factors alters the metabolism and growth of plant. These environmental factors include hot and cold temperatures, plant and weed interaction, waterlogging, drought, salinity, availability of soil nutrients, and metals availability. Any alteration in the balance of these factors produced hostile environment in which growth of plant was disturbed due to the production of ROS-based oxidative stress. The overproduction of ROS is harmful for different plant metabolic processes under biotic and abiotic stress. ROS production was also confined to the highly important cellular compartments like cell membrane, chloroplast, mitochondria, peroxisomes, and apoplast. These are vital organelles and loss of their function may result in the loss of plant. The breakdown or utilization of ROS at cellular level in plants is the basic of modulation and defence against ROS. Both enzymatic and nonenzymatic scavengers of ROS are used by plants to maintain the ROS and plant metabolism in equilibrium under stressful conditions. The plant species which are tolerant against a particular stress should be favoured to achieve better growth and yield of plant. The development in the field of genetic engineering also encourages the utilization of transgenic plants, which have better ROS scavenging system to overcome hostile conditions faced by crop plants. Also there is a need to find some single stress-tolerant genes to permanently overcome the stressful conditions faced by plants.

REFERENCES

Abbas T, Rizwan M, Ali S, Adrees M, Mahmood A, Zia-Ur-Rehman M, Ibrahim M, Arshad M, Qayyum MF (2018) Biochar application increased the growth and yield and reduced cadmium in drought stressed wheat grown in an aged contaminated soil. *Ecotoxicol Environ Saf* 148:825–833.

Abdul S, Muhammad AA, Shabir H, Hesham A, Sajjad H, Niaz A, Abdul G, Sayyed RZ, Fahad S, Subhan D, Rahul D (2021a) Zinc nutrition and arbuscular mycorrhizal symbiosis effects on maize (*Zea mays* L.) growth and productivity. *J Saudi Society Agricultural Sci.* https://doi.org/10.1016/j.sjbs.2021.06.096

Abdul S, Muhammad AA, Subhan D, Niaz A, Fahad S, Rahul D, Mohammad JA, Omaima N, Muhammad Habib Ur R, Bernard RG (2021b) Effect of arbuscular mycorrhizal fungi on the physiological functioning of maizeunder zinc-deficient soils. *Sci Rep* 11:18468.

Abideen Z, Koyro HW, Huchzermeyer B, Ansari R, Zulfiqar F, Gul B (2020) Ameliorating effects of biochar on photosynthetic e_ciency and antioxidant defence of Phragmites karka under drought stress. *Plant Biol* 22:259–266.

Adnan M, Fahad S, Muhammad Z, Shahen S, Ishaq AM, Subhan D, Zafar-ul-Hye M, Martin LB, Raja MMN, Beena S, Saud S, Imran A, Zhen Y, Martin B, Jiri H, Rahul D (2020) Coupling phosphate-solubilizing bacteria with phosphorus supplements improve maize phosphorus acquisition and growth under lime induced salinity stress. *Plants* 9(900). https://doi.org/10.3390/plants9070900

Adnan M, Fahad S, Khan IA, Saeed M, Ihsan MZ, Saud S, Riaz M, Wang D, Wu C. (2019). Integration of poultry manure and phosphate solubilizing bacteria improved availability of Ca bound P in calcareous soils. *3 Biotech* 9(10):368.

Adnan M, Shah Z, Sharif M, Rahman H. (2018b). Liming induces carbon dioxide (CO_2) emission in PSB inoculated alkaline soil supplemented with different phosphorus sources. *Environ Sci Poll Res* 25(10):9501–9509.

Adnan M, Zahir S, Fahad S, Arif M, Mukhtar A, Imtiaz AK, Ishaq AM, Abdul B, Hidayat U, Muhammad A, Inayat-Ur R, Saud S, Muhammad ZI, Yousaf J, Amanullah, Hafiz MH, Wajid N (2018a) Phosphate-solubilizing bacteria nullify the antagonistic effect of soil calcification on bioavailability of phosphorus in alkaline soils. *Sci Rep* 8:4339. https://doi.org/10.1038/s41598-018-22653-7

Afifi M, Swanton C (2012) Early physiological mechanisms of weed competition. *Weed Sci* 60:542–551.

Agostinetto D, Tarouco CP, Nohatto MA, Oliveira C, Fraga DS (2017) Metabolic activity of wheat and ryegrass plants in competition. *Planta Daninha* 35:017155463.

Ahanger MA, Tomar NS, Tittal M, Argal S, Agarwal RM (2017) Plant growth under water/salt stress: ROS production; antioxidants and significance of added potassium under such conditions. *Physiol Mol Biol Plants* 23(4):731–744.

Ahmad I, Jadoon SA, Said A, Adnan M, Mohammad F, Munsif F (2017) Response of sunflower varieties to NPK fertilization. *Pure Appl Biol* 6(1): 272–277.

Ahmad S, Kamran M, Ding R, Meng X, Wang H, Ahmad I, Fahad S, Han Q (2019) Exogenous melatonin confers drought stress by promoting plant growth, photosynthetic capacity and antioxidant defense system of maize seedlings. *PeerJ* 7:e7793. http://doi.org/10.7717/peerj.7793

Akbar H, Timothy JK, Jagadish T, Golam M, Apurbo KC, Muhammad F, Rajan B, Fahad S, Hasanuzzaman M (2020) Agricultural Land Degradation: Processes and Problems Undermining Future Food Security.

In: Fahad S, Hasanuzzaman M, Alam M, Ullah H, Saeed M, Khan AK, Adnan M (eds) *Environment, climate, plant and vegetation growth*. Springer Publ Ltd, Springer Nature Switzerland AG. Part of Springer Nature, pp. 17–62. https://doi.org/10.1007/978-3-030-49732-3

Akram M, Ashraf MY, Ahmad R, Waraich EA, Iqbal J, Mohsan M (2010) Screening for salt tolerance in maize (*Zea mays* L.) hybrids at an early seedling stage. *Pak J Bot* 42:141–154.

Akram R, Turan V, Hammad HM, Ahmad S, Hussain S, Hasnain A, Maqbool MM, Rehmani MIA, Rasool A, Masood N, Mahmood F, Mubeen M, Sultana SR, Fahad S, Amanet K, Saleem M, Abbas Y, Akhtar HM, Waseem F, Murtaza R, Amin A, Zahoor SA, ul Din MS, Nasim W (2018a) Fate of organic and inorganic pollutants in paddy soils. In: Hashmi MZ, Varma A (eds) *Environmental pollution of paddy soils, soil biology*. Springer International Publishing AG, Gewerbestrasse 11, Cham, CH-6330, Switzerland, pp. 197–214.

Akram R, Turan V, Wahid A, Ijaz M, Shahid MA, Kaleem S, Hafeez A, Maqbool MM, Chaudhary HJ, Munis, MFH, Mubeen M, Sadiq N, Murtaza R, Kazmi DH, Ali S, Khan N, Sultana SR, Fahad S, Amin A, Nasim W (2018b) Paddy land pollutants and their role in climate change. In: Hashmi, MZ, Varma, A (eds) *Environmental pollution of paddy soils, soil biology*. Springer International Publishing AG, Gewerbestrasse 11, Cham, CH-6330, Switzerland, pp. 113–124.

Al Mahmud J, Bhuyan MB, Anee TI, Nahar K, Fujita M, Hasanuzzaman M (2019) Reactive oxygen species metabolism and antioxidant defense in plants under metal/metalloid stress. In: Hasanuzzaman M, Hakeem KR, Nahar K, Alharby HF (eds) *Plant abiotic stress tolerance*. Springer, Cham, pp. 221–257.

Al Mahmud J, Hasanuzzaman M, Nahar K, Bhuyan MB, Fujita M (2018) Insights into citric acid-induced cadmium tolerance and phytoremediation in *Brassica juncea* L.: coordinated functions of metal chelation, antioxidant defense and glyoxalase systems. *Ecotoxicol Environ Saf* 147:990–1001.

AL-Aloosy YAM, AL-Tameemi AJ, Jumaa SS (2019) The role of enzymatic and non-enzymatic antioxidants in facing the environmental stresses on plant: a review. *Plant Arch* 19:1057–1060.

Amanullah, Muhammad I, Haider N, Shah K, Manzoor A, Asim M, Saif U, Izhar A, Fahad S, Adnan M et al. (2021) Integrated foliar nutrients application improve wheat (*Triticum Aestivum* L.) productivity under calcareous soils in drylands. *Commun Soil Sci Plant Anal*. https://doi.org/10.1080/00103624.2021.1956521

Amanullah, Fahad S (eds) (2017) Rice - technology and production. *IntechOpen Croatia 2017*. http://dx.doi.org/10.5772/64480

Amanullah, Fahad S (eds) (2018a) Corn - Production and human Health in changing climate. *IntechOpen United Kingdom 2018*. http://dx.doi.org/10.5772/intechopen.74074

Amanullah, Fahad S (eds) (2018b) Nitrogen in agriculture – updates. *IntechOpen Croatia 2018*. http://dx.doi.org/10.5772/65846

Amanullah, Shah K, Imran, Hamdan AK, Muhammad A, Abdel RA, Muhammad A, Fahad S, Azizullah S, Brajendra P (2020) Effects of climate change on irrigation water quality. In: Fahad S, Hasanuzzaman M, Alam M, Ullah H, Saeed M, Khan AK, Adnan M (eds) *Environment, climate, plant and vegetation growth*. Springer Publ Ltd, Springer Nature Switzerland AG. Part of Springer Nature, pp. 123–132. https://doi.org/10.1007/978-3-030-49732-3

Amir M, Muhammad A, Allah B, Sevgi Ç, Haroon ZK, Muhammad A, Emre A (2020) Bio fortification under climate change: the fight between quality and quantity. In: Fahad S, Hasanuzzaman M, Alam M, Ullah H, Saeed M, Khan AK, Adnan M (eds) *Environment, climate, plant and vegetation growth*. Springer Publ Ltd, Springer Nature Switzerland AG. Part of Springer Nature, pp. 173–228. https://doi.org/10.1007/978-3-030-49732-3

Amjad I, Muhammad H, Farooq S, Anwar H (2020) Role of plant bioactives in sustainable agriculture. In: Fahad S, Hasanuzzaman M, Alam M, Ullah H, Saeed M, Khan AK, Adnan M (eds) *Environment, climate, plant and vegetation growth*. Springer Publ Ltd, Springer Nature Switzerland AG. Part of Springer Nature, pp. 591–606. https://doi.org/10.1007/978-3-030-49732-3

Amjad SF, Mansoora N, Din IU, Khalid IR, Jatoi GH, Murtaza G, Yaseen S, Naz M, Danish S, Fahad S et al. (2021) Application of Zinc fertilizer and mycorrhizal inoculation on physio- biochemical parameters of wheat grown under water-stressed environment. *Sustainability* 13: 11007. https://doi.org/10.3390/su131911007

Anee TI, Nahar K, Rahman A, Mahmud JA, Bhuiyan TF, Alam MU, Fujita M, Hasanuzzaman M (2019) Oxidative damage and antioxidant defense in Sesamum indicum after di_erent waterlogging durations. *Plants* 8:196.

Antoniou C, Savvides A, Christou A, Fotopoulos V (2016) Unravelling chemical priming machinery in plants: the role of reactive oxygen–nitrogen–sulfur species in abiotic stress tolerance enhancement. *Curr Opin Plant Biol* 33:101–107.

Arafat Y, Shafi M, Khan MA, Adnan M, Basir A, Rahman IU, Arshad M, Khan A, Saleem N, Romman M, Rahman Z, Shah JA (2016) Yield response of Wheat cultivars to zinc application rates and methods. *Pure Appl Biol* 5(4):1260–1270.

Arif M, Talha J, Muhammad R, Fahad S, Muhammad A, Amanullah, Kawsar A, Ishaq AM, Bushra K, Fahd R (2020) Biochar; a Remedy for climate change. In: Fahad S, Hasanuzzaman M, Alam M, Ullah H, Saeed M, Khan AK, Adnan M (eds) *Environment, climate, plant and vegetation growth.* Springer Publ Ltd, Springer Nature Switzerland AG. Part of Springer Nature, pp. 151–172. https://doi.org/10.1007/978-3-030-49732-3

Ashfaq AR, Uzma Y, Niaz A, Muhammad AA, Fahad S, Haider S, Tayebeh Z, Subhan D, Süleyman T, Hesham AElE, Pramila T, Jamal MA, Sulaiman AA, Rahul D (2021) Toxicity of Cadmium and nickel in the context of applied activated carbon biochar for improvement in soil fertility. *Saudi Soc Agric Sci.* https://doi.org/10.1016/j.sjbs.2021.09.035

Ashraf M, Foolad MR (2007) Roles of glycinebetaine and proline in improving plant abiotic stress resistance. *Environ Exp Bot* 59:206–216.

Athar M, Masood IA, Sana S, Ahmed M, Xiukang W, Sajid F, Sher AK, Habib A, Faran M, Zafar H, Farhana G, Fahad S (2021) Bio-diesel production of sunflower through sulphur management in a semi-arid subtropical environment. *Environ Sci Pollution Res.* https://doi.org/10.1007/s11356-021-16688-z

Atif B, Hesham A, Fahad S (2021) Biochar coupling with phosphorus fertilization modifies antioxidant activity, osmolyte accumulation and reactive oxygen species synthesis in the leaves and xylem sap of rice cultivars under high-temperature stress. *Physiol Mol Biol Plants.* https://doi.org/10.1007/s12298-021-01062-7

Ayman ELS, Akbar H, Celaleddin B, Muhammad AI, Sohidul I, Shah F, Oksana S, Fatih Ç, Ram SM, Murat E (2020) Consequences of salinity stress on the quality of crops and its mitigation strategies for sustainable crop production: an outlook of arid and semi- arid regions. In: Fahad S, Hasanuzzaman M, Alam M, Ullah H, Saeed M, Khan AK, Adnan M (eds) *Environment, climate, plant and vegetation growth.* Springer Publ Ltd, Springer Nature Switzerland AG. Part of Springer Nature, pp. 503–534. https://doi.org/10.1007/978-3-030-49732-3

Aziz K, Daniel KYT, Fazal M, Muhammad ZA, Farooq S, Fan W, Fahad S, Ruiyang Z (2017a) Nitrogen nutrition in cotton and control strategies for greenhouse gas emissions: a review. *Environ Sci Pollut Res* 24:23471–23487. https://doi.org/10.1007/s11356-017-0131-y

Aziz K, Daniel KYT, Muhammad ZA, Honghai L, Shahbaz AT, Mir A, Fahad S (2017b) Nitrogen fertility and abiotic stresses management in cotton crop: a review. *Environ Sci Pollut Res* 24:14551–14566. https://doi.org/10.1007/s11356-017-8920-x

Bailly C, El-Maarouf-Bouteau H, Corbineau F (2008) From intracellular signaling networks to cell death: the dual role of reactive oxygen species in seed physiology. *C R Biol* 331(10): 806–814.

Banerjee A, Roychoudhury A (2018) Abiotic stress, generation of reactive oxygen species, and their consequences: an overview. In: Singh VP, Singh S, Tripathi D, Mohan Prasad S, Chauhan DK (eds) *Revisiting the role of reactive oxygen species (ROS) in plants: ROS Boon or bane for plants?* Wiley, New York, pp. 23–50.

Barroso AAM, Yamauti MS, Alves PLC (2010) Interference between weed species and two bean cultivars in two times of sowing. *Bragantia* 69:609–616.

Baseer M, Adnan M, Fazal M, Fahad S, Muhammad S, Fazli W, Muhammad A, Jr. Amanullah, Depeng W, Saud S, Muhammad N, Muhammad Z, Fazli S, Beena S, Mian AR, Ishaq AM (2019) Substituting urea by organic wastes for improving maize yield in alkaline soil. *J Plant Nutrition.* https://doi.org/10.1080/01904167.2019.1659344

Bayram AY, Seher Ö, Nazlican A (2020) Climate change forecasting and modeling for the year of 2050. In: Fahad S, Hasanuzzaman M, Alam M, Ullah H, Saeed M, Khan AK, Adnan M (eds) *Environment, climate, plant and vegetation growth.* Springer Publ Ltd, Springer Nature Switzerland AG. Part of Springer Nature, pp. 109–122. https://doi.org/10.1007/978-3-030-49732-3

Bienert GP, Schjoerring JK, Jahn TP (2006) Membrane transport of hydrogen peroxide. *Biochim Biophys Acta Biomembr* 1758(8):994–1003.

Bose J, Rodrigo-Moreno A, Shabala S (2014) ROS homeostasis in halophytes in the context of salinity stress tolerance. *J Exp Bot* 65:1241–1257.

Campos CN, Ávila RG, de Souza KRD, Azevedo LM, Alves JD (2019) Melatonin reduces oxidative stress and promotes drought tolerance in young *Co_ea arabica* L. plants. *Agric Water Manag* 211:37–47.

Carocho M, Ferreira IC (2013) A review on antioxidants, prooxidants and related controversy: Natural and synthetic compounds, screening and analysis methodologies and future perspectives. *Food Chem Toxicol* 51:15–25.

Caverzan A, Piasecki C, Chavarria G, Stewart CN, Vargas L (2019) Defenses against ROS in crops and weeds the effects of interference and herbicides. *Int J Mol Sci* 20(5):1086.

Chang W, Qiujuan J, Evgenios A, Haitao L, Gezi L, Jingjing Z, Fahad S, Ying J (2021) Hormetic effects of zinc on growth and antioxidant defense system of wheat plants. *Sci Total Environ.* https://doi.org/10.1016/j.scitotenv.2021.150992

Chen Y, Guo Z, Dong L, Fu Z, Zheng Q, Zhang G, Qin L, Sun X, Shi Z, Fahad S, Xie F, Saud S (2021) Turf performance and physiological responses of native Poa species to summer stress in Northeast China. *PeerJ* 9:e12252 http://doi.org/10.7717/peerj.12252

Choudhury FK, Rivero RM, Blumwald E, Mittler R (2017) Reactive oxygen species, abiotic stress and stress combination. *Plant J* 90:856–867.

Christou A, Filippou P, Manganaris GA, Fotopoulos V (2014) Sodium hydrosulfide induces systemic thermotolerance to strawberry plants through transcriptional regulation of heat shock proteins and aquaporin. *BMC Plant Biol* 14:1–11.

Corpas FJ (2015) what is the role of hydrogen peroxide in plant peroxisomes? *Plant Biol* 17:1099–1103.

Das K Roychoudhury A (2014) Reactive oxygen species (ROS) and response of antioxidants as ROS-scavengers during environmental stress in plants. *Front Environ Sci* 2:53.

Das P, Nutan KK, Singla-Pareek SL, Pareek A (2015) Oxidative environment and redox homeostasis in plants: dissecting out significant contribution of major cellular organelles. *Front Environ Sci* 2:70.

Deepranjan S, Ardith S, Siva D, Sonam S, Shikha, Manoj P, Amitava R, Sayyed RZ, Abdul G, Mohammad JA, Subhan D, Fahad S, Rahul D (2021) Optimizing nutrient use efficiency, productivity, energetics, and economics of red cabbage following mineral fertilization and biopriming with compatible rhizosphere microbes. *Sci Rep* 11:15680.

Del Río LA, Corpas FJ, López-Huertas E, Palma JM (2018) Plant superoxide dismutases: function under abiotic stress conditions. In: Gupta D, Palma J, Corpas F, (eds) *Antioxidants and antioxidant enzymes in higher plants*. Springer, Cham, Switzerland, pp. 1–26.

Depeng W, Fahad S, Saud S, Muhammad K, Aziz K, Mohammad NK, Hafiz MH, Wajid N (2018) Morphological acclimation to agronomic manipulation in leaf dispersion and orientation to promote “Ideotype” breeding: evidence from 3D visual modeling of “super” rice (*Oryza sativa* L.). *Plant Physiol Biochem* 135:499–510. https://doi.org/10.1016/j.plaphy.2018.11.010

Dietz KJ, Turkan I, Krieger-Liszkay A (2016) Redox and reactive oxygen species dependent signaling into and out of the photosynthesizing chloroplast. *Plant Physiol* 171:1541–1550.

Ding X, Jiang Y, He L, Zhou Q, Yu J, Hui D, Huang D (2016) Exogenous glutathione improves high root-zone temperature tolerance by modulating photosynthesis, antioxidant and osmolytes systems in cucumber seedlings. *Sci Rep* 6:35424.

El-Amier Y, Elhindi K, El-Hendawy S, Al-Rashed S, Abd-ElGawad A (2019) Antioxidant system and biomolecules alteration in *Pisum sativum* under heavy metal stress and possible alleviation by 5-aminolevulinic acid. *Molecules* 24:4194.

Emre B, Ömer SU, Martín LB, Andre D, Fahad S, Rahul D, Zafar-ul-Hye M, Ghulam SH, Subhan D (2021) Studying soil erosion by evaluating changes in physico-chemical properties of soils under different land-use types. *J Saudi Society Agricultural Sci* 20:190–197.

Fahad S, Abdul B, Adnan M (eds) (2018b) Global wheat production. *IntechOpen United Kingdom 2018*. http://dx.doi.org/10.5772/intechopen.72559

Fahad S, Bajwa AA, Nazir U, Anjum SA, Farooq A, Zohaib A, Sadia S, Nasim W, Adkins S, Saud S (2017a) Crop production under drought and heat stress: plant responses and management options. *Front Plant Sci* 8:1147.

Fahad S, Bajwa AA, Nazir U, Anjum SA, Farooq A, Zohaib A, Sadia S, Nasim W, Adkins S, Saud S, Ihsan MZ, Alharby H, Wu C, Wang D, Huang J (2017b) Crop production under drought and heat stress: plant responses and management options. *Front Plant Sci* 8:1147. https://doi.org/10.3389/fpls.2017.01147

Fahad S, Bano A (2012) Effect of salicylic acid on physiological and biochemical characterization of maize grown in saline area. *Pak J Bot* 44:1433–1438.

Fahad S, Chen Y, Saud S, Wang K, Xiong D, Chen C, Wu C, Shah F, Nie L, Huang J (2013) Ultraviolet radiation effect on photosynthetic pigments, biochemical attributes, antioxidant enzyme activity and hormonal contents of wheat. *J Food Agri Environ* 11(3&4):1635–1641.

Fahad S, Hussain S, Bano A, Saud S, Hassan S, Shan D, Khan FA, Khan F, Chen Y, Wu C, Tabassum MA, Chun MX, Afzal M, Jan A, Jan MT, Huang J (2014a) Potential role of phytohormones and plant growth-promoting rhizobacteria in abiotic stresses: consequences for changing environment. *Environ Sci Pollut Res* 22(7):4907–4921. https://doi.org/10.1007/s11356-014-3754-2

Fahad S, Hussain S, Matloob A, Khan FA, Khaliq A, Saud S, Hassan S, Shan D, Khan F, Ullah N, Faiq M, Khan MR, Tareen AK, Khan A, Ullah A, Ullah N, Huang J (2014b) Phytohormones and plant responses to salinity stress: a review. *Plant Growth Regul* 75(2):391–404. https://doi.org/10.1007/s10725-014-0013-y

Fahad S, Hussain S, Saud S, Hassan S, Chauhan BS, Khan F et al (2016a) Responses of rapid viscoanalyzer profile and other rice grain qualities to exogenously applied plant growth regulators under high day and high night temperatures. *PLoS One* 11(7):e0159590. https://doi.org/10.1371/journal.pone.0159590

Fahad S, Hussain S, Saud S, Hassan S, Ihsan Z, Shah AN, Wu C, Yousaf M, Nasim W, Alharby H, Alghabari F, Huang J (2016c) Exogenously applied plant growth regulators enhance the morphophysiological growth and yield of rice under high temperature. *Front Plant Sci* 7:1250. https://doi.org/10.3389/fpls.2016.01250

Fahad S, Hussain S, Saud S, Hassan S, Tanveer M, Ihsan MZ, Shah AN, Ullah A, Nasrullah KF, Ullah S, Alharby NW, Wu C, Huang J (2016d) A combined application of biochar and phosphorus alleviates heat-induced adversities on physiological, agronomical and quality attributes of rice. *Plant Physiol Biochem* 103:191–198.

Fahad S, Hussain S, Saud S, Khan F, Hassan S, Jr, Nasim W, Arif M, Wang F, Huang J (2016b) Exogenously applied plant growth regulators affect heat-stressed rice pollens. *J Agron Crop Sci* 202:139–150.

Fahad S, Hussain S, Saud S, Khan F, Hassan S, Nasim W, Arif M, Wang F, Huang J (2016) Exogenously applied plant growth regulators act heat-stressed rice pollens. *J Agron Crop Sci* 202:139–150.

Fahad S, Hussain S, Saud S, Tanveer M, Bajwa AA, Hassan S, Shah AN, Ullah A, Wu C, Khan FA, Shah F, Ullah S, Chen Y, Huang J (2015a) A biochar application protects rice pollen from high-temperature stress. *Plant Physiol Biochem* 96:281–287.

Fahad S, Muhammad ZI, Abdul K, Ihsanullah D, Saud S, Saleh A, Wajid N, Muhammad A, Imtiaz AK, Chao W, Depeng W, Jianliang H (2018a): Consequences of high temperature under changing climate optima for rice pollen characteristics-concepts and perspectives. *Arch Agron Soil Sci.* https://doi.org/10.1080/03650340.2018.1443213

Fahad S, Nie L, Chen Y, Wu C, Xiong D, Saud S, Hongyan L, Cui K, Huang J (2015b) Crop plant hormones and environmental stress. *Sustain Agric Rev* 15:371–400.

Fahad S, Saud S, Yajun C, Chao W, Depeng W (eds) (2021f) Abiotic stress in plants. *IntechOpen United Kingdom 2021.* http://dx.doi.org/10.5772/intechopen.91549

Fahad S, Hasanuzzaman M, Alam M, Ullah H, Saeed M, Ali Khan I, Adnan M. (eds) (2020) *Environment, climate, plant and vegetation growth.* Springer Nature Switzerland AG 2020. https://doi.org/10.1007/978-3-030-49732-3

Fahad S, Adnan M, Hassan S, Saud S, Hussain S, Wu C, Wang D, Hakeem KR, Alharby HF, Turan V, Khan MA, Huang J (2019b) Rice responses and tolerance to high temperature. In: Hasanuzzaman M, Fujita M, Nahar K, Biswas JK (eds) *Advances in rice research for abiotic stress tolerance.* Woodhead Publ Ltd, Abington Hall Abington, Cambridge CB1 6AH, Cambs, England, pp. 201–224.

Fahad S, Rehman A, Shahzad B, Tanveer M, Saud S, Kamran M, Ihtisham M, Khan SU, Turan V, Rahman MHU (2019a). Rice responses and tolerance to metal/metalloid toxicity. In: Hasanuzzaman M, Fujita M, Nahar K, Biswas JK (eds) *Advances in rice research for abiotic stress tolerance.* Woodhead Publ Ltd, Abington Hall Abington, Cambridge Cb1 6AH, CAMBS, England, pp. 299–312.

Fahad S, Sönmez O, Saud S, Wang D, Wu C, Adnan M, Arif M, Amanullah (eds) (2021e) *Engineering tolerance in crop plants against abiotic stress, First edition. ed, Footprints of climate variability on plant diversity.* CRC Press, Boca Raton.

Fahad S, Sonmez O, Saud S, Wang D, Wu C, Adnan M, Turan V (eds) (2021b) *Climate change and plants: biodiversity, growth and interactions, First edition. ed, Footprints of climate variability on plant diversity.* CRC Press, Boca Raton.

Fahad S, Sonmez O, Saud S, Wang D, Wu C, Adnan M, Turan V (eds) (2021c). *Developing climate resilient crops: improving global food security and safety, First edition. ed, Footprints of climate variability on plant diversity.* CRC Press, Boca Raton.

Fahad S, Sönmez O, Saud S, Wang D, Wu C, Adnan M, Turan V (eds) (2021a) *Plant growth regulators for climate-smart agriculture, First edition. ed, Footprints of climate variability on plant diversity.* CRC Press, Boca Raton, FL.

Fahad S, Sönmez O, Turan V, Adnan M, Saud S, Wu C, Wang D (eds) (2021d). *Sustainable soil and land management and climate change, First edition. ed, Footprints of climate variability on plant diversity.* CRC Press, Boca Raton.

Farah R, Muhammad R, Muhammad SA, Tahira Y, Muhammad AA, Maryam A, Shafaqat A, Rashid M, Muhammad R, Qaiser H, Afia Z, Muhammad AA, Muhammad A, Fahad S (2020) Alternative and non-conventional soil and crop management strategies for increasing water use efficiency. In: Fahad S, Hasanuzzaman M, Alam M, Ullah H, Saeed M, Khan AK, Adnan M (eds) *Environment, climate, plant and vegetation growth.* Springer Publ Ltd, Springer Nature Switzerland AG. Part of Springer Nature, pp. 323–338. https://doi.org/10.1007/978-3-030-49732-3

Farhana G, Ishfaq A, Muhammad A, Dawood J, Fahad S, Xiuling L, Depeng W, Muhammad F, Muhammad F, Syed AS (2020) Use of crop growth model to simulate the impact of climate change on yield of various wheat cultivars under different agro-environmental conditions in Khyber Pakhtunkhwa, *Pakistan. Arabian J Geosci* 13:112. https://doi.org/10.1007/s12517-020-5118-1

Farhat A, Hafiz MH, Wajid I, Aitazaz AF, Hafiz FB, Zahida Z, Fahad S, Wajid F, Artemi C (2020) A review of soil carbon dynamics resulting from agricultural practices. *J Environ Manage* 268:110319.

Fariduddin Q, Khan TA, Yusuf M (2014) Hydrogen peroxide mediated tolerance to copper stress in the presence of 28-homobrassinolide in Vigna radiata. *Acta Physiol Plant* 36:2767–2778.

Fazli W, Muhmmad S, Amjad A, Fahad S, Muhammad A, Muhammad N, Ishaq AM, Imtiaz AK, Mukhtar A, Muhammad S, Muhammad I, Rafi U, Haroon I, Muhammad A (2020) Plant-microbes interactions and functions in changing climate. In: Fahad S, Hasanuzzaman M, Alam M, Ullah H, Saeed M, Khan AK, Adnan M (eds) *Environment, climate, plant and vegetation growth.* Springer Publ Ltd, Springer Nature Switzerland AG. Part of Springer Nature, pp. 397–420. https://doi.org/10.1007/978-3-030-49732-3

Fotopoulos V, Tanou G, Ziogas V, Molassiotis A (2010) Involvement of AsA/DHA and GSH/GSSG ratios in gene and protein expression and in the activation of defense mechanisms under abiotic stress conditions. In: Anjum NA, Chan MT, Umar S (eds) *Ascorbate-glutathione pathway and stress tolerance in plants.* Springer, Dordrecht, the Netherlands, pp. 265–302.

Foyer CH, Noctor G (2011) Ascorbate and glutathione: The heart of the redox hub. *Plant Physiol* 155:2–18.

Fracasso A, Trindade L, Amaducci S (2016) Drought tolerance strategies highlighted by two Sorghum bicolor races in a drydown experiment. *J Plant Physiol* 190:1–14.

Gal J, Afifi, M, Lee E, Lukens L, Swanton CJ (2015) Detection of neighboring weeds alters soybean seedling roots and nodulation. *Weed Sci* 63:888–900.

Garg N, Manchanda G (2009) ROS generation in plants: boon or bane? *Plant Biosys* 143:8–96.

Gechev TS, Hille J (2005) Hydrogen peroxide as a signal controlling plant programmed cell death. *J Cell Biol* 168(1):17–20.

Ghulam M, Muhammad AA, Donald LS, Sajid M, Muhammad FQ, Niaz A, Ateeq Ur R, Shakeel A, Sajjad H, Muhammad A, Summia M, Aqib HAK, Fahad S, Rahul D, Mazhar I, Timothy DS (2021) Formalin fumigation and steaming of various composts differentially influence the nutrient release, growth and yield of muskmelon (*Cucumis melo* L.). *Sci Rep* 11:21057.

Gill SS, Tuteja N (2010) Reactive oxygen species and antioxidant machinery in abiotic stress tolerance in crop plants. *Physiol Bioch* 48:909–930.

Gohari G, Alavi Z, Esfandiari E, Panahirad S, Hajihoseinlou S, Fotopoulos V (2020) Interaction between hydrogen peroxide and sodium nitroprusside following chemical priming of *Ocimum basilicum* L. against salt stress. *Physiol Plant* 168:361–373.

Gopakumar L, Bernard NO, Donato V (2020) Soil microarthropods and nutrient cycling. In: Fahad S, Hasanuzzaman M, Alam M, Ullah H, Saeed M, Khan AK, Adnan M (eds) *Environment, climate, plant and vegetation growth.* Springer Publ Ltd, Springer Nature Switzerland AG. Part of Springer Nature, pp. 453–472. https://doi.org/10.1007/978-3-030-49732-3

Guler NS, Pehlivan N (2016) Exogenous low-dose hydrogen peroxide enhances drought tolerance of soybean (*Glycine max* L.) through inducing antioxidant system. *Acta Biol Hung* 267: 169–183.

Guofu L, Zhenjian B, Fahad S, Guowen C, Zhixin X, Hao G, Dandan L, Yulong L, Bing L, Guoxu J, Saud S (2021) Compositional and structural changes in soil microbial communities in response to straw

mulching and plant revegetation in an abandoned artificial pasture in Northeast China. *Glob Ecol Conserv* 31:e01871.
Gupta D, Pena LB, Romero-Puertas MC, Hernández A, Inouhe M, Sandalio LM (2017) NADPH oxidases differentially regulate ROS metabolism and nutrient uptake under cadmium toxicity. *Plant Cell Environ* 40:509–526.
Habib N, Ali Q, Ali S, Javed MT, Zulqurnain HM, Perveen R, Shahid MR, Rizwan M, Abdel-Daim MM, Elkelish A (2020) Use of Nitric oxide and hydrogen peroxide for better yield of wheat (*Triticum aestivum* L.) under water deficit conditions: growth, osmoregulation, and antioxidative defense mechanism. *Plants* 9:285.
Habib R, Ashfaq A, Aftab W, Manzoor H, Fahd R, Wajid I, Aminul I, Vakhtang S, Muhammad A, Asmat U, Abdul W, Syeda RS, Shah S, Shahbaz K, Fahad S, Manzoor H, Saddam H, Wajid N (2017) Application of CSM-CROPGRO-Cotton model for cultivars and optimum planting dates: evaluation in changing semi-arid climate. *Field Crops Res.* http://dx.doi.org/10.1016/j.fcr.2017.07.007
Hafeez M, Farman U, Muhammad MK, Xiaowei L, Zhijun Z, Sakhawat S, Muhammad I, Mohammed AA, Mandela F-G, Nicolas D, Muzammal R, Fahad S, Yaobin L (2021) Metabolic-based insecticide resistance mechanism and ecofriendly approaches for controlling of beet armyworm Spodoptera exigua: a review. *Environ Sci Pollution Res.* https://doi.org/10.1007/s11356-021-16974-w
Hafiz MH, Abdul K, Farhat A, Wajid F, Fahad S, Muhammad A, Ghulam MS, Wajid N, Muhammad M, Hafiz FB (2020b) Comparative effects of organic and inorganic fertilizers on soil organic carbon and wheat productivity under arid region. *Commun Soil Sci Plant Anal.* https://doi.org/10.1080/00103624.2020.1763385
Hafiz MH, Farhat A, Ashfaq A, Hafiz FB, Wajid F, Carol Jo W, Fahad S, Gerrit H (2020a) Predicting kernel growth of maize under controlled water and nitrogen applications. *Int J Plant Prod.* https://doi.org/10.1007/s42106-020-00110-8
Hafiz MH, Farhat A, Shafqat S, Fahad S, Artemi C, Wajid F, Chaves CB, Wajid N, Muhammad M, Hafiz FB (2018) Offsetting land degradation through nitrogen and water management during maize cultivation under arid conditions. *Land Degrad Dev* 1–10. https://doi.org/10.1002/ldr.2933
Hafiz MH, Muhammad A, Farhat A, Hafiz FB, Saeed AQ, Muhammad M, Fahad S, Muhammad A (2019) Environmental factors affecting the frequency of road traffic accidents: a case study of sub-urban area of Pakistan. *Environ Sci Pollut Res.* https://doi.org/10.1007/s11356-019-04752-8
Hafiz MH, Wajid F, Farhat A, Fahad S, Shafqat S, Wajid N, Hafiz FB (2016) Maize plant nitrogen uptake dynamics at limited irrigation water and nitrogen. *Environ Sci Pollut Res* 24(3):2549–2557. https://doi.org/10.1007/s11356-016-8031-0
Haider SA, Lalarukh I, Amjad SF, Mansoora N, Naz M, Naeem M, Bukhari SA, Shahbaz M, Ali SA, Marfo TD, Subhan D, Rahul D, Fahad S (2021) Drought stress alleviation by potassium-nitrate-containing chitosan/montmorillonite microparticles confers changes in *Spinacia oleracea* L. *Sustain* 13:9903. https://doi.org/10.3390/su13179903
Halliwell B (2006) Reactive species and antioxidants. Redox biology is a fundamental theme of aerobic life. *Plant Physiol.* 141(2):312–322.
Hamza SM, Xiukang W, Sajjad A, Sadia Z, Muhammad N, Adnan M, Fahad S et al. (2021) Interactive effects of gibberellic acid and NPK on morpho-physio-biochemical traits and organic acid exudation pattern in coriander (*Coriandrum sativum* L.) grown in soil artificially spiked with boron. *Plant Physiol Biochem* 167 (2021) 884–900.
Han QH, Huang B, Ding CB, Zhang, ZW, Chen YE, Hu C, Zhou LJ, Huang Y, Liao JQ, Yuan SE (2017) Effects of melatonin on anti-oxidative systems and photosystem II in cold-stressed rice seedlings. *Front Plant Sci* 8:785.
Hasanuzzaman M, Bhuyan MHM, Anee TI, Parvin K, Nahar K, Mahmud JA, Fujita M (2019) Regulation of ascorbate-glutathione pathway in mitigating oxidative damage in plants under abiotic stress. *Antioxidants* 8(9):384.
Hasanuzzaman M, Bhuyan MHM, Zulfiqar F, Raza A, Mohsin SM, Mahmud JA, Fotopoulos V (2020) Reactive oxygen species and antioxidant defense in plants under abiotic stress: revisiting the crucial role of a universal defense regulator. *Antioxidants* 9(8):681.
Hasanuzzaman M, Nahar K, Alam M, Roychowdhury R, Fujita M (2013) Physiological, biochemical, and molecular mechanisms of heat stress tolerance in plants. *Int J Mol Sci* 14:9643–9684.

Hasanuzzaman M, Nahar K, Gill SS, Alharby HF, Razafindrabe BH, Fujita M (2017) Hydrogen peroxide pretreatment mitigates cadmium-induced oxidative stress in *Brassica napus* L.: An intrinsic study on antioxidant defense and glyoxalase systems. *Front Plant Sci* 8:115.

Hasanuzzaman M, Nahar K, Rahman A, Mahmud JA, Alharby HF, Fujita M (2018) Exogenous glutathione attenuates lead-induced oxidative stress in wheat by improving antioxidant defense and physiological mechanisms. *J Plant Interact* 13:203–212.

Hesham FA, Fahad S (2020) Melatonin application enhances biochar efficiency for drought tolerance in maize varieties: modifications in physio-biochemical machinery. *Agron J* 112(4):1–22.

Higuchi ET (2006) Look back over the studies of lignin biochemistry. *J Wood Sci* 52:2–8.

Hmaeid N, Wali M, Mahmoud OM, Pueyo JJ, Ghnaya T, Abdelly C (2018) Efficient rhizobacteria promote growth and alleviate NaCl-induced stress in the plant species *Sulla carnosa*. *Appl Soil Ecol* 133:104–113.

Huang Li Y, Li Xiao-X, Zhang Yun-B, Fahad S, Wang F (2021) dep1 improves rice grain yield and nitrogen use efficiency simultaneously by enhancing nitrogen and dry matter translocation. *J Integrative Agri*. https://doi.org/10.1016/S2095-3119(21)63795-4

Huang S, Aken OV, Schwarzlander M, Belt K, Millar AH (2016) The roles of mitochondrial reactive oxygen species in cellular signaling and stress response in plants. *Plant Physiol* 171:1551–1559.

Hussain MA, Fahad S, Rahat S, Muhammad FJ, Muhammad M, Qasid A, Ali A, Husain A, Nooral A, Babatope SA, Changbao S, Liya G, Ibrar A, Zhanmei J, Juncai H (2020) Multifunctional role of brassinosteroid and its analogues in plants. *Plant Growth Regul*. https://doi.org/10.1007/s10725-020-00647-8

Ibrar K, Aneela R, Khola Z, Urooba N, Sana B, Rabia S, Ishtiaq H, Ur Rehman M, Salvatore M (2020) Microbes and environment: global warming reverting the frozen zombies. In: Fahad S, Hasanuzzaman M, Alam M, Ullah H, Saeed M, Khan AK, Adnan M (eds) *Environment, climate, plant and vegetation growth.* Springer Publ Ltd, Springer Nature Switzerland AG. Part of Springer Nature, pp. 607–634. https://doi.org/10.1007/978-3-030-49732-3

Ikram U, Khadim D, Muhammad T, Muhammad S, Fahad S (2021) Gibberellic acid and urease inhibitor optimize nitrogen uptake and yield of maize at varying nitrogen levels under changing climate. *Environ Sci Pollution Res*. https://doi.org/10.1007/s11356-021-16049-w

Ilyas M, Mohammad N, Nadeem K, Ali H, Aamir HK, Kashif H, Fahad S, Aziz K, Abid U, (2020) Drought tolerance strategies in plants: a mechanistic approach. *J Plant Growth Regul*. https://doi.org/10.1007/s00344-020-10174-5

Intergovernmental Panel on Climate Change (IPCC) (2007) *Climate change – the physical science basis*. In *Contribution of Working Group I to the Fourth Assessment Report of the Intergovernmental Panel on Climate Change*. Cambridge University Press, Cambridge, UK.

Iqra M, Amna B, Shakeel I, Fatima K, Sehrish L, Hamza A, Fahad S (2020) Carbon cycle in response to global warming. In: Fahad S, Hasanuzzaman M, Alam M, Ullah H, Saeed M, Khan AK, Adnan M (eds) *Environment, climate, plant and vegetation growth*. Springer Publ Ltd, Springer Nature Switzerland AG. Part of Springer Nature, pp. 1–16. https://doi.org/10.1007/978-3-030-49732-3

Irfan M, Muhammad M, Muhammad JK, Khadim MD, Dost M, Ishaq AM, Waqas A, Fahad S, Saud S et al. (2021) Heavy metals immobilization and improvement in maize (*Zea mays* L.) growth amended with biochar and compost. *Sci Rep* 11:18416.

Jabborova D, Sulaymanov K, Sayyed RZ, Alotaibi SH, Enakiev Y, Azimov A, Jabbarov Z, Ansari MJ, Fahad S, Danish S et al. (2021) Mineral fertilizers improves the quality of turmeric and soil. *Sustain* 13: 9437. https://doi.org/10.3390/su13169437

Jajoo A (2017) Effects of environmental pollutants polycyclic aromatic hydrocarbons (PAH) on photosynthetic processes. In: Hou H, Najafpour M, Moore G, Allakhverdiev S (eds) *Photosynthesis: structures, mechanisms, and applications*. Springer, Cham, 249–259.

Jajoo A, Mekala NR, Tomar RS Grieco M, Tikkanen M, Aro EM (2014) Inhibitory effects of polycyclic aromatic hydrocarbons (PAHs) on photosynthetic performance are not related to their aromaticity. *J Photochem Photobiol B* 137:151–155.

Jan M, Anwar-ul-Haq M, Adnan NS, Muhammad Y, Javaid I, Xiuling L, Depeng W, Fahad S (2019) Modulation in growth, gas exchange, and antioxidant activities of salt-stressed rice (*Oryza sativa* L.) genotypes by zinc fertilization. *Arabian J Geosci* 12:775. https://doi.org/10.1007/s12517-019-4939-2

Janska A, Marsik P, Zelenkova S, Ovesna J (2010) Cold stress and acclimation: what is important for metabolic adjustment? *Plant Biol* 12:395–405.

Javed MT, Akram MS, Habib N, Tanwir K, Ali Q, Niazi NK, Iqbal N (2018) Deciphering the growth, organic acid exudations, and ionic homeostasis of *Amaranthus viridis* L. and *Portulaca oleracea* L. under lead chloride stress. *Environ Sci Pollut Res* 25(3):2958–2971.

Javed MT, Akram MS, Tanwir K, Chaudhary HJ, Ali Q, Stoltz E, Lindberg S (2017) Cadmium spiked soil modulates root organic acids exudation and ionic contents of two differentially Cd tolerant maize (*Zea mays* L.) cultivars. *Ecotoxicol Environ Saf* 141:216–225.

Javed MT, Habib N, Akram MS, Ali Q, Haider MZ, Tanwir K, Chaudhary HJ (2019) The effect of lead pollution on nutrient solution pH and concomitant changes in plant physiology of two contrasting *Solanum melongena* L. cultivars. *Environ Sci Pollut Res* 26(33):34633–34644.

Kabała K, Zboinska M, Głowiak D, Reda M, Jakubowska D, Janicka M (2019) Interaction between the signaling molecules hydrogen sulfide and hydrogen peroxide and their role in vacuolarH^+-ATPase regulation in cadmium-stressed cucumber roots. *Physiol Plant* 166:688–704.

Kamarn M, Wenwen C, Irshad A, Xiangping M, Xudong Z, Wennan S, Junzhi C, Shakeel A, Fahad S, Qingfang H, Tiening L (2017) Effect of paclobutrazol, a potential growth regulator on stalk mechanical strength, lignin accumulation and its relation with lodging resistance of maize. *Plant Growth Regul* 84:317–332. https://doi.org/10.1007/s10725-017-0342-8

Kapoor D, Singh S, Kumar V, Romero R, Prasad R, Singh J (2019) Antioxidant enzymes regulation in plants in reference to reactive oxygen species (ROS) and reactive nitrogen species (RNS). *Plant Gene* 19:100182.

Kaur N, Kaur J, Grewal SK, Singh I (2019) Effect of Heat Stress on Antioxidative defense system and its amelioration by heat acclimation and salicylic acid pre-treatments in three pigeonpea genotypes. *Indian J Agric Biochem* 32:106–110.

Kawano T (2003) Roles of the reactive oxygen species generating peroxidase reactions in plant defense and growth induction. *Plant Cell Rep* 21:829–937.

Khadim D, Fahad S, Jahangir MMR, Iqbal M, Syed SA, Shah AK, Ishaq AM, Rahul D et al. (2021a) Biochar and urease inhibitor mitigate NH_3 and N_2O emissions and improve wheat yield in a urea fertilized alkaline soil. *Sci Rep* 11:17413.

Khadim D, Ur Saif R, Fahad S, Syed SA, Shah AK et al. (2021b) Influence of variable biochar concentration on yield-scaled nitrous oxide emissions, Wheat yield and nitrogen use efficiency. *Sci Rep* 11:16774.

Khan MMH, Niaz A, Umber G, Muqarrab A, Muhammad AA, Muhammad I, Shabir H, Shah F, Vibhor A, Shams HA-H, Reham A, Syed MBA, Nadiyah MA, Ali TKZ, Subhan D, Rahul D (2021) Synchronization of Boron application methods and rates is environmentally friendly approach to improve quality attributes of *Mangifera indica* L. On sustainable basis. *Saudi J Bio Sci*. https://doi.org/10.1016/j.sjbs.2021.10.036

Kim C, Dogra V (2019) Singlet oxygen metabolism: From genesis to signaling. *Front Plant Sci* 10:1640.

Kim MJ, Ciani S, Schachtman DP (2010) A peroxidase contributes to ROS production during Arabidopsis root response to potassium deficiency. *Mol Plant* 3:420–427.

Kreslavski VD, Brestic M, Zharmukhamedov SK, Lyubimov VY, Lankin AV, Jajoo A, Allakhverdiev SI (2017) Mechanisms of inhibitory effects of polycyclic aromatic hydrocarbons in photosynthetic primary processes in pea leaves and thylakoid preparations. *Plant Biol 1* 9(5):683–688.

Lashari MS, Liu Y, Li Y, Pan W, Fu J, Pan G, Zheng J, Zheng J, Zhang X, Yu X (2013) Effects of amendment of biochar-manure compost in conjunction with pyroligneous solution on soil quality and wheat yield of a salt-stressed cropland from Central China Great Plain. *Field Crop Res* 144:113–118.

Laxa M, Liebthal M, Telman W, Chibani K, Dietz KJ (2019) The role of the plant antioxidant system in drought tolerance. *Antioxidants* 8:94.

Liu J, Hasanuzzaman M, Wen H, Zhang J, Peng T, Sun H, Zhao Q (2019) High temperature and drought stress cause abscisic acid and reactive oxygen species accumulation and suppress seed germination growth in rice. *Protoplasma* 256:1217–1227.

Liu T, Ye X, Li M, Li J, Qi H, Hu X (2020) H_2O_2 and NO are involved in trehalose-regulated oxidative stress tolerance in cold-stressed tomato plants. *Environ Exp Bot* 171:103961.

Liu Y, He C (2016) Regulation of plant reactive oxygen species (ROS) in stress responses: learning from AtRBOHD. *Plant Cell Rep* 35:995–1007.

Lobell DB, Field CB (2007) Global scale climate–Crop yield relationships and the impacts of recent warming. *Environ Res Lett* 2(1):014002.

Lopez-Mungia A, Hernandez-Romero Y, Pedraza-Chaverri J, Miranda MA, Regla I, Martinez A, Castillo E (2011) Phenylpropanoid glycoside anatogues: enzymatic synthesis, antioxidant activity and theoretical study of their Free radical Scavenger mechanism. *Plos One* 6(6):201–215.

Loreti E, Van VH, Perata P (2016) Plant responses to flooding stress. *Curr Opin Plant Biol* 33:64–71.

Mahar A, Amjad A, Altaf HL, Fazli W, Ronghua L, Muhammad A, Fahad S, Muhammad A, Rafiullah, Imtiaz AK, Zengqiang Z (2020) Promising technologies for Cd-contaminated soils: drawbacks and possibilities. In: Fahad S, Hasanuzzaman M, Alam M, Ullah H, Saeed M, Khan AK, Adnan M (eds) *Environment, climate, plant and vegetation growth*. Springer Publ Ltd, Springer Nature Switzerland AG. Part of Springer Nature, pp. 63–92. https://doi.org/10.1007/978-3-030-49732-3

Mahmood UH, Tassaduq R, Chandni I, Adnan A, Muhammad A, Muhammad MA, Muhammad HR, Mehmood AN, Alam S, Fahad S (2021) Linking plants functioning to adaptive responses under heat stress conditions: a mechanistic review. *J Plant Growth Regul*. https://doi.org/10.1007/s00344-021-10493-1

Mahmud JA, Hasanuzzaman M, Nahar K, Bhuyan MHMB, Fujita M (2018) Insights into citric acid-induced cadmium tolerance and phytoremediation in *Brassica juncea* L. coordinated functions of metal chelation, antioxidant defense and glyoxalase systems. *Ecotoxicol Environ Saf* 147:990–1001.

Malhotra C, Kapoor RT, Ganjewala D, Singh N (2017) Sodium silicate mediated response of antioxidative defense system in *Lycopersicon esculentum* mill under water stress. *Int J Phytomed* 9:364–378.

Manzer HS, Saud A, Soumya M, Abdullah A, Qasi DA, Bander MA, Hayssam MA, Hazem MK, Fahad S, Vishnu DR, Om PN (2021) Molybdenum and hydrogen sulfide synergistically mitigate arsenic toxicity by modulating defense system, nitrogen and cysteine assimilation in faba bean (*Vicia faba* L.) seedlings. *Environ Pollut* 290:117953.

Md Jakir H, Allah B (2020) Development and applications of transplastomic plants; A way towards eco-friendly agriculture. In: Fahad S, Hasanuzzaman M, Alam M, Ullah H, Saeed M, Khan AK, Adnan M (eds) *Environment, climate, plant and vegetation growth*. Springer Publ Ltd, Springer Nature Switzerland AG. Part of Springer Nature, pp. 285–322. https://doi.org/10.1007/978-3-030-49732-3

Md. Enamul H, Shoeb AZM, Mallik AH, Fahad S, Kamruzzaman MM, Akib J, Nayyer S, Mehedi A, Swati AS, Md Yeamin A, Most SS (2020) Measuring vulnerability to environmental hazards: qualitative to quantitative. In: Fahad S, Hasanuzzaman M, Alam M, Ullah H, Saeed M, Khan AK, Adnan M (eds) *Environment, climate, plant and vegetation growth*. Springer Publ Ltd, Springer Nature Switzerland AG. Part of Springer Nature, pp. 421–452. https://doi.org/10.1007/978-3-030-49732-3

Mehla N, Sindhi V, Josula D, Bisht P, Wani SH (2017) An introduction to antioxidants and their roles in plant stress tolerance. In: Khan MIR, Khan NA (eds) *Reactive oxygen species and antioxidant systems in plants: role and regulation under abiotic stress*. Springer, Singapore, pp. 1–23.

Miller G, Suzuki N, Ciftci-Yilmaz S, Mittler R (2010) Reactive oxygen species homeostasis and signalling during drought and salinity stresses. *Plant Cell Environ* 33:453–467.

Mittler R (2017) ROS are good. *Trends Plant Sci* 22:11–19.

Mohammad IAW, Munir A, Adel RAU, Mutair A, Muhammad IR (2020) Advances in pyrolytic technologies with improved carbon capture and storage to combat climate change. In: Fahad S, Hasanuzzaman M, Alam M, Ullah H, Saeed M, Khan AK, Adnan M (eds) *Environment, climate, plant and vegetation growth*. Springer Publ Ltd, Springer Nature Switzerland AG. Part of Springer Nature, pp. 535–576. https://doi.org/10.1007/978-3-030-49732-3

Mubeen M, Ashfaq A, Hafiz MH, Muhammad A, Hafiz UF, Mazhar S, Muhammad SD, Asad A, Amjed A, Fahad S, Wajid N (2020) Evaluating the climate change impact on water use efficiency of cotton-wheat in semi-arid conditions using DSSAT model. *J Water Clim Change*. https://doi.org/10.2166/wcc.2019.179/622035/jwc2019179.pdf

Muhammad N, Muqarrab A, Khurram S, Fiaz A, Fahim N, Muhammad A, Shazia A, Omaima N, Sulaiman AA, Fahad S, Subhan D, Rahul D (2021) Kaolin and Jasmonic acid improved cotton productivity under water stress conditions. *J Saudi Soc Agric Sci* 28:6606–6614. https://doi.org/10.1016/j.sjbs.2021.07.043

Muhammad TQ, Amna F, Amna B, Barira Z, Xitong Z, Ling-Ling C (2020) Effectiveness of conventional crop improvement strategies vs. omics. In: Fahad S, Hasanuzzaman M, Alam M, Ullah H, Saeed M, Khan AK, Adnan M (eds) *Environment, climate, plant and vegetation growth*. Springer Publ Ltd, Springer Nature Switzerland AG. Part of Springer Nature, pp. 253–284. https://doi.org/10.1007/978-3-030-49732-3

Muhammad Z, Abdul MK, Abdul MS, Kenneth BM, Muhammad S, Shahen S, Ibadullah J, Fahad S (2019) Performance of *Aeluropus lagopoides* (mangrove grass) ecotypes, a potential turfgrass, under high saline conditions. *Environ Sci Pollut Res*. https://doi.org/10.1007/s11356-019-04838-3

Muzammal R, Fahad S, Guanghui D, Xia C, Yang Y, Kailei T, Lijun L, Fei-Hu L, Gang D (2021) Evaluation of hemp (*Cannabis sativa* L.) as an industrial crop: a review. *Environ Sci Pollution Res*. https://doi.org/10.1007/s11356-021-16264-5

Naseer F, Baig J, Din SU, Nafees MA, Alam A, Hameed N, Adnan M, Arshad M, Khan MA, Romman M, Mian IA, Shah SRA (2016) Impact of water quality on distribution of macro invertebrate in Jutialnallah, Gilgit-Baltistan, Pakistan. *Int J Biosci* 9(6):451–459.

Nath M, Bhatt D, Bhatt MD, Prasad R, Tuteja N (2018) Microbe-mediated enhancement of nitrogen and phosphorus content for crop improvement. In: Prasad R, Gill SS, Tuteja N (eds) *Crop improvement through microbial biotechnology*. Academic Press, Cambridge, UK, pp. 291–301.

Negrao S, Schmockel SM, Tester M (2017) Evaluating physiological responses of plants to salinity stress. *Ann Bot* 119:1–11.

Nishio K, Horie M, Akazawa Y, Shichiri M, Iwahashi H, Hagihara Y, Niki E (2013) Attenuation of lipopolysaccharide (LPS)-induced cytotoxicity by tocopherols and tocotrienols. *Redox Biol* 1(1):97–103.

Noor M, Naveed R, Ajmal J, Fahad S, Muhammad A, Fazli W, Saud S, Hassan S (2020) Climate change and costal plant lives. In: Fahad S, Hasanuzzaman M, Alam M, Ullah H, Saeed M, Khan AK, Adnan M (eds) *Environment, climate, plant and vegetation growth*. Springer Publ Ltd, Springer Nature Switzerland AG. Part of Springer Nature, pp. 93–108. https://doi.org/10.1007/978-3-030-49732-3

Paciolla C, De-Tullio MC, Chiappetta A, Innocenti AM, Bitonti MB, Liso R, Arrigoni O (2001) Short-and long-term effects of dehydroascorbate in *Lupinus albus* and *Allium cepa* roots. *Plant Cell Physiol* 42(8):857–863.

Panuccio MR, Jacobsen SE, Akhtar SS, Muscolo A (2014) Effect of saline water irrigation on seed germination and early seedling growth of the halophyte quinoa. *AoB Plants* 6:plu047.

Park JS, Lee, EJ (2019) Waterlogging induced oxidative stress and the mortality of the Antarctic plant, *Deschampsia antarctica*. *J Ecol Environ* 43:29.

Parvez S, Abbas G, Shahid M, Amjad M, Hussain M, Asad SA, Imran M, Naeem MA (2020) Effect of salinity on physiological, biochemical and photostabilizing attributes of two genotypes of quinoa (*Chenopodium quinoa* Willd.) exposed to arsenic stress. *Ecotoxicol Environ Saf* 187:109814.

Piasecki C, Rizzardi MA, Schons J, Caverzan A, Oliveira C (2018) Interference of volunteer corn on stress metabolism and yield of dry bean. *Planta Daninha* 36:e018176669.

Qadir M, Quillero E, Nangia V, Noble AD (2014) Economics of salt-induced and degradation and restoration. *Nat Res Forum* 38:282–295.

Qamar-Uz Z, Zubair A, Muhammad Y, Muhammad ZI, Abdul K, Fahad S, Safder B, Ramzani PMA, Muhammad N (2017) Zinc biofortification in rice: leveraging agriculture to moderate hidden hunger in developing countries. *Arch Agron Soil Sci* 64:147–161. https://doi.org/10.1080/03650340.2017.1338343

Qi J, Wang J, Gong Z, Zhou, JM (2017) Apoplastic ROS signaling in plant immunity. *Curr Opin Plant Biol* 38:92–100.

Radosevich SR, Holt JS, Ghersa CM (2007) *Ecology of weeds and invasive plants: relationship to agriculture and natural resource management*, 3rd ed., John Wiley & Sons, Hoboken, NJ, USA.

Rady MM, Belal HE, Gadallah FMS, Emida WM (2020) Selenium application in two methods promotes drought tolerance in *Solanum lycopersicum* plant by inducing the antioxidant defense system. *Sci Hortic* 266:109290.

Rahman I, Ali S, Alam M, Adnan M, Basir A, Ullah H, Malik MFA, Shah AS, Ibrahim M (2016) Effect of seed priming on germination performance and Yield of okra (abelmoschusesculentusl.). *Pakistan J Agric Res* 29(3):253–262.

Raja V, Majeed U, Kang H, Andrabi KI, John R (2017) Abiotic stress: Interplay between ROS, hormones and MAPKs. *Environ Exp Bot* 137:142–157.

Ramegowda V, Senthil-Kumar M (2015) The interactive effects of simultaneous biotic and abiotic stresses on plants: mechanistic understanding from drought and pathogen combination. *J Plant Physiol* 176: 47–54.

Rasheed R, Iqbal M, Ashraf A, Hussain I, Shafiq F, Yousaf A, Zaheer A (2018) Glycine betaine counteracts the inhibitory effects of waterlogging on growth, photosynthetic pigments, oxidative defence system, nutrient composition, and fruit quality in tomato. *J Hortic Sci Biotechnol* 93:385–391.

Rashid M, Qaiser H, Khalid SK, Mohammad IAW, Zhang A, Muhammad A, Shahzada SI, Rukhsanda A, Ghulam AS, Shahzada MM, Sarosh A, Muhammad FQ (2020) Prospects of biochar in alkaline soils to mitigate climate change. In: Fahad S, Hasanuzzaman M, Alam M, Ullah H, Saeed M, Khan AK, Adnan M (eds) *Environment, climate, plant and vegetation growth*. Springer Publ Ltd, Springer Nature Switzerland AG. Part of Springer Nature, pp. 133–150. https://doi.org/10.1007/978-3-030-49732-3

Rehman M, Fahad S, Saleem MH, Hafeez M, Muhammad HR, Liu F, Deng G (2020) Red light optimized physiological traits and enhanced the growth of ramie (*Boehmeria nivea* L.). *Photosynthetica* 58(4): 922–931.

Repetto M, Semprine J, Boveris A (2012) Lipid peroxidation: chemical mechanism, biological implications and analytical determination. In: Catala A (ed) *Lipid Peroxidation*. In Tech, Rijeka, Croatia, pp. 3–30.

Rezayian M, Ebrahimzadeh H, Niknam V (2020) Nitric oxide stimulates antioxidant system and osmotic adjustment in soybean under drought stress. *J Soil Sci Plant Nutr* 1–11. https://doi.org/10.1007/s42729-020-00198-x.

Richards SL, Wilikins KA, Swarbreck SM (2015) The hydroxyl radical in plants: from seed to seed. *J Exp Bot* 66:37–46.

Romero-Puertas MC, Corpas FJ, Sandalio LM, Leterrier M, Rodríguez-Serrano M, Del Río LA, Palma JM (2006) Glutathione reductase from pea leaves: response to abiotic stress and characterization of the peroxisomal isozyme. *New Phytologist* 170(1):43–52.

Roychoudhury A, Basu S, Sengupta DN (2012) Antioxidants and stress-related metabolites in the seedlings of two indica rice varieties exposed to cadmium chloride toxicity. *Acta Physiol Plant* 34(3):835–847.

Sadam M, Muhammad TQ, Ghulam M, Muhammad SK, Faiz AJ (2020) Role of biotechnology in climate resilient agriculture. In: Fahad S, Hasanuzzaman M, Alam M, Ullah H, Saeed M, Khan AK, Adnan M (eds) *Environment, climate, plant and vegetation growth*. Springer Publ Ltd, Springer Nature Switzerland AG. Part of Springer Nature, pp. 339–366. https://doi.org/10.1007/978-3-030-49732-3

Sade B, Soylu S, Soylu E (2011) Drought and oxidative stress. *Afr J Biotechnol* 10(54):11102–11109.

Safi UK, Ullah F, Mehmood S, Fahad S, Ahmad Rahi A, Althobaiti F, et al. (2021) Antimicrobial, antioxidant and cytotoxic properties of *Chenopodium glaucum* L. *PLoS One* 16(10):e0255502. https://doi.org/10.1371/journal.pone.0255502

Saha I, De AK, Sarkar B, Ghosh A, Dey N, Adak MK (2018) Cellular response of oxidative stress when sub1A QTL of rice receives water deficit stress. *Plant Sci Today* 5:84–94.

Sahrish N, Shakeel A, Ghulam A, Zartash F, Sajjad H, Mukhtar A, Muhammad AK, Ahmad K, Fahad S, Wajid N, Sezai E, Carol Jo W, Gerrit H (2022) Modeling the impact of climate warming on potato phenology. *European J AgroN* 132:126404.

Sajid H, Jie H, Jing H, Shakeel A, Satyabrata N, Sumera A, Awais S, Chunquan Z, Lianfeng Z, Xiaochuang C, Qianyu J, Junhua Z (2020) Rice production under climate change: adaptations and mitigating strategies. In: Fahad S, Hasanuzzaman M, Alam M, Ullah H, Saeed M, Khan AK, Adnan M (eds) *Environment, climate, plant and vegetation growth*. Springer Publ Ltd, Springer Nature Switzerland AG. Part of Springer Nature, pp. 659–686. https://doi.org/10.1007/978-3-030-49732-3

Sajjad H, Muhammad M, Ashfaq A, Waseem A, Hafiz MH, Mazhar A, Nasir M, Asad A, Hafiz UF, Syeda RS, Fahad S, Depeng W, Wajid N (2019) Using GIS tools to detect the land use/land cover changes during forty years in Lodhran district of Pakistan. *Environ Sci Pollut Res*. https://doi.org/10.1007/s11356-019-06072-3

Saleem MH, Fahad S, Adnan M, Mohsin A, Muhammad SR, Muhammad K, Qurban A, Inas AH, Parashuram B, Mubassir A, Reem MH (2020a) Foliar application of gibberellic acid endorsed phytoextraction of copper and alleviates oxidative stress in jute (*Corchorus capsularis* L.) plant grown in highly copper-contaminated soil of China. *Environ Sci Pollution Res*. https://doi.org/10.1007/s11356-020-09764-3

Saleem MH, Fahad S, Shahid UK, Mairaj D, Abid U, Ayman ELS, Akbar H, Analía L, Lijun L (2020c) Copper-induced oxidative stress, initiation of antioxidants and phytoremediation potential of flax (*Linum usitatissimum* L.) seedlings grown under the mixing of two different soils of China. *Environ Sci Poll Res*. https://doi.org/10.1007/s11356-019-07264-7

Saleem MH, Rehman M, Fahad S, Tung SA, Iqbal N, Hassan A, Ayub A, Wahid MA, Shaukat S, Liu L, Deng G (2020b) Leaf gas exchange, oxidative stress, and physiological attributes of rapeseed (*Brassica napus* L.) grown under different light-emitting diodes. *Photosynthetica* 58(3):836–845.

Saman S, Amna B, Bani A, Muhammad TQ, Rana MA, Muhammad SK (2020) QTL mapping for abiotic stresses in cereals. In: Fahad S, Hasanuzzaman M, Alam M, Ullah H, Saeed M, Khan AK, Adnan M (eds) *Environment, climate, plant and vegetation growth*. Springer Publ Ltd, Springer Nature Switzerland AG. Part of Springer Nature, pp. 229–252. https://doi.org/10.1007/978-3-030-49732-3

Sandalio LM, Romero-Puertas MC (2015) Peroxisomes sense and respond to environmental cues by regulating ROS and RNS signalling networks. *Ann Bot* 116:475–485.

Satish L, Rency AS, Ramesh M (2018) Spermidine sprays alleviate the water deficit-induced oxidative stress in finger millet (*Eleusine coracana* L. Gaertn.) plants. *3 Biotech* 8:63.

Saud S, Chen Y, Fahad S, Hussain S, Na L, Xin L, Alhussien SA (2016) Silicate application increases the photosynthesis and its associated metabolic activities in Kentucky bluegrass under drought stress and post-drought recovery. *Environ Sci Pollut Res* 23(17):17647–17655. https://doi.org/10.1007/s11356-016-6957-x

Saud S, Chen Y, Long B, Fahad S, Sadiq A (2013) The different impact on the growth of cool season turf grass under the various conditions on salinity and drought stress. *Int J Agric Sci Res* 3:77–84.
Saud S, Fahad S, Cui G, Chen Y, Anwar S (2020) Determining nitrogen isotopes discrimination under drought stress on enzymatic activities, nitrogen isotope abundance and water contents of Kentucky bluegrass. *Sci Rep* 10:6415. https://doi.org/10.1038/s41598-020-63548-w
Saud S, Fahad S, Yajun C, Ihsan MZ, Hammad HM, Nasim W, Amanullah Jr, Arif M and Alharby H (2017) Effects of nitrogen supply on water stress and recovery mechanisms in Kentucky bluegrass plants. *Front. Plant Sci* 8:983. https://doi.org/10.3389/fpls.2017.00983
Saud S, Li X, Chen Y, Zhang L, Fahad S, Hussain S, Sadiq A, Chen Y (2014) Silicon application increases drought tolerance of Kentucky bluegrass by improving plant water relations and morph physiological functions. *Sci World J* 2014:1–10. https://doi.org/10.1155/2014/368694
Senol C (2020) The effects of climate change on human behaviors. In: Fahad S, Hasanuzzaman M, Alam M, Ullah H, Saeed M, Khan AK, Adnan M (eds) *Environment, climate, plant and vegetation growth.* Springer Publ Ltd, Springer Nature Switzerland AG. Part of Springer Nature, pp. 577–590. https://doi.org/10.1007/978-3-030-49732-3
Shafi MI, Adnan M, Fahad S, Fazli W, Ahsan K, Zhen Y, Subhan D, Zafar-ul-Hye M, Martin B, Rahul D (2020) Application of single superphosphate with humic acid improves the growth, yield and phosphorus uptake of wheat (*Triticum aestivum* L.) in calcareous soil. *Agron* 10:1224. https://doi.org/10.3390/agronomy10091224
Shah F, Lixiao N, Kehui C, Tariq S, Wei W, Chang C, Liyang Z, Farhan A, Fahad S, Huang J (2013) Rice grain yield and component responses to near 2°C of warming. *Field Crop Res* 157:98–110.
Sharma C, Mathur S, Tomar RS, Jajoo A (2018) Investigating role of Triton X-100 in ameliorating deleterious effects of anthracene in wheat plants. *Photosynthetica* 56:652–659.
Sharma NYS (2016) Reactive oxygen species, oxidative stress and ROS scavenging system in plants. *J Chem Pharm Res* 8:595–604.
Sharma P, Jha AB, Dubey RS, Pessarakli M (2012) Reactive oxygen species, oxidative damage, and antioxidative defense mechanism in plants under stressful conditions. *J Bot* 2012:217037.
Sidra K, Javed I, Subhan D, Allah B, Syed IUSB, Fatma B, Khaled DA, Fahad S, Omaima N, Ali TKZ, Rahul D (2021) Physio-chemical characterization of indigenous agricultural waste materials for the development of potting media. *J Saudi Society Agricultural Sci.* https://doi.org/10.1016/j.sjbs.2021.08.058
Singh A, Kumar A, Yadav S, Singh IK (2019) Reactive oxygen species-mediated signaling during abiotic stress. *Plant Gene* 18:100173.
Sohag AAM, Tahjib UFM, Brestic M, Afrin S, Sakil MA, Hossain MT, Hossain MA, Hossain MA (2020) Exogenous salicylic acid and hydrogen peroxide attenuate drought stress in rice. *Plant Soil Environ* 66:7–13.
Subhan D, Zafar-ul-Hye M, Fahad S, Saud S, Martin B, Tereza H, Rahul D (2020) Drought stress alleviation by ACC deaminase producing *Achromobacter xylosoxidans* and *Enterobacter cloacae*, with and without timber waste biochar in maize. *Sustain* 12(6286). https://doi.org/10.3390/su12156286
Sun Y, Wang H, Liu S, Peng X (2016) Exogenous application of hydrogen peroxide alleviates drought stress in cucumber seedlings. *S Afr J Bot* 106:23–28.
Taiz L, Zeiger E (2006) *Plant physiology*, 4th ed. Sinauer Associates, Inc., Sunderland, MA.
Takagi D, Takumi S, Hashiguchi M, Sejima T, Miyake C (2016) Superoxide and singlet oxygen produced within the thylakoid membranes both cause photosystem I photoinhibition. *Plant Physiol* 171:1626–1634.
Tanou G, Filippou P, Belghazi M, Job D, Diamantidis G, Fotopoulos V, Molassiotis A (2012) Oxidative and nitrosative-based signaling and associated post-translational modifications orchestrate the acclimation of citrus plants to salinity stress. *Plant J* 72:585–599.
Tanwir K, Akram MS, Masood S, Chaudhary HJ, Lindberg S, Javed MT (2015) Cadmium-induced rhizospheric pH dynamics modulated nutrient acquisition and physiological attributes of maize (*Zea mays* L.). *Environ Sci Pollut Res* 22(12):9193–9203.
Tariq M, Ahmad S, Fahad S, Abbas G, Hussain S, Fatima Z, Nasim W, Mubeen M, Ur Rehman MH, Khan MA, Adnan M. (2018). The impact of climate warming and crop management on phenology of sunflower-based cropping systems in Punjab, Pakistan. *Agri Forest Met* 256:270–282.
Telfer A (2014) Singlet oxygen production by PSII under light stress: mechanism, detection and the protective role of β-carotene. *Plant Cell Physiol* 55:1216–1223.

Tomar RS, Singh B, Jajoo A (2019) Effects of organic pollutants on photosynthesis. In: Ahmad P, Ahanger MA, Alyemeni MN, Alam P (eds) *Photosynthesis, productivity and Environmental stress*. Wiley, Boca Raton, pp. 1–26. https://doi.org/10.1002/9781119501800.ch1

Trovato M, Mattioli R, Costantino P (2008) Multiple role of proline in plant stress tolerance and development. *Rendiconti Lincei* 19:325–346.

Unsar N-U, Muhammad R, Syed HMB, Asad S, Mirza AQ, Naeem I, Muhammad HR, Fahad S, Shafqat S (2020) Insect pests of cotton crop and management under climate change scenarios. In: Fahad S, Hasanuzzaman M, Alam M, Ullah H, Saeed M, Khan AK, Adnan M (eds) *Environment, climate, plant and vegetation growth*. Springer Publ Ltd, Springer Nature Switzerland AG. Part of Springer Nature, pp. 367–396. https://doi.org/10.1007/978-3-030-49732-3

Wahid F, Fahad S, Subhan D, Adnan M, Zhen Y, Saud S, Manzer HS, Martin B, Tereza H, Rahul D (2020) Sustainable management with mycorrhizae and phosphate solubilizing bacteria for enhanced phosphorus uptake in calcareous soils. *Agri* 10 (334). https://doi.org/10.3390/agriculture10080334

Wajid N, Ashfaq A, Asad A, Muhammad T, Muhammad A, Muhammad S, Khawar J, Ghulam MS, Syeda RS, Hafiz MH, Muhammad IAR, Muhammad ZH, Muhammad HR, Veysel T, Fahad S, Suad S, Aziz K, Shahzad A (2017) Radiation efficiency and nitrogen fertilizer impacts on sunflower crop in contrasting environments of Punjab. *Pakistan Environ Sci Pollut Res* 25:1822–1836. https://doi.org/10.1007/s11356-017-0592-z

Wang L, Apel K (2016) Singlet oxygen. In: Nonell S, Flors C (eds) *Applications in biosciences and nanosciences*. Royal Society of Chemistry, pp. 267–278.

Wang WB, Kim YH, Lee HS, Kim KY, Deng XP, Kwak SS (2009) Analysis of antioxidant enzyme activity during germination of alfalfa under salt and drought stresses. *Plant Physiol Biochem* 47:570–577.

Wang X, Li S, Liu Y, Ma C (2015) Redox regulated peroxisome homeostasis. *Redox Biol* 4:104–108.

Wu C, Kehui C, She T, Ganghua L, Shaohua W, Fahad S, Lixiao N, Jianliang H, Shaobing P, Yanfeng D (2020) Intensified pollination and fertilization ameliorate heat injury in rice (*Oryza sativa* L.) during the flowering stage. *Field Crops Res* 252:107795.

Wu C, Tang S, Li G, Wang S, Fahad S, Ding Y (2019) Roles of phytohormone changes in the grain yield of rice plants exposed to heat: a review. *PeerJ* 7:e7792. https://doi.org/10.7717/peerj.7792

Xue M, Guo T, Ren M, Wang Z, Tang K, Zhang W, Wang M (2019) Constitutive expression of chloroplast glycerol-3-phosphate acyltransferase from *Ammopiptanthus mongolicus* enhances unsaturation of chloroplast lipids and tolerance to chilling, freezing and oxidative stress in transgenic Arabidopsis. *Plant Physiol Biochem* 143:375–387.

Yamasaki H, Ogura MP, Kingjoe KA, Cohen MF (2019) D-Cysteine-induced rapid root abscission in the water fern *Azolla pinnata*: Implications for the linkage between d-amino acid and reactive sulfur species (RSS) in plant environmental responses. *Antioxidants* 8:411.

Yang Z, Zhang Z, Zhang T, Fahad S, Cui K, Nie L, Peng S, Huang J (2017) The effect of season-long temperature increases on rice cultivars grown in the central and southern regions of China. *Front Plant Sci* 8:1908. https://doi.org/10.3389/fpls.2017.01908

Yildiztekin M, Kaya C, Tuna AL, Ashraf M (2015). Oxidative stress and antioxidative mechanisms in tomato (*Solanum lycopersicum* l.) plants sprayed with different pesticides. *Pak J Bot* 47:717–721.

Zafar-ul-Hye M, Muhammad TH, Muhammad A, Fahad S, Martin B, Tereza D, Rahul D, Subhan D (2020b) Potential role of compost mixed biochar with rhizobacteria in mitigating lead toxicity in spinach. *Scientific Rep* 10:12159. https://doi.org/10.1038/s41598-020-69183-9

Zafar-ul-Hye M, Muhammad N, Subhan D, Fahad S, Rahul D, Mazhar A, Ashfaq AR, Martin B, Jiˇrí H, Zahid HT, Muhammad N (2020a) Alleviation of cadmium adverse effects by improving nutrients uptake in bitter gourd through cadmium tolerant rhizobacteria. *Environ* 7(54). https://doi.org/10.3390/environments7080054

Zafar-ul-Hye M, Akbar MN, Iftikhar Y, Abbas M, Zahid A, Fahad S, Datta R, Ali M, Elgorban AM, Ansari MJ et al. (2021) Rhizobacteria inoculation and caffeic acid alleviated drought stress in lentil plants. *Sustain* 13:9603. https://doi.org/10.3390/su13179603

Zahida Z, Hafiz FB, Zulfiqar AS, Ghulam MS, Fahad S, Muhammad RA, Hafiz MH, Wajid N, Muhammad S (2017) Effect of water management and silicon on germination, growth, phosphorus and arsenic uptake in rice. *Ecotoxicol Environ Saf* 144:11–18.

Zahir SM, Zheng HG, Ala Ud D, Amjad A, Ata Ur R, Kashif J, Shah F, Saud S, Adnan M, Fazli W, Saud A, Manzer HS, Shamsher A, Wajid N, Hafiz MH, Fahad S (2021) Synthesis of silver nanoparticles using Plantago lanceolata extract and assessing their antibacterial and antioxidant activities. *Sci Rep* 11:20754.
Zia-Ur-Rehman M (2020) Environment, climate change and biodiversity. In: Fahad S, Hasanuzzaman M, Alam M, Ullah H, Saeed M, Khan AK, Adnan M (eds) *Environment, climate, plant and vegetation growth*. Springer Publ Ltd, Springer Nature Switzerland AG. Part of Springer Nature, pp. 473–502. https://doi.org/10.1007/978-3-030-49732-3

9

Global Change Implications on Oxidative Status in Photosynthetic Organisms

G. Malanga, E. Robello, and S. Puntarulo
Universidad de Buenos Aires, Facultad de Farmacia y Bioquímica, Fisicoquímica, Buenos Aires, Argentina; CONICET-Universidad de Buenos Aires. Instituto de Bioquímica y Medicina Molecular (IBIMOL), Buenos Aires, Argentina

CONTENTS

9.1 Introduction

9.1.1 Climatic Change

The Intergovernmental Panel on Climate Change (IPCC) defines Climate Change as 'any variation in the state of the climate that can be identified (through statistical data) by observing changes in its properties that persist during a period of determined time, usually decades. The causes of it can be both natural and anthropogenic'. The 2007 IPCC Fourth Assessment Report pointed out that global temperature has increased by 0.6°C to 0.8°C over the past century (Liu et al. 2012). Global warming is becoming a very dangerous situation for all living creatures. Plants are sensitive to biological damage due to this increase in temperature, and the whole ecosystem is compromised (Fahad et al. 2016a, 2016b, 2016c, 2016d; Yang et al. 2017). The latest data from UN in 2013–2014 confirmed (95% probability) that the global warming is caused by anthropogenic activities, and is conditioned by the so-called *Greenhouse effect* (Tkemaladze and Makhashvili 2016). Climate change has the potential to affect, both positively and negatively, many features such as the location, time, and productivity of agricultural systems (Deepranjan et al. 2021; Haider et al. 2021; Huang Li et al. 2021; Ikram et al. 2021; Jabborova et al. 2021; Khadim et al. 2021a, 2021b; Manzer et al. 2021; Muzammal et al. 2021; Abdul et al. 2021a, 2021b; Ashfaq et al. 2021; Amjad et al. 2021; Atif et al. 2021; Athar et al. 2021; Adnan et al. 2018a, 2018b, 2019; Akram et al. 2018a, 2018b; Aziz et al. 2017a, 2017b; Chang et al. 2021; Chen et al. 2021; Emre et al. 2021; Habib et al. 2017; Hafiz et al. 2016, 2019, 2020a, 2020b; Ghulam et al. 2021; Guofu et al. 2021; Hafeez et al. 2021; Khan et al. 2021; Kamarn et al. 2017; Muhammad et al. 2019; Safi et al. 2021; Sajjad et al. 2019; Saud et al. 2013, 2014, 2016, 2017, 2020; Shah et al. 2013; Qamar et al. 2017; Hamza et al. 2021; Irfan et al.

DOI: 10.1201/9781003286417-9

2021; Wajid et al. 2017; Yang et al. 2017; Zahida et al. 2017; Depeng et al. 2018; Hussain et al. 2020; Shafi et al. 2020; Wahid et al. 2020; Subhan et al. 2020; Zafar-ul-Hye et al. 2020a, 2020b; Zafar-ul-Hye et al. 2021; Adnan et al. 2020; Ilyas et al. 2020; Saleem et al. 2020a, 2020b, 2020c; Rehman et al. 2020; Farhat et al. 2020; Wu et al. 2020; Mubeen et al. 2020; Farhana et al. 2020; Jan et al. 2019; Wu et al. 2019; Ahmad et al. 2019; Baseer et al. 2019; Hafiz et al. 2018; Tariq et al. 2018). The 4th and 5th IPCC reports clearly outlined the global and regional impacts of projected climate change on agriculture, water resources, natural ecosystems, and food security (Shah 2009; Fahad and Bano 2012; Fahad et al. 2013, 2014a, 2014b, 2015a, 2015b, 2016a, 2016b, 2016c, 2016d, 2017, 2018a, 2018b, 2019a, 2019b, 2020, 2021a, 2021b, 2021c, 2021d, 2021e, 2021f; Hesham and Fahad 2020; Iqra et al. 2020; Akbar et al. 2020; Mahar et al. 2020; Noor et al. 2020; Bayram et al. 2020; Fahad S 2017, 2018a, 2018b; Amanullah et al. 2020, 2021; Rashid et al. 2020; Arif et al. 2020; Amir et al. 2020; Saman et al. 2020; Muhammad et al. 2020; Md Jakir and Allah 2020; Mahmood et al. 2021; Farah et al. 2020; Sadam et al. 2020; Unsar et al. 2020; Fazli et al. 2020; Md. Enamul et al. 2020; Gopakumar et al. 2020; Zia-Ur-Rehman 2020; Ayman et al. 2020; Mohammad et al. 2020a, 2020b; Senol 2020; Amjad et al. 2020; Ibrar et al. 2020; Sajid et al. 2020; Muhammad et al. 2021; Sidra et al. 2021; Zahir et al. 2021; Sahrish et al. 2022). The process of global warming impacts coastal and marine environments and the organisms living in them (Harley et al. 2006). Variations in aquatic environmental conditions can cause extreme physiological stress, such us a moderate increase in temperature can cause changes in metabolic rate, causing modifications to biological processes in growth and development (Zahoor et al. 2016; Branca et al. 2013). In the past two decades the potential effects of climate change on different ecosystem have been investigated. Several studies have reported how climate change impacts on plant biodiversity distribution and adaptations (Singh et al. 2018).

The generation of reactive oxygen species (ROS) is a common event in many types of abiotic stress exposure, regardless of the plant species (Shah 2009). So, there is growing interest in the evaluation of oxidative stress in relation to the impact of global climate change on natural ecosystems in different trophic levels (Hernando et al. 2015). The main objective of this work is to briefly summarize the role of oxidative alterations depending on global changes on photosynthetic organisms.

9.2 Oxidative Stress

The formation of ROS) and other oxidants is a well-established physiological event in aerobic cells. Between 2% and 4% of O_2 consumed by living cells generates ROS. The mechanism of this generation is of fundamental relevance since several organelles (such as the mitochondria, the endoplasmic reticulum, and the peroxisomes) and the activity of cytoplasmic enzymes (such as xanthine oxidase) are capable of partially reducing O_2 to superoxide anion ($O_2^{\bullet}$) and hydrogen peroxide (H_2O_2), secondarily generating hydroxyl radical ($^{\bullet}OH$) (Boveris 1998). If the production of these species is not fully controlled, their reactions might lead to cell damage. Thus, an active antioxidant activity is required. This capacity includes the specific scavenger activity of antioxidant enzymes, such as superoxide dismutase (SOD), catalase (CAT), among others; and of nonenzymatic compounds, both hydrophilic and lipophilic nature. Also, the presence of complexing agents such as ferritin and ceruloplasmin that avoid the generation of reactive species by complexing metals, e.g. Fe and Cu, to compounds with low ability to catalyse dangerous formation of reactive species play a role in the overall antioxidant cellular ability. Moreover, the activity of repairing enzymes to improve the capacity of the antioxidant enzymes contributes to the cellular defence system (Figure 9.1). These cellular mechanisms allow the cells to maintain the concentration of reactive species in steady state controlled values by equalling the rate of use or disappearance to its rate of production (Boveris 1998).

An imbalance, whether due to an increase in the production of the oxidizing species or a decrease in the activity of the antioxidant species, defines 'oxidative stress.' However, the possibility of reaching this condition strongly depends on the magnitude of the change. The diagram shown in Figure 9.2 indicates that the condition, in terms of the effect on the redox balance in the cells, could be described as eustress or distress, that is, a positive stress or a deleterious stress (Niki 2016). These concepts were used in the biomedical sciences in the 'General Adaptation Syndrome' by the endocrinologist Hans Selye, and he

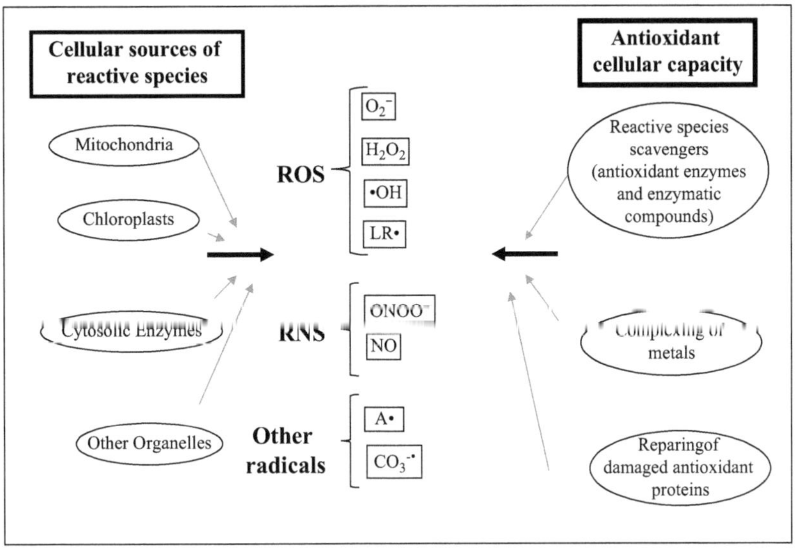

FIGURE 9.1 Diagram showing that both ROS and RNS generation by intracellular sites and cellular antioxidant systems regulate steady state concentration of reactive species.

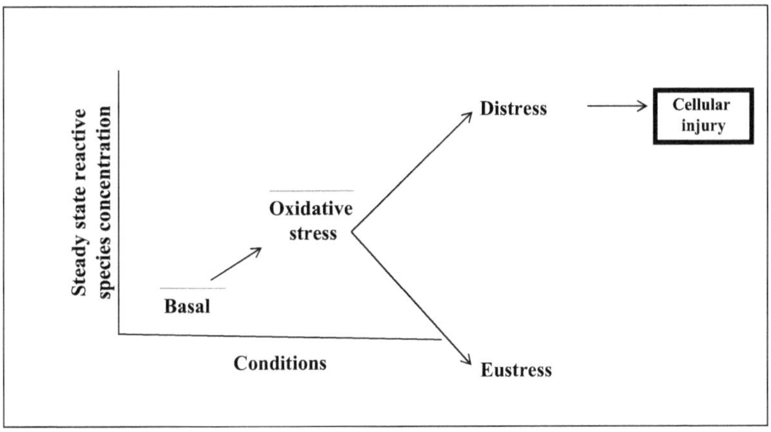

FIGURE 9.2 Diagram showing that changes in steady state concentration of reactive species affect the redox balance in the cells.

distinguished two types of stress, "eustress" and "distress," and these were later introduced into psychology (Lazarus 1966). Eustress enhance function, for example through training or challenging work, whereas distress refers to a persistent stress that is not resolved through coping or adaptation and may lead to illnesses. Later, this concept was extended to the balance between tolerance and sensitivity in many biological systems, and it was understood as how an alteration may determine whether a stress factor has a positive or negative effect (Kranner et al. 2010). The observation that slight increases in the steady state concentration of reactive species could be interpreted as a benefit for the cell, allowing the triggering of protective pathways when exposed to other challenges, is a new and important way of studying the biochemistry of reactive species in biological organisms. The imbalance in ROS production is considered as the cause of several pathological changes in photosynthetic organisms, like photoinhibition (Thiele and Krause 1994), pigment bleaching (Godde and Buchhold 1992), hypersensitive response against an invading pathogen (Inze and Van Montagu 2003), and plant tumour formation (Ji 1996). However, there is a growing piece of evidence showing that these active species are used as redox signalling messengers with important regulatory functions (Niki 2016; Moldogazieva et al. 2018).

9.3 Biological Implication

Over the past decades, global warming has increased exerting substantial effects on acidification, salinity, temperature, dissolved O_2, light penetration, nutrient conditions, biotoxins presence, and metal inputs due to higher sediment runoff specially impacting on coastal and benthic ecosystems, among others (Quartino et al. 2013; Gutt et al. 2016; Sahade et al. 2015, González et al. 2017). Besides, the increasing water temperatures and sand stratification in the sea, as a consequence of the global climate change, might be potential stressors for the growth and persistence of many harmful algal blooms (Sellner et al. 2003; Vadstein et al. 2004, Cabrera et al. 2019).

Moreover, the changes in climate, such as changes in temperature, precipitation, droughts, ultraviolet (UV) irradiation, are important factors to produce alterations in the normal physiology of plant growth. Additionally, forest fires, insect infestations, and disease outbreaks are the consequence of climate change that mainly affects crop plants (Shah 2009). The diagram shown in Figure 9.3 indicates that some of the most studied consequences of global changes can impact on the oxidative and nitrosative status of photosynthetic organisms.

9.3.1 Temperature Stress

Drought promotes oxidative stress on the photosynthetic apparatus. Besides, it produces inhibition of protein synthesis, inactivation of several chloroplast enzymes, deterioration of the electron chain transport on the thylakoids, and increases the membrane permeability (Guang and Sacandalios 1998). Also, it was reported an increase in several antioxidants enzymes such us glutathione reductase and dehydroascorbate reductase in the resurrection of the plant *Sporobolus stapfiana* (Sgherri and Navari Izzo 1995; Burke et al. 1985). Also, low temperatures combined with strong light exposures could produce photooxidative stress by weakening antioxidant enzymes and increasing the production of ROS, like in tomato and spinach in which illumination produced a depletion of Cu ions and a reduction of Cu/Zn-SOD activity, increasing oxidative injury (Wise and Naylor 1987; Michalski and Kaniuga 1982).

The increase in the average temperatures in a definite area promotes storms and the changes in the precipitation pattern. These changes can promote flooding of soils, which can lead to increase in soil acidity. The lower pH can promote an increase in availability of Fe^{2+} for terrestrial plants. This form of Fe can be

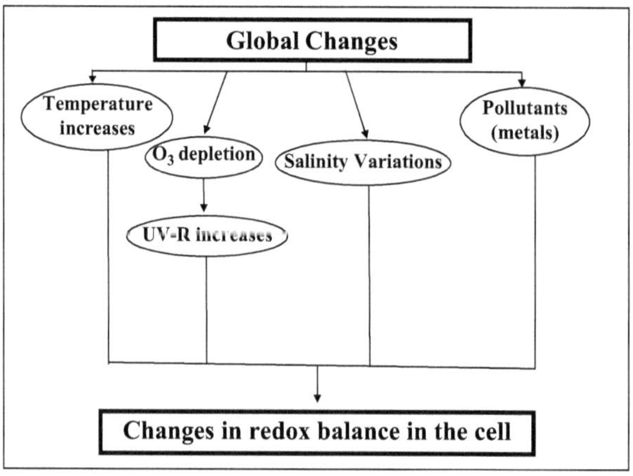

FIGURE 9.3 The most studied effects of global changes that impact on oxidative and nitrosative status of photosynthetic organisms.

toxic to organisms, since it can by a reaction with $O_2^{\bullet}$ catalyse the Fenton reaction and generate $^{\bullet}OH$ (Briat and Lobréaux 1997). These pH values could be responsible for a promotion in the Fe^{2+} concentrations in the soil solution from 10 to >2000 mg l^{-1} (180 µM to >36 mM) (Becker and Asch 2005). Studies carried out on embryonic soybean axes of 24 h and 48 h of development (Ilyas et al. 2016; Robello et al. 2007) showed that the overloading with 500 µM Fe produced an increase in total cellular Fe and in the labile iron pool (LIP, a pool of Fe, which can promote ROS generation). However, the alterations produced by that experimental situation did not affect the redox balance in the cellular hydrophilic medium, and did not damage the integrity of cell membranes in axes of 24 h and 48 h of development. On the other hand, it is well documented the appearance, after exposure to excess Fe, of the bronzing phenomenon in rice, which is a characteristic of Fe toxicity (Nikolic and Pavlovic 2018). Bronzing includes small reddish brown spots on the leaves, and it seems to occur at Fe concentrations above 500 mg Fe per kg leaf dry weight (Nikolic et al. 2007). When bronzing symptoms range from 15% to 30% there is yield reduction, and further increases lead to toxicity, and even to complete crop failure (Becker and Asch 2005).

9.3.2 UV Radiation

The stratospheric ozone depletion as a consequence of the certain human activities has been increasing the fluxes of UV-B (290–320 nm) radiation reaching Earth's surface (Kerr and McElroy 1993). Increased UV-B radiation, as a consequence of a decrease in the ozone layer, is potentially harmful to all forms of life by affecting a number of physiological and biochemical processes in plants, macro- and micro-algae, such as growth and survival, buoyancy, photosynthesis, pigmentation, CO_2 uptake, ribulose 1,5-bisphosphate carboxylase/oxygenase (RuBisCO) activity, and alterations in the native structure of proteins and DNA molecules (Rastogi and Incharoensakdi 2013). The plants can acclimatize themselves when the UV-B exposure is mild, but an exposure of a higher magnitude would allow relevant metabolic disorders in these organisms such as, alteration of the timing of flowering, number of flowers, pollen germination, pollen tube elongation, fruit yields, and subsequent seed mass and seed germination failure (Takshak and Agrawal 2019). Carotenoids, which are accessory photosynthetic pigments, could quench ROS (such as singlet oxygen, $^1O_2$), and provide structural stability to light-harvesting complex proteins and thylakoid membranes. They also react with lipid peroxidation products and terminate free radical chain reactions. These compounds can change their concentration in medicinal plants due to UV irradiation: among the medicinal plants, its concentration under increased UV-B dose was found to be increased in *Picea asperata*, reduced in *Withania somnifera*, and remained unaffected in *Turnera diffusa* (Takshak, and Agrawal 2019). Besides, it has been studied that UV-B radiation can not only damage the photosystem-II (PSII) but promotes the production of $^{\bullet}OH$ and carbon centre radicals in isolated thylakoid along with the accumulation of SOD (Hideg et al. 1993; Foyer et al. 1994; Hideg and Vass 1996). It is also seen that guaiacol peroxidase, which is found in organelles, especially in vacuoles, cytosol, and cell wall in plants (Sharma et al. 2012), can regulate H_2O_2 intracellular levels using different organic compounds as substrates, such as guaiacol or pyrogallol (Gill and Tuteja 2010). It has been observed an increase of the activity of this enzyme in plants growing under UV radiation exposure (Ibrahim et al. 2013; Soares et al. 2019).

The increase in UV-B radiation is also detrimental to cyanobacteria (Latifi et al. 2009). Hernando et al. (2015) showed that variations in environmental factors caused by climate change, such as UV-B exposure, generate a situation of oxidative damage in *Microcystis aeruginosa*, either as a direct or indirect consequence suggesting a high ability of *M. aeruginosa* to detect a potential stress situation due to the reactive species production rapidly triggering antioxidant defences. On the other hand, growing UV radiation is changing habitat conditions, and these changes are leading to alteration in the composition in terms of the presence of many diatom species in the marine ecosystem (Hernando et al. 2015).

Estevez et al. (2001) suggested that an increase in UV-B radiation was responsible for the development of oxidative stress conditions in the hydrophilic cellular media, assessed by the ascorbyl radical ($A^{\bullet}$)/ascorbate (AH^-) content ratio, in Antarctic *Chlorella* sp. cells. Moreover, Giordano et al. (2004) reported a similar increase in this ratio in *Gunnera magellanica*, suggesting that this effect was controlled by an increase in the AH^- pool. Also, the inhibition of leaf growth described here after UV-B exposure may result from accumulation of UV-B-induced damage to key macromolecules and cellular structures, by the production of oxidative stress.

9.3.3 Salinity Stress and Metal Contamination

As stated before, global changes could affect salinity of the environmental media of photosynthetic organisms. Ashraf et al. (2019) reported that when rice plants face salinity stress, water deficiency, or high concentrations of heavy metals (like Pb, Cd, Ni), the common response is that they undergo increasing generation of ROS, and the reduction of their growth and productivity, probably due to the oxidative stress. NaCl-treated pea plants enhanced $O_2^{\bullet}$ production and reduced the Mn-SOD activity (Hernandez et al. 1993). However, pea cultures sensitive to salt stress, exposed to 70mM NaCl, increased the content of H_2O_2 and lipid peroxidation, without changing their antioxidant capacity (Hernandez et al. 1995). Ozgur et al. (2013) reported that halophytes plants growing in high salt concentration developed high lipid peroxidation in shoots, and an increase in the antioxidant capacity. This response pattern of change in the total antioxidant ability helped these plants to deal with salt-dependent stress better than glycophytes plants.

Hypoosmotic conditions cause a water influx resulting in an increased turgor pressure that in addition to oxidative stress could be the reason for replacement of large centric by small pennate diatoms with high ability to adapt to changing osmotic conditions. Data presented by Hernando et al. (2015) showed that hypoosmotic conditions in phytoplankton favour water influx into the cells, which resulted in increased turgor pressure and oxidative stress. This stress was linked to a number of other cellular toxic processes, including damages to proteins, enzyme inactivation, and DNA breakage. The authors characterized the effects on a natural phytoplankton assemblage from a coastal Antarctic environment to the changes in salinity by seasonal and long-term influx of fresh water from glacier melting.

As previously stated, melt water increments from glaciers due to global warming have relevant effects on salinity among other factors such as light penetration, temperature, nutrient conditions, and metal inputs due to higher sediment runoff in coastal waters (Quartino et al. 2013; Sahade et al. 2015). In the King George Island region, Antarctica, high sedimentation reduces underwater photosynthetically active radiation (PAR, 400–700 nm) leading to a lower carbon balance by some macroalgal species (Deregibus et al. 2016). González et al. (2017) studied the effects of the exposure of some abiotic factors affected by the glacier impact on the physiological and oxidative profiles of Antarctic macroalgaes. The effects of the higher sediment load, which implies less light and higher Fe values in Island D as compared to Peñón de Pesca, Potter Cove, King George Island, Antarctica, on the oxidative status of the macroalgaes *Gigartina skottsbergii* and *Himantothallus grandifolius* were characterized. Although they are adapted to skip rough abiotic conditions *G. skottsbergii* seemed to be more sensitive to the external stress. The Fe present in the seawater was significantly incorporated in *G. skottsbergii* and may contribute, together with the other physicochemical changes, to produce oxidative stress and damage in this species. On the other hand, *H. grandifolius*, an Antarctic perennial macroalgae, responded as a well-adapted organism. The successful adaptation might respond to a wide spectrum of factors and mechanisms, including antioxidant ability to cope with oxidative stress, but further studies are required to have a clear picture of the complex network of factors leading to this successful survival and the relative contribution of each alteration in such an extreme environment.

9.4 Nitrosative Stress in Terrestrials and Aquatic Systems

NO generation has also been reported in the case of abiotic (e.g. high temperature, drought, etc.) and biotic (e.g. pathogen interaction) stress agents. Though most of the findings regard terrestrial plants, very few studies are reported for marine photosynthetic organisms, including microalgae, seaweeds, sea grasses, and mangroves (Kumar et al. 2015). High amounts of NO were produced at high temperatures in the zooxanthella *S. microadriaticum*, commonly found in symbiotic association with reef-building corals. It was hypothesized that NO acted as a cytotoxic molecule either by inactivating the metabolic pathways or by causing damage to macromolecules and irreversible inhibition of mitochondrial respiration. However, a positive correlation between increase in NO levels, caspase 3-like activity, and temperature indicates that NO triggers dinoflagellate cell death. NO at lower concentrations has been shown to be effective against the oxidative damages either by activating oxidative enzymes or by detoxifying superoxide ions. Also, a photoprotective effect of NO has been recorded in *Phaeodactylum tricornutum* (Kumar et al. 2015).

9.5 Conclusion

This brief summary clearly shows that the effect of global changes on the oxidative balance in the photosynthetic organisms is a common feature. Even though, the overall contribution of these effects strongly depends on the studied system and the magnitude of the challenge. It is now widely accepted that naturally triggered increases in the steady state concentration of reactive species, such as eustress, could be even beneficial to the survival of the species when exposed to further insults. However, when the antioxidant capacity is overwhelmed, cellular damage could be the consequence. The persistency in the generation of environmentally dangerous anthropogenic changes is severely affecting the communities, both in aquatic and terrestrial organisms, putting at risk the survival of many species. Oxidative biochemical reactions seem as constant factors involved in beneficial and damaging pathways. A better understanding of these mechanisms could contribute to the design of strategies that could be reflected not only in the preservation of the biodiversity but also in the improvement of the economic activity of the areas affected by the global changes.

REFERENCES

Abdul S, Muhammad AA, Shabir H, Hesham A, Sajjad H, Niaz A, Abdul G, Sayyed RZ, Fahad S, Subhan D, Rahul D (2021a) Zinc nutrition and arbuscular mycorrhizal symbiosis effects on maize (*Zea mays* L.) growth and productivity. *J Saudi Soc Agric Sci*. https://doi.org/10.1016/j.sjbs.2021.06.096

Abdul S, Muhammad AA, Subhan D, Niaz A, Fahad S, Rahul D, Mohammad JA, Omaima N, Muhammad Habib Ur R, Bernard RG (2021b) Effect of arbuscular mycorrhizal fungi on the physiological functioning of maizeunder zinc-deficient soils. *Sci Rep* 11:18468

Adnan M, Fahad S, Muhammad Z, Shahen S, Ishaq AM, Subhan D, Zafar-ul-Hye M, Martin LB, Raja MMN, Beena S, Saud S, Imran A, Zhen Y, Martin B, Jiri H, Rahul D (2020) Coupling phosphate-solubilizing bacteria with phosphorus supplements improve maize phosphorus acquisition and growth under lime induced salinity stress. *Plants* 9(900). https://doi.org/10.3390/plants9070900

Adnan M, Fahad S, Khan IA, Saeed M, Ihsan MZ, Saud S, Riaz M, Wang D, Wu C. (2019). Integration of poultry manure and phosphate solubilizing bacteria improved availability of Ca bound P in calcareous soils. *3 Biotech* 9(10):368.

Adnan M, Shah Z, Sharif M, Rahman H. (2018b). Liming induces carbon dioxide (CO_2) emission in PSB inoculated alkaline soil supplemented with different phosphorus sources. *Environ Sci Poll Res* 25(10):9501–9509.

Adnan M, Zahir S, Fahad S, Arif M, Mukhtar A, Imtiaz AK, Ishaq AM, Abdul B, Hidayat U, Muhammad A, Inayat-Ur R, Saud S, Muhammad ZI, Yousaf J, Amanullah, Hafiz MH, Wajid N (2018a) Phosphate-solubilizing bacteria nullify the antagonistic effect of soil calcification on bioavailability of phosphorus in alkaline soils. *Sci Rep* 8:4339. https://doi.org/10.1038/s41598-018-22653-7

Ahmad S, Kamran M, Ding R, Meng X, Wang H, Ahmad I, Fahad S, Han Q (2019) Exogenous melatonin confers drought stress by promoting plant growth, photosynthetic capacity and antioxidant defense system of maize seedlings. *PeerJ* 7:e7793. http://doi.org/10.7717/peerj.7793

Akbar H, Timothy JK, Jagadish T, Golam M, Apurbo KC, Muhammad F, Rajan B, Fahad S, Hasanuzzaman M (2020) Agricultural land degradation: processes and problems undermining future food security. In: Fahad S, Hasanuzzaman M, Alam M, Ullah H, Saeed M, Khan AK, Adnan M (eds) *Environment, climate, plant and vegetation growth*. Springer Publ Ltd, Springer Nature Switzerland AG. Part of Springer Nature, pp. 17–62. https://doi.org/10.1007/978-3-030-49732-3

Akram R, Turan V, Hammad HM, Ahmad S, Hussain S, Hasnain A, Maqbool MM, Rehmani MIA, Rasool A, Masood N, Mahmood F, Mubeen M, Sultana SR, Fahad S, Amanet K, Saleem M, Abbas Y, Akhtar HM, Waseem F, Murtaza R, Amin A, Zahoor SA, ul Din MS, Nasim W (2018a) Fate of organic and inorganic pollutants in paddy soils. In: Hashmi, MZ and Varma, A (eds) *Environmental pollution of paddy soils, soil biology*. Springer International Publishing AG, Gewerbestrasse 11, Cham, CH-6330, Switzerland, pp. 197–214.

Akram R, Turan V, Wahid A, Ijaz M, Shahid MA, Kaleem S, Hafeez A, Maqbool MM, Chaudhary HJ, Munis, MFH, Mubeen M, Sadiq N, Murtaza R, Kazmi DH, Ali S, Khan N, Sultana SR, Fahad S, Amin A, Nasim W (2018b) Paddy land pollutants and their role in climate change. In: Hashmi, MZ and Varma, A (eds) *Environmental pollution of paddy soils, soil biology*. Springer International Publishing AG, Gewerbestrasse 11, Cham, CH-6330, Switzerland, pp. 113–124.

Amanullah, Muhammad I, Haider N, Shah K, Manzoor A, Asim M, Saif U, Izhar A, Fahad S, Adnan M et al. (2021) Integrated foliar nutrients application improve wheat (*Triticum Aestivum* L.) productivity under calcareous soils in drylands. *Commun Soil Sci Plant Anal.* https://Doi.Org/10.1080/00103624.2021.1956521

Amanullah, Shah K, Imran, Hamdan AK, Muhammad A, Abdel RA, Muhammad A, Fahad S, Azizullah S, Brajendra P (2020) Effects of climate change on irrigation water quality. In: Fahad S, Hasanuzzaman M, Alam M, Ullah H, Saeed M, Khan AK, Adnan M (eds) *Environment, climate, plant and vegetation growth*. Springer Publ Ltd, Springer Nature Switzerland AG. Part of Springer Nature, pp. 123–132. https://doi.org/10.1007/978-3-030-49732-3

Amir M, Muhammad A, Allah B, Sevgi Ç, Haroon ZK, Muhammad A, Emre A (2020) Bio fortification under climate change: the fight between quality and quantity. In: Fahad S, Hasanuzzaman M, Alam M, Ullah H, Saeed M, Khan AK, Adnan M (eds) *Environment, climate, plant and vegetation growth*. Springer Publ Ltd, Springer Nature Switzerland AG. Part of Springer Nature, pp. 173–228. https://doi.org/10.1007/978-3-030-49732-3

Amjad I, Muhammad H, Farooq S, Anwar H (2020) Role of plant bioactives in sustainable agriculture. In: Fahad S, Hasanuzzaman M, Alam M, Ullah H, Saeed M, Khan AK, Adnan M (eds) *Environment, climate, plant and vegetation growth*. Springer Publ Ltd, Springer Nature Switzerland AG. Part of Springer Nature, pp. 591–606. https://doi.org/10.1007/978-3-030-49732-3

Amjad SF, Mansoora N, Din IU, Khalid IR, Jatoi GH, Murtaza G, Yaseen S, Naz M, Danish S, Fahad S et al. (2021) Application of Zinc fertilizer and mycorrhizal inoculation on physio- biochemical parameters of wheat grown under water-stressed environment. *Sustainability* 13:11007. https://doi.org/10.3390/su131911007

Arif M, Talha J, Muhammad R, Fahad S, Muhammad A, Amanullah, Kawsar A, Ishaq AM, Bushra K, Fahd R (2020) Biochar; a remedy for climate change. In: Fahad S, Hasanuzzaman M, Alam M, Ullah H, Saeed M, Khan AK, Adnan M (eds) *Environment, climate, plant and vegetation growth*. Springer Publ Ltd, Springer Nature Switzerland AG. Part of Springer Nature, pp. 151–172. https://doi.org/10.1007/978-3-030-49732-3

Ashfaq AR, Uzma Y, Niaz A, Muhammad AA, Fahad S, Haider S, Tayebeh Z, Subhan D, Süleyman T, Hesham AElE, Pramila T, Jamal MA, Sulaiman AA, Rahul D (2021) Toxicity of Cadmium and nickel in the context of applied activated carbon biochar for improvement in soil fertility. *Saudi Soc Agric Sci.* https://doi.org/10.1016/j.sjbs.2021.09.035

Ashraf MA, Iqbal M, Rasheed R, Hussain I, Salman Mubarik M, Saleem Arif M, Ria M (2019) Abiotic stress-induced oxidative stress in rice. *Advances in Rice Research for Abiotic Stress Tolerance* 2019:489–504.

Athar M, Masood IA, Sana S, Ahmed M, Xiukang W, Sajid F, Sher AK, Habib A, Faran M, Zafar H, Farhana G, Fahad S (2021) Bio-diesel production of sunflower through sulphur management in a semi-arid subtropical environment. *Environ Sci Pollution Res.* https://doi.org/10.1007/s11356-021-16688-z

Atif B, Hesham A, Fahad S (2021) Biochar coupling with phosphorus fertilization modifies antioxidant activity, osmolyte accumulation and reactive oxygen species synthesis in the leaves and xylem sap of rice cultivars under high-temperature stress. *Physiol Mol Biol Plants.* https://doi.org/10.1007/s12298-021-01062-7

Ayman ELS, Akbar H, Celaleddin B, Muhammad AI, Sohidul I, Shah F, Oksana S, Fatih Ç, Ram SM, Murat E (2020) Consequences of salinity stress on the quality of crops and its mitigation strategies for sustainable crop production: an outlook of arid and semi- arid regions. In: Fahad S, Hasanuzzaman M, Alam M, Ullah H, Saeed M, Khan AK, Adnan M (eds) *Environment, climate, plant and vegetation growth*. Springer Publ Ltd, Springer Nature Switzerland AG. Part of Springer Nature, pp. 503–534. https://doi.org/10.1007/978-3-030-49732-3

Aziz K, Daniel KYT, Fazal M, Muhammad ZA, Farooq S, Fan W, Fahad S, Ruiyang Z (2017a) Nitrogen nutrition in cotton and control strategies for greenhouse gas emissions: a review. *Environ Sci Pollut Res* 24:23471–23487. https://doi.org/10.1007/s11356-017-0131-y

Aziz K, Daniel KYT, Muhammad ZA, Honghai L, Shahbaz AT, Mir A, Fahad S (2017b) Nitrogen fertility and abiotic stresses management in cotton crop: a review. *Environ Sci Pollut Res* 24:14551–14566. https://doi.org/10.1007/s11356-017-8920-x

Baseer M, Adnan M, Fazal M, Fahad S, Muhammad S, Fazli W, Muhammad A, Jr. Amanullah, Depeng W, Saud S, Muhammad N, Muhammad Z, Fazli S, Beena S, Mian AR, Ishaq AM (2019) Substituting urea by organic wastes for improving maize yield in alkaline soil. *J Plant Nutr.* https://doi.org/10.1080/01904167.2019.1659344

Bayram AY, Seher Ö, Nazlican A (2020) Climate change forecasting and modeling for the year of 2050. In: Fahad S, Hasanuzzaman M, Alam M, Ullah H, Saeed M, Khan AK, Adnan M (eds) *Environment, climate, plant and vegetation growth.* Springer Publ Ltd, Springer Nature Switzerland AG. Part of Springer Nature, pp. 109–122. https://doi.org/10.1007/978-3-030-49732-3

Becker M, Asch F (2005) Iron toxicity in rice-conditions and management concepts. *J Plant Nutr Soil Sci* 168(4):558–573.

Boveris A (1998) Biochemistry of free radicals: from electrons to tissues. *Medicina* 58:350–356.

Branca G, Lipper L, McCarthy N, Jolejole MC (2013) Food security, climate change, and sustainable land management: a review. *Agron Sust Develop* 33(4):635–650.

Briat JF, Lobréaux S (1997) Iron transport and storage in plants. *Trends Plant Sci* 2(5):187–193.

Burke JJ, Gamble PE, Hatfield JJ, Quisenberry JE (1985) Plant morphological and biochemical response to field water deficit. I. Response of glutathione reductase activity and paraquat sensitivity. *Plant Physiol* 79:415–419.

Cabrera J, González PM, Puntarulo S (2019) Oxidative effects of the harmful algal blooms on primary organisms of the food web. *Biocell* 43(2):41–50.

Chang W, Qiujuan J, Evgenios A, Haitao L, Gezi L, Jingjing Z, Fahad S, Ying J (2021) Hormetic effects of zinc on growth and antioxidant defense system of wheat plants. *Sci Total Environ.* https://doi.org/10.1016/j.scitotenv.2021.150992

Chen Y, Guo Z, Dong L, Fu Z, Zheng Q, Zhang G, Qin L, Sun X, Shi Z, Fahad S, Xie F, Saud S. (2021) Turf performance and physiological responses of native Poa species to summer stress in Northeast China. *PeerJ* 9:e12252. http://doi.org/10.7717/peerj.12252

Deepranjan S, Ardith S, Siva D, Sonam S, Shikha, Manoj P, Amitava R, Sayyed RZ, Abdul G, Mohammad JA, Subhan D, Fahad S, Rahul D (2021) Optimizing nutrient use efficiency, productivity, energetics, and economics of red cabbage following mineral fertilization and biopriming with compatible rhizosphere microbes. *Sci Rep* 11:15680.

Depeng W, Fahad S, Saud S, Muhammad K, Aziz K, Mohammad NK, Hafiz MH, Wajid N (2018) Morphological acclimation to agronomic manipulation in leaf dispersion and orientation to promote "Ideotype" breeding: Evidence from 3D visual modeling of "super" rice (*Oryza sativa* L.). *Plant Physiol Biochem* 135:499–510. https://doi.org/10.1016/j.plaphy.2018.11.010

Deregibus D, Quartino ML, Campana GL, Momo FR, Wiencke C, Zacher K (2016) Photosynthetic light requirements and vertical distribution of macroalgae in newly ice-free areas in Potter Cove, South Shetland Islands, Antarctica. *Polar Biol* 39(1):153–166.

Emre B, Ömer SU, Martín LB, Andre D, Fahad S, Rahul D, Muhammad Z-ul-H, Ghulam SH, Subhan D (2021) Studying soil erosion by evaluating changes in physico-chemical properties of soils under different land-use types. *J Saudi Soc Agric Sci* 20:190–197.

Estevez MS, Malanga G, Puntarulo S (2001) UV-B effects on Antarctic *Chlorella sp* cells. *J Photochem Photobiol B Biol* 62(1–2):19–25.

Fahad S, Abdul B, Adnan M (eds) (2018b) Global wheat production. *IntechOpen United Kingdom 2018.* http://dx.doi.org/10.5772/intechopen.72559

Fahad S, Bajwa AA, Nazir U, Anjum SA, Farooq A, Zohaib A, Sadia S, Nasim W, Adkins S, Saud S, Ihsan MZ, Alharby H, Wu C, Wang D, Huang J (2017) Crop production under drought and heat stress: plant responses and Management Options. *Front Plant Sci* 8:1147. https://doi.org/10.3389/fpls.2017.01147

Fahad S, Bano A (2012) Effect of salicylic acid on physiological and biochemical characterization of maize grown in saline area. *Pak J Bot* 44:1433–1438.

Fahad S, Chen Y, Saud S, Wang K, Xiong D, Chen C, Wu C, Shah F, Nie L, Huang J (2013) Ultraviolet radiation effect on photosynthetic pigments, biochemical attributes, antioxidant enzyme activity and hormonal contents of wheat. *J Food Agri Environ* 11(3&4):1635–1641.

Fahad S, Hussain S, Bano A, Saud S, Hassan S, Shan D, Khan FA, Khan F, Chen Y, Wu C, Tabassum MA, Chun MX, Afzal M, Jan A, Jan MT, Huang J (2014a) Potential role of phytohormones and plant growth-promoting rhizobacteria in abiotic stresses: consequences for changing environment. *Environ Sci Pollut Res* 22(7):4907–4921. https://doi.org/10.1007/s11356-014-3754-2

Fahad S, Hussain S, Matloob A, Khan FA, Khaliq A, Saud S, Hassan S, Shan D, Khan F, Ullah N, Faiq M, Khan MR, Tareen AK, Khan A, Ullah A, Ullah N, Huang J (2014b) Phytohormones and plant responses to salinity stress: a review. *Plant Growth Regul* 75(2):391–404. https://doi.org/10.1007/s10725-014-0013-y

Fahad S, Hussain S, Saud S, Hassan S, Chauhan BS, Khan F et al. (2016a) Responses of rapid viscoanalyzer profile and other rice grain qualities to exogenously applied plant growth regulators under high day and high night temperatures. *PLoS One* 11(7):e0159590. https://doi.org/10.1371/journal.pone.0159590

Fahad S, Hussain S, Saud S, Hassan S, Ihsan Z, Shah AN, Wu C, Yousaf M, Nasim W, Alharby H, Alghabari F, Huang J (2016c) Exogenously applied plant growth regulators enhance the morphophysiological growth and yield of rice under high temperature. *Front Plant Sci* 7:1250. https://doi.org/10.3389/fpls.2016.01250

Fahad S, Hussain S, Saud S, Hassan S, Tanveer M, Ihsan MZ, Shah AN, Ullah A, Nasrullah KF, Ullah S, Alharby HNW, Wu C, Huang J (2016d) A combined application of biochar and phosphorus alleviates heat-induced adversities on physiological, agronomical and quality attributes of rice. *Plant Physiol Biochem* 103:191–198.

Fahad S, Hussain S, Saud S, Khan F, Hassan S, Nasim W, Arif M, Wang F, Huang J (2016b) Exogenously applied plant growth regulators affect heat-stressed rice pollens. *J Agron Crop Sci* 202:139–150.

Fahad S, Hussain S, Saud S, Tanveer M, Bajwa AA, Hassan S, Shah AN, Ullah A, Wu C, Khan FA, Shah F, Ullah S, Chen Y, Huang J (2015a) A biochar application protects rice pollen from high-temperature stress. *Plant Physiol Biochem* 96:281–287.

Fahad S, Muhammad ZI, Abdul K, Ihsanullah D, Saud S, Saleh A, Wajid N, Muhammad A, Imtiaz AK, Chao W, Depeng W, Jianliang H (2018a) Consequences of high temperature under changing climate optima for rice pollen characteristics-concepts and perspectives. *Archives Agron Soil Sci.* https://doi.org/10.1080/03650340.2018.1443213

Fahad S, Nie L, Chen Y, Wu C, Xiong D, Saud S, Hongyan L, Cui K, Huang J (2015b) Crop plant hormones and environmental stress. *Sustain Agric Rev* 15:371–400.

Fahad S, Saud S, Yajun C, Chao W, Depeng W (Eds.), (2021f) Abiotic stress in plants. *IntechOpen United Kingdom 2021.* http://dx.doi.org/10.5772/intechopen.91549

Fahad S, Hasanuzzaman M, Alam M, Ullah H, Saeed M, Ali Khan I, Adnan M (eds) (2020) Environment, climate, plant and vegetation growth. *Springer Nature Switzerland AG 2020.* https://doi.org/10.1007/978-3-030-49732-3

Fahad S, Adnan M, Hassan S, Saud S, Hussain S, Wu C, Wang D, Hakeem KR, Alharby HF, Turan V, Khan MA, Huang J (2019b) Rice responses and tolerance to high temperature. In: Hasanuzzaman M, Fujita M, Nahar K, Biswas, JK (eds) *Advances in rice research for abiotic stress tolerance.* Woodhead Publ Ltd, Abington Hall Abington, Cambridge CB1 6AH, Cambs, England, pp. 201–224.

Fahad S, Rehman A, Shahzad B, Tanveer M, Saud S, Kamran M, Ihtisham M, Khan SU, Turan V, Rahman MHU (2019a) Rice responses and tolerance to metal/metalloid toxicity. In: Hasanuzzaman M, Fujita M, Nahar K, Biswas JK (eds) *Advances in rice research for abiotic stress tolerance.* Woodhead Publ Ltd, Abington Hall Abington, Cambridge CB1 6AH, Cambs, England, pp. 299–312.

Fahad S, Sönmez O, Saud S, Wang D, Wu C, Adnan M, Arif M, Amanullah (eds) (2021e) *Engineering tolerance in crop plants against abiotic stress, First edition. ed, Footprints of climate variability on plant diversity.* CRC Press, Boca Raton.

Fahad S, Sonmez O, Saud S, Wang D, Wu C, Adnan M, Turan V (eds) (2021b) *Climate change and plants: biodiversity, growth and interactions, First edition. ed, Footprints of climate variability on plant diversity.* CRC Press, Boca Raton.

Fahad S, Sonmez O, Saud S, Wang D, Wu C, Adnan M, Turan V (eds) (2021c) *Developing climate resilient crops: improving global food security and safety, First edition. ed, Footprints of climate variability on plant diversity.* CRC Press, Boca Raton.

Fahad S, Sönmez O, Saud S, Wang D, Wu C, Adnan M, Turan V (eds) (2021a) *Plant growth regulators for climate-smart agriculture, First edition. ed, Footprints of climate variability on plant diversity.* CRC Press, Boca Raton, FL.

Fahad S, Sönmez O, Turan V, Adnan M, Saud S, Wu C, Wang D (eds) (2021d) *Sustainable soil and land management and climate change, First edition. ed, Footprints of climate variability on plant diversity.* CRC Press, Boca Raton.

Farah R, Muhammad R, Muhammad SA, Tahira Y, Muhammad AA, Maryam A, Shafaqat A, Rashid M, Muhammad R, Qaiser H, Afia Z, Muhammad AA, Muhammad A, Fahad S (2020) Alternative and non-conventional soil and crop management strategies for increasing water use efficiency. In: Fahad S,

Hasanuzzaman M, Alam M, Ullah H, Saeed M, Khan AK, Adnan M (eds) *Environment, climate, plant and vegetation growth*. Springer Publ Ltd, Springer Nature Switzerland AG. Part of Springer Nature, pp. 323–338. https://doi.org/10.1007/978-3-030-49732-3

Farhana G, Ishfaq A, Muhammad A, Dawood J, Fahad S, Xiuling L, Depeng W, Muhammad F, Muhammad F, Syed AS (2020) Use of crop growth model to simulate the impact of climate change on yield of various wheat cultivars under different agro-environmental conditions in Khyber Pakhtunkhwa, Pakistan. *Arabian J Geosci* 13:112. https://doi.org/10.1007/s12517-020-5118-1

Farhat A, Hafiz MH, Wajid I, Aitazaz AF, Hafiz FB, Zahida Z, Fahad S, Wajid F, Artemi C (2020) A review of soil carbon dynamics resulting from agricultural practices. *J Environ Manage* 268(2020):110319.

Fazli W, Muhmmad S, Amjad A, Fahad S, Muhammad A, Muhammad N, Ishaq AM, Imtiaz AK, Mukhtar A, Muhammad S, Muhammad I, Rafi U, Haroon I, Muhammad A (2020) *Plant-microbes interactions and functions in changing climate*. In: Fahad S, Hasanuzzaman M, Alam M, Ullah H, Saeed M, Khan AK, Adnan M (eds) *Environment, climate, plant and vegetation growth*. Springer Publ Ltd, Springer Nature Switzerland AG. Part of Springer Nature, pp. 397–420. https://doi.org/10.1007/978-3-030-49732-3

Foyer CH, Lelandais M, Kunert KJ (1994) Photooxidative stress in plants. *Physiol Plant* 92:669–717.

Ghulam M, Muhammad AA, Donald LS, Sajid M, Muhammad FQ, Niaz A, Ateeq Ur R, Shakeel A, Sajjad H, Muhammad A, Summia M, Aqib HAK, Fahad S, Rahul D, Mazhar I, Timothy DS (2021) Formalin fumigation and steaming of various composts differentially influence the nutrient release, growth and yield of muskmelon (*Cucumis melo* L.). *Sci Rep* 11:21057.

Gill SS, Tuteja N (2010) Reactive oxygen species and antioxidant machinery in abiotic stress tolerance in crop plants. *Plant Physiol Biochem* 48(12):909–930.

Giordano CV, Galatro A, Puntarulo S, Ballaré CL (2004) The inhibitory effects of UV-B radiation (280-315 nm) on *Gunnera magellanica* growth correlate with increased DNA damage but not with oxidative damage to lipids. *Plant Cell Environm* 27(11):1415–1423.

Godde D, Buchhold J (1992) Effect of longterm fumigation with ozone on the turnover of the D-1 reaction center polypeptide of photosystem II in spruce (Piceaabies). *Physiol Plant* 86:568–574.

González PM, Deregibus D, Malanga G, Campana GL, Zacher K, Quartino ML, Puntarulo S (2017) Oxidative balance in macroalgae from Antarctic waters. Possible role of Fe. *J Exp Mar Biol Ecol* 486:379–386.

Gopakumar L, Bernard NO, Donato V (2020) Soil microarthropods and nutrient cycling. In: Fahad S, Hasanuzzaman M, Alam M, Ullah H, Saeed M, Khan AK, Adnan M (eds) *Environment, climate, plant and vegetation growth*. Springer Publ Ltd, Springer Nature Switzerland AG. Part of Springer Nature, pp. 453–472. https://doi.org/10.1007/978-3-030-49732-3

Guang L, Sacandalios JG (1998) Two structurally similar maize cytosolic superoxide dismutase genes, Sod4 and Sod4A, respond differentially to abscisic and high osmoticum. *Plant Physiol* 117:217–224.

Guofu L, Zhenjian B, Fahad S, Guowen C, Zhixin X, Hao G, Dandan L, Yulong L, Bing L, Guoxu J, Saud S (2021) Compositional and structural changes in soil microbial communities in response to straw mulching and plant revegetation in an abandoned artificial pasture in Northeast China. *Glob Ecol Conserv* 31(2021):e01871.

Gutt J, Constable A, Cummings V, Hosie G, McIntyre T, Mintenbeck K, Murray A, Peck L, Ropert-Coudert Y, Saba GK, Schofield O, Schloss I, Stefels J, Takahashi K (2016) Vulnerability of Southern Ocean biota to climate change. *Antarctic Environments Portal*. https://doi.org/10.18124/D46P48

Habib R, Ashfaq A, Aftab W, Manzoor H, Fahd R, Wajid I, Md. Aminul I, Vakhtang S, Muhammad A, Asmat U, Abdul W, Syeda RS, Shah S, Shahbaz K, Fahad S, Manzoor H, Saddam H, Wajid N (2017) Application of CSM-CROPGRO-Cotton model for cultivars and optimum planting dates: Evaluation in changing semi-arid climate. *Field Crops Res*. http://dx.doi.org/10.1016/j.fcr.2017.07.007

Hafeez M, Farman U, Muhammad MK, Xiaowei L, Zhijun Z, Sakhawat S, Muhammad I, Mohammed AA, Mandela F-G, Nicolas D, Muzammal R, Fahad S, Yaobin L (2021) Metabolic-based insecticide resistance mechanism and ecofriendly approaches for controlling of beet armyworm Spodoptera exigua: a review. *Environ Sci Pollution Res*. https://doi.org/10.1007/s11356-021-16974-w

Hafiz MH, Abdul K, Farhat A, Wajid F, Fahad S, Muhammad A, Ghulam MS, Wajid N, Muhammad M, Hafiz FB (2020b): Comparative effects of organic and inorganic fertilizers on soil organic carbon and wheat productivity under arid region. *Commun Soil Sci Plant Anal*. https://doi.org/10.1080/00103624.2020.1763385

Hafiz MH, Farhat A, Ashfaq A, Hafiz FB, Wajid F, Carol Jo W, Fahad S, Gerrit H (2020a) Predicting kernel growth of maize under controlled water and nitrogen applications. *Int J Plant Prod*. https://doi.org/10.1007/s42106-020-00110-8

Hafiz MH, Farhat A, Shafqat S, Fahad S, Artemi C, Wajid F, Chaves CB, Wajid N, Muhammad M, Hafiz FB (2018) Offsetting land degradation through nitrogen and water management during maize cultivation under arid conditions. *Land Degrad Dev* 1–10. https://doi.org/10.1002/ldr.2933

Hafiz MH, Muhammad A, Farhat A, Hafiz FB, Saeed AQ, Muhammad M, Fahad S, Muhammad A (2019) Environmental factors affecting the frequency of road traffic accidents: a case study of sub-urban area of Pakistan. *Environ Sci Pollut Res*. https://doi.org/10.1007/s11356-019-04752-8

Hafiz MH, Wajid F, Farhat A, Fahad S, Shafqat S, Wajid N, Hafiz FB (2016) Maize plant nitrogen uptake dynamics at limited irrigation water and nitrogen. *Environ Sci Pollut Res* 24(3):2549–2557. https://doi.org/10.1007/s11356-016-8031-0

Haider SA, Lalarukh I, Amjad SF, Mansoora N, Naz M, Naeem M, Bukhari SA, Shahbaz M, Ali SA, Marfo TD, Subhan D, Rahul D, Fahad S (2021) Drought stress alleviation bypotassium-nitrate-containingchitosan/montmorillonite microparticles confers changes in *Spinacia oleracea* L. *Sustain* 13: 9903. https://doi.org/10.3390/su13179903

Hamza SM, Xiukang W, Sajjad A, Sadia Z, Muhammad N, Adnan M, Fahad S et al. (2021) Interactive effects of gibberellic acid and NPK on morpho-physio-biochemical traits and organic acid exudation pattern in coriander (*Coriandrum sativum* L.) grown in soil artificially spiked with boron. *Plant Physiol Biochem* 167(2021):884–900.

Harley CD, Randall Hughes A, Hultgren KM, Miner BG, Sorte CJ, Thornber CS, Rodriguez LF, Tomanek L, Williams SL (2006) The impacts of climate change in coastal marine systems. *Ecol Lett* 9(2):228–241.

Hernandez JA, Corpas FJ, Gomez LA, del Río LA, Sevilla F (1993) Salt induced oxidative stress mediated by active oxygen species in pea leaf mitochondria. *Plant Physiol* 89:103–110.

Hernandez JA, Olmos E, Corpas FJ, Sevilla F, del Río LA (1995) Salt-induced oxidative stress in chloroplasts of pea plants. *Plant Sci* 105:151–167.

Hernando M, Schloss IR, Malanga G, Almandoz GO, Ferreyra GA, Aguiar MB, Puntarulo S (2015) Effects of salinity changes on coastal Antarctic phytoplankton physiology and assemblage composition. *J Exp Mar Biol Ecol* 466:110–119.

Hesham FA and Fahad S (2020) Melatonin application enhances biochar efficiency for drought tolerance in maize varieties: Modifications in physio-biochemical machinery. *Agron J* 112(4):1–22.

Hideg É, Sass L, Barbato R, Vass I (1993) Inactivation of photosynthetic oxygen evolution by UV-B irradiation: A thermoluminescence study. *Photosynth Res* 38(3):455–462.

Hideg É, Vass I (1996) UV-B induced free radical production in plant leaves and isolated thylakoid membranes. *Plant Sci* 115(2):251–260.

Huang Li Y, Li X-X, Zhang Y-B, Fahad S, Wang F (2021) dep1 improves rice grain yield and nitrogen use efficiency simultaneously by enhancing nitrogen and dry matter translocation. *J Integrative Agri*. https://doi.org/10.1016/S2095-3119(21)63795-4

Hussain MA, Fahad S, Rahat S, Muhammad FJ, Muhammad M, Qasid A, Ali A, Husain A, Nooral A, Babatope SA, Changbao S, Liya G, Ibrar A, Zhanmei J, Juncai H (2020) Multifunctional role of brassinosteroid and its analogues in plants. *Plant Growth Regul*. https://doi.org/10.1007/s10725-020-00647-8

Ibrahim NA, El-Zairy EMR, Abdalla WA, Khalil HM (2013) Combined UV-protecting and reactive printing of Cellulosic/wool blends. *Carbohydr Polym* 92(2):1386–1394.

Ibrar K, Aneela R, Khola Z, Urooba N, Sana B, Rabia S, Ishtiaq H, Mujaddad UR, Salvatore M (2020) Microbes and environment: global warming reverting the frozen zombies. In: Fahad S, Hasanuzzaman M, Alam M, Ullah H, Saeed M, Khan AK, Adnan M (eds) *Environment, climate, plant and vegetation growth*. Springer Publ Ltd, Springer Nature Switzerland AG. Part of Springer Nature, pp. 607–634. https://doi.org/10.1007/978-3-030-49732-3

Ikram U, Khadim D, Muhammad T, Muhammad S, Fahad S (2021) Gibberellic acid and urease inhibitor optimize nitrogen uptake and yield of maize at varying nitrogen levels under changing climate. *Environ Sci Pollution Res*. https://doi.org/10.1007/s11356-021-16049-w

Ilyas H, Alam SS, Khan I, Shah B, Naeem A, Khan N, Ullah W, Adnan M, Shah SRA, Junaid K, Ahmed N, Iqbal M (2016) Medicinal plants rhizosphere exploration for the presence of potential biocontrol fungi. *J Ento Zool Stud* 4(3):108–113.

Ilyas M, Mohammad N, Nadeem K, Ali H, Aamir HK, Kashif H, Fahad S, Aziz K, Abid U, (2020) Drought tolerance strategies in plants: a mechanistic approach. *J Plant Growth Regul*. https://doi.org/10.1007/s00344-020-10174-5

Inze D, Van Montagu M (2003) Oxidative stress: an introduction. In: Inzé D, Van Montagu M (eds) *Oxidative stress in plants*. CRC Press, London, pp 1–33.

Iqra M, Amna B, Shakeel I, Fatima K, Sehrish L, Hamza A, Fahad S (2020) Carbon cycle in response to global warming. In: Fahad S, Hasanuzzaman M, Alam M, Ullah H, Saeed M, Khan AK, Adnan M (eds) *Environment, climate, plant and vegetation growth*. Springer Publ Ltd, Springer Nature Switzerland AG. Part of Springer Nature, pp. 1–16. https://doi.org/10.1007/978-3-030-49732-3

Irfan M, Muhammad M, Muhammad JK, Khadim MD, Dost M, Ishaq AM, Waqas A, Fahad S, Saud S et al. (2021) Heavy metals immobilization and improvement in maize (*Zea mays* L.) growth amended with biochar and compost. *Sci Rep* 11:18416.

Jabborova D, Sulaymanov K, Sayyed RZ, Alotaibi SH, Enakiev Y, Azimov A, Jabbarov Z, Ansari MJ, Fahad S, Danish S et al. (2021) Mineral fertilizers improves the quality of turmeric and soil. *Sustain* 13:9437. https://doi.org/10.3390/su13169437

Jan M, Muhammad AH, Adnan NS, Muhammad Y, Javaid I, Xiuling L, Depeng W, Fahad S (2019) Modulation in growth, gas exchange, and antioxidant activities of salt-stressed rice (*Oryza sativa* L.) genotypes by zinc fertilization. *Arabian J Geosci* 12:775. https://doi.org/10.1007/s12517-019-4939-2

Ji LL (1996) Exercise, oxidative stress, and antioxidants. *Am J Sport Medic* 24(6 suppl):S20–S24.

Kamarn M, Wenwen C, Irshad A, Xiangping M, Xudong Z, Wennan S, Junzhi C, Shakeel A, Fahad S, Qingfang H, Tiening L (2017) Effect of paclobutrazol, a potential growth regulator on stalk mechanical strength, lignin accumulation and its relation with lodging resistance of maize. *Plant Growth Regul* 84:317–332. https://doi.org/10.1007/s10725-017-0342-8

Kerr JB, McElroy CT (1993) Evidence for large upward trends of ultraviolet-B radiation linked to ozone depletion. *Science* 262(5136):1032–1034.

Khadim D, Fahad S, Jahangir MMR, Iqbal M, Syed SA, Shah AK, Ishaq AM, Rahul D et al. (2021a) Biochar and urease inhibitor mitigate NH_3 and N_2O emissions and improve wheat yield in a urea fertilized alkaline soil. *Sci Rep* 11:17413.

Khadim D, Saif R, Fahad S, Syed SA, Shah AK et al. (2021b) Influence of variable biochar concentration on yield-scaled nitrous oxide emissions, Wheat yield and nitrogen use efficiency. *Sci Rep* 11:16774.

Khan MMH, Niaz A, Umber G, Muqarrab A, Muhammad AA, Muhammad I, Shabir H, Shah F, Vibhor A, Shams H. A-H, Reham A, Syed MBA, Nadiyah MA, Ali TKZ, Subhan D, Rahul D (2021) Synchronization of Boron application methods and rates is environmentally friendly approach to improve quality attributes of *Mangifera indica* L. On sustainable basis. *Saudi J Bio Sci*. https://doi.org/10.1016/j.sjbs.2021.10.036

Kranner I, Minibayeva FV, Beckett RP, Seal CE (2010.) What is stress? Concepts, definitions and applications in seed science. *New Phytol* 188(3):655–673.

Kumar A, Castellano I, Patti FP, Palumbo A, Buia MC (2015) Nitric oxide in marine photosynthetic organisms. *Nitric Oxide* 4:34–39.

Latifi A, Ruiz M, Zhang CC (2009) Oxidative stress in cyanobacteria. *FEMS Microb Rev* 33(2):258–278.

Lazarus, RS (1966) Some principles of psychological stress and their relation to dentistry. *J Dental Res* 45(6):1620–1626.

Liu Z, Yang X, Hubbard KG, Lin X (2012) Maize potential yields and yield gaps in the changing climate of northeast China. *Glob Chang Biol* 18:3441–3454.

Mahar A, Amjad A, Altaf HL, Fazli W, Ronghua L, Muhammad A, Fahad S, Muhammad A, Rafiullah, Imtiaz AK, Zengqiang Z (2020) Promising technologies for Cd-Contaminated soils: drawbacks and possibilities. In: Fahad S, Hasanuzzaman M, Alam M, Ullah H, Saeed M, Khan AK, Adnan M (eds) *Environment, climate, plant and vegetation growth*. Springer Publ Ltd, Springer Nature Switzerland AG. Part of Springer Nature, pp. 63–92. https://doi.org/10.1007/978-3-030-49732-3

Mahmood Ul H, Tassaduq R, Chandni I, Adnan A, Muhammad A, Muhammad MA, Muhammad HR, Mehmood AN, Alam S, Fahad S (2021) *Linking plants functioning to adaptive responses under heat stress conditions: a mechanistic review. J Plant Growth Regul.* https://doi.org/10.1007/s00344-021-10493-1

Manzer HS, Saud A, Soumya M, Abdullah A, Qasi DA, Bander MA, Hayssam MA, Hazem MK, Fahad S, Vishnu DR, Om PN (2021) Molybdenum and hydrogen sulfide synergistically mitigate arsenic toxicity by modulating defense system, nitrogen and cysteine assimilation in faba bean (*Vicia faba* L.) seedlings. *Environ Pollut* 290:117953.

Md Jakir H, Allah B (2020) Development and applications of transplastomic plants; A way towards eco-friendly agriculture. In: Fahad S, Hasanuzzaman M, Alam M, Ullah H, Saeed M, Khan AK, Adnan M (eds) *Environment, climate, plant and vegetation growth*. Springer Publ Ltd, Springer Nature Switzerland AG. Part of Springer Nature, pp. 285–322. https://doi.org/10.1007/978-3-030-49732-3

Md. Enamul H, Shoeb AZM, Mallik AH, Fahad S, Kamruzzaman MM, Akib J, Nayyer S, Mehedi A, Swati AS, Md Yeamin A, Most SS (2020) Measuring vulnerability to environmental hazards: qualitative to quantitative. In: Fahad S, Hasanuzzaman M, Alam M, Ullah H, Saeed M, Khan AK, Adnan M (eds) *Environment, climate, plant and vegetation growth*. Springer Publ Ltd, Springer Nature Switzerland AG. Part of Springer Nature, pp. 421–452. https://doi.org/10.1007/978-3-030-49732-3

Michalski WP, Kaniuga Z (1982) Photosynthetic apparatus of chilling-sensitive plants. XI. Reversibility by light of cold- and dark-induced inactivation of cyanide-sensitive superoxide dismutase activity in tomato leaf chloroplasts. *Biochim Biophys Acta (BBA)-Bioenerget* 680(3):250–257.

Mohammad IAW, Abdelazeem S, Munir A, Khalid E, Adel RAU (2020b) Extent of climate change in Saudi Arabia and its impacts on agriculture: a case study from Qassim region. In: Fahad S, Hasanuzzaman M, Alam M, Ullah H, Saeed M, Khan AK, Adnan M (eds) *Environment, climate, plant and vegetation growth*. Springer Publ Ltd, Springer Nature Switzerland AG. Part of Springer Nature, pp. 635–658. https://doi.org/10.1007/978-3-030-49732-3

Mohammad IAW, Munir A, Adel RAU, Mutair A, Muhammad IR (2020a) Advances in pyrolytic technologies with improved carbon capture and storage to combat climate change. In: Fahad S, Hasanuzzaman M, Alam M, Ullah H, Saeed M, Khan AK, Adnan M (eds) *Environment, climate, plant and vegetation growth*. Springer Publ Ltd, Springer Nature Switzerland AG. Part of Springer Nature, pp. 535–576. https://doi.org/10.1007/978-3-030-49732-3

Moldogazieva NT, Mokhosoev IM, Feldman NB, Lutsenko SV (2018) ROS and RNS signaling: adaptive redox switches through oxidative/nitrosative protein modifications. *Free Rad Res* 52(5):507–543.

Mubeen M, Ashfaq A, Hafiz MH, Muhammad A, Hafiz UF, Mazhar S, Muhammad SD, Asad A, Amjed A, Fahad S, Wajid N (2020) Evaluating the climate change impact on water use efficiency of cotton-wheat in semi-arid conditions using DSSAT model. *J Water Clim Change*. https://doi.org/10.2166/wcc.2019.179/622035/jwc2019179.pdf

Muhammad N, Muqarrab A, Khurram S, Fiaz A, Fahim N, Muhammad A, Shazia A, Omaima N, Sulaiman AA, Fahad S, Subhan D, Rahul D (2021) Kaolin and Jasmonic acid improved cotton productivity under water stress conditions. *J Saudi Soc Agric Sci* 28(2021):6606–6614. https://doi.org/10.1016/j.sjbs.2021.07.043

Muhammad TQ, Amna F, Amna B, Barira Z, Xitong Z, Ling-Ling C (2020) Effectiveness of conventional crop improvement strategies vs. omics. In: Fahad S, Hasanuzzaman M, Alam M, Ullah H, Saeed M, Khan AK, Adnan M (eds) *Environment, climate, plant and vegetation growth*. Springer Publ Ltd, Springer Nature Switzerland AG. Part of Springer Nature, pp. 253–284. https://doi.org/10.1007/978-3-030-49732-3

Muhammad Z, Abdul MK, Abdul MS, Kenneth BM, Muhammad S, Shahen S, Ibadullah J, Fahad S (2019) Performance of *Aeluropus lagopoides* (mangrove grass) ecotypes, a potential turfgrass, under high saline conditions. *Environ Sci Pollut Res*. https://doi.org/10.1007/s11356-019-04838-3

Muzammal R, Fahad S, Guanghui D, Xia C, Yang Y, Kailei T, Lijun L, Fei-Hu L, Gang D (2021) Evaluation of hemp (*Cannabis sativa* L.) as an industrial crop: a review. *Environ Sci Pollution Res*. https://doi.org/10.1007/s11356-021-16264-5

Niki E (2016) Oxidative stress and antioxidants: distress or eustress? *Arch Biochem Biophys* 595:19–24.

Nikolic M, Nikolic N, Liang Y, Kirkby EA, Römheld V (2007) Germanium-68 as an adequate tracer for silicon transport in plants. Characterization of silicon uptake in different crop species. *Plant Physiol* 143(1):495–503.

Nikolic M, Pavlovic J (2018) Plant responses to iron deficiency and toxicity and iron use efficiency in plants. In *Plant Micronutrient Use Efficiency*. Academic Press, pp. 55–69.

Noor M, Naveed Ur R, Ajmal J, Fahad S, Muhammad A, Fazli W, Saud S, Hassan S (2020) Climate change and costal plant lives. In: Fahad S, Hasanuzzaman M, Alam M, Ullah H, Saeed M, Khan AK, Adnan M (eds) *Environment, climate, plant and vegetation growth*. Springer Publ Ltd, Springer Nature Switzerland AG. Part of Springer Nature, pp. 93–108. https://doi.org/10.1007/978-3-030-49732-3

Ozgur R, Uzilday B, Sekmen AH, Turkan I (2013) Reactive oxygen species regulation and antioxidant defence in halophytes. *Funct Plant Biol* 40:832–847.

Qamar Z, Zubair A, Muhammad Y, Muhammad ZI, Abdul K, Fahad S, Safder B, Ramzani PMA, Muhammad N (2017) Zinc biofortification in rice: leveraging agriculture to moderate hidden hunger in developing countries. *Arch Agron Soil Sci* 64:147–161. https://doi.org/10.1080/03650340.2017.1338343

Quartino ML, Deregibus D, Campana GL, Latorre GEJ, Momo FR (2013) Evidence of macroalgal colonization on newly ice-free areas following glacial retreat in Potter Cove (South Shetland Islands), Antarctica. *PLoS One* 8(3):e58223.

Rashid M, Qaiser H, Khalid SK, Mohammad IAW, Zhang A, Muhammad A, Shahzada SI, Rukhsanda A, Ghulam AS, Shahzada MM, Sarosh A, Muhammad FQ (2020) Prospects of biochar in alkaline soils to mitigate climate change. In: Fahad S, Hasanuzzaman M, Alam M, Ullah H, Saeed M, Khan AK, Adnan M (eds) *Environment, climate, plant and vegetation growth.* Springer Publ Ltd, Springer Nature Switzerland AG. Part of Springer Nature, pp. 133–150. https://doi.org/10.1007/978-3-030-49732-3

Rastogi RP, Incharoensakdi A (2013) UV radiation-induced accumulation of photoprotective compounds in the green alga *Tetraspora sp.* CU2551. *Plant Physiol Biochem* 70: 7–13.

Rehman M, Fahad S, Saleem MH, Hafeez M, Muhammad Habib Ur Rahman, Liu F, Deng G (2020) Red light optimized physiological traits and enhanced the growth of ramie (*Boehmeria nivea* L.). *Photosynthetica* 58(4):922–931.

Robello E, Galatro A, Puntarulo S (2007) Iron role in oxidative metabolism of soybean axes upon growth: effect of iron overload. *Plant Sci* 172(5):939–947.

Sadam M, Muhammad TQ, Ghulam M, Muhammad SK, Faiz AJ (2020) Role of biotechnology in climate resilient agriculture. In: Fahad S, Hasanuzzaman M, Alam M, Ullah H, Saeed M, Khan AK, Adnan M (eds) *Environment, climate, plant and vegetation growth.* Springer Publ Ltd, Springer Nature Switzerland AG. Part of Springer Nature, pp. 339–366. https://doi.org/10.1007/978-3-030-49732-3

Safi UK, Ullah F, Mehmood S, Fahad S, Ahmad Rahi A, Althobaiti F et al. (2021) Antimicrobial, antioxidant and cytotoxic properties of *Chenopodium glaucum* L. *PLoS One* 16(10):e0255502. https://doi.org/10.1371/journal.pone.0255502

Sahade R, Lagger C, Torre L, Momo F, Monien P, Schloss I, Barnes DKA, Serveto N, Tarantelli S, Tatián M, Zamboni N, Abele D (2015) Climate change and glacier retreat drive shifts in an Antarctic benthic ecosystem. *Sci Adv* 1(10):e1500050.

Sahrish N, Shakeel A, Ghulam A, Zartash F, Sajjad H, Mukhtar A, Muhammad AK, Ahmad K, Fahad S, Wajid N, Sezai E, Carol Jo W, Gerrit H (2022) Modeling the impact of climate warming on potato phenology. *European J AgroN* 132:126404.

Sajid H, Jie H, Jing H, Shakeel A, Satyabrata N, Sumera A, Awais S, Chunquan Z, Lianfeng Z, Xiaochuang C, Qianyu J, Junhua Z (2020) Rice production under climate change: adaptations and mitigating strategies. In: Fahad S, Hasanuzzaman M, Alam M, Ullah H, Saeed M, Khan AK, Adnan M (eds) *Environment, climate, plant and vegetation growth.* Springer Publ Ltd, Springer Nature Switzerland AG. Part of Springer Nature, pp. 659–686. https://doi.org/10.1007/978-3-030-49732-3

Sajjad H, Muhammad M, Ashfaq A, Waseem A, Hafiz MH, Mazhar A, Nasir M, Asad A, Hafiz UF, Syeda RS, Fahad S, Depeng W, Wajid N (2019) Using GIS tools to detect the land use/land cover changes during forty years in Lodhran district of Pakistan. *Environ Sci Pollut Res.* https://doi.org/10.1007/s11356-019-06072-3

Saleem MH, Fahad S, Adnan M, Mohsin A, Muhammad SR, Muhammad K, Qurban A, Inas AH, Parashuram B, Mubassir A, Reem MH (2020a) Foliar application of gibberellic acid endorsed phytoextraction of copper and alleviates oxidative stress in jute (*Corchorus capsularis* L.) plant grown in highly copper-contaminated soil of China. *Environ Sci Pollution Res.* https://doi.org/10.1007/s11356-020-09764-3

Saleem MH, Fahad S, Shahid UK, Mairaj D, Abid U, Ayman ELS, Akbar H, Analía L, Lijun L (2020c) Copper-induced oxidative stress, initiation of antioxidants and phytoremediation potential of flax (*Linum usitatissimum* L.) seedlings grown under the mixing of two different soils of China. *Environ Sci Poll Res.* https://doi.org/10.1007/s11356-019-07264-7

Saleem MH, Rehman M, Fahad S, Tung SA, Iqbal N, Hassan A, Ayub A, Wahid MA, Shaukat S, Liu L, Deng G (2020b) Leaf gas exchange, oxidative stress, and physiological attributes of rapeseed (*Brassica napus* L.) grown under different light-emitting diodes. *Photosynthetica* 58(3):836–845.

Saman S, Amna B, Bani A, Muhammad TQ, Rana MA, Muhammad SK (2020) QTL mapping for abiotic stresses in cereals. In: Fahad S, Hasanuzzaman M, Alam M, Ullah H, Saeed M, Khan AK, Adnan M (eds) *Environment, climate, plant and vegetation growth.* Springer Publ Ltd, Springer Nature Switzerland AG. Part of Springer Nature, pp. 229–252. https://doi.org/10.1007/978-3-030-49732-3

Saud S, Chen Y, Fahad S, Hussain S, Na L, Xin L, Alhussien SA (2016) Silicate application increases the photosynthesis and its associated metabolic activities in Kentucky bluegrass under drought stress and post-drought recovery. *Environ Sci Pollut Res* 23(17):17647–17655. https://doi.org/10.1007/s11356-016-6957-x

Saud S, Chen Y, Long B, Fahad S, Sadiq A (2013) The different impact on the growth of cool season turf grass under the various conditions on salinity and drought stress. *Int J Agric Sci Res* 3:77–84.
Saud S, Fahad S, Cui G, Chen Y, Anwar S (2020) Determining nitrogen isotopes discrimination under drought stress on enzymatic activities, nitrogen isotope abundance and water contents of Kentucky bluegrass. *Sci Rep* 10:6415. https://doi.org/10.1038/s41598-020-63548-w
Saud S, Fahad S, Yajun C, Ihsan MZ, Hammad HM, Nasim W, Amanullah Jr, Arif M and Alharby H (2017) Effects of nitrogen supply on water stress and recovery mechanisms in Kentucky bluegrass Plants. *Front Plant Sci* 8:983. https://doi.org/10.3389/fpls.2017.00983
Saud S, Li X, Chen Y, Zhang L, Fahad S, Hussain S, Sadiq A, Chen Y (2014) Silicon application increases drought tolerance of Kentucky bluegrass by improving plant water relations and morph physiological functions. *SciWorld J* 2014:1–10. https://doi.org/10.1155/2014/368694
Sellner KG, Doucette GJ, Kirkpatrick GJ (2003) Harmful algal blooms: causes, impacts and detection. *J Industrial Microbiol Biotechnol* 30(7):383–406.
Senol C (2020) *The effects of climate change on human behaviors.* In: Fahad S, Hasanuzzaman M, Alam M, Ullah H, Saeed M, Khan AK, Adnan M (eds) *Environment, climate, plant and vegetation growth.* Springer Publ Ltd, Springer Nature Switzerland AG. Part of Springer Nature, pp. 577–590. https://doi.org/10.1007/978-3-030-49732-3
Sgherri CLM, Navari Izzo F (1995) Sunflower seedlings subjected to increasing water deficit stress: oxidative stress and defence mechanisms. *Physiol Plantar* 93(1):25–30.
Shafi MI, Adnan M, Fahad S, Fazli W, Ahsan K, Zhen Y, Subhan D, Zafar-ul-Hye M, Martin B, Rahul D (2020) Application of single superphosphate with humic acid improves the growth, yield and phosphorus uptake of wheat (*Triticum aestivum* L.) in calcareous soil. *Agron* (10):1224. https://doi.org/10.3390/agronomy10091224
Shah F, Lixiao N, Kehui C, Tariq S, Wei W, Chang C, Liyang Z, Farhan A, Fahad S, Huang J (2013) Rice grain yield and component responses to near 2°C of warming. *Field Crop Res* 157:98–110.
Shah T (2009) Climate change and groundwater: India's opportunities for mitigation and adaptation. *Environm Res Lett* 4(3):035005.
Sharma P, Jha AB, Dubey RS, Pessarakli M (2012). Reactive oxygen species, oxidative damage, and antioxidative defense mechanism in plants under stressful conditions. *J Bot* 2012:1–26.
Sidra K, Javed I, Subhan D, Allah B, Syed IUSB, Fatma B, Khaled DA, Fahad S, Omaima N, Ali TKZ, Rahul D (2021) Physio-chemical characterization of indigenous agricultural waste materials for the development of potting media. *J Saudi Soc Agric Sci.* https://doi.org/10.1016/j.sjbs.2021.08.058
Singh J, Singh RP, Khare R (2018) Influence of climate change on Antarctic flora. *Polar Sci* 18:94–101.
Soares C, Carvalho ME, Azevedo RA, Fidalgo F (2019) Plants facing oxidative challenges: a little help from the antioxidant networks. *Environm Experim Bot* 161: 4–25.
Subhan D, Zafar-ul-Hye M, Fahad S, Saud S, Martin B, Tereza H, Rahul D (2020) Drought stress alleviation by ACC deaminase producing *Achromobacter xylosoxidans* and *Enterobacter cloacae*, with and without timber waste biochar in maize. *Sustain* 12(6286). https://doi.org/10.3390/su12156286
Takshak S, Agrawal SB (2019) Defense potential of secondary metabolites in medicinal plants under UV-B stress. *J Photochem Photobiol B Biol* 193:51–88.
Tariq M, Ahmad S, Fahad S, Abbas G, Hussain S, Fatima Z, Nasim W, Mubeen M, Ur Rehman MH, Khan MA, Adnan M (2018). The impact of climate warming and crop management on phenology of sunflower-based cropping systems in Punjab, Pakistan. *Agric Forest Met* 256:270–282.
Thiele A, Krause GHH (1994). Xanthophyll cycle and thermal energy dissipation in Photosystem II: relationship between Zeaxanthin formation, energy-dependent fluorescence quenching and photoinhibition. *J Plant Physiol* 144: 324–332.
Tkemaladze GSH, Makhashvili KA (2016) Climate changes and photosynthesis. *Ann Agrar Sci* 14:1119–1126.
Unsar N-U, Muhammad R, Syed HMB, Asad S, Mirza AQ, Naeem I, Muhammad HR, Fahad S, Shafqat S (2020) Insect pests of cotton crop and management under climate change scenarios. In: Fahad S, Hasanuzzaman M, Alam M, Ullah H, Saeed M, Khan AK, Adnan M (eds) *Environment, climate, plant and vegetation growth.* Springer Publ Ltd, Springer Nature Switzerland AG. Part of Springer Nature, pp. 367–396. https://doi.org/10.1007/978-3-030-49732-3
Vadstein O, Stibor H, Lippert B, Løseth K, Roederer W, Sundt-Hansen L, Olsen Y (2004) Moderate increase in the biomass of omnivorous copepods may ease grazing control of planktonic algae. *Mar Ecol Progr Series* 270:199–207.

Wahid F, Fahad S, Subhan D, Adnan M,, Zhen Y, Saud S, Manzer HS, Martin B, Tereza H, Rahul D (2020) Sustainable management with mycorrhizae and phosphate solubilizing bacteria for enhanced phosphorus uptake in calcareous soils. *Agri* 10(334). https://doi.org/10.3390/agriculture10080334

Wajid N, Ashfaq A, Asad A, Muhammad T, Muhammad A, Muhammad S, Khawar J, Ghulam MS, Syeda RS, Hafiz MH, Muhammad IAR, Muhammad ZH, Muhammad HR, Veysel T, Fahad S, Suad S, Aziz K, Shahzad A (2017) Radiation efficiency and nitrogen fertilizer impacts on sunflower crop in contrasting environments of Punjab. *Pakistan Environ Sci Pollut Res* 25:1822–1836. https://doi.org/10.1007/s11356-017-0592-z

Wise RR, Naylor AW (1987) Chilling-enhanced photooxidation: evidence for the role of singlet oxygen and superoxide in the breakdown of pigments and endogenous antioxidants. *Plant Physiol* 83(2):278–282.

Wu C, Kehui C, She T, Ganghua L, Shaohua W, Fahad S, Lixiao N, Jianliang H, Shaobing P, Yanfeng D (2020) Intensified pollination and fertilization ameliorate heat injury in rice (*Oryza sativa* L.) during the flowering stage. *Field Crops Res* 252:107795.

Wu C, Tang S, Li G, Wang S, Fahad S, Ding Y (2019) Roles of phytohormone changes in the grain yield of rice plants exposed to heat: a review. *PeerJ* 7:e7792. https://doi.org/10.7717/peerj.7792

Yang Z, Zhang Z, Zhang T, Fahad S, Cui K, Nie L, Peng S, Huang J (2017) The effect of season-long temperature increases on rice cultivars grown in the central and southern regions of China. *Front Plant Sci* 8:1908. https://doi.org/10.3389/fpls.2017.01908

Zafar-ul-Hye M, Muhammad TH, Muhammad A, Fahad S, Martin B, Tereza D, Rahul D, Subhan D (2020b) Potential role of compost mixed biochar with rhizobacteria in mitigating lead toxicity in spinach. *Scientific Rep* 10:12159. https://doi.org/10.1038/s41598-020-69183-9.

Zafar-ul-Hye M, Muhammad N, Subhan D, Fahad S, Rahul D, Mazhar A, Ashfaq AR, Martin B, Jiˇrí H, Zahid HT, Muhammad N (2020a) Alleviation of Cadmium adverse effects by improving nutrients uptake in bitter gourd through cadmium tolerant rhizobacteria. *Environ* 7(54). https://doi.org/10.3390/environments7080054

Zafar-ul-Hye M, Akbar MN, Iftikhar Y, Abbas M, Zahid A, Fahad S, Datta R, Ali M, Elgorban AM, Ansari MJ et al. (2021) Rhizobacteria Inoculation and caffeic acid alleviated drought stress in lentil plants. *Sustain* 13:9603. https://doi.org/10.3390/su13179603

Zahida Z, Hafiz FB, Zulfiqar AS, Ghulam MS, Fahad S, Muhammad RA, Hafiz MH, Wajid N, Muhammad S (2017) Effect of water management and silicon on germination, growth, phosphorus and arsenic uptake in rice. *Ecotoxicol Environ Saf* 144:11–18.

Zahir SM, Zheng HG, Ala UD, Amjad A, Ata UR, Kashif J, Shah F, Saud S, Adnan M, Fazli W, Saud A, Manzer HS, Shamsher A, Wajid N, Hafiz MH, Fahad S (2021) Synthesis of silver nanoparticles using Plantago lanceolata extract and assessing their antibacterial and antioxidant activities. *Sci Rep* 11:20754.

Zahoor M, Afzal M, Ali M, Mohammad W, Khan N, Adnan M, Ali A, Saeed M (2016) Effect of organic waste and NPK fertilizer on potato yield and soil fertility. *Pure Appl Biol* 5(3):439–445.

Zia-Ur-Rehman M (2020) Environment, climate change and biodiversity. In: Fahad S, Hasanuzzaman M, Alam M, Ullah H, Saeed M, Khan AK, Adnan M (eds) *Environment, climate, plant and vegetation growth.* Springer Publ Ltd, Springer Nature Switzerland AG. Part of Springer Nature, pp. 473–502. https://doi.org/10.1007/978-3-030-49732-3

10

Postharvest Sucrose Losses under Water Stress Conditions in Sugarcane

Varucha Misra, Ashutosh Kumar Mall, and C. P. Prajapati
ICAR-Indian Institute of Sugarcane Research, Lucknow, India

S. Solomon
CSA University of Agriculture & Technology, Kanpur, India

CONTENTS

10.1 Introduction

Sugarcane, *Saccharum officinarum* L., belonging to family Poaceae is closely related to grasses. *Saccharum spontaneum*, *S. sinese*, *S. barberi*, *S. robustrum*, *S. officinarum* are the five species of sugarcane (Salunke and Desai 1988). Of all the portions of sugarcane plant, the economically important part is the sugarcane stalk, which is filled with sucrose. Sucrose is the principal form of transient carbohydrate, which might be synthesized, resynthesized, inverted many times before being accumulated in the sugarcane stalks (Alexander 1973). In this crop, process of sucrose storage involves inversion of sucrose into hexose moieties, interconversion of sucrose into glucose and fructose, and sucrose gets accumulated by gradient concentration (Salunke and Desai 1988). Sugarcane is an important commercially economic crop for our county, attaining the second highest producer position in the world. In 2018–2019, it was cultivated in 5,159 thousand hectares and produced 3,83,892 thousand tonnes of cane with a yield of 74.42 tonnes/ha (Anonymous 2019). In tropical regions of India, Maharashtra is the highest producer of sugarcane (93,040 tonnes), covering an area of 1,163 ha while Tamil Nadu has the highest productivity (98.98 tonnes/ha) as per 2018–2019 year. In subtropical zone, Uttar Pradesh is the highest state producer of sugarcane (1,71,834 tonnes) covering an area of 2,390 ha and Punjab has the highest productivity state of sugar (71.90 tonnes/ha) as per 2018–2019 reports.

Sugarcane is a perishable commodity and gets deteriorated as soon as it is harvested (Solomon 2009). After harvest, sucrose losses are further accentuated, which causes reduction in cane weight and low sugar recovery. In India, postharvest sucrose losses in this crop are always an important aspect, and several studies have been conducted to know the impact it causes (Solomon et al. 1997; Solomon 2009; Misra et al. 2016a; Qadeem et al. 2015). As a natural corollary, all the sugars, especially sucrose, are prone to biochemical and microbial degradation. Besides, sugarcane attracts a large number of insects

DOI: 10.1201/9781003286417-10

and pests as well as pathogens, which affect its growth and productivity. The most important effect is displayed by the vagaries of nature, which devour the sucrose content leading to low sugar recovery and therefore make it a perishable and loss-prone commodity. Most of the commercially grown sugarcane varieties in India have medium sucrose content (about 12.0%–14.0% of the cane) at peak maturity/ ripening, but due to excessive pre- and postmilling losses, the average sugar recovery remains in the range of 9.5%–10.2%. Thus, around 25%–35% of the sugar synthesized in the sugarcane plant system is lost during various stages of cane handling and processing. This is a serious situation which affects sugarcane growers as well as performance and profitability of the sugar industry, per se. Postharvest or staling losses of sucrose are an important component of these. Deterioration in sucrose content in well-ripened sugarcane crop is due to several reasons such as microbial infestation, harvest and transportation injuries, preharvest burning, time lag between harvest and milling, etc. (Solomon 2000). In this respect, an important factor that plays a role herein is frequent occurring of abiotic stresses due to climate change circumstances (Deepranjan et al. 2021; Haider et al. 2021; Huang et al. 2021; Ikram et al. 2021; Jabborova et al. 2021; Khadim et al. 2021a, 2021b; Manzer et al. 2021; Muzammal et al. 2021; Abdul et al. 2021a, 2021b; Ashfaq et al. 2021; Amjad et al. 2021; Atif et al. 2021; Athar et al. 2021; Adnan et al. 2018a, 2018b, 2019; Akram et al. 2018a, 2018b; Aziz et al. 2017a, 2017b; Chang et al. 2021; Chen et al. 2021; Emre et al. 2021; Habib Ur et al. 2017; Hafiz et al. 2016, 2018, 2019, 2020a, 2020b; Ghulam et al. 2021; Guofu et al. 2021; Hafeez et al. 2021; Khan et al. 2021; Kamarn et al. 2017; Muhammad et al. 2019; Safi et al. 2021; Sajjad et al. 2019; Saud et al. 2013, 2014, 2016, 2017, 2020; Shah et al. 2013; Qamar-Uz et al. 2017; Hamza et al. 2021; Irfan et al. 2021; Wajid et al. 2017; Yang et al. 2017; Zahida et al. 2017; Depeng et al. 2018; Hussain et al. 2020; Shafi et al. 2020; Wahid et al. 2020; Subhan et al. 2020; Zafar-ul-Hye et al. 2020a, 2020b, 2021; Adnan et al. 2020; Ilyas et al. 2020; Saleem et al. 2020a, 2020b, 2020c; Rehman et al. 2020; Farhat et al. 2020; Wu et al. 2020; Mubeen et al. 2020; Farhana et al. 2020; Jan et al. 2019; Wu et al. 2019; Ahmad et al. 2019; Baseer et al. 2019; Tariq et al. 2018; Fahad and Bano 2012; Fahad et al. 2013, 2014a, 2014b, 2015a, 2015b, 2016a, 2016b, 2016c, 2016d, 2017, 2018a, 2018b, 2019a, 2019b, 2020, 2021a, 2021b, 2021c, 2021d, 2021e, 2021f; Hesham and Fahad 2020. Iqra et al. 2020; Akbar et al. 2020; Mahar et al. 2020; Noor et al. 2020; Bayram et al. 2020; Amanullah and Fahad 2017, 2018a, 2018b; Amanullah et al. 2020, 2021; Rashid et al. 2020; Arif et al. 2020; Amir et al. 2020; Saman et al. 2020; Muhammad et al. 2020; Md Jakir and Allah 2020; Mahmood Ul et al. 2021; Farah et al. 2020; Sadam et al. 2020; Unsar et al. 2020; Fazli et al. 2020; Md. Enamul et al. 2020; Gopakumar et al. 2020; Zia-Ur-Rehman 2020; Ayman et al. 2020; Mohammad et al. 2020a, 2020b; Senol 2020; Amjad et al. 2020; Ibrar et al. 2020; Sajid et al. 2020; Muhammad et al. 2021; Sidra et al. 2021; Zahir et al. 2021; Sahrish et al. 2022). It has been reported that in India, various losses occur due to abiotic and biotic stresses which cause reduction in production (to as low as 2%–4% to as high as 25%–45%) and productivity of sugarcane (to as low as 3%–5% to as high as 25%–50%) (Shrivastava and Srivastava 2012). These abiotic and biotic losses have been illustrated in Table 10.1. Besides abiotic stress acting as a contributor to postharvest sucrose losses in sugarcane, there are other factors which in combination to abiotic stress further accentuate this problem and results in low recovery of sugar, viz. ambient temperature, humidity, activities of soluble invertases, cane variety which are mostly overlooked by sugar mills in management process (Solomon 2009). Solomon had reported that approximately one-fourth of the total supplied canes to Indian mills are of stale quality that causes a huge loss of approx. Rs. 1,600 crores of Indian sugar mills. The time lag between harvesting to milling leads to major postharvest sucrose losses because at that time possibility of attack of microbes like *Leuconostoc mesentroides* or *L. dextranicum* on sugarcane rises, and this is relatively more increased in case of abiotic stress–affected canes. This leads to crucial sucrose loss in cane and milled juice (Sharma et al. 1994; Solomon et al. 1990, 1997, 2001, 2003, 2006). The chance of attack of these microbes, *L. mesentroides* or *L. dextranicum,* increases when they attain favourable conditions, viz. in stacked cane or the cane that is in transport vehicles for more than 4–7 days or even left at cane centers for many days. These microbes convert sucrose into polysaccharides (Misra et al. 2016c). Furthermore, there are ample evidences that have revealed that weather is always of prime concern for postharvest losses, especially sucrose recovery; higher the temperature and humidity, more is the deterioration and so is the case of water logging (Solomon 2000).

TABLE 10.1

Decline in the Yield of Plant and Ratoon Crop(s) in Drought Years (1981, 1983, 1984 and 1988) as Compared to Normal Years (1982, 1986, and 1987) in Mauritius

	Sectors					
Per cent Yield Decline	**West**	**North**	**East**	**South**	**Centre**	**Whole**
Plant Crop						
(a) Main season	18.24	26.27	19.17	12.08	8.84	17.90
(b) Petite season	32.12	29.31	2.30	11.86	12.96	12.30
Ratoon Crop(s)						
(a) 1st ratoon	22.37	24.40	9.40	9.79	8.33	12.30
(b) 2nd ratoon	20.70	27.81	6.83	6.45	7.21	11.00
(c) 3rd ratoon	19.55	26.77	6.56	4.40	2.88	9.36
(d) 4th ratoon	21.33	25.95	9.96	4.81	2.56	10.39
(e) 5th ratoon	23.63	27.56	4.36	3.83	0.74	8.74
(f) 6th ratoon	23.99	30.91	7.17	3.97	4.74	11.17
(g) Older ratoons	26.78	21.86	12.43	4.15	5.08	11.00

Overall mean (a) Plant crops 15.10 + 3.96 (b) Ratoon crops 10.57 + 1.19
Source: Shrivastava et al. 1992).

Thus, this chapter focuses on water stress (drought and water logging) and its problems associated with sugarcane, particularly with respect to postharvest sucrose losses and the ways by which these losses could be minimized in postharvest canes.

10.2 Postharvest Sucrose Losses in Sugarcane

Postharvest sucrose losses in sugarcane are basically the sucrose losses occurring in sugarcane after harvest due to perishable nature. The first scientific experiment on postharvest deterioration was conducted by M. van Czerniecki in Java (now Indonesia), wherein in a period of five days, after harvest, 38% of the available sugar was lost. Thereafter, several studies have been conducted on this aspect to know the cause and amount of losses in harvested canes (Singh et al. 2015a; Misra et al. 2016a, 2016b; Rahman et al. 2016). Alexander (1973) had identified two major areas of postharvest quality decline. These are (i) inversion processes (ii) losses at factory level. The inversion process in sugarcane is accompanied after harvest and succeeding delays in cane delivering to factory for processing and are often termed as "field losses" or "primary losses" while losses at factory levels occur due to inefficient processing methods that cause sugar inversion and conversion reactions in milled juice (primary and mixed juice) and are termed as "secondary losses" (Figure 10.1). Went and Geerligs (1884) observed that main cause of postharvest deterioration is drying of the cane. Study of Weinberg (1903) conducted in Indonesia and India indicated that when the canes were kept for 4 days, 32% of the available sugar is lost. These studies led him to conclude that the time lag between harvesting and milling as well as storage environment plays a key role in postharvest deterioration process of sugar, and to have maximum sugar output, that cane should be milled within 24–36 hours after harvest. Earlier studies have revealed that sugar yield and profits could be increased by storing the sugarcane in heaps for one week or more and was considered to be 'after ripening,' but losses in recoverable sugar have been reliant on weight loss and purity change (Browne and Blouin 1907). Guilbeau et al. (1956) and Guilbeau et al. (1955) had showed that losses occurring at field are higher than factory losses. Sugarcane loses its sucrose content within few days at field after harvest irrespective of the care and proper cultural practices followed to make it ripen (Alexander 1973). There are various causes behind postharvest sucrose losses in sugarcane such as storage method, physical condition of cane, efficiency of processing unit, time lag between harvesting and milling, nature of variety, etc. One of the important causes on which research is being focused now are the losses occurring due to water stress under climate change scenario.

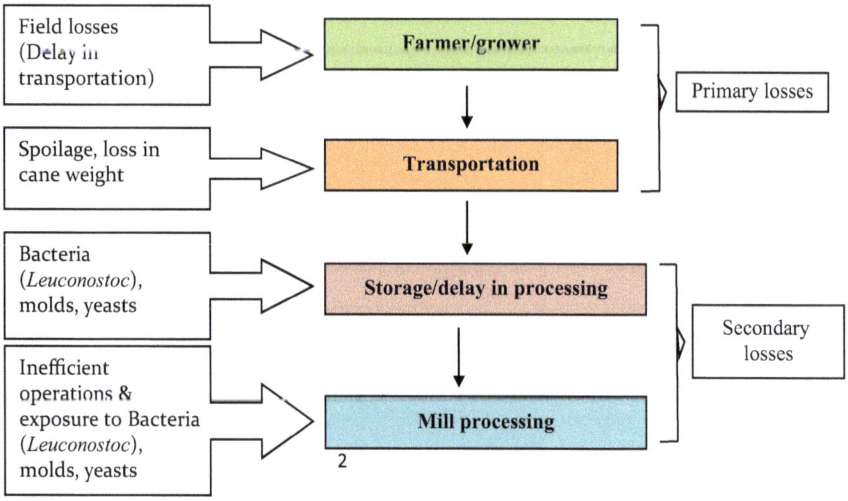

FIGURE 10.1 Losses in food chain in sugarcane crop at different stages.

10.3 Water Stress and Its Problem Associated with Sugarcane

Water stress is the major limiting factor in crop production worldwide (Vasantha 2016; Kumar et al. 2019; Fahad et al. 2019a, 2019b; Saud et al. 2013, 2014, 2017). Water stress is of two types: one where there is deficiency of water to crop and other where there is excess of water. Deficiency of water along with less/nil rainfall causes drought while extreme of water causes either water logging or flooding conditions (VanToai et al. 2001; Rasheed et al. 2015) due to poor drainage system and heavy rainfall (Islam et al. 2011). Nearly 60% of the total sugarcane agriculture suffers from lack of adequate supply, mainly because of limited availability of water for irrigation in lift irrigated areas, canal closure during summer in many of canal irrigated tracts, and droughts that occur in a cyclic manner (Vasantha 2016). Total water requirement for sugarcane crop annually varies from 1,850 mm to 2,500 mm. For one tonne of sugarcane water requirement has been estimated to be 250 tonnes with daily evaporation rate of 8–10 mm (Vasantha 2016). Effects of water stress on crop depend on severity, duration, and timing of stress (Farooq et al. 2009). Also, it depends on the way plants respond after removal of stress as well as other external factors which play a role in it (Plaut 2003).

In India, from 1890 to 2015 twenty-four droughts have been faced by agricultural crops, of which 2002 was worst (in last 100 years) while 2015 was unique (half country in drought and half in flood) in drought scale. As per India Water Portal.org., the year 2015 has been the driest year in India since 2009, and 40% of the country was deficient in rains by about 12%. Anand (2004) reported that the drought that occurred in 2004 affected 122 districts with moistures stress (rainfall < 40% of normal) and 49 districts with scanty rainfall (66%–99% below normal rate), and this had caused loss of productivity and decrease in yield of kharif crops along with sugarcane crop (productivity by 10%) (Zarabi 2004). Khurana (2019) had illustrated that larger area of India suffers from problem of drought. Drought has severely affected sugarcane crop in the year 2015 in various states, i.e. Maharashtra, Andhra Pradesh in tropical India; and Punjab in subtropical India (Shrivastava et al. 2016). In Maharashtra, one-third lies in semiarid zone, whereas 80%–84% of agriculture was in rainfed condition. Marathawada and Vidarbh are the regions of Maharashtra that were rigorously affected by drought and had caused huge losses to sugarcane (Kelkar-Khambete 2014). Central water commission had identified 51.13 million ha spread over 315 talukas out of 725 talukas in 13 states as drought-prone areas (Shrivastava and Srivastava 2006).

With respect to water logging, another type of water stress is, in India, approx. 11.6 million ha area has been under water logging due to soil physical degradation. Anonymous (2000) had revealed that 48.9 million ha area is exposed to soil water erosion. In India, total flood-prone area is 40 million ha, which includes 10 million ha of the area protected from floods. Of cropped area the one affected by the floods is 3.8 million ha (Shrivastava and Srivastava 2006). About 10%–30% of the sugarcane acreage faces problem of water logging. This is due to heavy rainfall. About 73.3% of the annual rainfall is caused by the southwest monsoon during June to September, which appears to be primarily responsible for floods/submergence and water-logging problems. In sugarcane growing areas, north Bihar (particularly Gandak, Budhi Gandak, Bagmati), Punjab (Firozpur and Faridkot districts), and some parts of U.P. (Ghagra Ghat area in eastern U.P.) are highly affected by such conditions. Nearly 50% of the cultivable land of north Bihar (about 1.7 million ha) is affected by flooding during rains. In flood-prone areas of Bihar and east U.P., for almost 2 to 3 months, rain water stands in fields at a level of 3 to 4 feet (Shrivastava and Srivastava 2006; Misra 2018).

10.4 Drought: Causes, Changes, Adaptations and Losses in Sugarcane

Sugarcane crop is well known for high water requirement that varies from 0.09 inches/day (minimum level in June) to 0.24 inches/day (maximum level in January) for full bloom of the crop (Shrivastava et al. 2011; Mall et al. 2019). Drought has been a common problem in current condition of climate change due to global warming. Occurrence of drought is unpredictable and can occur at any stage of crop, which contributes to low cane yield and productivity (Shrivastava et al. 2016). It is known that droughts occur due to several reasons, major one being long periods of high temperatures and less rainfall, late onset or early departure of monsoon, and even the decreasing water level (Shrivastava and Srivastava 2012). Drought can be classified on the basis of its occurrence as occasional, frequent, and rainfed drought (Shrivastava and Srivastava 2012). Furthermore Wilhite and Glantz had classified it into four broad categories as meteorological, agricultural, hydrological, and socioeconomic droughts. Irrespective of whichever type of drought the crop faces, it causes a strong impact on cane yield and productivity. Drought influences sugarcane and sugar productivity in many parts of the world (Misra 2018). As sugarcane is a long period crop, it requires more water supplies, and so there is requirement of more irrigation management. Gascho and Shih (1983) had showed that the formative stage of cane growth requires surplus water for proper germination and establishment; however, occurrence of low humidity causes negative impact on growth. Studies had reported that water stress or soil moisture stress at this stage had more negative impact on growth and yield, which further accentuated with high temperature and evaporation (Srinivasan 1993; Inman-Bamber and Smith 2005; Rao et al. 2007). The first and foremost response that a plant gives under water stress is the morphological and physiological responses (Rachmilevitch et al. 2006). Occurrence of drought at formative stage causes changes in these aspects, which in turn accentuates secondary stresses such as ionic changes, nutritional disorder, oxidative stresses, etc. This further causes difficulties in membrane disorganization, photosynthesis delay, and metabolic toxicity. Studies had showed that due to water stress ionic imbalance, hyperosmotic stress may also occur, causing disruption and interference in metabolic activities and at times plant also dies (Zhu 2001; Karuppanapandian et al. 2011; Liu et al. 2011). Less moisture content in the soil adversely affects cane yield. After 75 to 90 days of moisture stress soil, cane yield loss were revealed to be 0.39 mt/day while after 90 to 105 days of moisture stress soil, loss was of 0256 mt/day (Anonymous 1959). Cane yield reduced more when drought conditions were preceded by normal growing seasons rather than drought followed by drought seasons in Mauritius. However, under both these conditions it has been reported that there would be decline in cane yield in both ratoon and plant crop. Contrastingly, productivity of sugarcane has been more affected in ratoon crops under such conditions (viz, drought-normal-drought season than normal-normal-drought season) (Tables 10.1 and 10.2). Misra et al. (2016a) had revealed that postharvest sucrose losses in sugarcane grown under drought conditions cause higher sucrose losses and poor juice quality after 240 hours of harvest. Findings showed that there was a loss of 0.58 units and 3.58 units in sucrose content in postharvest canes grown under drought condition in comparison to canes grown under normal conditions in November and March harvest. Parashar (1960) also compared juice quality and sugar recovery in drought and normal grown canes at different times of crushing season in Bihar (in 1957–1959). Initially drought-affected cane had higher Brix and purity coefficient than the normal cane, but as the crushing season advanced, the drought-affected cane maintained its progress in Brix. However, purity increases in normal canes. Furthermore, sugar recovery from drought ones does not support to high pol percent in juice due to the presence of

TABLE 10.2

Influence of Drought Conditions Following Normal Year (Preceded by Normal and Drought Conditions) on Sugarcane Productivity in Mauritius

		Plant Crop		Ratoon Crop						
Crop Season	Conditions	Main Season	Petite Season	1st	2nd	3rd	4th	5th	6th	Older Ratoons
1981	Drought	96.6	85.4	82.8	78.7	76.6	74.5	73.1	72.0	73.3
1982	Normal	122.0	100.4	97.1	90.9	89.0	88.2	84.6	84.7	84.3
1983	Drought	110.3	84.2	85.6	76.6	75.6	72.4	70.6	79.0	70.1
% decline in yield in 1983 vs. 1982		9.6	16.1	11.8	15.7	15.1	17.7	12.3	18.5	16.8
1986	Normal	124.3	97.8	98.3	88.1	82.2	79.8	75.7	78.9	76.7
1987	Normal	126.2	96.5	99.8	92.6	85.1	81.1	79.2	76.6	80.5
1988	Drought	108.1	91.2	94.3	89.5	85.3	78.0	76.7	73.7	75.0
% decline in yield in 1988 vs. 1987		14.3	5.5	5.5	2.7	+0.2	3.8	3.1	3.8	6.8

Source: Shrivastava et al. 1992.

TABLE 10.3

Juice Quality Sugar Recovery and Losses in Drought Year (1957–1958) and Normal Year (1958–1959)

Quality Attribute	Drought Year (1957–1958)	Normal Year (1958–1959)
Pol% juice	13.024	12.706
Sugar recovery	10.006	10.008
Unknown losses	0.283	0.156
Pol losses in final molasses	1.480	1.320
Molasses % cane	4.24	4.00

Source: Parashar 1960.

some optically active bodies in the juice. Also, sugar losses and molasses percent cane are relatively higher in drought-affected canes as compared to normal ones (Table 10.3).

For adapting to such a condition, various changes have also been observed such as relative water content, membrane stability, changes in permeability of cell, etc. Researchers had revealed that nontoxic compatible solutes accumulation like proline, betaines, etc. are the responses to which plants respond in such a case (Abbasi et al. 2014; dos-Santos and de-Almeida Silva 2015). Of all the amino acid, most common one is proline which accumulates in huge amounts in such condition (Abraham et al. 2003; Hemaprabha et al. 2013; Munawarti et al. 2014).

10.5 Water Logging: Causes, Changes, Adaptations, and Losses in Sugarcane

Water logging has been defined as a state when water gets stagnated for a short or long duration due to changes in hydrologic regime, landscape, silting up of river beds, increased sedimentation, and reduced capacity of the drainage systems (Anonymous 2000). Water logging limits sugarcane productivity in many parts of the world. Crop competition, increasing pressure on land and various socioeconomic factors appear to have compelled sugarcane to be grown on poorly drained marginal lands prone to seasonal inundation, which affect its productivity (Carter 1976). Although the productivity of sugarcane crop is impaired by prolonged water logging, this may withstand such conditions better than most of the other crops. It is one of the reasons for its preferential cultivation in northern India, in spite of the short growing season. Sugarcane though susceptible to water logging conditions, the extent of damage to crop depends on the stage of growth at which it occurred, the duration of it, and tolerability by various varieties. Water logging reduces shoot and root growth, dry matter production, and total crop yield (Ploschuk et al. 2018). At germination, such conditions are most damaging for establishment of crop resulting in complete loss while if it occurred at tillering stage, the loss was comparatively less but severe. Maximum impact on the initial development of crop was observed during short-term flooding on rapid germinating varieties, viz. CP 80-1827 and CLP-239 (Venkataraman 1929). At grand growth phase, water logging affected more on quality than on yield (Raid 1992). As soon as duration of water logging conditions increases, the reduction in yield also increases (Srivastava et al. 1988). Hurney (1983) had showed that if soil saturated for more than three days, germination of canes was reduced. Even flooding conditions occurring at planting phase affect crop emergence. Emergence of crop was also affected by flooding at planting phase. There are various changes (morphological and physiological) observed in canes exposed to water-logging condition that causes effects on yield and productivity of sugarcane. These are reduced diffusion of oxygen to the root system affecting respiration growth and development of plants (Levitt 1980; Marschner 1986; Veen 1988), formation of arenchyma (air filled intercellular spaces) tissues due to ethylene accumulation in the roots (Feldman 1984), decreased active ion uptake and nutrient deficiency (Leyshon and Sheared 1974), low redox potential, development of pithiness (Naidu and Bhagyalakshmi 1973) and cavity formation (Dilley et al. 1975), reduced rate of photosynthesis (due to slow diffusion of CO_2 and reduced light availability), reduced chlorophyll content and nutrients like N, P and K, and aerenchyma formation. Reduction in chlorophyll content varies from variety to variety. For instance, CoS 91269 had lower chlorophyll content than CoPant 2011 and BO 91 (Anonymous 1980). Anonymous (1980) had even reported that the chlorophyll content varies in tolerant and susceptible water-logged canes. Furthermore, leaves yellowing had also been seen in such cases due to nutrient imbalance, which gives cane a scorched appearance. Dilley et al. (1975) had revealed that at times water logging also causes nonpathogenic large red brown lesions on leaves. In nutrient imbalance, Rege and Mascarenhas (1956) had shown increase in Ca and Mg levels in leaves under water-logging condition. Furthermore, there are number of reasons which cause these changes. These are air displacement from pore spaces in soil; occurrence of anaerobic environment; increase in pH and specific conductance; suppression of soil organic matter nitrates; replacement of soil ions in reduced forms, viz. Fe^{++}, Mn^{++}, NH^{4+}, S^{--}; increase in available phosphorus as phosphates (Levitt 1980; Marschner 1986). As stated earlier, water logging causes loss in yield of sugarcane but extent of loss depends on duration, severity, crop stage, and management practices adopted before and after occurrence of such abiotic stress. Besides, there are several other reasons for yield loss in water-logging conditions. These are stalk mortality, reduced crop growth, lodging, cane breakage, etc. Monoharan et al. (1990) revealed that water stagnation for two months resulted in reduction in cane yield in various varieties by 26%–36%. Gomathi and Chandran (2009) and Gomathi et al. (2010) had showed that water logging caused 22.4% reduction in number of millable canes (NMC), 45.6% reduction in single cane weight, 30.0% reduction in cane length, 15.9% reduction in internodal length, 17.8% reduction in cane thickness, and 40.1% reduction in cane yield. During the late grand growth phase of sugarcane, under artificial water-logging condition about 5%–30% loss in yield has been observed (Shrivastava and Srivastava 2006). Similarly, Deren et al. (1992) had also reported yield loss in water-logging sugarcane, but this loss ranges between 30% and 100% on 160 sugarcane genotypes when flooding condition was for 5 months. Another study showed that in Australia, canes exposed to such condition causes reduction in cane yield and dry matter by 17%–52% and 20%–56%, respectively, in sugarcane clones grown in Australia (Basnayake et al. 2012). In perspective of juice quality of such canes, it is of poor quality and with increase in time after harvest more poor juice quality has been reported (Misra et al. 2016b). Several studies had showed that extracted juice of flooded canes in July to September had poor sucrose content reducing sugars (Rege and Mascarenhas 1956; Madan et al. 1987) and phosphorus (Anonymous (Anakapalle, India), 1973), whereas high quantity of gums, glucose, total amount of nitrogen, nonprotein nitrogen ash (Anonymous (Anakapalle, India), 1973). It has been known that juice quality deterioration is very rapid in postwater logging period; quality of juice remained maintained till the time canes were under water-logged condition. The decline in juice quality is rapid when water level recedes with low sucrose content. Furthermore, under such condition sucrose inversion is also another factor, which contributes to enhancement of these losses (Solomon 2000, 2009). Gomathi and Chandran (2009) had revealed that 93 WL 1297, 98 WL 1357, 88 WL 2137, 92 WL 1029, NCO 310, and 57 NG 136 showed the least resistance to inversion of sucrose as compared to other clones in such condition. Madan et al. (1987) had also reported that invertase activity and enzymes of carbohydrate metabolism increase under such condition along with decrease in amylase activity which further contributes to poor juice quality and low sugar recovery. To cope up with such situation at times, sugarcane tries to adapt to such condition. The ways by which it does are increased levels of alcohol dehydrogenase (ADH) activity and ethanol production during anaerobic conditions, high level of fermentative metabolism in the roots for energy requirement, acetaldehyde accumulation (toxic metabolite), development of aerenchyma system in the cortex region facilitates the transportation of gases.

10.6 Management of Postharvest Sucrose Losses under Water Stress Condition

Efforts to manage postharvest quality losses of sugarcane at field and factory have been undertaken, but only partial success has been achieved (Figure 10.2). Furthermore, not much information has been known under water stress condition. Ray et al. (2009) reported that flooding at depth of 15 cm from soil surface and intermittent flooding at interval of 15 days (water level of 15–45 cm) showed no negative effect on cane yield. This may be due to the various management practices adopted for such condition. There are various parameters of juice quality which act as indicators such as dextran, gum, oligosaccharides, ethanol, mannitol, reducing sugars, titrable acidity, invertase content, juice viscosity, purity drop, etc. which help in knowing the cane quality being supplied to sugar mills. Under water stress condition, certain management practices should be adopted in order to minimize these losses. Following are the ways by which postharvest sucrose losses could be minimized in canes exposed to water stress conditions:

1. Prevention of losses from cut ends of sugarcane: Moisture loss after harvest takes place rapidly. To prevent these losses waxing of cut ends with paraffin for 24 hours was only effective (Barnes 1918).
2. Sugarcane should be placed in shade so as to reduce the further heat load of water-stressed canes. Hadi had showed that loss of available sugars in juice of S48 and Co 221 was 0.32% and 0.35%, respectively, when placed in shade, whereas this loss was higher when placed under sun (0.6%). Krishnamurthy Rao had reported the use of occasional sprinkling of water over harvested canes kept in shade so as to further reduce the heat load losses occurring after harvest. Besides, trash covering of harvested canes also helps in minimizing postharvest sucrose losses

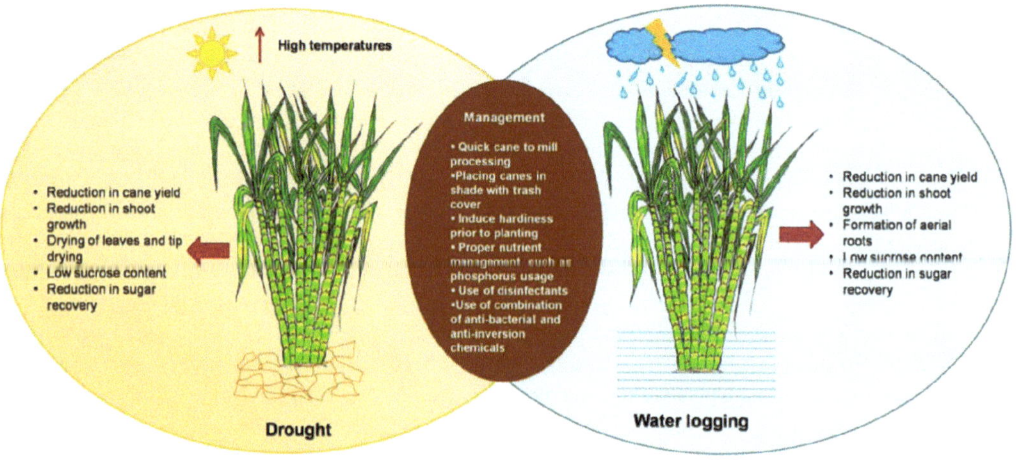

FIGURE 10.2 Effect and management of water stress in sugarcane. The arrows depict the changes and losses occurring in two different conditions of water stress (drought and water logging). Common minimization of postharvest sucrose losses in such sorts of canes has been depicted in brown colour.

in sugarcane. Dhanpal et al. (2019) had showed sugarcane trash to be an effective water management method.

3. Nutrient management should be done such as application of phosphorus so as to minimize the sucrose losses after harvest in water stress affected canes. Rege and Sannabhadti had showed that use of phosphorus as fertilizer helped in reducing deterioration in cane after harvest. Singh et al. (1992) had illustrated that drought-affected canes grown in calciorthent soils had relatively lesser sucrose deterioration than in canes grown in potassium fertilized soil. This is further supported by Solomon that sugarcane grown in soil containing high potassium and magnesium produced less dextrans after harvest. Anonymous (1987) had also showed that application of 2% of nitrogen under water-logging conditions in trench planting is also helpful in improving cane yields and in case of ratoon crop, 100 kg P_2O_5 along with 250 kg N per hectare is helpful in sustaining productivity and also minimizes the postharvest sucrose losses in sugarcane.
4. Inducing hardiness to sugarcane in drought-prone areas at the time of planting also helps in reducing the sucrose losses after harvest of cane. Soaking sets of sugarcane in saturated lime water (80 kg in 1000 L) for two hours a day before planting was found to be effective in improving cane yield and also minimizing the sucrose losses occurring after harvest in drought canes (Shrivastava and Srivastava 2012). Furthermore, trash mulching is also effective in such case during emergence in plant and initiation in ratoon crop, as it prevents soil moisture loss. This also helps in minimizing those losses in harvested canes that occur due to soil moisture loss.
5. Quick and efficient transport of harvested canes to mills for earlier crushing helps in minimizing postharvest losses in water stress–affected canes (Saleem et al. 2015; Solomon 2009).
6. Use of disinfectants over harvested canes is beneficial for minimizing the postharvest sucrose losses. Several disinfectants have been recommended for usage over harvested canes to reduce postharvest sucrose losses. These disinfectants also work in canes exposed to water stress conditions. Desai et al. (1985) had reported that spraying of benzoic acid (100 ppm) and formaldehyde (100 ppm) over harvested canes had significantly retarded postharvest losses. But use of formaldehyde has been banned as it is carcinogenic in nature (Hunt and Dale 2018). Furthermore, many bactericides such as Polycide, Bactrinol-100, BD Mill sanitizer, DBAC, IFOPOL, DNDT, ABF, Actin-ID, potassium permanganate and sodium metasilicate, Tsunami-100, Kcide 800, Sucroguard, perla soap solution (1%), allyl isothiocyanate (AITC), etc. have been recommended to check cane and juice deterioration. Coote (1984) had illustrated that application of bactericides on to the knives of mechanical harvesters or continuous loaders directly is advantageous in mills, while at field elimination of dextran up to 40% in juice is obtained by application of these biocides over harvested canes.
7. Use of combination of antibacterial and antiinversion chemicals over harvested canes should be done so as to minimize the sucrose losses in canes after harvest. Solomon et al. (2001) prepared some formulations containing antibacterial (quaternary ammonium compounds/thiocarbamate), anti-inversion chemicals along with some wetting agent. Application of these formulations over harvested canes helps in reducing loss of sucrose from harvested stored cane up to a period of one week, depending on the ambient temperature, variety, and storage conditions. Furthermore, another formulation containing 500–1,000 mg/L of sodium lauryl sulphate 500–10,000 mg/L benzalkonium chloride/thiocarbamate + 100–2,000 mg/L glutaraldehyde when sprayed on harvested canes were effective in minimizing postharvest sugarcane deterioration both at field (left for transportation) and mills (staling at the cane centres) due to synergistic action of antibacterial and antiinversion chemicals. Solomon et al. (2001) have also reported mist spraying of a formulation containing benzalkonium chloride 1–2 g/L + sodium metasilicate 5–10 g/L over harvested canes to be effective due to antideteriorating properties.
8. Application of neem extract (5%) may be used for water-stressed canes, as it was also effective in minimizing postharvest sucrose losses (Singh et al. 2015) under normal conditions. Further, the study showed that it is as effective as 1% sodium azide and picric acid application.
9. To avoid water-logging effect on canes, early planting should be preferred as well as proper drainage system (Shrivastava and Srivastava 2012).

10.7 Conclusion and Future Prospects

Water stress is an important problem in terms of changing climate. Sudden occurrence of drought and water-logging condition, due to less or extreme rainfall, affects the sugarcane crop irrespective of the growth stage at which it occurs. The impact of such abiotic stress depends on severity, duration, and timing of stress. Water stress causes a number of changes in sugarcane crop (physiological and morphological), which ultimately affects cane yield and sugar recovery. In case of drought, it causes membrane disorganization, photosynthesis delay, metabolic toxicity, ionic imbalance, and hyperosmotic stress resulting in low cane yield, sucrose content, and sugar recovery while in case of water logging, it causes reduction in shoot and root growth, dry matter production, and total crop yield. Postharvest sucrose losses in such sort of canes further accentuate this problem and cause more losses in sucrose content and sugar recovery. There are number of factors which contribute to it such as time lag between harvesting and milling, more infestation of microbes, etc. Management of these losses in sugarcane is far more difficult than the problem. Though there are management techniques available in reducing the postharvest sucrose losses in sugarcane affected by water stress condition, ignorance and less availability of literature and knowledge have made it more complicated.

There is a need to work on postharvest sucrose losses under different varieties grown under water stress conditions to know how much loss it causes to sugar recovery in mills. Climate change in future will further be worst due to various environmental disturbances, and so there is a need to develop certain tolerant varieties of sugarcane, which cause less loss in sucrose content after harvest. With the weather alterations, development and infestation of new microbes are also occurring in environment, which may be harmful for the harvested canes of such sorts. There is also need of knowing such harmful microbes attacking on such canes. Further, research needs to be done on its management by which it could be minimized to a large extent. Use of ecofriendly compounds and nonhazardous compounds such as pine oil, neem extract, etc. over harvested canes grown under water stress conditions should be studied so as to reduce these losses under such conditions.

ACKNOWLEDGMENT

Authors are thankful to Director, ICAR-IISR, Lucknow, for conducting various studies on postharvest sucrose losses in sugarcane for the past many years.

REFERENCES

Abbasi AR, Sarvestani R, Mohammadi B, Baghery A (2014) Drought stress-induced changes at physiological and biochemical levels in some common vetch (*Vicia sativa* L.) genotypes. *J Agric Sci Tech* 16:505–516.

Abdul S, Muhammad AA, Shabir H, Hesham AEE, Sajjad H, Niaz A, Abdul G, Sayyed RZ, Fahad S, Subhan D, Rahul D (2021a) Zinc nutrition and arbuscular mycorrhizal symbiosis effects on maize (*Zea mays* L.) growth and productivity. *J Saudi Soc Agric Sci* https://doi.org/10.1016/j.sjbs.2021.06.096

Abdul S, Muhammad AA, Subhan D, Niaz A, Fahad S, Rahul D, Mohammad JA, Omaima N, Muhammad HR, Bernard RG (2021b) Effect of arbuscular mycorrhizal fungi on the physiological functioning of maizeunder zinc-deficient soils. *Sci Rep* 11:18468.

Abraham E, Rigo G, Szekely G, Nagy R, Koncz C, Szabados L (2003) Light-dependent induction of proline biosynthesis by abscisic acid and salt stress is inhibited by brassinosteroid in Arabidopsis. *Plant Mol Biol* 51:363–372.

Adnan M, Fahad S, Muhammad Z, Shahen S, Ishaq AM, Subhan D, Zafar-ul-Hye M, Martin LB, Raja MMN, Beena S, Saud S, Imran A, Zhen Y, Martin B, Jiri H, Rahul D (2020) Coupling phosphate-solubilizing bacteria with phosphorus supplements improve maize phosphorus acquisition and growth under lime induced salinity stress. *Plants* 9(900). https://doi.org/10.3390/plants9070900

Adnan M, Fahad S, Khan IA, Saeed M, Ihsan MZ, Saud S, Riaz M, Wang D, Wu C (2019) Integration of poultry manure and phosphate solubilizing bacteria improved availability of Ca bound P in calcareous soils. *3 Biotech* 9(10):368.

Adnan M, Shah Z, Sharif M, Rahman H (2018b). Liming induces carbon dioxide (CO_2) emission in PSB inoculated alkaline soil supplemented with different phosphorus sources. *Environ Sci Poll Res* 25(10):9501–9509.

Adnan M, Zahir S, Fahad S, Arif M, Mukhtar A, Imtiaz AK, Ishaq AM, Abdul B, Hidayat U, Muhammad A, Inayat-Ur R, Saud S, Muhammad ZI, Yousaf J, Amanullah, Hafiz MH, Wajid N (2018a) Phosphate-solubilizing bacteria nullify the antagonistic effect of soil calcification on bioavailability of phosphorus in alkaline soils. *Sci Rep* 8:4339. https://doi.org/10.1038/s41598-018-22653-7

Ahmad S, Kamran M, Ding R, Meng X, Wang H, Ahmad I, Fahad S, Han Q (2019) Exogenous melatonin confers drought stress by promoting plant growth, photosynthetic capacity and antioxidant defense system of maize seedlings. *PeerJ* 7:e7793. http://doi.org/10.7717/peerj.7793

Akbar H, Timothy JK, Jagadish T, Golam M, Apurbo KC, Muhammad F, Rajan B, Fahad S, Hasanuzzaman M (2020) Agricultural land degradation: processes and problems undermining future food security. In: Fahad S, Hasanuzzaman M, Alam M, Ullah H, Saeed M, Khan AK, Adnan M (eds) *Environment, climate, plant and vegetation growth.* Springer Publ Ltd, Springer Nature Switzerland AG. Part of Springer Nature, pp. 17–62. https://doi.org/10.1007/978-3-030-49732-3

Akram R, Turan V, Hammad HM, Ahmad S, Hussain S, Hasnain A, Maqbool MM, Rehmani MIA, Rasool A, Masood N, Mahmood F, Mubeen M, Sultana SR, Fahad S, Amanet K, Saleem M, Abbas Y, Akhtar HM, Waseem F, Murtaza R, Amin A, Zahoor SA, ul Din MS, Nasim W (2018a) Fate of organic and inorganic pollutants in paddy soils. In: Hashmi MZ, Varma A (eds) *Environmental pollution of paddy soils, soil biology.* Springer International Publishing Ag, Cham, Switzerland, pp. 197–214.

Akram R, Turan V, Wahid A, Ijaz M, Shahid MA, Kaleem S, Hafeez A, Maqbool MM, Chaudhary HJ, Munis, MFH, Mubeen M, Sadiq N, Murtaza R, Kazmi DH, Ali S, Khan N, Sultana SR, Fahad S, Amin A, Nasim W (2018b) Paddy land pollutants and their role in climate change. In: Hashmi MZ, Varma A (eds) *Environmental pollution of paddy soils, soil biology.* Springer International Publishing Ag, Cham, Switzerland, pp. 113–124.

Alexander AG (1973) *Sugarcane physiology.* Elsevier Scientific Pub Comp, Amsterdam, pp. 573–609.

Amanullah, Muhammad I, Haider N, Shah K, Manzoor A, Asim M, Saif U, Izhar A, Fahad S, Adnan M et al. (2021) Integrated foliar nutrients application improve wheat (*Triticum aestivum* L.) productivity under calcareous soils in drylands. *Commun Soil Sci Plant Anal.* https://doi.org/10.1080/00103624.2021.1956521

Amanullah, Fahad S (eds) (2017) Rice - technology and production. IntechOpen Croatia 2017. http://dx.doi.org/10.5772/64480

Amanullah, Fahad S (eds) (2018a) Corn - Production and human Health in changing climate. IntechOpen United Kingdom 2018. http://dx.doi.org/10.5772/intechopen.74074

Amanullah, Fahad S (eds) (2018b) Nitrogen in agriculture – updates. IntechOpen Croatia 2018. http://dx.doi.org/10.5772/65846

Amanullah, Shah K, Imran, Hamdan AK, Muhammad A, Abdel RA, Muhammad A, Fahad S, Azizullah S, Brajendra P (2020) Effects of climate change on irrigation water quality. In: Fahad S, Hasanuzzaman M, Alam M, Ullah H, Saeed M, Khan AK, Adnan M (eds) *Environment, climate, plant and vegetation growth.* Springer Publ Ltd, Springer Nature Switzerland AG. Part of Springer Nature, pp. 123–132. https://doi.org/10.1007/978-3-030-49732-3

Amir M, Muhammad A, Allah B, Sevgi Ç, Haroon ZK, Muhammad A, Emre A (2020) Bio fortification under climate change: the fight between quality and quantity. In: Fahad S, Hasanuzzaman M, Alam M, Ullah H, Saeed M, Khan AK, Adnan M (eds) *Environment, climate, plant and vegetation growth.* Springer Publ Ltd, Springer Nature Switzerland AG. Part of Springer Nature, pp. 173–228. https://doi.org/10.1007/978-3-030-49732-3

Amjad I, Muhammad H, Farooq S, Anwar H (2020) Role of plant bioactives in sustainable agriculture. In: Fahad S, Hasanuzzaman M, Alam M, Ullah H, Saeed M, Khan AK, Adnan M (eds) *Environment, climate, plant and vegetation growth.* Springer Publ Ltd, Springer Nature Switzerland AG. Part of Springer Nature, pp. 591–606. https://doi.org/10.1007/978-3-030-49732-3

Amjad SF, Mansoora N, Din IU, Khalid IR, Jatoi GH, Murtaza G, Yaseen S, Naz M, Danish S, Fahad S et al. (2021) Application of zinc fertilizer and mycorrhizal inoculation on physio-biochemical parameters of wheat grown under water-stressed environment. *Sustainability* 13:11007. https://doi.org/10.3390/su131911007

Anand A (2004) Drought-like situation in 122 districts. *The Statesman* (New Delhi), July 17. https://www.livehindustan.com/tags/drought-like-situation-in-four-districts-uttarakhand. Accessed 26 July 2019.
Anonymous (1959) *Survey of Sugarcane Research and Development in India: Andhra Pradesh Indian Central Sugarcane Committee*, New Delhi, India, p. 126.
Anonymous (1980) Informe de Operaciones de la Industria Azucarera Peruana Anos (Peru).
Anonymous (1987) *Research achievements: 1912–1987*, SBI, Coimbatore, India, p. 121.
Anonymous (2000) *Vision 2000: Natural Resource Management Research*, Division of NRM, ICAR, Krishi Bhavan, New Delhi, p. 46.
Anonymous (2019) *Cooperative Sugar* 50(6):32–36.
Anonymous (Anakapalle, India) (1973) Sixty years of Agricultural Research at Anakapalle Sugarcane Research Station, Anakapalle, APAU, Hyderabad, p. 178.
Arif M, Talha J, Muhammad R, Fahad S, Muhammad A, Amanullah, Kawsar A, Ishaq AM, Bushra K, Fahd R (2020) Biochar; a remedy for climate change. In: Fahad S, Hasanuzzaman M, Alam M, Ullah H, Saeed M, Khan AK, Adnan M (eds) *Environment, climate, plant and vegetation growth*. Springer Publ Ltd, Springer Nature Switzerland AG. Part of Springer Nature, pp. 151–172. https://doi.org/10.1007/978-3-030-49732-3
Ashfaq AR, Uzma Y, Niaz A, Muhammad AA, Fahad S, Haider S, Tayebeh Z, Subhan D, Süleyman T, Hesham AELE, Pramila T, Jamal MA, Sulaiman AA, Rahul D (2021) Toxicity of Cadmium and nickel in the context of applied activated carbon biochar for improvement in soil fertility. *Saudi Soc Agric Sci*. https://doi.org/10.1016/j.sjbs.2021.09.035
Athar M, Masood IA, Sana S, Ahmed M, Xiukang W, Sajid F, Sher AK, Habib A, Faran M, Zafar H, Farhana G, Fahad S (2021) Bio-diesel production of sunflower through sulphur management in a semi-arid subtropical environment. *Environ Sci Pollution Res*. https://doi.org/10.1007/s11356-021-16688-z
Atif B, Hesham A, Fahad S (2021) Biochar coupling with phosphorus fertilization modifies antioxidant activity, osmolyte accumulation and reactive oxygen species synthesis in the leaves and xylem sap of rice cultivars under high-temperature stress. *Physiol Mol Biol Plants*. https://doi.org/10.1007/s12298-021-01062-7
Ayman ELS, Hossain A, Barutçular C, Iqbal MA, Islam MS, Fahad S, Sytar O, Çig F, Meena RS, Erman M (2020) Consequences of salinity stress on the quality of crops and its mitigation strategies for sustainable crop production: an outlook of arid and semi-arid regions. In: Fahad S, Hasanuzzaman M, Alam M, Ullah H, Saeed M, Khan AK, Adnan M (eds) *Environment, climate, plant and vegetation growth*. Springer Publ Ltd, Springer Nature Switzerland AG. Part of Springer Nature, pp. 503–534. https://doi.org/10.1007/978-3-030-49732-3
Aziz K, Daniel KYT, Fazal M, Muhammad ZA, Farooq S, Fan W, Fahad S, Ruiyang Z (2017a) Nitrogen nutrition in cotton and control strategies for greenhouse gas emissions: a review. *Environ Sci Pollut Res* 24:23471–23487. https://doi.org/10.1007/s11356-017-0131-y
Aziz K, Daniel KYT, Muhammad ZA, Honghai L, Shahbaz AT, Mir A, Fahad S (2017b) Nitrogen fertility and abiotic stresses management in cotton crop: a review. *Environ Sci Pollut Res* 24:14551–14566. https://doi.org/10.1007/s11356-017-8920-x
Baseer M, Adnan M, Fazal M, Fahad S, Muhammad S, Fazli W, Muhammad A, Jr. Amanullah, Depeng W, Saud S, Muhammad N, Muhammad Z, Fazli S, Beena S, Mian AR, Ishaq AM (2019) Substituting urea by organic wastes for improving maize yield in alkaline soil. *J Plant Nutr*. https://doi.org/10.1080/01904167.2019.1659344
Basnayake JPA, Inman-Bamber NG, Lakshmanan P (2012) Sugarcane for water-limited environments. Genetic variation in cane yield and sugar content in response to water stress. *J Exp Bot* 63(16):6023–6033.
Bayram AY, Seher Ö, Nazlican A (2020) Climate change forecasting and modeling for the year of 2050. In: Fahad S, Hasanuzzaman M, Alam M, Ullah H, Saeed M, Khan AK, Adnan M (eds) *Environment, climate, plant and vegetation growth*. Springer Publ Ltd, Springer Nature Switzerland AG. Part of Springer Nature, pp. 109–122. https://doi.org/10.1007/978-3-030-49732-3
Browne CA, Blouin RF (1907) The Chemistry of Sugarcane and its Products in Louisiana, La Agric Exp Stn. Bull No. 91.
Carter CE (1976) Excess water decrease cane and sugar yields. *Proc Amer Soc Sugar Cane Technol* 6:44–51.
Chang W, Qiujuan J, Evgenios A, Haitao L, Gezi L, Jingjing Z, Fahad S, Ying J (2021) Hormetic effects of zinc on growth and antioxidant defense system of wheat plants. *Sci Total Environ*. https://doi.org/10.1016/j.scitotenv.2021.150992

Chen Y, Guo Z, Dong L, Fu Z, Zheng Q, Zhang G, Qin L, Sun X, Shi Z, Fahad S, Xie F, Saud S (2021) Turf performance and physiological responses of native Poa species to summer stress in Northeast China. *PeerJ* 9:e12252 http://doi.org/10.7717/peerj.12252

Coote CJ (1984) *Proceedings of the International Dextran Workshop.* Sugar Processing Research Institute. New Orleans, LA, pp. 70–72.

Deepranjan S, Ardith S, Siva D, Sonam S, Manoj P, Amitava R, Sayyed RZ, Abdul G, Mohammad JA, Subhan D, Fahad S, Rahul D (2021) Optimizing nutrient use efficiency, productivity, energetics, and economics of red cabbage following mineral fertilization and biopriming with compatible rhizosphere microbes. *Sci Rep* 11:15680.

Depeng W, Fahad S, Saud S, Muhammad K, Aziz K, Mohammad NK, Hafiz MH, Wajid N (2018) Morphological acclimation to agronomic manipulation in leaf dispersion and orientation to promote "Ideotype" breeding: evidence from 3D visual modeling of "super" rice (*Oryza sativa* L.). *Plant Physiol Biochem.* 135:499–510 https://doi.org/10.1016/j.plaphy.2018.11.010

Deren CW, Cherry RH, Snyder GH (1992) Effect of flooding in selected sugarcane clones and soil borne insects. Proc. Amer. Soc. Sugar Cane Technol. 22 (June 17-19). (Abst. Sugar Y. Azucar 87(6): 2).

Desai BB, Sangle PB, Gaur SL (1985) Chemical control of post harvest losses in sugar cane. *Curt Res Rep* 1(1):33.

Dhanpal R, Tayade AS, Bhaskaran A, Geetha P (2019) Efficient water management in sugarcane with composted coir pith and sugarcane trash under tropical Indian conditions. *Sugar Tech* 21(2):256–264.

Dilley DR, Heggestad HE, Powers WL, Weiser CJ (1975) Environmental stress. In: Brown AWA, Byerly TC, Gibbs M, Pietro AS (eds) *Crop productivity – research imperatives*, Michigan State University, Agril. Exp. Stn. and C.F. Kettering Foundation, Yellow Springs, OH, pp. 309–355.

dos-Santos CM, de-Almeida Silva M (2015) Physiological and biochemical responses of sugarcane to oxidative stress induced by water deficit and paraquat. *Acta Physiol Plant* 37:172.

Emre B, Ömer SU, Martín LB, Andre D, Fahad S, Rahul D, Muhammad Z-ul-H, Ghulam SH, Subhan D (2021) Studying soil erosion by evaluating changes in physico-chemical properties of soils under different land-use types. *J Saudi Soc Agric Sci* 20:190–197.

Fahad S, Abdul B, Adnan M (eds) (2018b) Global wheat production. IntechOpen United Kingdom 2018. http://dx.doi.org/10.5772/intechopen.72559

Fahad S, Bajwa AA, Nazir U, Anjum SA, Farooq A, Zohaib A, Sadia S, Nasim W, Adkins S, Saud S, Ihsan MZ, Alharby H, Wu C, Wang D, Huang J (2017) Crop production under drought and heat stress: plant responses and management options. *Front Plant Sci* 8:1147. https://doi.org/10.3389/fpls.2017.01147

Fahad S, Bano A (2012) Effect of salicylic acid on physiological and biochemical characterization of maize grown in saline area. *Pak J Bot* 44:1433–1438.

Fahad S, Chen Y, Saud S, Wang K, Xiong D, Chen C, Wu C, Shah F, Nie L, Huang J (2013) Ultraviolet radiation effect on photosynthetic pigments, biochemical attributes, antioxidant enzyme activity and hormonal contents of wheat. *J Food Agri Environ* 11(3&4):1635–1641.

Fahad S, Hussain S, Bano A, Saud S, Hassan S, Shan D, Khan FA, Khan F, Chen Y, Wu C, Tabassum MA, Chun MX, Afzal M, Jan A, Jan MT, Huang J (2014a) Potential role of phytohormones and plant growth-promoting rhizobacteria in abiotic stresses: consequences for changing environment. *Environ Sci Pollut Res* 22(7):4907–4921. https://doi.org/10.1007/s11356-014-3754-2

Fahad S, Hussain S, Matloob A, Khan FA, Khaliq A, Saud S, Hassan S, Shan D, Khan F, Ullah N, Faiq M, Khan MR, Tareen AK, Khan A, Ullah A, Ullah N, Huang J (2014b) Phytohormones and plant responses to salinity stress: a review. *Plant Growth Regul* 75(2):391–404. https://doi.org/10.1007/s10725-014-0013-y

Fahad S, Hussain S, Saud S, Hassan S, Chauhan BS, Khan F et al (2016a) Responses of rapid viscoanalyzer profile and other rice grain qualities to exogenously applied plant growth regulators under high day and high night temperatures. *PLoS One* 11(7):e0159590. https://doi.org/10.1371/journal.pone.0159590

Fahad S, Hussain S, Saud S, Hassan S, Ihsan Z, Shah AN, Wu C, Yousaf M, Nasim W, Alharby H, Alghabari F, Huang J (2016c) Exogenously applied plant growth regulators enhance the morphophysiological growth and yield of rice under high temperature. *Front Plant Sci* 7:1250. https://doi.org/10.3389/fpls.2016.01250

Fahad S, Hussain S, Saud S, Hassan S, Tanveer M, Ihsan MZ, Shah AN, Ullah A, Nasrullah KF, Ullah S, Alharby HNW, Wu C, Huang J (2016d) A combined application of biochar and phosphorus alleviates heat-induced adversities on physiological, agronomical and quality attributes of rice. *Plant Physiol Biochem* 103:191–198.

Fahad S, Hussain S, Saud S, Khan F, Hassan S, Jr, Nasim W, Arif M, Wang F, Huang J (2016b) Exogenously applied plant growth regulators affect heat-stressed rice pollens. *J Agron Crop Sci* 202:139 150.

Fahad S, Hussain S, Saud S, Tanveer M, Bajwa AA, Hassan S, Shah AN, Ullah A, Wu C, Khan FA, Shah F, Ullah S, Chen Y, Huang J (2015a) A biochar application protects rice pollen from high-temperature stress. *Plant Physiol Biochem* 96:281–287.

Fahad S, Muhammad ZI, Abdul K, Ihsanullah D, Saud S, Saleh A, Wajid N, Muhammad A, Imtiaz AK, Chao W, Depeng W, Jianliang H (2018a) Consequences of high temperature under changing climate optima for rice pollen characteristics-concepts and perspectives. *Archives Agron Soil Sci.* https://doi.org/10.1080/03650340.2018.1443213

Fahad S, Nie L, Chen Y, Wu C, Xiong D, Saud S, Hongyan L, Cui K, Huang J (2015b) Crop plant hormones and environmental stress. *Sustain Agric Rev* 15:371–400.

Fahad S, Saud S, Yajun C, Chao W, Depeng W (eds) (2021f) Abiotic stress in plants. IntechOpen United Kingdom 2021. http://dx.doi.org/10.5772/intechopen.91549

Fahad S, Hasanuzzaman M, Alam M, Ullah H, Saeed M, Ali Khan I, Adnan M. (Eds.) (2020) *Environment, climate, plant and vegetation growth.* Springer Nature Switzerland AG. https://doi.org/10.1007/978-3-030-49732-3

Fahad S, Adnan M, Hassan S, Saud S, Hussain S, Wu C, Wang D, Hakeem KR, Alharby HF, Turan V, Khan MA, Huang J (2019b) Rice responses and tolerance to high temperature. In: Hasanuzzaman M, Fujita M, Nahar K, Biswas JK (eds) *Advances in rice research for abiotic stress tolerance*. Woodhead Publ Ltd, Abington, Cambridge, UK, pp. 201–224.

Fahad S, Rehman A, Shahzad B, Tanveer M, Saud S, Kamran M, Ihtisham M, Khan SU, Turan V, Rahman MHU (2019a) Rice responses and tolerance to metal/metalloid toxicity. In: Hasanuzzaman M, Fujita M, Nahar K, Biswas JK (eds) *Advances in rice research for abiotic stress tolerance*. Woodhead Publ Ltd, Abington, Cambridge, UK, pp. 299–312.

Fahad S, Sönmez O, Saud S, Wang D, Wu C, Adnan M, Arif M, Amanullah (eds) (2021e) Engineering tolerance in crop plants against abiotic stress. In: *Footprints of climate variability on plant diversity*, 1st edn. CRC Press, Boca Raton.

Fahad S, Sonmez O, Saud S, Wang D, Wu C, Adnan M, Turan V (eds) (2021b) Climate change and plants: biodiversity, growth and interactions. In: *Footprints of climate variability on plant diversity*, 1st edn. CRC Press, Boca Raton.

Fahad S, Sonmez O, Saud S, Wang D, Wu C, Adnan M, Turan V (eds) (2021c) Developing climate resilient crops: improving global food security and safety. In: *Footprints of climate variability on plant diversity*, 1st edn. CRC Press, Boca Raton.

Fahad S, Sönmez O, Saud S, Wang D, Wu C, Adnan M, Turan V (eds) (2021a) Plant growth regulators for climate-smart agriculture. In: *Footprints of climate variability on plant diversity*, 1st edn. CRC Press, Boca Raton.

Fahad S, Sönmez O, Turan V, Adnan M, Saud S, Wu C, Wang D (eds) (2021d) Sustainable soil and land management and climate change. In: *Footprints of climate variability on plant diversity*, 1st edn. CRC Press, Boca Raton.

Farah R, Muhammad R, Muhammad SA, Tahira Y, Muhammad AA, Maryam A, Shafaqat A, Rashid M, Muhammad R, Qaiser H, Afia Z, Muhammad AA, Muhammad A, Fahad S (2020) Alternative and non-conventional soil and crop management strategies for increasing water use efficiency. In: Fahad S, Hasanuzzaman M, Alam M, Ullah H, Saeed M, Khan AK, Adnan M (eds) *Environment, climate, plant and vegetation growth.* Springer Publ Ltd, Springer Nature Switzerland AG. Part of Springer Nature, pp. 323–338. https://doi.org/10.1007/978-3-030-49732-3

Farhana G, Ishfaq A, Muhammad A, Dawood J, Fahad S, Xiuling L, Depeng W, Muhammad F, Muhammad F, Syed AS (2020) Use of crop growth model to simulate the impact of climate change on yield of various wheat cultivars under different agro-environmental conditions in Khyber Pakhtunkhwa, Pakistan. *Arabian J Geosci* 13:112. https://doi.org/10.1007/s12517-020-5118-1

Farhat A, Hafiz MH, Wajid I, Aitazaz AF, Hafiz FB, Zahida Z, Fahad S, Wajid F, Artemi C (2020) A review of soil carbon dynamics resulting from agricultural practices. *J Environ Manage* 268:110319.

Farooq M, Wahid A, Kobayashi N, Fujita D, Basra SMA (2009) Plant drought stress: effects, mechanisms and management. *Agron Sustain Dev* 29(1):185–212.

Fazli W, Muhmmad S, Amjad A, Fahad S, Muhammad A, Muhammad N, Ishaq AM, Imtiaz AK, Mukhtar A, Muhammad S, Muhammad I, Rafi U, Haroon I, Muhammad A (2020) Plant-microbes interactions and functions in changing climate. In: Fahad S, Hasanuzzaman M, Alam M, Ullah H, Saeed M, Khan AK,

Adnan M (eds) *Environment, climate, plant and vegetation growth*. Springer Publ Ltd, Springer Nature Switzerland AG. Part of Springer Nature, pp. 397–420. https://doi.org/10.1007/978-3-030-49732-3

Feldman LJ (1984) Regulation of root development. *Ann Rev Plant Physiol* 35:323–342.

Gascho GJ, Shih SF (1983) Sugarcane. In: Teare ID, Peet MM (eds) *Crop water relations*. John Wiley & Sons, New York, pp. 445–479.

Ghulam M, Muhammad AA, Donald LS, Sajid M, Muhammad FQ, Niaz A, Ateeq Ur R, Shakeel A, Sajjad H, Muhammad A, Summia M, Aqib HAK, Fahad S, Rahul D, Mazhar I, Timothy DS (2021) Formalin fumigation and steaming of various composts differentially influence the nutrient release, growth and yield of muskmelon (*Cucumis melo* L.). *Sci Rep* 11:21057.

Gomathi R, Chandran K (2009) Effect of waterlogging on growth and yield of sugarcane clones. Sugarcane Breeding Institute (SBI-ICAR). *Quart NewsLett* 29(4):1–2

Gomathi R, Chandran K, Gururaja Rao PN, Rakkiyappan P (2010) Effect of waterlogging in sugarcane and its management. Published by The Director, Sugarcane Breeding Institute (SBI-ICAR), Coimbatore. Extension Pub. No. 185.

Gopakumar L, Bernard NO, Donato V (2020) Soil microarthropods and nutrient cycling. In: Fahad S, Hasanuzzaman M, Alam M, Ullah H, Saeed M, Khan AK, Adnan M (eds) *Environment, climate, plant and vegetation growth*. Springer Publ Ltd, Springer Nature Switzerland AG. Part of Springer Nature, pp. 453–472. https://doi.org/10.1007/978-3-030-49732-3

Guilbeau WF, Coll FE, Matin LF (1955) Effects of delay in grinding on value and processing quality of sugarcane juice. *Sugar J* 18:30.

Guilbeau WF, Coll FE, Matin LF (1956) Losses from delay in delivering harvested cane. *Sugar Bull* 34:28.

Guofu L, Zhenjian B, Fahad S, Guowen C, Zhixin X, Hao G, Dandan L, Yulong L, Bing L, Guoxu J, Saud S (2021) Compositional and structural changes in soil microbial communities in response to straw mulching and plant revegetation in an abandoned artificial pasture in Northeast China. *Glob Ecol Conserv* 31:e01871.

Habib Ur R, Ashfaq A, Aftab W, Manzoor H, Fahd R, Wajid I, Md. Aminul I, Vakhtang S, Muhammad A, Asmat U, Abdul W, Syeda RS, Shah S, Shahbaz K, Fahad S, Manzoor H, Saddam H, Wajid N (2017) Application of CSM-CROPGRO-cotton model for cultivars and optimum planting dates: evaluation in changing semi-arid climate. *Field Crops Res.* http://dx.doi.org/10.1016/j.fcr.2017.07.007

Hafeez M, Farman U, Muhammad MK, Xiaowei L, Zhijun Z, Sakhawat S, Muhammad I, Mohammed AA, Mandela F-G, Nicolas D, Muzammal R, Fahad S, Yaobin L (2021) Metabolic-based insecticide resistance mechanism and ecofriendly approaches for controlling of beet armyworm *Spodoptera exigua*: a review. *Environ Sci Pollution Res.* https://doi.org/10.1007/s11356-021-16974-w

Hafiz MH, Abdul K, Farhat A, Wajid F, Fahad S, Muhammad A, Ghulam MS, Wajid N, Muhammad M, Hafiz FB (2020b) Comparative effects of organic and inorganic fertilizers on soil organic carbon and wheat productivity under arid region. *Commun Soil Sci Plant Anal.* https://doi.org/10.1080/00103624.2020.1763385

Hafiz MH, Farhat A, Ashfaq A, Hafiz FB, Wajid F, Carol Jo W, Fahad S, Gerrit H (2020a) Predicting kernel growth of maize under controlled water and nitrogen applications. *Int J Plant Prod.* https://doi.org/10.1007/s42106-020-00110-8

Hafiz MH, Farhat A, Shafqat S, Fahad S, Artemi C, Wajid F, Chaves CB, Wajid N, Muhammad M, Hafiz FB (2018) Offsetting land degradation through nitrogen and water management during maize cultivation under arid conditions. *Land Degrad Dev* 1–10. https://doi.org/10.1002/ldr.2933

Hafiz MH, Muhammad A, Farhat A, Hafiz FB, Saeed AQ, Muhammad M, Fahad S, Muhammad A (2019) Environmental factors affecting the frequency of road traffic accidents: a case study of sub-urban area of Pakistan. *Environ Sci Pollut Res.* https://doi.org/10.1007/s11356-019-04752-8

Hafiz MH, Wajid F, Farhat A, Fahad S, Shafqat S, Wajid N, Hafiz FB (2016) Maize plant nitrogen uptake dynamics at limited irrigation water and nitrogen. *Environ Sci Pollut Res* 24(3):2549–2557. https://doi.org/10.1007/s11356-016-8031-0

Haider SA, Lalarukh I, Amjad SF, Mansoora N, Naz M, Naeem M, Bukhari SA, Shahbaz M, Ali SA, Marfo TD, Subhan D, Rahul D, Fahad S (2021) Drought stress alleviation by potassium-nitrate-containing chitosan/montmorillonite microparticles confers changes in *Spinacia oleracea* L. *Sustainability* 13:9903. https://doi.org/10.3390/su13179903

Hamza SM, Xiukang W, Sajjad A, Sadia Z, Muhammad N, Adnan M, Fahad S et al. (2021) Interactive effects of gibberellic acid and NPK on morpho-physio-biochemical traits and organic acid exudation pattern in

coriander (*Coriandrum sativum* L.) grown in soil artificially spiked with boron. *Plant Physiol Biochem* 167(2021):884–900.
Hemaprabha G, Swapna S, Lavanya DL, Sajitha B, Venkataramana S (2013) Evaluation of drought tolerance potential of elite genotypes and progenies of sugarcane (Saccharum sp. hybrids). *Sugar Tech* 15(1):9–16.
Hesham FA, Fahad S (2020) Melatonin application enhances biochar efficiency for drought tolerance in maize varieties: modifications in physio-biochemical machinery. *Agron J* 112(4):1–22.
Huang L-Y, Li X-X, Zhang Y-B, Fahad S, Wang F (2021) dep1 improves rice grain yield and nitrogen use efficiency simultaneously by enhancing nitrogen and dry matter translocation. *J Integrative Agri.* https://doi.org/10.1016/S2095-3119(21)63795-4
Hunt A, Dale N (2018). Economic valuation in formaldehyde regulation. Organization for economic cooperation and development pp 39. http://www.oecd.org/officialdocuments/publicdisplaydocumentpdf/?cote=ENV/WKP(2018)8&docLanguage=En
Hurney AP (1983) Effect of flooding on cane yield. BSES Bull No.3:11.
Hussain MA, Fahad S, Rahat S, Muhammad FJ, Muhammad M, Qasid A, Ali A, Husain A, Nooral A, Babatope SA, Changbao S, Liya G, Ibrar A, Zhanmei J, Juncai H (2020) Multifunctional role of brassinosteroid and its analogues in plants. *Plant Growth Regul* https://doi.org/10.1007/s10725-020-00647-8
Ibrar K, Aneela R, Khola Z, Urooba N, Sana B, Rabia S, Ishtiaq H, Mujaddad R, Salvatore M (2020) Microbes and environment: global warming reverting the frozen zombies. In: Fahad S, Hasanuzzaman M, Alam M, Ullah H, Saeed M, Khan AK, Adnan M (eds) *Environment, climate, plant and vegetation growth.* Springer Publ Ltd, Springer Nature Switzerland AG. Part of Springer Nature, pp. 607–634. https://doi.org/10.1007/978-3-030-49732-3
Ikram U, Khadim D, Muhammad T, Muhammad S, Fahad S (2021) Gibberellic acid and urease inhibitor optimize nitrogen uptake and yield of maize at varying nitrogen levels under changing climate. *Environ Sci Pollution Res.* https://doi.org/10.1007/s11356-021-16049-w
Ilyas M, Mohammad N, Nadeem K, Ali H, Aamir HK, Kashif H, Fahad S, Aziz K, Abid U (2020) Drought tolerance strategies in plants: a mechanistic approach. *J Plant Growth Regul.* https://doi.org/10.1007/s00344-020-10174-5
Inman-Bamber NG, Smith DM (2005) Water relations in sugarcane and response to water deficits. *Field Crops Res* 92:185–202.
Iqra M, Amna B, Shakeel I, Fatima K, Sehrish L, Hamza A, Fahad S (2020) Carbon cycle in response to global warming. In: Fahad S, Hasanuzzaman M, Alam M, Ullah H, Saeed M, Khan AK, Adnan M (eds) *Environment, climate, plant and vegetation growth.* Springer Publ Ltd, Springer Nature Switzerland AG. Part of Springer Nature, pp. 1–16. https://doi.org/10.1007/978-3-030-49732-3
Irfan M, Muhammad M, Muhammad JK, Khadim MD, Dost M, Ishaq AM, Waqas A, Fahad S, Saud S et al. (2021) Heavy metals immobilization and improvement in maize (*Zea mays* L.) growth amended with biochar and compost. *Sci Rep* 11:18416.
Islam MS, Miah MAS, Begum MK, Alam MR, Arefin, MS (2011) Growth, yield and juice quality of some selected sugarcane clones under water-logging stress condition. *World J Agric Sci* 7(4):504–509.
Jabborova D, Sulaymanov K, Sayyed RZ, Alotaibi SH, Enakiev Y, Azimov A, Jabbarov Z, Ansari MJ, Fahad S, Danish S et al. (2021) Mineral fertilizers improves the quality of turmeric and soil. *Sustainability* 13:9437. https://doi.org/10.3390/su13169437
Jan M, Muhammad AH, Adnan NS, Muhammad Y, Javaid I, Xiuling L, Depeng W, Fahad S (2019) Modulation in growth, gas exchange, and antioxidant activities of salt-stressed rice (*Oryza sativa* L.) genotypes by zinc fertilization. *Arabian J Geosci* 12:775. https://doi.org/10.1007/s12517-019-4939-2
Kamarn M, Wenwen C, Irshad A, Xiangping M, Xudong Z, Wennan S, Junzhi C, Shakeel A, Fahad S, Qingfang H, Tiening L (2017) Effect of paclobutrazol, a potential growth regulator on stalk mechanical strength, lignin accumulation and its relation with lodging resistance of maize. *Plant Growth Regul* 84:317–332. https://doi.org/10.1007/s10725-017-0342-8
Karuppanapandian T, Moon JH, Kim C, Manoharan K, Kim W (2011) Reactive oxygen species in plants: their generation, signal transduction, and scavenging mechanisms. *Aust J Crop Sci* 5(6):709–725.
Kelkar-Khambete A (2014) Drought in Maharashtra: Lack of management or vagaries of climate change? http://www.indiawaterportal.org. Accessed 26 September 2015.
Khadim D, Fahad S, Jahangir MMR, Iqbal M, Syed SA, Shah AK, Ishaq AM, Rahul D et al. (2021a) Biochar and urease inhibitor mitigate NH_3 and N_2O emissions and improve wheat yield in a urea fertilized alkaline soil. *Sci Rep* 11:17413.

Khadim D, Saif R, Fahad S, Syed SA, Shah AK et al. (2021b) Influence of variable biochar concentration on yield-scaled nitrous oxide emissions, wheat yield and nitrogen use efficiency. *Sci Rep* 11:16774.

Khan MMH, Niaz A, Umber G, Muqarrab A, Muhammad AA, Muhammad I, Shabir H, Shah F, Vibhor A, Shams HA-H, Reham A, Syed MBA, Nadiyah MA, Ali TKZ, Subhan D, Rahul D (2021) Synchronization of Boron application methods and rates is environmentally friendly approach to improve quality attributes of *Mangifera indica* L. On sustainable basis. *Saudi J Bio Sci.* https://doi.org/10.1016/j.sjbs.2021.10.036

Khurana I (2019) *Dealing with droughts Water governance and management in India.* Springer Publications, pp. 1–45.

Kumar D, Malik N, Sengar RS (2019) Physio-biochemical insights into sugarcane genotypes under water stress. *Biol Rhythm Res.* https://doi.org/10.1080/09291016.2019.1587838

Levitt J (1980) *Responses of plants to environmental stresses Vol. II. Water, salt and other stresses*, Academic Press, New York, p. 606.

Leyshon AJ, Sheared RW (1974) Influence of short term flooding on the growth and plant nutrient composition of barley. *Can J Soil Sci* 54:463–473.

Liu C, Liu Y, Guo K, Fan D, Li G, Zheng Y, Yu L, Yang R (2011) Effect of drought on pigments, osmotic adjustment and antioxidant enzymes in six woody plant species in karst habitats of Southwestern China. *Environ Exp Bot* 71:174–183.

Madan VK, Singh K, Pande HP, Soni N, Saxena YR (1987) Foliar enzymes of sugarcane II. Biochemical studies of waterlogged plants. *Recent Adv Sci Technol* 1(1):79–85.

Mahar A, Amjad A, Altaf HL, Fazli W, Ronghua L, Muhammad A, Fahad S, Muhammad A, Rafiullah, Imtiaz AK, Zengqiang Z (2020) Promising technologies for Cd-contaminated soils: drawbacks and possibilities. In: Fahad S, Hasanuzzaman M, Alam M, Ullah H, Saeed M, Khan AK, Adnan M (eds) *Environment, climate, plant and vegetation growth.* Springer Publ Ltd, Springer Nature Switzerland AG. Part of Springer Nature, pp. 63–92. https://doi.org/10.1007/978-3-030-49732-3

Mahmood Ul H, Tassaduq R, Chandni I, Adnan A, Muhammad A, Muhammad MA, Muhammad H-R, Mehmood AN, Alam S, Fahad S (2021) Linking plants functioning to adaptive responses under heat stress conditions: a mechanistic review. *J Plant Growth Regul.* https://doi.org/10.1007/s00344-021-10493-1

Mall AK, Misra V, Kumar A, Pathak AD (2019) Drought: status, Losses and management in sugarcane crop. In: Bharti PK, Ray J (eds) *Farm and development: technological perspectives*, Discovery Publishing House Pvt. Ltd., New Delhi, pp. 165–187.

Manzer HS, Saud A, Soumya M, Abdullah AA, Qasi DA, Bander MAM, Hayssam MA, Hazem MK, Fahad S, Vishnu DR, Om PN (2021) Molybdenum and hydrogen sulfide synergistically mitigate arsenic toxicity by modulating defense system, nitrogen and cysteine assimilation in faba bean (*Vicia faba* L.) seedlings. *Environ Pollut* 290:117953.

Marschner H (1986) *Mineral nutrition of higher plants*, Academic Press, New York, p. 674.

Md Jakir H, Allah B (2020) Development and applications of transplastomic plants; a way towards eco-friendly agriculture. In: Fahad S, Hasanuzzaman M, Alam M, Ullah H, Saeed M, Khan AK, Adnan M (eds) *Environment, climate, plant and vegetation growth.* Springer Publ Ltd, Springer Nature Switzerland AG. Part of Springer Nature, pp. 285–322. https://doi.org/10.1007/978-3-030-49732-3

Md. Enamul H, Shoeb AZM, Mallik AH, Fahad S, Kamruzzaman MM, Akib J, Nayyer S, Mehedi KM, Swati AS, Md Yeamin A, Most SS (2020) Measuring vulnerability to environmental hazards: qualitative to quantitative. In: Fahad S, Hasanuzzaman M, Alam M, Ullah H, Saeed M, Khan AK, Adnan M (eds) *Environment, climate, plant and vegetation growth.* Springer Publ Ltd, Springer Nature Switzerland AG. Part of Springer Nature, pp. 421–452. https://doi.org/10.1007/978-3-030-49732-3

Misra V (2018) Post-harvest quality of sugarcane affected by prolonged water stress condition. Amity University, Thesis, p. 229.

Misra V, Solomon S, Shrivastava AK, Shukla SP, Ansari MI (2016c) Post harvest sugarcane deterioration: Leuconostoc and its effect. *J Funct Environ Bot* 6(1):1–7.

Misra V, Solomon S, Singh P, Prajapati CP, Ansari MI (2016a) Impact of drought on post-harvest quality of sugarcane crop. *Adv Life Sci* 20(5):9496–9505.

Misra V, Solomon S, Singh P, Prajapati CP, Ansari MI (2016b) Effect of water logging on post-harvest sugarcane deterioration. *Agrica* 5(2):119–132.

Mohammad IA-W, Abdelazeem S, Munir A, Khalid E, Adel RAU (2020a) Extent of climate change in Saudi Arabia and its impacts on agriculture: a case study from Qassim Region. In: Fahad S, Hasanuzzaman

M, Alam M, Ullah H, Saeed M, Khan AK, Adnan M (eds) *Environment, climate, plant and vegetation growth*. Springer Publ Ltd, Springer Nature Switzerland AG. Part of Springer Nature, pp. 635–658. https://doi.org/10.1007/978-3-030-49732-3

Mohammad IA-W, Ahmad M, Usman ARA, Akanji M, Rafique MI (2020b) Advances in pyrolytic technologies with improved carbon capture and storage to combat climate change. In: Fahad S, Hasanuzzaman M, Alam M, Ullah H, Saeed M, Khan AK, Adnan M (eds) *Environment, climate, plant and vegetation growth*. Springer Publ Ltd, Springer Nature Switzerland AG. Part of Springer Nature, pp. 535–576. https://doi.org/10.1007/978-3-030-49732-3

Monoharan ML, Duraisamy K, Krishnamurthy SV, Vijayaraghan H, Muthkrishnan K (1990) Performance of sugarcane varieties in waterlogged condition. *Maharashtra Sugar* 15(11):39–45.

Mubeen M, Ashfaq A, Hafiz MH, Muhammad A, Hafiz UF, Mazhar S, Muhammad SD, Asad A, Amjed A, Fahad S, Wajid N (2020) Evaluating the climate change impact on water use efficiency of cotton-wheat in semi-arid conditions using DSSAT model. *J Water Clim Change*. https://doi.org/10.2166/wcc.2019.179/622035/jwc2019179.pdf

Muhammad N, Muqarrab A, Khurram S, Fiaz A, Fahim N, Muhammad A, Shazia A, Omaima N, Sulaiman AA, Fahad S, Subhan D, Rahul D (2021) Kaolin and Jasmonic acid improved cotton productivity under water stress conditions. *J Saudi Society Agricultural Sci* 28:6606–6614. https://doi.org/10.1016/j.sjbs.2021.07.043

Muhammad TQ, Amna F, Amna B, Barira Z, Xitong Z, Ling-Ling C (2020) Effectiveness of conventional crop improvement strategies vs. omics. In: Fahad S, Hasanuzzaman M, Alam M, Ullah H, Saeed M, Khan AK, Adnan M (eds) *Environment, climate, plant and vegetation growth*. Springer Publ Ltd, Springer Nature Switzerland AG. Part of Springer Nature, pp. 253–284. https://doi.org/10.1007/978-3-030-49732-3

Muhammad Z, Abdul MK, Abdul MS, Kenneth BM, Muhammad S, Shahen S, Ibadullah J, Fahad S (2019) Performance of *Aeluropus lagopoides* (mangrove grass) ecotypes, a potential turfgrass, under high saline conditions. *Environ Sci Pollut Res*. https://doi.org/10.1007/s11356-019-04838-3

Munawarti A, Taryono Semiarti E, Sismindari S (2014) Morphological and biochemical responses of glagah (*Saccharum spontaneum* L.) accessions to drought stress. *J Trop Life Sci* 4(1):61–66.

Muzammal R, Fahad S, Guanghui D, Xia C, Yang Y, Kailei T, Lijun L, Fei-Hu L, Gang D (2021) Evaluation of hemp (*Cannabis sativa* L.) as an industrial crop: a review. *Environ Sci Pollution Res*. https://doi.org/10.1007/s11356-021-16264-5

Naidu KM, Bhagyalakshmi KV (1973) Relative turgidity and stomatal movement of sugarcane varieties in relation to their drought resistance. *Indian J Agric Sci* 43(11):1016–1018.

Noor M, Naveed Ur R, Ajmal J, Fahad S, Muhammad A, Fazli W, Saud S, Hassan S (2020) Climate change and costal plant lives. In: Fahad S, Hasanuzzaman M, Alam M, Ullah H, Saeed M, Khan AK, Adnan M (eds) *Environment, climate, plant and vegetation growth*. Springer Publ Ltd, Springer Nature Switzerland AG. Part of Springer Nature, pp. 93–108. https://doi.org/10.1007/978-3-030-49732-3

Parashar DR (1960) Behaviour of drought affected cane during the process of sugar manufacture. *Ind Sugar* 10(1):55–58.

Plaut Z (2003) Plant exposure to water stress during specific growth stages, Encyclopedia of Water Science, Taylor & Francis Group, pp. 673–675.

Ploschuk RA, Miralles DJ, Colmer TD, Ploschuk EL, Striker GG (2018) Waterlogging of winter crops at early and late stages: impacts on leaf physiology, growth and yield. *Front Plant Sci* 20:1–15.

Qadeem W, Sattar S, Adnan M, Zaman M, Ali I, Shah SRA, Junaid K, Said F, Rahman IU, Saleem N (2015) Efficacy of botanical and microbial extracts against Angoumois grain moth Sitotrogacerealella (Olivier) (Lepidoptera: Gelechiidae) under laboratory conditions. *J Ento Zool Stud* 3(5):451–454.

Qamar-Uz Z, Zubair A, Muhammad Y, Muhammad ZI, Abdul K, Fahad S, Safder B, Ramzani PMA, Muhammad N (2017) Zinc biofortification in rice: leveraging agriculture to moderate hidden hunger in developing countries. *Arch Agron Soil Sci* 64:147–161. https://doi.org/10.1080/03650340.2017.1338343

Rachmilevitch S, Costa MD, Huang B (2006) Physiological and biochemical indicator for stress tolerance. In: Huang B (ed) *Plant–environment interactions*. 3rd edn, New York, CRC Press, pp. 322–338.

Rahman UI, Ali S, Rahman LU, Adnan M, Ibrahim M, Saleem N, Irshad M (2016) Effect of seed priming on growth parameters of Okra (*Abelmoschus esculentus* L.). *Pure Appl Biol* 5(1):165–171.

Raid RN (1992) Influence of short term flooding on initial stand establishment of sugarcane. Proc. American Soc. Sugar Cane Technol. 22: (June 17-19), (Abst. Sugar Y. Azucar 87(6):26.

Rao CM, Kumar UV, Naik RB, Sekha D, Katti GS (2007) Relative performance of sugarcane clones/varieties under soil moisture stress/drought during formative stage. *Ind Sugar* 56(11):23–28.

Rasheed MT, Inayatullah M, Shah B, Ahmed N, Khan A, Ali M, Ahmed S, Junaid K, Adnan M, Huma Z (2015) Relative abundance of insect pollinators on two cultivars of sunflower in Islamabad. *J Ento Zool Studies* 3(6):164–165.

Rashid M, Qaiser H, Khalid SK, Mohammad I-W, Zhang A, Muhammad A, Shahzada SI, Rukhsanda A, Ghulam AS, Shahzada MM, Sarosh A, Muhammad FQ (2020) Prospects of biochar in alkaline soils to mitigate climate change. In: Fahad S, Hasanuzzaman M, Alam M, Ullah H, Saeed M, Khan AK, Adnan M (eds) *Environment, climate, plant and vegetation growth.* Springer Publ Ltd, Springer Nature Switzerland AG. Part of Springer Nature, pp. 133–150. https://doi.org/10.1007/978-3-030-49732-3

Roy JD, Thomas RS, Berry G (2009) Sugarcane response to high water intermittent flooding. *J Crop Improv* 24(1):12–27.

Rege RD, Mascarenhas JP (1956) Studies on the influence of floods on cane growth and quality in the Pamba River valley. *Proc Inter Soc Sugar Cane Technol* 9:375–389.

Rehman M, Fahad S, Saleem MH, Hafeez M, Rahman MH, Liu F, Deng G (2020) Red light optimized physiological traits and enhanced the growth of ramie (*Boehmeria nivea* L.). *Photosynthetica* 58(4):922–931.

Sadam M, Muhammad TQ, Ghulam M, Muhammad SK, Faiz AJ (2020) Role of biotechnology in climate resilient agriculture. In: Fahad S, Hasanuzzaman M, Alam M, Ullah H, Saeed M, Khan AK, Adnan M (eds) *Environment, climate, plant and vegetation growth.* Springer Publ Ltd, Springer Nature Switzerland AG. Part of Springer Nature, pp. 339–366. https://doi.org/10.1007/978-3-030-49732-3

Safi UK, Ullah F, Mehmood S, Fahad S, Ahmad Rahi A, Althobaiti F, et al. (2021) Antimicrobial, antioxidant and cytotoxic properties of *Chenopodium glaucum* L. *PLoS ONE* 16(10):e0255502. https://doi.org/10.1371/journal.pone.0255502

Sahrish N, Shakeel A, Ghulam A, Zartash F, Sajjad H, Mukhtar A, Muhammad AK, Ahmad K, Fahad S, Wajid N, Sezai E, Carol Jo W, Gerrit H (2022) Modeling the impact of climate warming on potato phenology. *Eur J AgroN* 132:126404.

Sajid H, Jie H, Jing H, Shakeel A, Satyabrata N, Sumera A, Awais S, Chunquan Z, Lianfeng Z, Xiaochuang C, Qianyu J, Junhua Z (2020) Rice production under climate change: adaptations and mitigating strategies. In: Fahad S, Hasanuzzaman M, Alam M, Ullah H, Saeed M, Khan AK, Adnan M (eds) *Environment, climate, plant and vegetation growth.* Springer Publ Ltd, Springer Nature Switzerland AG. Part of Springer Nature, pp. 659–686. https://doi.org/10.1007/978-3-030-49732-3

Sajjad H, Muhammad M, Ashfaq A, Waseem A, Hafiz MH, Mazhar A, Nasir M, Asad A, Hafiz UF, Syeda RS, Fahad S, Depeng W, Wajid N (2019) Using GIS tools to detect the land use/land cover changes during forty years in Lodhran district of Pakistan. *Environ Sci Pollut Res.* https://doi.org/10.1007/s11356-019-06072-3

Saleem MH, Fahad S, Adnan M, Mohsin A, Muhammad SR, Muhammad K, Qurban A, Inas AH, Parashuram B, Mubassir A, Reem MH (2020a) Foliar application of gibberellic acid endorsed phytoextraction of copper and alleviates oxidative stress in jute (*Corchorus capsularis* L.) plant grown in highly copper-contaminated soil of China. *Environ Sci Pollution Res.* https://doi.org/10.1007/s11356-020-09764-3

Saleem MH, Fahad S, Shahid UK, Mairaj D, Abid U, Ayman ELS, Akbar H, Analía L, Lijun L (2020c) Copper-induced oxidative stress, initiation of antioxidants and phytoremediation potential of flax (*Linum usitatissimum* L.) seedlings grown under the mixing of two different soils of China. *Environ Sci Poll Res.* https://doi.org/10.1007/s11356-019-07264-7

Saleem MH, Rehman M, Fahad S, Tung SA, Iqbal N, Hassan A, Ayub A, Wahid MA, Shaukat S, Liu L, Deng G (2020b) Leaf gas exchange, oxidative stress, and physiological attributes of rapeseed (*Brassica napus* L.) grown under different light-emitting diodes. *Photosynthetica* 58(3):836–845.

Saleem N, Adnan M, Khan NA, Zaheer S, Jalal F, Amin M, Khan WM, Arif M, Rahman I, Ibrahim M, Jamal Y, Shah SRA, Junaid K, Ali M (2015) Dual purpose canola: grazing and grains options. *Pak J Weed Sci Res* 21(2):295–304.

Salunke DK, Desai BB (1988) Sugarcane In: *Post-harvest biotechnology of sugar crops.* CRC Press, USA, pp. 35–74.

Saman S, Amna B, Bani A, Muhammad TQ, Rana MA, Muhammad SK (2020) QTL mapping for abiotic stresses in cereals. In: Fahad S, Hasanuzzaman M, Alam M, Ullah H, Saeed M, Khan AK, Adnan M (eds) *Environment, climate, plant and vegetation growth.* Springer Publ Ltd, Springer Nature Switzerland AG. Part of Springer Nature, pp. 229–252. https://doi.org/10.1007/978-3-030-49732-3

Saud S, Chen Y, Fahad S, Hussain S, Na L, Xin L, Alhussien SA (2016) Silicate application increases the photosynthesis and its associated metabolic activities in Kentucky bluegrass under drought stress and post-drought recovery. *Environ Sci Pollut Res* 23(17):17647–17655. https://doi.org/10.1007/s11356-016-6957-x

Saud S, Chen Y, Long B, Fahad S, Sadiq A (2013) The different impact on the growth of cool season turf grass under the various conditions on salinity and drought stress. *Int J Agric Sci Res* 3:77–84.

Saud S, Fahad S, Cui G, Chen Y, Anwar S (2020) Determining nitrogen isotopes discrimination under drought stress on enzymatic activities, nitrogen isotope abundance and water contents of Kentucky bluegrass. *Sci Rep* 10:6415. https://doi.org/10.1038/s41598-020-63548-w

Saud S, Fahad S, Yajun C, Ihsan MZ, Hammad HM, Nasim W, Amanullah, Arif M, Alharby H (2017) Effects of nitrogen supply on water stress and recovery mechanisms in Kentucky bluegrass plants. *Front Plant Sci* 8:983. https://doi.org/10.3389/fpls.2017.00983

Saud S, Li X, Chen Y, Zhang L, Fahad S, Hussain S, Sadiq A, Chen Y (2014) Silicon application increases drought tolerance of Kentucky bluegrass by improving plant water relations and morph physiological functions. *SciWorld J* 2014:1–10. https://doi.org/10.1155/2014/368694

Senol C (2020) The effects of climate change on human behaviors. In: Fahad S, Hasanuzzaman M, Alam M, Ullah H, Saeed M, Khan AK, Adnan M (eds) *Environment, climate, plant and vegetation growth.* Springer Publ Ltd, Springer Nature Switzerland AG. Part of Springer Nature, pp. 577–590. https://doi.org/10.1007/978-3-030-49732-3

Shafi MI, Adnan M, Fahad S, Fazli W, Ahsan K, Zhen Y, Subhan D, Zafar-ul-Hye M, Martin B, Rahul D (2020) Application of single superphosphate with humic acid improves the growth, yield and phosphorus uptake of wheat (*Triticum aestivum* L.) in calcareous soil. *Agronomy* 10:1224. https://doi.org/10.3390/agronomy10091224

Shah F, Lixiao N, Kehui C, Tariq S, Wei W, Chang C, Liyang Z, Farhan A, Fahad S, Huang J (2013) Rice grain yield and component responses to near 2°C of warming. *Field Crop Res* 157:98–110.

Sharma KP, Batta SK, Singh R (1994) Studies on minimizing dextran problems in sugarcane under subtropical conditions. *Trop Agric* (Trinidad) 71:119–122.

Shrivastava AK, Ghosh AK, Agnihotri VP (1992) *Sugar cane ratoons*, Oxford & IBH Publishing Co. Pvt. Ltd., New Delhi, p. 184.

Shrivastava AK, Srivastava AK, Solomon S (2011) Sustaining sugarcane productivity under depleting water resources. *Curr Sci* 101(6):748–754.

Shrivastava AK, Srivastava S (2012) *Improving crop resistant to abiotic stress.* In: Tuteja N, Gill SS, Tiburcio AF, Tuteja R (eds) Vol. II, Wiley-V.C.H. Verlag GmbH & Co. KGaA, pp. 885–921.

Shrivastava AK, Srivastava TK, Srivastava AK, Misra V, Shrivastava S, Singh VK, Shukla SP (2016). Climate change induced abiotic stresses affecting sugarcane and their mitigation. ICAR-Indian Institute of Sugarcane Research, Lucknow, p. 108. ISBN:978-93-5265-983-8.

Shrivastava AK, Srivastava MK (2006) *Abiotic stresses affecting sugarcane: Sustaining productivity*, International Book distributing Co., Lucknow, India, p. 322.

Sidra K, Javed I, Subhan D, Allah B, Syed IUSB, Fatma B, Khaled DA, Fahad S, Omaima N, Ali TKZ, Rahul D (2021) Physio-chemical characterization of indigenous agricultural waste materials for the development of potting media. *J Saudi Society Agricultural Sci.* https://doi.org/10.1016/j.sjbs.2021.08.058

Singh J, Kumar A, Abbas Z (2015) Post harvest quantity and quality losses of sugarcane crop. *J Biol Chem Res* 32(2):711–719.

Singh KDN, Jha BK, Misra GK, Samad A (1992) Effects of potassium on yield and quality of sugarcane on calciorthent soils under drought condition. *J Indian Soc Soil Sci* 40(1):105–110.

Singh P, Solomon S, Prajapati CP, Kumar S, Misra V, Chandra A (2015a) Deterioration of fresh and stale cane juice at high ambient temperature in relation to expression of invertases and the growth of Leuconostoc sp. *Agrica* 4(2):79–85.

Solomon S (2000) Post-harvest cane deterioration and its milling consequences. *Sugar Tech* 1:1–18.

Solomon S (2009) Post harvest deterioration. *Sugar Tech* 11(2):109–123.

Solomon S, Banerji R, Shrivastava AK, Singh P, Singh I, Verma M, Prajapati CP, Sawnani A (2006) Post-harvest deterioration of sugarcane and chemical methods to minimize sucrose losses. *Sugar Tech* 8(1):74–78.

Solomon S, Ramadurai R, Shanmugnathan S, Shrivastava AK, Deb S, Singh I (2003) Management of biological losses in milling tandom to improve sugar recovery. *Sugar Tech* 5(3):137–142.

Solomon S, Shahi HN, Suman A, Gaur A, Deb S, Singh I (2001) A survey of post- harvest biological losses in Indian sugar factories: an emerging challenge. *Proc Int Soc Sugar Cane Technol* 24:380–381.

Solomon S, Shrivastava AK, Srivastava BL, Madan VK (1997) Pre-milling sugar losses and their management in sugarcane. Technical Bulletin No. 37. Indian Institute of Sugarcane Research, Lucknow, p. 217.

Solomon S, Srivastava KK, Bhatnagar S, Madan VK (1990) Post-harvest changes in invertase activity and juice quality in sugarcane. *Indian Sugar* 39(12):895–899.

Srinivasan TR (1993) Management practices for drought situation lead paper presented in the 8th Sugarcane Research and Development Workers Meeting of Northern Karnataka held at MKN. Hubli.

Srivastava SC, Johari DP, Gill PS (1988) *Manual of sugarcane production in India*, Indian Council of Agricultural Research, New Delhi, p. 194.

Subhan D, Zafar-ul-Hye M, Fahad S, Saud S, Martin B, Tereza H, Rahul D (2020) Drought stress alleviation by ACC deaminase producing *Achromobacter xylosoxidans* and *Enterobacter cloacae*, with and without timber waste biochar in maize. *Sustainability* 12(6286). https://doi.org/10.3390/su12156286

Tariq M, Ahmad S, Fahad S, Abbas G, Hussain S, Fatima Z, Nasim W, Mubeen M, Rehman MH, Khan MA, Adnan M (2018) The impact of climate warming and crop management on phenology of sunflower-based cropping systems in Punjab, Pakistan. *Agri Forest Met* 256:270–282.

Unsar N-U, Muhammad R, Syed HMB, Asad S, Mirza AQ, Naeem I, Muhammad HR, Fahad S, Shafqat S (2020) Insect pests of cotton crop and management under climate change scenarios. In: Fahad S, Hasanuzzaman M, Alam M, Ullah H, Saeed M, Khan AK, Adnan M (eds) *Environment, climate, plant and vegetation growth*. Springer Publ Ltd, Springer Nature Switzerland AG. Part of Springer Nature, pp. 367–396. https://doi.org/10.1007/978-3-030-49732-3

VanToai TT, St. Martin SK, Chase K, Boru, G, Schnipke V, Schmitthenner AF, Lark KG (2001) Identification of a QTL associated with tolerance of soybean to soil waterlogging. *Crop Sci* 41:1247–1252.

Vasantha S (2016) Mechanisms of salinity and drought tolerance in sugarcane. In: Ram B, Govindraj P, Sundar AR, Bhaskaran A (eds) *Biotechnological and conventional tools for biotic and abiotic stress management in sugarcane*. Winter School, 7–27 November, pp. 144–150.

Veen BW (1988) Influence of oxygen deficiency on growth and function of plant roots. *Plant Soil* 111:259–266.

Venkataraman TS (1929) Studies of sugarcane roots at different stages of growth. *Mem Dept Agric India Bot Ser* 16(5):145–157.

Wahid F, Fahad S, Subhan D, Adnan M, Zhen Y, Saud S, Manzer HS, Martin B, Tereza H, Rahul D (2020) Sustainable management with mycorrhizae and phosphate solubilizing bacteria for enhanced phosphorus uptake in calcareous soils. *Agriculture* 10(334). https://doi.org/10.3390/agriculture10080334

Wajid N, Ashfaq A, Asad A, Muhammad T, Muhammad A, Muhammad S, Khawar J, Ghulam MS, Syeda RS, Hafiz MH, Muhammad IAR, Muhammad ZH, Muhammad HR, Veysel T, Fahad S, Suad S, Aziz K, Shahzad A (2017) Radiation efficiency and nitrogen fertilizer impacts on sunflower crop in contrasting environments of Punjab. *Pakistan Environ Sci Pollut Res* 25:1822–1836. https://doi.org/10.1007/s11356-017-0592-z

Weinberg J (1903) The deterioration of cut sugarcane. *Agric Ledger* 6:1–8.

Went FAFC, Geerligs HCP (1884) On the deterioration of the saccharose content of sugarcane after being cut. *Sugar Cane* 24:482–485.

Wu C, Kehui C, She T, Ganghua L, Shaohua W, Fahad S, Lixiao N, Jianliang H, Shaobing P, Yanfeng D (2020) Intensified pollination and fertilization ameliorate heat injury in rice (*Oryza sativa* L.) during the flowering stage. *Field Crops Res* 252:107795.

Wu C, Tang S, Li G, Wang S, Fahad S, Ding Y (2019) Roles of phytohormone changes in the grain yield of rice plants exposed to heat: a review. *PeerJ* 7:e7792. https://doi.org/10.7717/peerj.7792

Yang Z, Zhang Z, Zhang T, Fahad S, Cui K, Nie L, Peng S, Huang J (2017) The effect of season-long temperature increases on rice cultivars grown in the central and southern regions of China. *Front Plant Sci* 8:1908. https://doi.org/10.3389/fpls.2017.01908

Zafar-ul-Hye M, Muhammad TH, Muhammad A, Fahad S, Martin B, Tereza D, Rahul D, Subhan D (2020b) Potential role of compost mixed biochar with rhizobacteria in mitigating lead toxicity in spinach. *Scientific Rep* 10:12159. https://doi.org/10.1038/s41598-020-69183-9

Zafar-ul-Hye M, Muhammad N, Subhan D, Fahad S, Rahul D, Mazhar A, Ashfaq AR, Martin B, Ji ˇrí H, Zahid HT, Muhammad N (2020a) Alleviation of cadmium adverse effects by improving nutrients uptake in bitter gourd through cadmium tolerant rhizobacteria. *Environments* 7(54). https://doi.org/10.3390/environments7080054

Zafar-ul-Hye M, Akbar MN, Iftikhar Y, Abbas M, Zahid A, Fahad S, Datta R, Ali M, Elgorban AM, Ansari MJ et al. (2021) Rhizobacteria inoculation and caffeic acid alleviated drought stress in lentil plants. *Sustain* 13:9603. https://doi.org/10.3390/su13179603

Zahida Z, Hafiz FB, Zulfiqar AS, Ghulam MS, Fahad S, Muhammad RA, Hafiz MH, Wajid N, Muhammad S (2017) Effect of water management and silicon on germination, growth, phosphorus and arsenic uptake in rice. *Ecotoxicol Environ Saf* 144:11–18.

Zahir SM, Zheng HG, Ala Ud D, Amjad A, Ata Ur R, Kashif J, Shah F, Saud S, Adnan M, Fazli W, Saud A, Manzer HS, Shamsher A, Wajid N, Hafiz MH, Fahad S (2021) Synthesis of silver nanoparticles using *Plantago lanceolata* extract and assessing their antibacterial and antioxidant activities. *Sci Rep* 11:20754.

Zarabi S (2004) Bad monsoon may upset growth prospects. *Hindustan Times* (Lucknow, edn.), July 22, p. 13.

Zhu JK (2001) Cell signaling under salt, water and cold stresses. *Curr Opin Plant Biol* 4:401–406.

Zia-Ur-Rehman M (2020) Environment, climate change and biodiversity. In: Fahad S, Hasanuzzaman M, Alam M, Ullah H, Saeed M, Khan AK, Adnan M (eds) *Environment, climate, plant and vegetation growth.* Springer Publ Ltd, Springer Nature Switzerland AG. Part of Springer Nature, pp. 473–502. https://doi.org/10.1007/978-3-030-49732-3

11

Legume Production and Climate Change

Abdel Rahman M. Al-Tawaha
Al Hussein Bin Talal University, Ma'an, Jordan

David McNeil
University of Tasmania LPO, Sandy Bay, Tasmania, Australia

Yahia A. Othman
The University of Jordan, Amman, Jordan

Ibrahim M. AlRawashdeh
Al Hussein Bin Talal University, Ma'an, Jordan

Amanullah and Imran
The University of Agriculture Peshawar, Peshawar, Pakistan

Muhammad Adnan and Muhammad Zamin
University of Swabi, Anbar, Pakistan

Ali M. Qaisi
The University of Jordan, Amman, Jordan

Abdelrazzaq Al-Tawaha
University Putra Malaysia, Serdang, Malaysia

Nusrat Jahan
Aligarh Muslim University, Aligarh, India

Mohd Abas Shah
ICAR-Central Potato Research Station, Jalandhar, India

Shah Khalid
The University of Agriculture Peshawar, Pakistan

Rokayya Sami
Taif University, Taif, Kingdom of Saudi Arabia

Abdur Rauf
University of Swabi, Anbar, Pakistan

Devarajan Thangadurai
Karnatak University, Dharwad, India

Jeyabalan Sangeetha
Central University of Kerala, Kasaragod, India

Shah Fahad
Hainan University, Haikou, China; The University of Haripur, Haripur, Pakistan

Refat A. Youssef
National Research Centre, Giza, Egypt

DOI: 10.1201/9781003286417-11

Wafa'a A. Al-Taisan
Imam Abdulrahman Bin Faisal University, Dammam, Saudi Arabia

Duraid K. A. Al-Taey
Al-Qasim Green University, Baghdad, Iraq

CONTENTS

11.1 Introduction

Legume family (Fabaceae, syn. Leguminosae) is the third largest family of angiosperms, including over 750 genera and 19,000 species ranging from small herbs to large trees. Mediterranean environment is wealthy in legumes, both as woody shrubs of the late progression phases and as short-lived, initiate grassy plants (Arianoutsou and Thanos 1996).

Pulses including two sorts depend on growing period: firstly winter season crops such as chickpeas, lentils, lathyrus, broad bean, and vetch; secondly summer season types, for example, soybeans and peas. They have many weals in our globe; they supply humans with food; feed for animals; and also provide the soil with nitrogen fixation through bacterial nodes and increase the CO_2 concentration of the ambiance (Khan et al. 2016; Mousavi Derazmahalleh et al. 2019). Studying the pulses species under stress conditions helps in choosing the right pulses for pasture breeding and soil preservation programmes, mainly in arid and semiarid regions. A study was undertaken to create biogeographical manners in the pulses flora of southern Africa so as to assist the selection of species with agricultural potency (Trytsman et al. 2016).

Pulse crops play an important part in human nourishment, mammal feeding, and soil fecundity improvement; they play a big role in decreasing paucity and contributing to persistence of economic progress in the world. Embracing new varieties of pulses enables you with latent power to endure stress situations and aid in expanding the planting zones, which leads to better revenue for the countries. Leaves from pulses trees and shrubs can be a precious supplement for nourishment of livestock, particularly through the desiccated time when other nurture raw materials are insufficient or their alimentary value is low. The agronomic assessment at two agro-ecologically discrete regions in eastern DRC verified that diverse forage trees and shrub species are tailored distinctly. The species with top forage superiority, *L. diversifolia* (especially accession ILRI 15551) and *L. leucocephala*, execute preferably at the site (Keamanyola) with advanced soil productiveness (Katunga 2014).

11.2 Ecology and Adaptation of Legumes Crop

Legumes are generally developed rainfed and are showing to distinct kinds of drought zones from terminal drought to sporadic drought (Purushothaman et al. 2013). Acclimation technologies are endeavours to reinstate the synchrony or to generate new management between ambiance and plant communities. Both chickpeas and pigeonpeas are important grain legumes developed for their protein-rich seeds used for human utilization, for their capacity to renovate and conserve soil fecundity by nitrogen fixation, and for their convenience to adjust very well to different cropping manners (Upadhyaya et al. 2010). Heat stress through venereal progress can issue noteworthy yield loss. Chickpea (*Cicer arietinum* L.) is an important food legume, and heat stress influences chickpea ontogeny over a range of milieu (Devasirvatham et al. 2012; Ahmad et al. 2016). Four accessions of *C. reticulatum* and one accession of *C. pinnatifidum* were found to be an impedance to water deficiency and heat stress (up to 41.8°C) (Canci and Toker 2009).

In chickpea, total root biomass in precocious development stages has been revealed to appreciably participate to seed product under incurable water deficit in middle and South India (Kashiwagi 2015). Deeper root coordination and superior root desiccated mass could be allied with enhanced prevention to soil water restriction in drought tolerant legume than in drought-sensitive ones (Abdelhamid 2010). Reproduction programmes are ongoing to promote cultivars tailored to a wide range of latitudes and with disease resistance (Siddique et al. 2000).

Conformation strategies that are debated include prior seeding of pulse crops, use of cold pulses, crop succession within crop rotation, and modulations to the microclimate such as forthright seeding into standing straw (Cutforth 2007).

Prospective pulse genotypes will require to preserve or smooth to be greater under the expected climate changes. The most significant way to enhance the granule yield of cool period food legumes per unit area under stress niches will be the enlargement of cultivars with conflict to numerous abiotic stresses in the circumstance of endurable incorporated pulses output, insect and pest eradication, nutrients organization, and irrigation use if accessible (Toker and Yadav 2010). Faba bean has a lot of proper imputes for sustainable recent cropping patterns, but these outcomes propose that yield will be inadequate in case of anticipated climate change, presuppose the expansion of heat-tolerant cultivars, or enhanced flexibility by other mechanisms such as earlier flowering periods (Bishop 2016).

Faba bean yield will turn into ever more uneven with anticipated climate change and underline the requirement to recover heat stress resilience of faba bean floral advance and anthesis to endorse future (Bishop 2016). Faba bean yield was minimized by heat stress within the hotness range recognized to contrive yield decreasing in other output species, and as in other species. Replacement of imported soybean food by locally created pulses could have a convenient effect, mainly for pulses produced and used on-farm (Köpke and Nemecek 2010). Modest is identified about the natural adjustment of faba bean germplasm, or about the regions and impacts of genes that effect traits associated to drought adjustment, chiefly stomatal morphology and role as key manners for gas reciprocity between plant and ambiance (Khazaei 2014). Katerji (2011) drought at flowering, early podding and grain-filling phases decrease both grain and straw yields.

Promoting new varieties that can be tolerated or conflict for environment invasion mainly climate change will be sharing in raising the prime creation of pulses during minimize the stomata conductance, CO_2 improvement, universal warming tolerates. Enhancing the leaf area index also take part in a large role in adjustment the legumes which offer to the CO_2 raising. On the other, the nitrogen fixation outfitted by legume and C:N ratio can be enhanced.

In order to enhance crop execution, it is necessary to be aware of how crop species react and adjust to water-deficit circumstances (Khazaei 2014). Due to variability in quantity, interval, strength, and allocation of rainfall globally, new varieties should be enhanced to evade drought incident. Adjustment based on the genetic makeup and phenotype elasticity, and species with high flexibility are more adapted to stress environment and can survive for a long period and generate high yield ingredients.

Drought is a very imperative ecological cause restraining the output of pulses crops as well other field crops globally. Climate change patterns have led to variability in rainfall patterns, and increased periods of winter and spring drought in continents will affect areas of pulses planting. The planting required for varieties showing prior drought bearing and water-use efficiency is fabricated in the enhancing stressed within pulses plant improving programmes on selection of varieties preferable to tolerate prolonged times of water deficit.

Modern molecular markers help in limiting quantitative trait loci that participated in selection of the favourable traits used in breeding programmes. Amelioration demanded for pulses depended on conductors of water deficit, for example, chlorophyll fluorescence, stomatal density, and root density. Genetic resource gathering and classification can have an important part in distinguishing pulses genotypes adapted to water-deficit-prone environments. In legumes, the influences of water deficit or other stresses on nitrogen fixation require to be quantified alongside the influences on the plant itself (Abberton et al. 2008).

Molecular analysis stressed that many genes are concerned in plant acclimation to all stresses that exposed pulses as well as other crops. Direct gene transfer can be sharing in enhancing the yield and yield ingredients during genetic stability of legume crops under various ecology microclimates. Molecular biology has a wide range of capacity enforcements in pasture pulses breeding, including marker-assisted and genomics-assisted selection and the recognition of quantitative trait loci and elect genes for necessitate characters (Ullah et al. 2016; Nichols et al. 2012).

11.3 Physiological Responses of Grain Legumes to Stress Environments

Crop growth and productivity are affected by many biotic and abiotic factors (Turk and Tawaha 2001, 2002a, 2002b; Tawaha and Turk 2002, 2004; Tawaha et al. 2003; Turk et al. 2003a, 2003b, 2004; Khan et al. 2017; Al-Rifaee et al. 2004; Musallam et al. 2004; Supanjani et al. 2005; Sulpanjani et al. 2005; Abu-Darwish and Abu Dieyeh 2009; Al-Tawaha et al. 2010; Al-Tawaha and Seguin 2006; Al-Tawaha et al. 2006a, 2006b, 2007, 2018a, 2018b, 2018c, 2018d; Abu Obaid et al. 2018; Deepranjan et al. 2021; Haider et al. 2021; Huang et al. 2021; Ikram et al. 2021; Jabborova et al. 2021; Khadim et al. 2021a, 2021b; Manzer et al. 2021; Muzammal et al. 2021; Abdul et al. 2021a, 2021b; Ashfaq et al. 2021; Amjad et al. 2021; Atif et al. 2021; Athar et al. 2021; Adnan et al. 2018a, 2018b, 2019, 2020; Akram et al. 2018a, 2018b; Aziz et al. 2017a, 2017b; Chang et al. 2021; Chen et al. 2021; Emre et al. 2021; Habib Ur et al. 2017; Hafiz et al. 2016, 2018, 2019, 2020a, 2020b; Ghulam et al. 2021; Guofu et al. 2021; Hafeez et al. 2021; Khan et al. 2021; Kamarn et al. 2017; Muhammad et al. 2019; Safi et al. 2021; Sajjad et al. 2019; Saud et al. 2013, 2014, 2016, 2017, 2020; Shah et al. 2013; Qamar-Uz et al. 2017; Hamza et al. 2021; Irfan et al. 2021;Wajid et al. 2017; Yang et al. 2017; Zahida et al. 2017; Depeng et al. 2018; Hussain et al. 2020; Shafi et al. 2020; Wahid et al. 2020; Subhan et al. 2020a, 2020b; Zafar-ul-Hye et al. 2020a, 2020b, 2021; Ilyas et al. 2020; Saleem et al. 2020a, 2020b, 2020c; Rehman et al. 2020; Farhat et al. 2020; Wu et al. 2020; Mubeen et al. 2020; Farhana et al. 2020; Jan et al. 2019; Wu et al. 2019; Ahmad et al. 2019; Baseer et al. 2019; Fahad and Bano 2012; Fahad et al. 2013, 2014a, 2014b, 2015a, 2015b, 2016a, 2016b, 2016c, 2016d, 2017, 2018a, 2018b, 2019a, 2019b, 2020, 2021a, 2021b, 2021c, 2021d, 2021e, 2021f; Hesham and Fahad 2020; Iqra et al. 2020; Akbar et al. 2020; Mahar et al. 2020; Noor et al. 2020; Bayram et al. 2020; Amanullah and Fahad 2017, 2018a, 2018b; Amanullah et al. 2020, 2021; Rashid et al. 2020; Arif et al. 2020; Amir et al. 2020; Saman et al. 2020; Muhammad et al. 2020; Md Jakir and Allah 2020; Mahmood Ul et al. 2021; Farah et al. 2020; Sadam et al. 2020; Unsar et al. 2020; Fazli et al. 2020; Md. Enamul et al. 2020; Gopakumar et al. 2020; Zia-Ur-Rehman 2020; Ayman et al. 2020; Mohammad et al. 2020a, 2020b; Senol 2020; Amjad et al. 2020; Ibrar et al. 2020; Sajid et al. 2020; Muhammad et al. 2021; Sidra et al. 2021; Zahir et al. 2021; Sahrish et al. 2022). On the other hand, physiological approaches of plants including grain legumes provide general outlines of plant responses and finding stress tolerance–related traits or elite cultivars (Araújo et al. 2015). During the past decade, several physiological variables have been evaluated and recommended to understand the influence of environmental stress on grain legumes such as nutrient uptake and translocation, shoot architecture, leaf gas exchange (photosynthesis, transpiration, and stomatal conductance), chlorophyll content and fluorescence, yield, seed quality, root dynamic and its interaction with soil biota (Dadresan et al. 2015; Ferchichi et al. 2016; Kurban et al. 1998; McDonald and Paulsen 1997). Shoot and root architecture can significantly affect yield and biological control of diseases and insects and lead to reduction in pesticide use (Duc et al. 2015; Leskovar and Othman 2016). Photosynthesis rate depend on light intensity that reaches chloroplasts (Yang et al. 2018).

Shade conditions in close-planting and intercropping systems directly affect the photosynthetic capacity of grain legumes (e.g. soybean, *Glycine max*) by regulating the expression of photosynthetic proteins (Yang et al. 2018). In addition, environmental stress such as water deficit and salt stress lead to direct changes in gas exchange performance. McDonald and Paulsen (1997) assessed the influence of day/night temperature (20°C/15°C, 30°C/15°C, or 30°C/25°C day/night) on leaf chlorophyll and photosynthesis in common bean (*Phaseolus vulgaris*), cowpea (*Vigna unguiculata*), faba bean (*Vicia faba*), and field pea (*Pisum sativum*). They found that 30°C/25°C day/night decreased chlorophyll fluorescence in all species except cowpea which was tolerant to high temperature stress. Conversely, high day temperature (but not high night temperature) increased leaf chlorophyll, N content, as well as photosynthetic rates (McDonald and Paulsen 1997). Mycorrhizal association in leguminous crops (e.g. soybean) improves N fixation and photosynthesis, especially in phosphorus-starved soils (Bulgarelli et al. 2017). The symbiotic relationship of arbuscular mycorrhizal fungi (AMF) with legumes in poor soil (specifically, low P) improves nitrogenase activity, carboxylation rate of Rubisco, P content in nodules, and lowers the metabolic limitation of photosynthesis (Bulgarelli et al. 2017).

Environmental stresses such as drought and salinity (soil and irrigated water) are commonly responsible for partial or complete failure of grain legumes, especially in arid and semiarid regions. In fact, those stresses significantly affect the physiological functions of leguminous crops and their interaction with soil biota including AMF and rhizobia (Almeida et al. 2000; Farooq et al. 2017; Ghanbari et al. 2015).

11.3.1 Salinity Stress

11.3.1.1 Growth and Physiology

About 98% of the salt in the soil are excluded by most plants allowing only 2% to be transported to the shoots. Under salinity stress, roots must exclude the excess Na^+ and Cl^- dissolved in the soil, or the salt will gradually build up to toxic levels. Soil salinity reduces plant growth and development by reducing the plant's potential to absorb water (osmotic or water-deficit effect), and it may moves in the transpiration stream and injure the transpiring leaf cells (salt-specific effects) (Munns 2005).

High soil salinity leads to changes in the membrane permeability, cell osmotic potential, as well as leaf water potential and leads to reduction in plant height, leaf number, and leaf area (Kurban et al. 1998; López-Aguilar et al. 2003; Othman et al. 2006). In addition, salinity stress has an adverse effect on seed germination, shoot growth, and antioxidant activities. For example, salinity stress significantly decreased root length, water-use efficiency, and increased root-to-shoot ratio of chickpea (*Cicer arietinum*) (Jan et al. 2013). In alfalfa, root mycorrhizal colonization, shoot nutrient (K^+, Ca^{+2}, P, Zn, Cu, Mg^{+2}), protein and proline content were significantly lower under salinity stress compared to control (Ashrafi et al. 2014). Soybean 'Manokin' grown in sandy soil and irrigated with different levels of saline water (2.0, 5.0, 7.7, 10.0, 11.9, and 13.7 dS m^{-1}) showed that plants received 11.1 and 13.0 dS m^{-1} electrical conductivity (EC) treatments had lower height, leaf area, and shoot dry weight (Wang et al. 2005). In addition, shoot-to-root ratios decreased with increasing Cl^- salinity. Interestingly, moderate saline conditions (5.0 and 7.7 dS m^{-1}) had higher plant height, leaf area, and shoot dry weight than those grown in lower (2.0 dS m^{-1}) or higher (10.0, 11.9, and 13.7 dS m^{-1}) salinity level treatments (Wang et al. 2005). Growing alfalfa (*Medicago sativa* cv. Magali) in 100 mM NaCl significantly decreased shoot relative water content and water potential (Serraj and Drevon 1998). Irrigation with saline water (2 dS m^{-1}) negatively affected the grain-filling stage and decreased soybean grain production by 80% compared to control (Bustingorri and Lavado 2013; Moussa and Hassan 2016). In alfalfa 'Gongong No.1,' soil salinity reduced photosynthesis rate, transpiration, and water-use efficiency (Guo et al. 2016). In addition, fine root growth and surface area of alfalfa declined across soil profile (0–30 cm soil depth) when grown under salinity stress (Guo et al. 2016). At moderate salinity stress (50mM NaCl), leaf-level photosynthesis rate and stomatal conductance of *Alhagi pseudoalhagi* (leguminous plant) were about 150% of that of the control (0.0mM NaCl), while both variables were about 60% lower of that the control in the 200mM NaCl salinity treatment (Kurban et al. 1999). High salinity levels (8, and 10 ds m^{-1}NaCl) reduced photosynthetic pigments (chlorophyll and carotenoid) content and increased peroxidase and polyphenol oxidase activity in fenugreek (*Trigonella foenum*) leaves compared to control (Dadresan et al. 2015).

Salinity-tolerant legumes have the potential to limit salt uptake and transport throughout the plant tissues, regulate the osmotic and ionic balance of root and shoot cells, and adjust leaf development and senescence (Munns 2005). The adaptation of some grain legumes to salt stress (e.g. faba bean) can be attributed to the accumulation of nutrient (N, P, K^+, Ca^{+2}, and Mg^{+2}) and exclusion of Na^+ and Cl^- (Abdelhamid et al. 2010). During germination process, the tolerance of legume plants to NaCl stress depends on the antioxidant activity of the seeds. Higher levels of antioxidant activities such as ascorbate peroxidase (plays a key role in the scavenging of reactive oxygen species) improve the salinity tolerance mechanisms and protect the seeds against radicals that leads to metabolic damage (Morais et al. 2012).

Generally, cool season grain legumes are less tolerant to soil abiotic stresses than cereals (Jayasundara et al. 1998). However, considerable variation exists among grain legumes species in their sensitivity to soil abiotic stresses (Jayasundara et al. 1998). For example, faba bean is relatively less sensitive to salinity and alkalinity than chickpea and lentil (*Lens culinaris*). In addition, field pea is relatively more adaptive to salinity and alkalinity than chickpea and lentil. Lupins (*Lupinus angustifolius*, *L. abtus*, and *L. luteus*) are the most tolerant grain legumes (cool season) to acidic soils compared to faba bean, field pea, chickpea, and lentil (Jayasundara et al. 1998). While *Vigna radiata* growth was significantly and gradually reduced by increasing salinity by 9.1, 16.2 and 28.2 dS m^{-1}, the growth of *Alhagi pseudoalhagi* was promoted at 9.1 and 16.2 dS m^{-1} and then slightly decreased at 28.2 dS m^{-1} (Kurban et al. 1998). López-Aguilar et al. (2003) grew tepary bean *(Phaseolus acutifolius*), cowpea, and wild bean *(Phaseolus jiliformis*) in a nutrient solution salinized with 80 mmol L^{-1} NaCl for 14 days. They found that the tolerance of cowpea to saline conditions was less than tepary bean and wild bean. This is because Na^+ ions moved easily to the leaves in cowpea, while in tepary bean and wild bean, Na^+ mobility was restricted through the capacity of tepary bean and wild bean stem and petiole to hold Na^+. In addition, the uptake and transport of Ca^{+2} and Mg^{+2} in cowpea and wild bean were significantly lower than tepary bean (López-Aguilar et al. 2003). Morais et al. (2012) investigated the impact of different NaCl levels (0mM, 50mM, 100mM, and 200mM) for 15 days on seed germination and on the activity of antioxidant enzymes of *Acacia longifolia* and *Ulex europaeus*. Salinity stress reduced germination rate percentages in *U. europaeus* (native species) by 97% and 100% for the 100mM and 200mM NaCl, respectively. In the invasive species (*A. longifolia*), the reduction was 16% at 200mM NaCl. For *A. longifolia*, salinity stress reduced antioxidant enzyme catalase (CAT) and glutathione reductase (GR) activity and increased ascorbate peroxidase (APX) significantly. For *U. europaeus*, lower activity of CAT and APX and higher GR and POX were found at 50mM NaCl. However, protein content increased in both species as NaCl concentration increased.

Several studies indicated that the presence of specific soil minerals can improve the tolerance of leguminous crops to salt stress such as Si and N. On the other hand, the presence of NO_3^- was slightly mitigated in the NaCl impact on alfalfa 'Magali' grown in 100mM NaCl (Serraj and Drevon 1998). Root elongation and root multiplication rate of alfalfa 'Batini' and 'Trifecta' decreased significantly by increased salinity (0, 3.0, 6.4, and 12.2 dS m^{-1} NaCl) but N (NO_3^-–N or NH_4^+–N) addition decreased the deleterious effect of salt stress (Esechie et al. 2002).

11.3.1.2 Nodulation and Microorganisms Association

Most legumes as well as the N fixation process by microorganisms in the nodules of the leguminous crops are sensitive to salinity (Bruning and Rozema 2013). Soil salinity adversely affects the nodulation and N fixation capacities of rhizobia, resulting in lower productivity (Kurdali and Al-Ain 2002). In addition, nitrogenase activity and nodule permeability to O_2 diffusion are significantly inhibited by salinity (Serraj and Drevon 1998). Mashhady et al. (1998) assessed the influence of different salinity levels on the growth and survival of *Rhizobium meliloti* strains and their symbiotic efficiencies with lucerne *(Medicago sativa* var. Maxidore*)* under greenhouse conditions for 72 h. Growth of all rhizobial strains (nodulation) and symbiotic N_2-fixation of lucerne plants were negatively affected by high salinity (100mM –250mM NaCl). In fact, growing lucerne in salty sand (250mM NaCl) completely inhibited the formation of nodules. Irrigation with high level of saline water (ECw 33 dS m^{-1}) completely inhibited indigenous rhizobial nodulation and N_2 fixation of *Sesbania aculeate* (Kurdali and Al-Ain 2002). Conversely, low salinity levels (ECw = 4.03 dSm^{-1}) significantly enhanced N_2 fixation.

Nodulation process response to salt stress in grain legumes depends on cultivar, nutrient levels in the soil rhizosphere (especially P), soil microorganism (e.g. AMF), and root interaction. Under saline conditions, double inoculation of alfalfa seed with AMF (*Glomus mosseae*) and alfalfa rhizobia (*Sinorhizobium meliloti*) improved nutrient uptake and resulted in a considerable increase in yield (Ashrafi et al. 2014). Phosphorus supply (37mM P) to salt-stressed (150mM NaCl) chickpea enhanced nodulation and rhizobial symbiosis efficiency, as well as salinity tolerance (Kaci et al. 2018).

11.3.1.3 Nutrients

Soil salinity disturbs hormonal regulation, nutritional imbalance, and photosynthesis and induces ion toxicity and osmotic effects cause a reduction in grain yield and quality of legumes (Farooq et al. 2017). Salt stress decreased macro- (P, K, Ca^{+2}, Mg^{+2}) and micronutrients (Cu, B, Zn, Mn, and Fe) (Esechie and Rodriguez 1998). Under salinity stress (100 mM NaCl), Na^+ was the essential ion responsible for the inhibition of growth and shoot biomass in faba bean (Bulut et al. 2011). Salt stress (150mM NaCl) increased Na^+ and P and decreased K^+ content in the leaves and root of chickpea (Kaci et al. 2018). In addition, soil salinity (150 mM NaCl) induced ionic imbalance and reduced P-use efficiency (Kaci et al. 2018). When *Alhagi pseudoalhagi* (leguminous plant) was subjected to different salinity stresses (0 mM, 50 mM, 100 mM, and 200 mM NaCl) for 30 days, nutrient (Ca^{+2}, Mg^{+2}, N, and K^+) uptake were induced by the 50 mM NaCl, whereas the uptake and transportation rates of those nutrients reduced significantly under higher levels of salinity treatments (Kurban et al. 1999). Increasing salt levels (0, 4, 6, 8, and 10 ds m^{-1}) lead to higher Na, P, B, Cu, and Zn content of fenugreek (*Trigonella foenum*) shoot and reduced the accumulation of Ca^{+2}, Fe, Mg^{+2}, and K^+ (Dadresan et al. 2015). However, under salinity conditions (75 mM NaCl), *Medicago ciliaris* was able to mitigate this impact by increasing nutrient use efficiency, specifically, K^+ and Fe (Ferchichi et al. 2016). In fact, adequate soil N can mitigate the sensitivity of grain legumes to salinity stress. For example, growing chickpea in a range of salt levels (control, 2.5, 5.0, and 7.5 dS m^{-1}) and three N sources (NO_3^-, NH_4^+, and NH_4NO_3) in greenhouse conditions showed that the application of N (NO_3^- or NH_4^+) reduced the reduction of chickpea growth rate, water-use efficiency, and grain yield at 7.5 dS m^{-1} salinity level (Jan et al. 2013).

Overall, the principal inferences of reviewing the impact and tolerance mechanisms of grain legumes to salt stress are: (1) salinity stress decreased seed germination of grain legumes by restricting water uptake and/or ion toxicity in the embryo, (2) salt stress reduce mineral uptake, growth and yield due to ion imbalance and toxicity and lower photosynthesis rate, (3) tolerance mechanisms of grain legumes to salt stress might develop through hormonal regulation, compartmentalization, and/or excretion of toxic ions, accumulation of compatible osmolytes and expanding the antioxidant capacity in the shoots, (4) the symbiotic association of grain legumes and soil biota (e.g. AMF) might lead to significant improvement of salinity tolerance by increases in the available nutrients (Farooq et al. 2017).

11.3.2 Drought Stress

Drought stress adversely affects grain legumes growth and productivity during all developmental stages. Water-deficit stress diminishes the ability of legumes roots to extract soil water and this significantly impacts the morphophysiological functions such as, photosynthetic rate, nutrient acquisition (e.g. Fe, N, and P) and N fixation. In fact, water-deficit stress has higher negative impact on growth and water-use efficiency than high temperature (30°C /25°C day/night) in common bean (*Phaseolus vulgaris*), cowpea, and field pea (*Pisum sativum*) (McDonald and Paulsen 1997). However, under drought stress, Fe content in common bean (*Phaseolus vulgaris* L.) seeds was affected the most, while N and protein contents were affected the least under the same stress condition (Ghanbari et al. 2015).

Photosynthesis rate and stomatal conductance are among the key plant processes to be affected by drought (Othman et al. 2014). In faba bean, water stress negatively impacts chlorophyll fluorescence (F_v/F_m) and maximal photochemical efficiency of photosystem-II (Abid et al. 2017). Water-deficit soybean had lower photosynthesis, symbiotic N_2 fixation, and yield than well-watered plants (Djekoun and Planchon 1991). In common bean (*Phaseolus vulgaris* L.), genotypes grown under drought conditions had lower grain yield (39%–69%) than well-watered genotypes (Ghanbari et al. 2015). Nitrogen fixation of soybean decreased steadily throughout the water deficiency period. Lower fixation process attributed to a reduction in nodule mass, which affected nodule structural constituents after a severe water-deficit stress. However, the fixation process depended more on the water status of the nodules than on the changes in nodule mass (Djekoun and Planchon 1991). Raffinose family oligosaccharides play an essential role in stress tolerance (drought and salinity) in grain legumes (e.g. *Macrotyloma uniflorum*) possibly through osmoprotection and/or ROS scavenging (George et al. 2018). Cellular analyses of *Lotus Japonica* roots using confocal microscopy showed that drought stress triggers the metabolism of reactive oxygen and nitrogen species (ROS and RNS) in cortical cells through the increase of protein nitration and the accumulation of nitric oxide and *S*-nitrosothiols (Signorelli et al. 2019).

11.3.3 Nutrient Stress

Nutrient levels including N and P in the root zone significantly affect the physiological responses of grain legumes to stress environments. Root architecture and its ability to establish effective symbiotic relationship with soil biota (rhizobia and mycorrhiza) can lead to efficient N, P, and water management (Duc et al. 2015). However, improving P acquisition and use efficiency in leguminous crops are potential challenges in the face of high P demand of legumes for N_2-fixing nodule, decreasing rock phosphate resources, rising world population, and climate change (Pang et al. 2018; Sulieman and Tran 2015).

The ability of leguminous crops to nodulate is a substantial characteristic especially in arid and semiarid environments. Generally, all legumes including grain leguminous crops have the potential to nodulate (Sprent and Gehlot 2010). Legumes have been known to fix atmospheric N_2 through the symbiotic relationship with rhizobia and then give the fixed N to soils and plant N budgets (Valentine et al. 2017). Phosphorus plays key roles in nodule development and N acquisition and metabolism in legumes (Valentine et al. 2017). In fact, severe P deficiency can lead to inhibition of nodule growth and formation process (Almeida et al. 2000). Therefore, nodules have developed several strategies to mitigate the immediate impact of P deficiency and to respond to prolonged P limitation, such as the recycling of amino acids (organic N) glutamate dehydrogenase and aspartate aminotransferase (Valentine et al. 2017). At high soil P levels, legume plants (e.g. *Virgilia divaricata*) transport more amino acids relative to inorganic N and ureides (final products of N_2 fixation) in their xylem sap to shoots, whereas at low P supply the plants transport higher amounts of ureides compared to amino acids and NH_4 (Magadlela et al. 2017). In white clover (*Trifolium repens*), P deficiency induced organic N feedback regulation (symbiotic N_2 fixation) by high production of amino acid asparagine in the nodules and roots (Almeida et al. 2000).

Nodules are the unique structures that develop on legume roots because of the symbiotic association with rhizobia. Generally, both biotic and abiotic stresses induce indigenous ethylene production, which is known for its negative influence in nodulation. Some rhizobial strains nodulate the host plant (legumes) by reducing endogenous ethylene production in the roots through their 1-aminocyclopropane-1-carboxylate deaminase activity (Shaharoona et al. 2011).

Legumes have developed adaptive strategies at physiological level to mitigate P deficiency including (1) the conservation of the P-homeostasis in nodules as an essential approach for rhizobia–legume symbiosis under low P levels and (2) the stabilization of P levels in the symbiotic tissues through the formation of a strong P sink in nodules and the P remobilization from organic-P containing substances (Sulieman and Tran 2015). The bacterial adaptation of legumes to fluctuation in P and N supply via nodule metabolites and xylem transport process to shoot might be an essential factor that supports the growth of this leguminous plant under these inconstant soil conditions (Magadlela et al. 2017).

11.4 Agronomic Approaches to Management of Stress Induced by Climate Change

Increased global atmospheric CO_2 (eCO_2) has been shown to have a wide range of potential environmental effects (IPCC 2018). Not all effects will express themselves equally in every region but rather in a mix that varies from region to region. The primary driver is increased global atmospheric CO_2 concentration, which creates a greenhouse effect and retains more heat in the Earth's atmosphere. In C3 plants such as legumes, this can directly affect, and often increase, plant growth (Taub 2010) via changes in stomatal conductance (Parvin et al. 2019), which can increase CO_2 uptake, reduce ozone sensitivity (Morgan et al. 2003), and reduce water use under drought conditions (Parvin et al. 2019). Increased atmospheric CO_2 concentration will also directly affect pH of any water bodies it dissolves in (rain, lakes, oceans; Cross et al. 2014). The result of these pH changes can alter nutrient availability including dissolution of carbonates. Secondary effects from increased trapping of energy in the atmosphere will result in increased global temperatures (IPCC 2018), overall increased rainfall, locally reduced rainfall, increased evaporative demand (Bengtsson et al. 2009), increased climate and weather event variability (Bao et al. 2017), increased water demand, potential changes in climate boundaries, and altered cloud cover, which may in turn affect frost frequencies (IPCC 2018). Tertiary changes will also arise from altered mixes of these environmental changes. Examples include changes in ranges of pests and diseases, altered breakdown rates of manures, and complex interactions of diseases, pests, CO_2 levels and water, and nutrient availability (Dáder et al. 2016). Complex cropping adaptation strategies will be required involving government, social, varietal, and agronomic changes (Debela et al. 2019). Generally, discussion will hinge more on areas where secondary plant stress losses exceed the benefits of elevated CO_2 (e.g. Ullah et al. 2018). Agronomic adaptation, including breeding legumes suited to the new conditions, will be a major means of meeting these complex interacting environmental changes.

The best available varieties and agronomic practices for a region will normally be known for historical climates, and these will have been implemented for a crop to the extent available, financially possible, and accepted by local growers. For example, Pulse Australia provides production guides across Australia for a range of pulses (http://www.pulseaus.com.au/growing-pulses/publications) and Pooniya et al. (2015) provide a summary of pulse management principles for India. However, climate change will alter a complex mix of environmental parameters and modified agronomic responses will be needed to take advantage of beneficial changes or manage complex interacting stresses. Even if environmental changes are potentially advantageous for a crop, the new environment may induce stresses if management does not change. For example, increased CO_2 will mean more carbon can be fixed leading to larger plants which may then suffer water, N, or P stress due to inadequate application if application rates are not raised to target the increased yield potential. Agronomic management will often require shift of the cropping options to higher, cooler altitudes or higher or lower latitudes along natural environmental gradients. However, this means that there will be some disconnection between daylength and UV exposure and other environmental variables which affect varietal phenology and suitability with elevated CO_2 adding further confusion potentially requiring new optimal planting dates, fertility, and pest management practices.

Yadav et al. (2010) have extensively reviewed the agronomic management of cool season pulses in order to mitigate the adverse effects of climate change. Therefore, this section will stress recent studies under specific agronomic management practices aimed at specific individual or interacting stresses. Much of the information will come from Free Air CO_2 Enrichment (FACE) research (Jones et al. 2014), open top chamber research (OTC; Li et al. 2017), enrichment chambers (EC), and simulation studies (Irigoyen et al. 2014). The Australian AgFace (https://www.piccc.org.au/research/AGFACE) has looked at stresses in peas, chickpeas, and lentils in addition to nonlegume grain crops. SoyFACE (https://soyface.illinois.edu/) has investigated soybeans. Overall however, grain legumes have not been as well studied in field stress studies at eCO_2 as other C3 crops (Bishop et al. 2014 list 6 of 41 C3 studies were grain legumes). Grain legumes have a particular set of characteristics that sets them apart as crops. Generally they are large seeded with high seed protein, they fix nitrogen, and are thus more limited by carbohydrate availability and stresses that adversely affect their symbiosis (e.g. waterlogging, phosphorus), in general they have had less breeding and thus are less improved than cereals (though soybeans are an exception), they are often high in nitrogen and thus prime targets for pests and diseases, frequently they are less determinate, they all have C3 physiology (which photosynthesize more with elevated CO_2; Bishop et al. 2014) and are frequently directly consumed by people and thus seed food quality is particularly important. Agronomic management decisions thus need to be made within these constraints a summary of which is given in Table 11.1.

TABLE 11.1

Summary of Environmental Effects Resulting from Climate Change, Potential Cropping Effects and Agronomic Responses for Grain Legume Production

Stress	Distribution	Mechanism	Negative Effects	Agronomic Response	Benefit	Agronomic Response	Legume Specific Effects
Elevated CO_2 creating Photosynthesis. Acclimation (PSA)	Universally eCO_2 up equally	eCO_2 increases C fixation (PS) increases C:N plus growth and demand for other nutrients. PSA can result.	Dilution of available minerals; extra water use due to extra growth, may lower harvest index.	Breed for improved eCO_2 use and transfer to grain. Lower planting rates. Increase nutrient application.	Higher growth and yield if PSA low. Support higher C and N fixation and P uptake, reduced sensitivity to other moderate stresses.	Ensure good nodulation and VA presence. Apply minerals to support higher growth.	Less PSA in C3. Other sinks, (N fixation, VA infection) reduce PSA.
eCO_2 in water; lower pH	Universally pH down equally	eCO_2 results in greater CO_2 in water. Small decrease in pH.	Minor soil acidification? Minor altered mineral solubilities.	Potentially increase application of minerals mobilised at high pH.	Increase low pH available nutrients.	Reduce application of affected minerals.	-
Water stress (W)	Variable ± Interacts with temp, P other stresses	± due to changed climate zones. Up as higher T increases evapotranspiration.	Greater growth, high T increases evapotranspiration and water stress.	Irrigate, lower planting rate, breed for better suited varieties.	Less susceptible to drought stress due to eCO_2 reducing water demand.	Increase yield expectation thus increase other inputs.	Less effect of drought on N fixation.
Temp. stress (T)	Universally up but variable extent	eCO_2 greenhouse effect increases global T (~1-2 degrees presently).	If above optimal lower seed quality, germination, plant growth, increase water demand.	Breed for better T response lines, change planting times, change crops, increase irrigation.	Below optimal T areas increased germination, growth and seed quality.	Modify variety, modify yield expectation and agronomy.	Can affect symbiosis as well as plant.
Mineral Nutrient stress (M)	Variable often increased stress	Greater PS increases demand may reduce sensitivity to moderate stress. Dilution of grain minerals.	High stress (P,K) can result in PSA. Moderate stress less sensitive due to eCO_2. Low mineral quality of grain.	Breed for increased mineral uptake. Increase application of nutrients and micronutrients. Improve VA and nodulation.	Greater PS increases N fixation and VA mineral uptake. Faster mineralization of N. Less sensitive to moderate stress.	Breed for increased N fixation and VA infection, modify nutrient application.	Can reduce NO_3^- sensitivity of fixation.
Storm Stress (S)	Variable often increased	Increased high intensity events due to extra greenhouse energy.	Increased erosion, waterlogging, soil crusting and hypoxia.	Raised bed cropping, manage for increased soil surface protection.	More rain reduces drought.	Select genetics, nutrients suited to extra water.	Waterlogging harms nodules.
Pests and Diseases (P)	Variable up, down, unaffected	Altered climates change pest distribution, altered plant biochemistry changes pest nutrition.	Can increase pest and disease spread and intensity of effect.	Re-evaluate disease, pest severity. Breed-resistant varieties. Modify and raise protection schedules.	Higher C:N ratio may reduce feeding. Higher temps, altered humidity may stress pests.	Reduce chemical control based on reduced incidences.	Higher C:N may reduce feeding behaviour.
Soil response to rotation (R)	Variable often elevated C:N	Higher PS, higher C:N, biomass. Interact with T, W, S, and M.	less release of N to soil more tied up with carbon.	Make greater use of mulches. Increase fertilizer rates.	Greater organic matter. Higher T gives faster N mineralization.	Greater use of mulches, modify N application.	May reduce legume N benefit of rotation.

11.4.1 CO_2 Stress

Low atmospheric CO_2 concentration has limiting effects on most crops and thus constitutes an environmental stress. However, in most crops the full benefits of enhanced atmospheric CO_2 reducing this limitation cannot be realized due to photosynthetic acclimation (Cai et al. 2018). Photosynthetic acclimation (PSA) arises from other factors not being able to increase at the same rate as photosynthesis (PS) and as a consequence C:N ratios will increase impeding further photosynthesis. However, in legumes their ability to convert fixed C to fixed N in the nodules generates an additional sink that eliminates the main source of photosynthetic acclimation. Under different circumstances modelling indicates other factors (e.g. P, K, water) may become limiting and lead to photosynthetic acclimation (Irigoyen et al. 2014). Thus, modification (i.e. increased application) of these factors from the previous optimal agronomy may be required as discussed in the next sections. Additionally, enhancement of arbuscular mycorrhizal fungi colonizing legume roots may increase the C sink (Kaschuk et al. 2010) and increase nutrient uptake thereby overcoming nutrient (e.g. P, K) limitations. Defoliation of feed legumes may also increase C demand and thus prevent photosynthetic acclimation. To be of benefit to grain growers the enhanced growth under eCO_2 must lead to enhanced grain yield. However, for peanuts grown at eCO_2 (Bannayan et al. 2009) with the present varieties, harvest index was reduced generating large, low-yield plants. Direct varietal selection for enhanced yield under eCO_2 via a range of physiological mechanisms (Tausz et al. 2013) is necessary and has been shown possible. Current cultivars of pea (Bourgault et al. 2016), field bean (Bunce, 2008), and soybean (Li et al. 2017) have shown a range of responses across cultivars to eCO_2. In soybean the range was 5%–105% grain increase under elevated CO_2 for spaced plants, but only 0%–62% for plants grown at normal spacing. Thus, some of the benefit that arises from additional growth under eCO_2 would be impeded in highly competitive close spaced cropping situations.

Breeding and selection of new legume varieties suited to the altered environments arising from climate change can be expected to continue to be a major agronomic response for maintaining and increasing yields. To achieve full breeding benefits under eCO_2 conditions, selection for physiological traits and benefits in high CO_2 environments similar to the new field situation will be needed. Indeed assuming 10 years to create and release a new variety the responses of the crop will need to be aimed at conditions expected 20 years in the future. A difficult task. Other field changes that may have occurred include elevated temperature, reduced rainfall, more rapid organic matter turnover, increased rainfall intensity, and sowing at a different latitude or altitude. Additionally, modified or new limitations induced by the effects of eCO_2, (e.g. water, temperature, sowing date, N, P, K, etc.) will require agronomic changes to maximally overcome CO_2 limitation to the extent possible. Only under these circumstances will full benefits be obtained in legumes under climate change scenarios.

11.4.2 Water Stress

Theoretically C3 crops should be able to take up more CO_2 per unit loss of water with eCO_2. This is well supported by the evidence from FACE and OTC results (Bishop et al. 2014). In their metaanalysis the prediction that the relative stimulation of yield by eCO_2 would be greatest in dry conditions was generally supported. However, there are four limitations to this conclusion. Firstly, while relative stimulation fell, the total yield stimulation from added moisture did not, indicating return for irrigation would be the same under eCO_2 conditions. Secondly, the data only had limited input from C3 legumes (soybeans) and the effect was not obvious if only the legumes were included. More recent data with peas (Jin et al. 2015) confirms that with eCO_2 drought stress was reduced. Additionally, the elevated temperatures associated with climate change would add to water stress by increasing evapotranspiration and potentially reducing much of the benefit. Finally, while global precipitation changes are small, long-term precipitation maps indicate numerous prime legume growing areas where rainfall has decreased (Adler et al. 2017; Jin et al. 2015). When all these effects are taken together water stress will remain of similar concern to the present. Breeding is needed for; (1) good germination under water stress and (2) phenology suited to changing climates with altered planting times, water availability, and temperature profiles. Other agronomic changes include modifying plant population to better match water availability and evaporative demand, selecting of optimal species mix for the new environment, managing rotations for conservation of water, and minimizing water loss through weed control and surface management (e.g. mulching). These actions will all remain as core agronomic activities in the eCO_2, higher temperature, variable rainfall, future

Breeding for improved physiology may also benefit yields. While no research has been conducted on legumes, Tausz-Posch et al. (2012) showed high Transpiration Use Efficient wheat (TUE) had an additive effect with eCO_2 (2% yield increase at ambient CO_2 against 19% increase at eCO_2). This is consistent with the idea that overcoming multiple limitations at eCO_2 gives the greatest gains. Gray et al. (2013) showed eCO_2 minimized the effect of drought on nitrogen fixation in soybeans. Achieving high seed yield and quality will remain as key factors in crop management to ensure that growers achieve maximum returns. In addition to these direct agronomic management methods aimed specifically at drought alleviation under eCO_2 reduction of other limitations such as P (Jin et al. 2015) would also reduce the effect of drought. This is consistent with the work of Bannayan et al. (2009) with peanuts at higher eCO_2. They showed substantial increases in top growth in all lines but reduced seed yields in some, suggesting potential new limiting factors had arisen (e.g. nutrients, water, competitive sinks) as a result requiring either breeding or agronomic modification in the environment to eliminate the new effects.

11.4.3 Temperature Stress

The earliest manifestation of temperature stress is at sowing affecting both germination rate and percentage (Raveneau et al. 2011). They indicated, at temperatures above the optimal germination temperature for pea and field bean lines, percentage of germination could fall by ~5%–20% per degree, potentially adversely affecting establishment. In the limited number of lines they used, they also found a wide range in optimal germination temperatures of between 20°C–27°C for pea and 30°C–36°C for beans. Below the optimal temperature percentage, germination decreased (more slowly than above optimal) and rate of germination also significantly decreased. Under these suboptimal conditions the increased temperatures could lead to increased and faster germination and hence greater early vegetative growth. The consequence of this is that in an increasing temperature environment there will be a need to reselect lines for the new germination temperatures or modify optimal planting depth and time to either avoid or take advantage of the changing germination temperatures. Indeed with a 2 degree temperature rise, areas marginal for peas may become better suited to beans and changing crops becomes a real possibility.

Potentially, increased growing season temperatures could enhance crop growth benefit from eCO_2. The greater availability of CO_2 would mean the increase in photorespiration with increased temperature would not be as great. However, Bishop et al. (2014) found growing season average temperature was not a good predictor of the magnitude of biomass and yield responses to eCO_2. Thus, regionally the increase in temperatures could be of benefit in cool environments and deleterious in hot environments. In particular, it would increase evapotranspiration and interact with drought. Agronomically this could lead to changes needed in planting dates and consequent breeding for changed phenology as well as utilization of typical drought management agronomic methods such as lower sowing rates, mulching, lower yield target nutrient inputs etc. Increased soil temperatures could allow earlier planting and increase soil N mineralization (Guntiñas et al. 2012) and thus modify timing and amount of starter N fertilization.

Bannayan et al. (2009) grew peanuts at elevated temperatures and eCO_2. However, at higher temperatures the biomass response was reduced to eCO_2. Thus, the sensitivity of individual cultivars to temperature will modify the response to eCO_2. Therefore, to achieve maximum yields it will be necessary to reselect cultivars based on the new temperature regime at the higher CO_2 level.

11.4.4 Nutritional Stress

Under severe K deficiency soybeans showed little growth or yield enhancement with elevated atmospheric CO_2 levels. Thus, optimum K fertilization is needed for maximum crop productivity under eCO_2. However, eCO_2 partially compensated for growth inhibition and seed yield with low rather than severe limitations for K (Singh and Reddy 2017).

Even in non-N fixing eucalyptus woodlands limited P and N availability can be increased under eCO_2 conditions, presumably through increased soil available carbohydrates from the root system increasing soil N fixation and P mobilization (Hasegawa et al. 2016). He et al. (2014) provide direct evidence for changes in soybean soil microbial profile in eCO_2 with enhanced presence of P and N fixing and mobilizing microbes. Within N fixing, VA infected legume crops, it would be reasonable to expect stimulation of N and P availability to be greater. This greater stimulation seems to be the case for N in peas. Bourgault et al. (2016) found 26% and 28% stimulation of pea seed and biomass yield with eCO_2 in an irrigated/dryland FACE experiment. Across the five varieties used some maintained seed N and plant N levels in the eCO_2 treatment indicating similar increased N uptake with similar harvest indices. Li et al. (2017) found similar variety dependent increases in yield (mean 40%) in soybeans grown at eCO_2. Seed and plant N levels increased equally primarily via fixed N. These data suggest that the greater C fixation fed greater nodular N fixation and additional N fertilization is not needed but additional fertilization with other minerals may be necessary.

One of the issues in many legumes is the initial N stress they may suffer if there is inadequate seed and soil availability to establish the symbiosis. This is often overcome by some early starter N fertilization, but care is needed not to apply so much nitrogen as to inhibit the establishment of effective nodular fixation of N. Potentially with greater C fixation at eCO_2 the usual balance could be upset. Butterly et al. (2016) investigated the inhibitory effect of soil NO_3^- on N fixation of peas in eCO_2. They found the inhibitory effect was significantly reduced and thus early N fertilization to stimulate plant growth could maximally benefit legume N fixation and yield under eCO_2.

N fixation under eCO_2 seems to be more robust to a range of stresses. Nitrogen fixation is generally highly sensitive to water stress. However, Parvin et al. (2019) showed that pea N fixation was less affected by water stress conditions under eCO_2. Gray et al. (2013) showed similar reductions of drought stress on N fixation in soybean. Jin et al. (2015) showed that combined P application and eCO_2 enhanced drought tolerance. Overall there appears to be a reduction in drought, N and P stress under eCO_2 enabling legumes to take advantage of greater C fixation to enhance yields. However, selection of the most suitable varieties is critical, and even though stress may be reduced there are still benefits to be obtained due to the enhanced photosynthesis by reducing stress by agronomic practices such as increased P and early N fertilization and irrigation.

Overall, nutritional stresses can be induced by increased C fixation generating increased demand for the nutrients in the greater biomass and yield as a result of eCO_2. However, at the same time the ability of the legumes tested so far indicate the plant has a greater ability to deal with these stresses and increase uptake or fixation. The full benefit of the increased C fixation, however, only becomes available if greater application of nutrients takes place and plants are bred specifically for the new higher yield potential situation.

11.4.5 Rotational

Lam et al. (2013) found that peas grown under eCO_2 had substantially increased biomass and nitrogen though with a higher C:N ratio. While the pea rotation benefitted the following wheat crop in all circumstances the nitrogen benefit from the eCO_2 pea crop (even though it had greater total N) was less than from the lower productivity ambient CO_2 pea crop. Increased temperature and higher C:N ratios may have tied up more of the N in the soil resulting in less availability in the rotation. Rotations will therefore need to be modified to take into account the increase in total biomass and increased C:N ratio. Potentially this will mean changes are needed in fertilization regimes and trash management.

11.4.6 Storm Intensity

Both modelling and direct measurement support the existence of increased rainfall intensity as a result of climate change (Bao et al. 2017). Thus, agronomic management will need to change to incorporate potential effects of the increased intensity. The main effects will be increased potential for surface erosion, potential hypoxia from increased waterlogging (further exacerbated by increased global temperatures), and physical destruction of plant yield via lodging, shattering, or grain damage. Hypoxia may have severe adverse effects on N fixation in legumes as well as damaging seeds at planting. Systems for managing high rainfall intensity events exist already in wet warm production areas such as raised bed cropping. This form of cropping removes water quickly from the surface roots and has been shown to increase legume yields (e.g. faba beans, Armstrong et al. 2015). Breeding for increased resistance to waterlogging is possible and extensively undertaken in soybean (Valliyodan et al. 2017). Surface erosion protection can be improved using existing methods such as maintaining groundcover, surface mulching, no till cropping, and contour ploughing. Generally breeding for quality, lodging, and shattering resistance are already goals within most legume breeding programmes. Lodging in peas can also be improved by intercropping with a support crop such as barley (Podgorska-Lesiak and Sobkowicz 2013). Heavy rainfall events can lead to surface crusting which can reduce field establishment of legumes (particularly those with epigeal emergence) significantly both by direct increased physical impedance and by interactions with herbicides (Wilson and Sbatella 2014). This effect is likely to be greater with increased soil temperatures due to climate change.

Thus, surface protection is important at germination as well as during later growth stages and between crops. Legumes must increasingly be grown with genetic resistance to heavy rain event stresses as well as having the environment managed to reduce the impact of these higher intensity rainfall events.

11.4.7 Pests and Diseases

Pests and diseases are considered in detail elsewhere; however, their agronomic management can be expected to change with eCO_2, increased temperatures, and modified environments. Pest ranges will alter, and plant–pest interactions will change (Luck et al. 2011). Little research has been carried out on pest interactions with legumes under eCO_2 but, as an example, in bell peppers attacked by aphids and Cucumber mosaic virus (CMV) Dáder et al. (2016) found several effects that could be expected to reduce aphid damage and spray regimes. A twofold decrease on transmission of CMV was observed when receptor plants were exposed to eCO_2 before aphid inoculation when compared to ambient CO_2. However, it seems at least in Lucerne that the direction of the effect, increase or decrease in aphid intensity, can be variety specific (Johnson et al. 2014). Hughes and Bazzaz (2001) found similar reductions in aphids on *Vicia Faba*. In the SoyFACE, Eastburn et al. (2010) found $eCO_2 \pm eO_3$ reduced downy mildew by 39%–66%, increased brown spot severity slightly but did not affect incidence of Sudden Death Syndrome (SDS). These trials did not include interactions with temperature, rainfall, nutrition, and varietal changes expected to also occur with climate change. Thus, agronomic management of these pests and diseases will need to be examined and may change. Timing and need for sprays, rotations, scouting, and thresholds will all need to be adapted to the new environment with little universality in the agronomic changes needed.

11.5 Conclusions

When agronomic management of stresses of legumes is considered with respect to climate change, a few overall trends seem to stand out. However, there will be no quick alternative to the need to develop new cultivars for the new environments and systematically reoptimize agronomic management associated with the crop. Most of the agronomic options presently available will continue to be used; their optimal levels will simply need reassessment. Some trends seem to be (1) the ability of fixing legumes to make good use of the elevated CO_2 to increase growth by utilizing excess C for N fixation and VA mediated nutrient uptake. This will require varieties and treatments that allow conversion of this growth to grain. (2) The reduction of some moderate, but not severe, stresses (e.g. drought, early N, P, and K) under eCO_2 but a need to manage the increased demand put on these factors by the greater growth. (3) The potential need to rework the phenology of varieties for regions as temperature and water profiles change. (4) A swathe of new interactions will occur as the multiple environmental, pest and disease stresses are all modified together. The same types of agronomic modifications will be needed as in the past but with ongoing climate change their optimal values will be continuously shifting.

REFERENCES

Abberton MT et al. (2008) The genetic improvement of forage grasses and legumes to enhance adaptation of grasslands to climate change. FAO, May 2008.

Abdelhamid M, Shokr M, Bekheta M (2010) Growth, root characteristics, and leaf nutrients accumulation of four faba bean (*Vicia faba* L.) cultivars differing in their broomrape tolerance and the soil properties in relation to salinity. *Commun Soil Sci Plant Anal* 41(22):2713–2728.

Abdelhamid MT (2010) Efficient root system in legume crops to stress environments. In: Yadav S., Redden R. (eds) *Climate change and management of cool season grain legume crops*. Springer, Dordrecht. pp. 229–242.

Abdul S, Muhammad AA, Shabir H, Hesham AEE, Sajjad H, Niaz A, Abdul G, Sayyed RZ, Fahad S, Subhan D, Rahul D (2021a) Zinc nutrition and arbuscular mycorrhizal symbiosis effects on maize (*Zea mays* L.) growth and productivity. *J Saudi Society Agricultural Sci.* https://doi.org/10.1016/j.sjbs.2021.06.096

Abdul S, Muhammad AA, Subhan D, Niaz A, Fahad S, Rahul D, Mohammad JA, Omaima N, Muhammad HR, Bernard RG (2021b) Effect of arbuscular mycorrhizal fungi on the physiological functioning of maizeunder zinc-deficient soils. *Sci Rep* 11:18468.

Abid G, M'Hamdi M, Mingeot D et al. (2017) Effect of drought stress on chlorophyll fluorescence, antioxidant enzyme activities and gene expression patterns in faba bean (*Vicia faba* L.). *Arch Agron Soil Sci* 63:536–552.

Abu Obaid AM, Melnyk AV, Onichko VI et al. (2018) Evaluation of six sunflower cultivars for forage productivity under salinity conditions. *Adv Environ Biol* 12(7):13–15.

Abu-Darwish MS, Abu Dieyeh ZH, Batarseh et al. (2009) Trace element contents and essential oil yields from wild thyme plant (*Thymus serpyllum* L.) grown at different natural-variable environments, Jordan. *J Food Agric Environ* 7(3&4):920–924.

Adler RF, Gu G, Sapiano M et al. (2017) Global precipitation: means, variations and trends during the satellite era (1979–2014). *Surv Geophys* 38:679–699. https://doi.org/10.1007/s10712-017-9416-4

Adnan M, Fahad S, Muhammad Z, Shahen S, Ishaq AM, Subhan D, Zafar-ul-Hye M, Martin LB, Raja MMN, Beena S, Saud S, Imran A, Zhen Y, Martin B, Jiri H, Rahul D (2020) Coupling phosphate-solubilizing bacteria with phosphorus supplements improve maize phosphorus acquisition and growth under lime induced salinity stress. *Plants* 9(900). https://doi.org/10.3390/plants9070900

Adnan M, Fahad S, Khan IA, Saeed M, Ihsan MZ, Saud S, Riaz M, Wang D, Wu C (2019) Integration of poultry manure and phosphate solubilizing bacteria improved availability of Ca bound P in calcareous soils. *3 Biotech* 9(10):368.

Adnan M, Shah Z, Sharif M, Rahman H (2018b). Liming induces carbon dioxide (CO_2) emission in PSB inoculated alkaline soil supplemented with different phosphorus sources. *Environ Sci Poll Res* 25(10):9501–9509.

Adnan M, Zahir S, Fahad S, Arif M, Mukhtar A, Imtiaz AK, Ishaq AM, Abdul B, Hidayat U, Muhammad A, Inayat-Ur R, Saud S, Muhammad ZI, Yousaf J, Amanullah, Hafiz MH, Wajid N (2018a) Phosphate-solubilizing bacteria nullify the antagonistic effect of soil calcification on bioavailability of phosphorus in alkaline soils. *Sci Rep* 8:4339. https://doi.org/10.1038/s41598-018-22653-7

Ahmad S, Kamran M, Ding R, Meng X, Wang H, Ahmad I, Fahad S, Han Q (2019) Exogenous melatonin confers drought stress by promoting plant growth, photosynthetic capacity and antioxidant defense system of maize seedlings. *PeerJ* 7:e7793. http://doi.org/10.7717/peerj.7793

Ahmad W, Hassan G, Ali M, Khan N, Ishaq M, Afridi K, Shah IA, Adnan M, Tajudin R, Ullah L, Rehman AU, Hussain B (2016) Estimates of heritability, genetic advance and correlation in F3 populations of wheat. *Pure Appl Biol* 5(4):1142–1150.

Akbar H, Timothy JK, Jagadish T, Golam M, Apurbo KC, Muhammad F, Rajan B, Fahad S, Hasanuzzaman M (2020) Agricultural land degradation: processes and problems undermining future food security. In: Fahad S, Hasanuzzaman M, Alam M, Ullah H, Saeed M, Khan AK, Adnan M (eds) *Environment, climate, plant and vegetation growth.* Springer Publ Ltd, Springer Nature Switzerland AG. Part of Springer Nature, pp. 17–62. https://doi.org/10.1007/978-3-030-49732-3

Akram R, Turan V, Hammad HM, Ahmad S, Hussain S, Hasnain A, Maqbool MM, Rehmani MIA, Rasool A, Masood N, Mahmood F, Mubeen M, Sultana SR, Fahad S, Amanet K, Saleem M, Abbas Y, Akhtar HM, Waseem F, Murtaza R, Amin A, Zahoor SA, ul Din MS, Nasim W (2018a) Fate of organic and inorganic pollutants in paddy soils. In: Hashmi MZ, Varma A (eds) *Environmental pollution of paddy soils, soil biology*. Springer International Publishing Ag, Cham, Switzerland, pp. 197–214.

Akram R, Turan V, Wahid A, Ijaz M, Shahid MA, Kaleem S, Hafeez A, Maqbool MM, Chaudhary HJ, Munis, MFH, Mubeen M, Sadiq N, Murtaza R, Kazmi DH, Ali S, Khan N, Sultana SR, Fahad S, Amin A, Nasim W (2018b) Paddy land pollutants and their role in climate change. In: Hashmi MZ, Varma A (eds) *Environmental pollution of paddy soils, soil biology*. Springer International Publishing Ag, Cham, Switzerland, pp. 113–124.

Almeida J, Hartwig U, Freshner M et al. (2000) Evidence that P deficiency induces N feedback regulation of symbiotic N_2 fixation in white clover (*Trifolium repens* L.). *J Exp Bot* 51:1289–1297.

Al-Rifaee M, Turk MA, Tawaha AM (2004) Effect of seed size and plant population density on yield and yield components of local faba bean (*Vicia faba* L.) major. *Int J Agric Biol* 6(2):294–299.

Al-Tawaha A, Al-Karaki G, Al Tawaha A et al. (2018d) Effects water flow rate on quantity and quality of lettuce (*Lactuca sativa* L.) in nutrient film technique (Nft) under hydroponics conditions. *Bulg J Agric Sci* 24(5):791–798.

Al-Tawaha AM, Seguin P (2006) Effects of seeding date, row spacing, and weeds on soybean seed isoflavone concentration. *Can J Plant Sci* 86:1079–1082.

Al-Tawaha AM, Al-Tawaha AM, Alu'datt MH et al. (2018c) Effects of soil type and rainwater harvesting treatments in the growth, productivity and morphological trains of barley plants cultivated in semi-arid environment. *Aust J Crop Sci* 12(6):975–979.

Al-Tawaha AM, Seguin P, Smith DL et al. (2006a) Foliar application of elicitors alters isoflavone concentrations and other seed characteristics of field-grown soybean. *Can J Plant Sci* 86:677–684.

Al-Tawaha AM, Seguin P, Smith DL et al. (2007) Effects of irrigation on isoflavone concentrations of soybean grown in southwestern Québec. *J Agron Crop Sci* 193(4):238–246.

Al-Tawaha AM, Yadav SS, Turk M et al. (2010) Crop production and management technologies for drought prone environments. Chapter in Book. Climate Change and Drought Management in Cool Season Grain Legume Crops. (Springer).

Al-Tawaha AR, Al-Karaki GN, Al-Tawaha AR et al. (2018a) Effect of water flow rate on quantity and quality of lettuce (*Lactuca sativa* L.) in nutrient film technique (NFT) under hydroponics conditions. *Bulg J Agric Sci* 24(5):791–798.

Al-Tawaha AR, Turk MA, Al-Tawaha AR et al. (2018b) Using chitosan to improve growth of maize cultivars under salinity conditions. *Bulg J Agric Sci* 24(3):437–442.

Al-Tawaha AM, Seguin P, Smith DL et al. (2006b) Foliar application of elicitors alters isoflavone concentrations and other seed characteristics of field-grown soybean. *Can J Plant Sci* 86:677–684.

Amanullah, Muhammad I, Haider N, Shah K, Manzoor A, Asim M, Saif U, Izhar A, Fahad S, Adnan M et al. (2021) Integrated foliar nutrients application improve wheat (*Triticum aestivum* L.) productivity under calcareous soils in drylands. *Commun Soil Sci Plant Anal.* https://doi.org/10.1080/00103624.2021.1956521

Amanullah, Fahad S (eds) (2017) Rice – technology and production. IntechOpen Croatia 2017. http://dx.doi.org/10.5772/64480

Amanullah, Fahad S (eds) (2018a) Corn – Production and human Health in changing climate. IntechOpen United Kingdom 2018. http://dx.doi.org/10.5772/intechopen.74074

Amanullah, Fahad S (eds) (2018b) Nitrogen in agriculture – updates. IntechOpen Croatia 2018. http://dx.doi.org/10.5772/65846

Amanullah, Shah K, Imran, Hamdan AK, Muhammad A, Abdel RA, Muhammad A, Fahad S, Azizullah S, Brajendra P (2020) Effects of climate change on irrigation water quality. In: Fahad S, Hasanuzzaman M, Alam M, Ullah H, Saeed M, Khan AK, Adnan M (eds) *Environment, climate, plant and vegetation growth.* Springer Publ Ltd, Springer Nature Switzerland AG. Part of Springer Nature, pp. 123–132. https://doi.org/10.1007/978-3-030-49732-3

Amir M, Muhammad A, Allah B, Sevgi Ç, Haroon ZK, Muhammad A, Emre A (2020) Bio fortification under climate change: the fight between quality and quantity. In: Fahad S, Hasanuzzaman M, Alam M, Ullah H, Saeed M, Khan AK, Adnan M (eds) *Environment, climate, plant and vegetation growth.* Springer Publ Ltd, Springer Nature Switzerland AG. Part of Springer Nature, pp. 173–228. https://doi.org/10.1007/978-3-030-49732-3

Amjad I, Muhammad H, Farooq S, Anwar H (2020) Role of plant bioactives in sustainable agriculture. In: Fahad S, Hasanuzzaman M, Alam M, Ullah H, Saeed M, Khan AK, Adnan M (eds) *Environment, climate, plant and vegetation growth.* Springer Publ Ltd, Springer Nature Switzerland AG. Part of Springer Nature, pp. 591–606. https://doi.org/10.1007/978-3-030-49732-3

Amjad SF, Mansoora N, Din IU, Khalid IR, Jatoi GH, Murtaza G, Yaseen S, Naz M, Danish S, Fahad S et al. (2021) Application of zinc fertilizer and mycorrhizal inoculation on physio-biochemical parameters of wheat grown under water-stressed environment. *Sustainability* 13:11007. https://doi.org/10.3390/su131911007

Araújo S, Beebe S, Crespi M et al. (2015) Abiotic stress responses in legumes: strategies used to cope with environmental challenges. *Crit Rev Plant Sci* 34(1–3):237–280.

Arianoutsou M, Thanos CA (1996) Legumes in the fire-prone mediterranean regions: an example from Greece. *Int J Wildland Fire* 6(2):77–82.

Arif M, Talha J, Muhammad R, Fahad S, Muhammad A, Amanullah, Kawsar A, Ishaq AM, Bushra K, Fahd R (2020) Biochar; a remedy for climate change. In: Fahad S, Hasanuzzaman M, Alam M, Ullah H, Saeed M, Khan AK, Adnan M (eds) *Environment, climate, plant and vegetation growth*. Springer Publ Ltd, Springer Nature Switzerland AG. Part of Springer Nature, pp. 151–172. https://doi.org/10.1007/978-3-030-49732-3

Armstrong RD, Eagle C, Flood R (2015) Improving grain yields on a sodic clay soil in a temperate, medium-rainfall cropping environment. *Crop Pasture Sci* 66(5):492–505. https://doi.org/10.1071/CP14210

Ashfaq AR, Uzma Y, Niaz A, Muhammad AA, Fahad S, Haider S, Tayebeh Z, Subhan D, Süleyman T, Hesham AELE, Pramila T, Jamal MA, Sulaiman AA, Rahul D (2021) Toxicity of Cadmium and nickel in the context of applied activated carbon biochar for improvement in soil fertility. *Saudi Soc Agric Sci*. https://doi.org/10.1016/j.sjbs.2021.09.035

Ashrafi E, Zahedi M, Razmjoo J (2014) Co-inoculations of arbuscular mycorrhizal fungi and rhizobia under salinity in alfalfa. *Soil Sci Plant Nutr* 60(5):619–629.

Athar M, Masood IA, Sana S, Ahmed M, Xiukang W, Sajid F, Sher AK, Habib A, Faran M, Zafar H, Farhana G, Fahad S (2021) Bio-diesel production of sunflower through sulphur management in a semi-arid subtropical environment. *Environ Sci Pollution Res*. https://doi.org/10.1007/s11356-021-16688-z

Atif B, Hesham A, Fahad S (2021) Biochar coupling with phosphorus fertilization modifies antioxidant activity, osmolyte accumulation and reactive oxygen species synthesis in the leaves and xylem sap of rice cultivars under high-temperature stress. *Physiol Mol Biol Plants*. https://doi.org/10.1007/s12298-021-01062-7

Ayman ELS, Hossain A, Barutçular C, Iqbal MA, Islam MS, Fahad S, Sytar O, Çig F, Meena RS, Erman M (2020) Consequences of salinity stress on the quality of crops and its mitigation strategies for sustainable crop production: an outlook of arid and semi-arid regions. In: Fahad S, Hasanuzzaman M, Alam M, Ullah H, Saeed M, Khan AK, Adnan M (eds) *Environment, climate, plant and vegetation growth*. Springer Publ Ltd, Springer Nature Switzerland AG. Part of Springer Nature, pp. 503–534. https://doi.org/10.1007/978-3-030-49732-3

Aziz K, Daniel KYT, Fazal M, Muhammad ZA, Farooq S, Fan W, Fahad S, Ruiyang Z (2017a) Nitrogen nutrition in cotton and control strategies for greenhouse gas emissions: a review. *Environ Sci Pollut Res* 24:23471–23487. https://doi.org/10.1007/s11356-017-0131-y

Aziz K, Daniel KYT, Muhammad ZA, Honghai L, Shahbaz AT, Mir A, Fahad S (2017b) Nitrogen fertility and abiotic stresses management in cotton crop: a review. *Environ Sci Pollut Res* 24:14551–14566. https://doi.org/10.1007/s11356-017-8920-x

Bannayan M, Tojo Soler CM, Garcia Y et al. (2009) Interactive effects of elevated [CO_2] and temperature on growth and development of a short- and long-season peanut cultivar. *Clim Change* 93(3–4):389–406.

Bao J, Sherwood SC, Alexander LV et al. (2017) Future increases in extreme precipitation exceed observed scaling rates. *Nature Clim Change Lett*, January pp. 1–6. https://doi.org/10.1175/JCLI-D-19-0920.1

Baseer M, Adnan M, Fazal M, Fahad S, Muhammad S, Fazli W, Muhammad A, Jr. Amanullah, Depeng W, Saud S, Muhammad N, Muhammad Z, Fazli S, Beena S, Mian AR, Ishaq AM (2019) Substituting urea by organic wastes for improving maize yield in alkaline soil. *J Plant Nutr.* https://doi.org/10.1080/01904167.2019.1659344

Bayram AY, Seher Ö, Nazlican A (2020) Climate change forecasting and modeling for the year of 2050. In: Fahad S, Hasanuzzaman M, Alam M, Ullah H, Saeed M, Khan AK, Adnan M (eds) *Environment, climate, plant and vegetation growth*. Springer Publ Ltd, Springer Nature Switzerland AG. Part of Springer Nature, pp. 109–122. https://doi.org/10.1007/978-3-030-49732-3

Bengtsson L, Hodges KI, Keenlyside N (2009) Will extratropical storms intensify in a warmer climate? *J Clim* 22:2276–2301.

Bishop KA, Leakey ADB, Ainsworth EA (2014) How seasonal temperature or water inputs affect the relative response of C3 crops to elevated [CO_2]: a global analysis of open top chamber and free air CO_2 enrichment studies. *Food Energy Secur* 3(1):33–45.

Bishop J (2016) Susceptibility of Faba Bean (*Vicia faba* L.) to heat stress during floral development and anthesis. *J Agron Crop Sci* 202(6):508–516.
Bourgault M, Brand J, Tausz M, Fitzgerald GJ (2016) Yield, growth and grain nitrogen response to elevated CO_2 of five field pea (*Pisum sativum* L.) cultivars in a low rainfall environment. *Field Crops Res* 196:1–9.
Bruning B, Rozema J (2013) Symbiotic nitrogen fixation in legumes: perspectives for saline agriculture. *Environ Exp Bot* 92:134–143.
Bulgarelli RG, Marcos FCC, Ribeiro RV et al. (2017) Mycorrhizae enhance nitrogen fixation and photosynthesis in phosphorus-starved soybean (*Glycine max* L. Merrill). *Environ Exp Bot* 140:26–33.
Bulut F, Akıncı S, Eroğlu A (2011) Growth and uptake of sodium and potassium in broad bean (*vicia faba*) under salinity stress. *Commun Soil Sci Plant Anal* 42(8):945–961.
Bunce JA (2008) Contrasting responses of seed yield to elevated carbon dioxide under field conditions within *Phaseolus vulgaris*. *Agric Ecosyst Environ* 128:219–224.
Bustingorri C, Lavado R (2013) Soybean response and ion accumulation under sprinkler irrigation with sodium-rich saline water. *J Plant Nutr* 36(11):1743–1766.
Butterly CR, Armstrong R, Chen D, Tang C (2016) Free-air CO_2 enrichment (FACE) reduces the inhibitory effect of soil nitrate on N2 fixation of *Pisum sativum*. *Ann Bot* 117:177–185.
Cai C, Li G, Yang H et al. (2018) Do all leaf photosynthesis parameters of rice acclimate to elevated CO_2, elevated temperature, and their combination, in FACE environments? *Glob Change Biol* 24(4):1685–1707.
Canci H, Toker C (2009) Evaluation of annual wild Cicer species for drought and heat resistance under field conditions. *Genet Resour Crop Evol* 56:1. https://doi.org/10.1007/s10722-008-9335-9
Chang W, Qiujuan J, Evgenios A, Haitao L, Gezi L, Jingjing Z, Fahad S, Ying J (2021) Hormetic effects of zinc on growth and antioxidant defense system of wheat plants. *Sci Total Environ.* https://doi.org/10.1016/j.scitotenv.2021.150992
Chen Y, Guo Z, Dong L, Fu Z, Zheng Q, Zhang G, Qin L, Sun X, Shi Z, Fahad S, Xie F, Saud S (2021) Turf performance and physiological responses of native Poa species to summer stress in Northeast China. *PeerJ* 9:e12252 http://doi.org/10.7717/peerj.12252
Cross JN, Monacci NM, Musielewicz S et al. (2014) High-resolution ocean and atmosphere pCO_2 timeseries measurements from mooring GAKOA_149W_60N in the Gulf of Alaska from 2011-05-19 to 2017-01-20 (NODC Accession 0116714), Version 5.5, National Oceanographic Data Center, NOAA, Dataset, https://doi.org/10.3334/CDIAC/OTG. TSM_GAKOA_149W_60N, 2014.
Cutforth HW (2007) Adaptation of pulse crops to the changing climate of the Northern Great Plains. *Agron J* 99:1684–1699.
Dáder B, Fereres A, Moreno A et al. (2016) Elevated CO_2 impacts bell pepper growth with consequences to *Myzus persicae* life history, feeding behaviour and virus transmission ability. *Sci Rep* 6:19–120.
Dadresan M, Luthe D, Reddivari L et al. (2015) Effect of salinity stress and surfactant treatment on physiological traits and nutrient absorption of fenugreek plant. *Commun Soil Sci Plant Anal* 46(22):2807–2820.
Debela N, McNeil DL, Bridle K et al. (2019) Adaptation to climate change in the pastoral and agropastoral systems of Borana, South Ethiopia: options and barriers. *Am J Clim Change* 8(1):40–60.
Deepranjan S, Ardith S, Siva D, Sonam S, Manoj P, Amitava R, Sayyed RZ, Abdul G, Mohammad JA, Subhan D, Fahad S, Rahul D (2021) Optimizing nutrient use efficiency, productivity, energetics, and economics of red cabbage following mineral fertilization and biopriming with compatible rhizosphere microbes. *Sci Rep* 11:15680.
Depeng W, Fahad S, Saud S, Muhammad K, Aziz K, Mohammad NK, Hafiz MH, Wajid N (2018) Morphological acclimation to agronomic manipulation in leaf dispersion and orientation to promote "Ideotype" breeding: evidence from 3D visual modeling of "super" rice (*Oryza sativa* L.). *Plant Physiol Biochem.* 135:499–510 https://doi.org/10.1016/j.plaphy.2018.11.010
Devasirvatham V et al. (2012) High temperature tolerance in chickpea and its implications for plant improvement. *Crop Pasture Sci* 63(5):419–428.
Djekoun A, Planchon C (1991) Water status effect on dinitrogen fixation and photosynthesis in soybean. *Agron J* 83:316–322.
Duc D, Agrama H, Bao S et al. (2015) Breeding annual grain legumes for sustainable agriculture: new methods to approach complex traits and target new cultivar ideotypes. *Crit Rev Plant Sci* 34(1–3):381–411.
Eastburn D, Degennaro M, Delucia E et al. (2010) Elevated atmospheric carbon dioxide and ozone alter soybean diseases at SoyFACE. *Glob Change Biol* 16:320–330.

Emre B, Ömer SU, Martín LB, Andre D, Fahad S, Rahul D, Muhammad Z-H, Ghulam SH, Subhan D (2021) Studying soil erosion by evaluating changes in physico-chemical properties of soils under different land-use types. *J Saudi Soc Agric Sci* 20:190–197.

Esechie H, Al-Barhi B, Al-Gheity S et al. (2002). Root and shoot growth in salinity-stressed alfalfa in response to nitrogen source. *J Plant Nutr* 25:2559–2569.

Esechie H, Rodriguez V, (1998) Ion compartmentation in salinity-stressed alfalfa seedlings growing under different temperature regimes. *Commun Soil Sci Plant Anal* 29(17–18): 2607–2618.

Fahad S, Abdul B, Adnan M (eds) (2018b) Global wheat production. IntechOpen United Kingdom 2018. http://dx.doi.org/10.5772/intechopen.72559

Fahad S, Bajwa AA, Nazir U, Anjum SA, Farooq A, Zohaib A, Sadia S, Nasim W, Adkins S, Saud S, Ihsan MZ, Alharby H, Wu C, Wang D, Huang J (2017) Crop production under drought and heat stress: plant responses and management options. *Front Plant Sci* 8:1147. https://doi.org/10.3389/fpls.2017.01147

Fahad S, Bano A (2012) Effect of salicylic acid on physiological and biochemical characterization of maize grown in saline area. *Pak J Bot* 44:1433–1438.

Fahad S, Chen Y, Saud S, Wang K, Xiong D, Chen C, Wu C, Shah F, Nie L, Huang J (2013) Ultraviolet radiation effect on photosynthetic pigments, biochemical attributes, antioxidant enzyme activity and hormonal contents of wheat. *J Food Agri Environ* 11(3&4):1635–1641.

Fahad S, Hussain S, Bano A, Saud S, Hassan S, Shan D, Khan FA, Khan F, Chen Y, Wu C, Tabassum MA, Chun MX, Afzal M, Jan A, Jan MT, Huang J (2014a) Potential role of phytohormones and plant growth-promoting rhizobacteria in abiotic stresses: consequences for changing environment. *Environ Sci Pollut Res* 22(7):4907–4921. https://doi.org/10.1007/s11356-014-3754-2

Fahad S, Hussain S, Matloob A, Khan FA, Khaliq A, Saud S, Hassan S, Shan D, Khan F, Ullah N, Faiq M, Khan MR, Tareen AK, Khan A, Ullah A, Ullah N, Huang J (2014b) Phytohormones and plant responses to salinity stress: a review. *Plant Growth Regul* 75(2):391–404. https://doi.org/10.1007/s10725-014-0013-y

Fahad S, Hussain S, Saud S, Hassan S, Chauhan BS, Khan F et al (2016a) Responses of rapid viscoanalyzer profile and other rice grain qualities to exogenously applied plant growth regulators under high day and high night temperatures. *PLoS One* 11(7):e0159590. https://doi.org/10.1371/journal.pone.0159590

Fahad S, Hussain S, Saud S, Hassan S, Ihsan Z, Shah AN, Wu C, Yousaf M, Nasim W, Alharby H, Alghabari F, Huang J (2016c) Exogenously applied plant growth regulators enhance the morphophysiological growth and yield of rice under high temperature. *Front Plant Sci* 7:1250. https://doi.org/10.3389/fpls.2016.01250

Fahad S, Hussain S, Saud S, Hassan S, Tanveer M, Ihsan MZ, Shah AN, Ullah A, Nasrullah KF, Ullah S, Alharby HNW, Wu C, Huang J (2016d) A combined application of biochar and phosphorus alleviates heat-induced adversities on physiological, agronomical and quality attributes of rice. *Plant Physiol Biochem* 103:191–198.

Fahad S, Hussain S, Saud S, Khan F, Hassan S, Jr, Nasim W, Arif M, Wang F, Huang J (2016b) Exogenously applied plant growth regulators affect heat-stressed rice pollens. *J Agron Crop Sci* 202:139–150.

Fahad S, Hussain S, Saud S, Tanveer M, Bajwa AA, Hassan S, Shah AN, Ullah A, Wu C, Khan FA, Shah F, Ullah S, Chen Y, Huang J (2015a) A biochar application protects rice pollen from high-temperature stress. *Plant Physiol Biochem* 96:281–287.

Fahad S, Muhammad ZI, Abdul K, Ihsanullah D, Saud S, Saleh A, Wajid N, Muhammad A, Imtiaz AK, Chao W, Depeng W, Jianliang H (2018a) Consequences of high temperature under changing climate optima for rice pollen characteristics-concepts and perspectives. *Archives Agron Soil Sci.* https://doi.org/10.1080/03650340.2018.1443213

Fahad S, Nie L, Chen Y, Wu C, Xiong D, Saud S, Hongyan L, Cui K, Huang J (2015b) Crop plant hormones and environmental stress. *Sustain Agric Rev* 15:371–400.

Fahad S, Saud S, Yajun C, Chao W, Depeng W (eds) (2021f) Abiotic stress in plants. IntechOpen United Kingdom 2021. http://dx.doi.org/10.5772/intechopen.91549

Fahad S, Hasanuzzaman M, Alam M, Ullah H, Saeed M, Ali Khan I, Adnan M (eds) (2020) *Environment, climate, plant and vegetation growth.* Springer Nature Switzerland AG. https://doi.org/10.1007/978-3-030-49732-3

Fahad S, Adnan M, Hassan S, Saud S, Hussain S, Wu C, Wang D, Hakeem KR, Alharby HF, Turan V, Khan MA, Huang J (2019b) Rice responses and tolerance to high temperature. In: Hasanuzzaman M, Fujita M, Nahar K, Biswas JK (eds) *Advances in rice research for abiotic stress tolerance.* Woodhead Publ Ltd, Abington, Cambridge, UK, pp. 201–224.

Fahad S, Rehman A, Shahzad B, Tanveer M, Saud S, Kamran M, Ihtisham M, Khan SU, Turan V, Rahman MHU (2019a) Rice responses and tolerance to metal/metalloid toxicity. In: Hasanuzzaman M, Fujita M, Nahar K, Biswas JK (eds) *Advances in rice research for abiotic stress tolerance*. Woodhead Publ Ltd, Abington, Cambridge, UK, pp. 299–312.

Fahad S, Sönmez O, Saud S, Wang D, Wu C, Adnan M, Arif M, Amanullah (eds) (2021e) Engineering tolerance in crop plants against abiotic stress. In: *Footprints of climate variability on plant diversity*, 1st edn. CRC Press, Boca Raton.

Fahad S, Sonmez O, Saud S, Wang D, Wu C, Adnan M, Turan V (eds) (2021b) Climate change and plants: biodiversity, growth and interactions. In: *Footprints of climate variability on plant diversity*, 1st edn. CRC Press, Boca Raton.

Fahad S, Sonmez O, Saud S, Wang D, Wu C, Adnan M, Turan V (eds) (2021c) Developing climate resilient crops: improving global food security and safety. In: *Footprints of climate variability on plant diversity*, 1st edn. CRC Press, Boca Raton.

Fahad S, Sönmez O, Saud S, Wang D, Wu C, Adnan M, Turan V (eds) (2021a) Plant growth regulators for climate-smart agriculture. In: *Footprints of climate variability on plant diversity*, 1st edn. CRC Press, Boca Raton.

Fahad S, Sönmez O, Turan V, Adnan M, Saud S, Wu C, Wang D (eds) (2021d) Sustainable soil and land management and climate change. In: *Footprints of climate variability on plant diversity*, 1st edn. CRC Press, Boca Raton.

Farah R, Muhammad R, Muhammad SA, Tahira Y, Muhammad AA, Maryam A, Shafaqat A, Rashid M, Muhammad R, Qaiser H, Afia Z, Muhammad AA, Muhammad A, Fahad S (2020) Alternative and non-conventional soil and crop management strategies for increasing water use efficiency. In: Fahad S, Hasanuzzaman M, Alam M, Ullah H, Saeed M, Khan AK, Adnan M (eds) *Environment, climate, plant and vegetation growth*. Springer Publ Ltd, Springer Nature Switzerland AG. Part of Springer Nature, pp. 323–338. https://doi.org/10.1007/978-3-030-49732-3

Farhana G, Ishfaq A, Muhammad A, Dawood J, Fahad S, Xiuling L, Depeng W, Muhammad F, Muhammad F, Syed AS (2020) Use of crop growth model to simulate the impact of climate change on yield of various wheat cultivars under different agro-environmental conditions in Khyber Pakhtunkhwa, Pakistan. *Arabian J Geosci* 13:112. https://doi.org/10.1007/s12517-020-5118-1

Farhat A, Hafiz MH, Wajid I, Aitazaz AF, Hafiz FB, Zahida Z, Fahad S, Wajid F, Artemi C (2020) A review of soil carbon dynamics resulting from agricultural practices. *J Environ Manage* 268:110319.

Farooq M, Gogoi N, Hussain M et al. (2017) Effects, tolerance mechanisms and management of salt stress in grain legumes. *Plant Physiol Biochem* 118:199–217.

Fazli W, Muhmmad S, Amjad A, Fahad S, Muhammad A, Muhammad N, Ishaq AM, Imtiaz AK, Mukhtar A, Muhammad S, Muhammad I, Rafi U, Haroon I, Muhammad A (2020) Plant-microbes interactions and functions in changing climate. In: Fahad S, Hasanuzzaman M, Alam M, Ullah H, Saeed M, Khan AK, Adnan M (eds) *Environment, climate, plant and vegetation growth*. Springer Publ Ltd, Springer Nature Switzerland AG. Part of Springer Nature, pp. 397–420. https://doi.org/10.1007/978-3-030-49732-3

Ferchichi S, El Khouni A, Zorrig W et al. (2016) Nutrient uptake and use efficiencies in *Medicago ciliaris* under salinity. *J Plant Nutr* 39:932–941.

George S, Manoharan D, Li J et al. (2018) Drought and salt stress in *Macrotyloma uniflorum* leads to common and specific transcriptomic responses and reveals importance of raffinose family oligosaccharides in stress tolerance. *Gene Rep* 10:7–16.

Ghanbari A, Mousavi S, Pessarakli M (2015) Accumulation of reserve compounds in common bean seeds under drought stress. *J Plant Nutr* 38(4):609–623.

Ghulam M, Muhammad AA, Donald LS, Sajid M, Muhammad FQ, Niaz A, Ateeq Ur R, Shakeel A, Sajjad H, Muhammad A, Summia M, Aqib HAK, Fahad S, Rahul D, Mazhar I, Timothy DS (2021) Formalin fumigation and steaming of various composts differentially influence the nutrient release, growth and yield of muskmelon (*Cucumis melo* L.). *Sci Rep* 11:21057.

Gopakumar L, Bernard NO, Donato V (2020) Soil microarthropods and nutrient cycling. In: Fahad S, Hasanuzzaman M, Alam M, Ullah H, Saeed M, Khan AK, Adnan M (eds) *Environment, climate, plant and vegetation growth*. Springer Publ Ltd, Springer Nature Switzerland AG. Part of Springer Nature, pp. 453–472. https://doi.org/10.1007/978-3-030-49732-3

Gray SB, Strellner RS, Puthuval KK et al. (2013) Nodulation of field-grown soybean is enhanced by Free-Air CO2 Enrichment only when combined with drought stress. *Funct Plant Biol* 40(2):137–147.

Guntiñas ME, Leirós MC, Trasar-Cepeda C et al. (2012) Effects of moisture and temperature on net soil nitrogen mineralization: A laboratory study. *Eur J Soil Biol* 48:73–80.

Guo P, Jin H, Wei H et al. (2016) Fine root growth and water use efficiency in alfalfa (*Medicago sativa* L. cv. Gongong No. 1) planted along a salinity gradient in coastal area of Dalian, Northeast China. *Soil Sci Plant Nutr* 62(2):164–172.

Guofu L, Zhenjian B, Fahad S, Guowen C, Zhixin X, Hao G, Dandan L, Yulong L, Bing L, Guoxu J, Saud S (2021) Compositional and structural changes in soil microbial communities in response to straw mulching and plant revegetation in an abandoned artificial pasture in Northeast China. *Glob Ecol Conserv* 31:e01871.

Habib Ur R, Ashfaq A, Aftab W, Manzoor H, Fahd R, Wajid I, Md. Aminul I, Vakhtang S, Muhammad A, Asmat U, Abdul W, Syeda RS, Shah S, Shahbaz K, Fahad S, Manzoor H, Saddam H, Wajid N (2017) Application of CSM-CROPGRO-cotton model for cultivars and optimum planting dates: evaluation in changing semi-arid climate. *Field Crops Res.* http://dx.doi.org/10.1016/j.fcr.2017.07.007

Hafeez M, Farman U, Muhammad MK, Xiaowei L, Zhijun Z, Sakhawat S, Muhammad I, Mohammed AA, Mandela F-G, Nicolas D, Muzammal R, Fahad S, Yaobin L (2021) Metabolic-based insecticide resistance mechanism and ecofriendly approaches for controlling of beet armyworm *Spodoptera exigua*: a review. *Environ Sci Pollution Res.* https://doi.org/10.1007/s11356-021-16974-w

Hafiz MH, Abdul K, Farhat A, Wajid F, Fahad S, Muhammad A, Ghulam MS, Wajid N, Muhammad M, Hafiz FB (2020b) Comparative effects of organic and inorganic fertilizers on soil organic carbon and wheat productivity under arid region. *Commun Soil Sci Plant Anal.* https://doi.org/10.1080/00103624.2020.1763385

Hafiz MH, Farhat A, Ashfaq A, Hafiz FB, Wajid F, Carol Jo W, Fahad S, Gerrit H (2020a) Predicting kernel growth of maize under controlled water and nitrogen applications. *Int J Plant Prod.* https://doi.org/10.1007/s42106-020-00110-8

Hafiz MH, Farhat A, Shafqat S, Fahad S, Artemi C, Wajid F, Chaves CB, Wajid N, Muhammad M, Hafiz FB (2018) Offsetting land degradation through nitrogen and water management during maize cultivation under arid conditions. *Land Degrad Dev* 1–10. https://doi.org/10.1002/ldr.2933

Hafiz MH, Muhammad A, Farhat A, Hafiz FB, Saeed AQ, Muhammad M, Fahad S, Muhammad A (2019) Environmental factors affecting the frequency of road traffic accidents: a case study of sub-urban area of Pakistan. *Environ Sci Pollut Res.* https://doi.org/10.1007/s11356-019-04752-8

Hafiz MH, Wajid F, Farhat A, Fahad S, Shafqat S, Wajid N, Hafiz FB (2016) Maize plant nitrogen uptake dynamics at limited irrigation water and nitrogen. *Environ Sci Pollut Res* 24(3):2549–2557. https://doi.org/10.1007/s11356-016-8031-0

Haider SA, Lalarukh I, Amjad SF, Mansoora N, Naz M, Naeem M, Bukhari SA, Shahbaz M, Ali SA, Marfo TD, Subhan D, Rahul D, Fahad S (2021) Drought stress alleviation by potassium-nitrate-containing chitosan/montmorillonite microparticles confers changes in *Spinacia oleracea* L. *Sustainability* 13:9903. https://doi.org/10.3390/su13179903

Hamza SM, Xiukang W, Sajjad A, Sadia Z, Muhammad N, Adnan M, Fahad S et al. (2021) Interactive effects of gibberellic acid and NPK on morpho-physio-biochemical traits and organic acid exudation pattern in coriander (*Coriandrum sativum* L.) grown in soil artificially spiked with boron. *Plant Physiol Biochem* 167(2021):884–900.

Hasegawa S, Macdonald CA, Power SA (2016) Elevated carbon dioxide increases soil nitrogen and phosphorus availability in a phosphorus-limited Eucalyptus woodland. *Glob Change Biol* 22:1628–1643.

He Z, Xiong J, Kent AD et al. (2014) Distinct responses of soil microbial communities to elevated CO_2 and O_3 in a soybean agro-ecosystem. *ISME J* 8:714–726.

Hesham FA, Fahad S (2020) Melatonin application enhances biochar efficiency for drought tolerance in maize varieties: modifications in physio-biochemical machinery. *Agron J* 112(4):1–22.

Huang L-Y, Li X-X, Zhang Y-B, Fahad S, Wang F (2021) dep1 improves rice grain yield and nitrogen use efficiency simultaneously by enhancing nitrogen and dry matter translocation. *J Integrative Agri.* https://doi.org/10.1016/S2095-3119(21)63795-4

Hughes L, Bazzaz FA (2001) Effects of elevated CO_2 on five plant-aphid interactions. *Entomol Exp Appl* 99:87–96.

Hussain MA, Fahad S, Rahat S, Muhammad FJ, Muhammad M, Qasid A, Ali A, Husain A, Nooral A, Babatope SA, Changbao S, Liya G, Ibrar A, Zhanmei J, Juncai H (2020) Multifunctional role of brassinosteroid and its analogues in plants. *Plant Growth Regul.* https://doi.org/10.1007/s10725-020-00647-8

Ibrar K, Aneela R, Khola Z, Urooba N, Sana B, Rabia S, Ishtiaq H, Mujaddad R, Salvatore M (2020) Microbes and environment: global warming reverting the frozen zombies. In: Fahad S, Hasanuzzaman M, Alam M, Ullah H, Saeed M, Khan AK, Adnan M (eds) *Environment, climate, plant and vegetation growth.* Springer Publ Ltd, Springer Nature Switzerland AG. Part of Springer Nature, pp. 607–634. https://doi.org/10.1007/978-3-030-49732-3

Ikram U, Khadim D, Muhammad T, Muhammad S, Fahad S (2021) Gibberellic acid and urease inhibitor optimize nitrogen uptake and yield of maize at varying nitrogen levels under changing climate. *Environ Sci Pollution Res.* https://doi.org/10.1007/s11356-021-16049-w

Ilyas M, Mohammad N, Nadeem K, Ali H, Aamir HK, Kashif H, Fahad S, Aziz K, Abid U (2020) Drought tolerance strategies in plants: a mechanistic approach. *J Plant Growth Regul.* https://doi.org/10.1007/s00344-020-10174-5

IPCC (2018) Global warming of 1.5°C. An IPCC Special Report on the impacts of global warming of 1.5°C above pre-industrial levels and related global greenhouse gas emission pathways, in the context of strengthening the global response to the threat of climate change, sustainable development, and efforts to eradicate poverty. 32pp, ISBN 978-92-9169-151-7.

Iqra M, Amna B, Shakeel I, Fatima K, Sehrish L, Hamza A, Fahad S (2020) Carbon cycle in response to global warming. In: Fahad S, Hasanuzzaman M, Alam M, Ullah H, Saeed M, Khan AK, Adnan M (eds) *Environment, climate, plant and vegetation growth.* Springer Publ Ltd, Springer Nature Switzerland AG. Part of Springer Nature, pp. 1–16. https://doi.org/10.1007/978-3-030-49732-3

Irfan M, Muhammad M, Muhammad JK, Khadim MD, Dost M, Ishaq AM, Waqas A, Fahad S, Saud S et al. (2021) Heavy metals immobilization and improvement in maize (*Zea mays* L.) growth amended with biochar and compost. *Sci Rep* 11:18416.

Irigoyen JJ, Goicoechea N, Antolín MC et al. (2014) Growth, photosynthetic acclimation and yield quality in legumes under climate change simulations: an updated survey. *Plant Sci* 226:22–29.

Jabborova D, Sulaymanov K, Sayyed RZ, Alotaibi SH, Enakiev Y, Azimov A, Jabbarov Z, Ansari MJ, Fahad S, Danish S et al. (2021) Mineral fertilizers improves the quality of turmeric and soil. *Sustainability* 13:9437. https://doi.org/10.3390/su13169437

Jan A, Osman M, Amanullah (2013). Response of chickpea to nitrogen sources under salinity stress. *J Plant Nutr* 36:1373–1382.

Jan M, Muhammad AH, Adnan NS, Muhammad Y, Javaid I, Xiuling L, Depeng W, Fahad S (2019) Modulation in growth, gas exchange, and antioxidant activities of salt-stressed rice (*Oryza sativa* L.) genotypes by zinc fertilization. *Arabian J Geosci* 12:775 https://doi.org/10.1007/s12517-019-4939-2

Jayasundara H, Thomsom B, Tang C (1998) Responses of cool season grain legumes to soil abiotic stresses. *Adv Agron* 63:77–151.

Jin J, Lauricella D, Armstrong R et al. (2015) Phosphorus application and elevated CO_2 enhance drought tolerance in field pea grown in a phosphorus-deficient vertisol. *Ann Bot* 116:975–985.

Johnson SN, Ryalls JMW, Karley AJ (2014) Global climate change and crop resistance to aphids: contrasting responses of lucerne genotypes to elevated atmospheric carbon dioxide. *Ann Appl Biol* 165:62–72.

Jones AG, Scullion J, Ostle N et al. (2014) Completing the FACE of elevated CO_2 research. *Environ Int* 73:252–258.

Kaci H, Chaker-Haddadj A, Aid F (2018) Enhancing of symbiotic efficiency and salinity tolerance of chickpea by phosphorus supply. *Acta Agric Scand Section B Soil Plant Sci* 68(6):534–540.

Kamarn M, Wenwen C, Irshad A, Xiangping M, Xudong Z, Wennan S, Junzhi C, Shakeel A, Fahad S, Qingfang H, Tiening L (2017) Effect of paclobutrazol, a potential growth regulator on stalk mechanical strength, lignin accumulation and its relation with lodging resistance of maize. *Plant Growth Regul* 84:317–332. https://doi.org/10.1007/s10725-017-0342-8

Kaschuk G, Leffelaar PA, Giller KE et al. (2010) Responses of legumes to rhizobia and arbuscular mycorrhizal fungi: a meta-analysis of potential photosynthate limitation of symbioses. *Soil Biol Biochem* 42:125–127.

Kashiwagi J (2015) Scope for improvement of yield under drought through the root traits in chickpea (*Cicer arietinum* L.). *Field Crops Res* 175:74.

Katerji N (2011) Faba bean productivity in saline–drought conditions. *Eur J Agron* 35(1):2–12.

Katunga MMD (2014) Agro-ecological adaptation and participatory evaluation of multipurpose tree and shrub legumes in mid altitudes of Sud-Kivu, D. R. Congo. *Am J Plant Sci* 5:2031–2039.

Khadim D, Fahad S, Jahangir MMR, Iqbal M, Syed SA, Shah AK, Ishaq AM, Rahul D et al. (2021a) Biochar and urease inhibitor mitigate NH_3 and N_2O emissions and improve wheat yield in a urea fertilized alkaline soil. *Sci Rep* 11:17413.

Khadim D, Saif R, Fahad S, Syed SA, Shah AK et al. (2021b) Influence of variable biochar concentration on yield-scaled nitrous oxide emissions, wheat yield and nitrogen use efficiency. *Sci Rep* 11:16774.

Khan A, Gul F, Jamal Y, Ali J, Adnan M, Khan MA, Naeem I, Bacha Z (2017) Weeds management in maize hybrids with night plowing. *Pak J Weed Sci Res* 22(2):179–191.

Khan A, Shah N, Gul A, Sahar NU, Ismail A, Muhammad, Aziz F, Farooq M, Adnan M, Rizwan M (2016) Comparative study of toxicological impinge of glyphosate and atrazine (herbicide) on stress biomarkers; blood biochemical and hematological parameters of the freshwater common carp (*Cyprinuscarpio*). *Pol J Environ Stud* 25(5):1993–1999.

Khan MMH, Niaz A, Umber G, Muqarrab A, Muhammad AA, Muhammad I, Shabir H, Shah F, Vibhor A, Shams HA-H, Reham A, Syed MBA, Nadiyah MA, Ali TKZ, Subhan D, Rahul D (2021) Synchronization of Boron application methods and rates is environmentally friendly approach to improve quality attributes of *Mangifera indica* L. On sustainable basis. *Saudi J Bio Sci*. https://doi.org/10.1016/j.sjbs.2021.10.036

Khazaei H (2014) Leaf traits associated with drought adaptation in faba bean (Vicia faba L.) Doctoral thesis. Helsinki 2014. p 59.

Köpke U, Nemecek T (2010) Ecological services of faba bean. *Field Crop Res* 115(3):217–233.

Kurban H, Saneoka H, Nehira K et al. (1998) Effect of salinity on growth and accumulation of organic and inorganic solutes in the leguminous plants *Alhagi pseudoalhagi* and *Vigna radiata*. *Soil Sci Plant Nutr* 44:589–597.

Kurban H, Saneoka H, Nehira K et al. (1999) Effect of salinity on growth, photosynthesis and mineral composition in leguminous plant *Alhagi pseudoalhagi* (Bieb.). *Soil Sci Plant Nutr* 45:851–862.

Kurdali F, Al-Ain F (2002) Effect of different water salinity levels on growth, nodulation, and N_2-fixation by dhaincha and on growth of sunflower using a ^{15}N-tracer technique. *J Plant Nutr* 25(11):2483–2498.

Lam SK, Chen D, Norton R et al. (2013) The effect of elevated atmospheric carbon dioxide concentration on the contribution of residual legume and fertilizer nitrogen to a subsequent wheat crop. *Plant Soil*, 364:81–91.

Leskovar D, Othman Y (2016) Low nitrogen fertigation promotes root development and transplant quality in globe artichoke. *HortScience* 51:567–572.

Li Y, Yu Z, Liu X et al. (2017) Elevated CO_2 increases nitrogen fixation at the reproductive phase contributing to various yield responses of soybean cultivars. *Front Plant Sci* 8:1546.

López-Aguilar R, Orduño-Cruz A, Lucero-Arce A et al (2003) Response to salinity of three grain legumes for potential cultivation in arid areas. *Soil Sci Plant Nutr* 49(3):329–336.

Luck J, Spackman M, Freeman A et al (2011) Climate change and diseases of food crops. *Plant Pathol* 60:113–121.

Magadlela A, Beukes C, Venter F et al (2017) Does P deficiency affect nodule bacterial composition and N source utilisation in a legume from nutrient-poor Mediterranean-type ecosystems?. *Soil Biol Biochem* 104:164–174.

Mahar A, Amjad A, Altaf HL, Fazli W, Ronghua L, Muhammad A, Fahad S, Muhammad A, Rafiullah, Imtiaz AK, Zengqiang Z (2020) Promising technologies for Cd-contaminated soils: drawbacks and possibilities. In: Fahad S, Hasanuzzaman M, Alam M, Ullah H, Saeed M, Khan AK, Adnan M (eds) *Environment, climate, plant and vegetation growth*. Springer Publ Ltd, Springer Nature Switzerland AG. Part of Springer Nature, pp. 63–92. https://doi.org/10.1007/978-3-030-49732-3

Mahmood Ul H, Tassaduq R, Chandni I, Adnan A, Muhammad A, Muhammad MA, Muhammad H-R, Mehmood AN, Alam S, Fahad S (2021) Linking plants functioning to adaptive responses under heat stress conditions: a mechanistic review. *J Plant Growth Regul*. https://doi.org/10.1007/s00344-021-10493-1

Manzer HS, Saud A, Soumya M, Abdullah AA, Qasi DA, Bander MAM, Hayssam MA, Hazem MK, Fahad S, Vishnu DR, Om PN (2021) Molybdenum and hydrogen sulfide synergistically mitigate arsenic toxicity by modulating defense system, nitrogen and cysteine assimilation in faba bean (*Vicia faba* L.) seedlings. *Environ Pollut* 290:117953.

Mashhady A, Salem S, Barakah F (1998) Effect of salinity on survival and symbiotic performance between *rhizobium meliloti* and *medicago sativa* l. in Saudi Arabian soils. *Arid Land Res Manag* 12(1):3–14.

McDonald GK, Paulsen GM (1997) High temperature effects on photosynthesis and water relations of grain legumes. *Plant Soil* 196:47–58.

Md Jakir H, Allah B (2020) Development and applications of transplastomic plants; a way towards eco-friendly agriculture. In: Fahad S, Hasanuzzaman M, Alam M, Ullah H, Saeed M, Khan AK, Adnan M (eds) *Envi-*

ronment, climate, plant and vegetation growth. Springer Publ Ltd, Springer Nature Switzerland AG. Part of Springer Nature, pp. 285–322. https://doi.org/10.1007/978 3 030 49732 3
Md. Enamul H, Shoeb AZM, Mallik AH, Fahad S, Kamruzzaman MM, Akib J, Nayyer S, Mehedi KM, Swati AS, Md Yeamin A, Most SS (2020) Measuring vulnerability to environmental hazards: qualitative to quantitative. In: Fahad S, Hasanuzzaman M, Alam M, Ullah H, Saeed M, Khan AK, Adnan M (eds) *Environment, climate, plant and vegetation growth.* Springer Publ Ltd, Springer Nature Switzerland AG. Part of Springer Nature, pp. 421–452. https://doi.org/10.1007/978-3-030-49732-3
Mohammad IA-W, Abdelazeem S, Munir A, Khalid E, Adel RAU (2020b) Extent of climate change in Saudi Arabia and its impacts on agriculture: a case study from Qassim Region. In: Fahad S, Hasanuzzaman M, Alam M, Ullah H, Saeed M, Khan AK, Adnan M (eds) *Environment, climate, plant and vegetation growth.* Springer Publ Ltd, Springer Nature Switzerland AG. Part of Springer Nature, pp. 635–658. https://doi.org/10.1007/978-3-030-49732-3
Mohammad IA-W, Ahmad M, Usman ARA, Akanji M, Rafique MI (2020a) Advances in pyrolytic technologies with improved carbon capture and storage to combat climate change. In: Fahad S, Hasanuzzaman M, Alam M, Ullah H, Saeed M, Khan AK, Adnan M (eds) *Environment, climate, plant and vegetation growth.* Springer Publ Ltd, Springer Nature Switzerland AG. Part of Springer Nature, pp. 535–576. https://doi.org/10.1007/978-3-030-49732-3
Morais MC, Panuccio MR, Muscolo A et al (2012) Does salt stress increase the ability of the exotic legume *Acacia longifolia* to compete with native legumes in sand dune ecosystems. *Environ Exp Bot* 82: 74–79.
Morgan PB, Ainsworth EA, Long SP (2003) How does elevated ozone impact soybean? A meta-analysis of photosynthesis, growth and yield. *Plant Cell Environ*, 26:1317–1328.
Mousavi Derazmahalleh M et al (2019) Adapting legume crops to climate change using genomic approaches. *Plant Cell Environ* 42:6–19.
Moussa H, Hassan M (2016) Growth enhancers to mitigate salinity stress in *Vicia faba. Int J Veg Sci* 22(3):243–250.
Mubeen M, Ashfaq A, Hafiz MH, Muhammad A, Hafiz UF, Mazhar S, Muhammad SD, Asad A, Amjed A, Fahad S, Wajid N (2020) Evaluating the climate change impact on water use efficiency of cotton-wheat in semi-arid conditions using DSSAT model. *J Water Clim Change.* https://doi.org/10.2166/wcc.2019.179/622035/jwc2019179.pdf
Muhammad N, Muqarrab A, Khurram S, Fiaz A, Fahim N, Muhammad A, Shazia A, Omaima N, Sulaiman AA, Fahad S, Subhan D, Rahul D (2021) Kaolin and Jasmonic acid improved cotton productivity under water stress conditions. *J Saudi Soc Agric Sci* 28:6606–6614. https://doi.org/10.1016/j.sjbs.2021.07.043
Muhammad TQ, Amna F, Amna B, Barira Z, Xitong Z, Ling-Ling C (2020) Effectiveness of conventional crop improvement strategies vs. omics. In: Fahad S, Hasanuzzaman M, Alam M, Ullah H, Saeed M, Khan AK, Adnan M (eds) *Environment, climate, plant and vegetation growth.* Springer Publ Ltd, Springer Nature Switzerland AG. Part of Springer Nature, pp. 253–284. https://doi.org/10.1007/978-3-030-49732-3
Muhammad Z, Abdul MK, Abdul MS, Kenneth BM, Muhammad S, Shahen S, Ibadullah J, Fahad S (2019) Performance of *Aeluropus lagopoides* (mangrove grass) ecotypes, a potential turfgrass, under high saline conditions. *Environ Sci Pollut Res.* https://doi.org/10.1007/s11356-019-04838-3
Munns R (2005) Genes and salt tolerance: bringing them together. *New Phytologist* 167:645–663.
Musallam IW, Al-Karaki G, Ereifej K et al. (2004) Yield and yield components of faba bean genotypes under rainfed and irrigation conditions. *Asian J Plant Sci* 3(4):439–448.
Muzammal R, Fahad S, Guanghui D, Xia C, Yang Y, Kailei T, Lijun L, Fei-Hu L, Gang D (2021) Evaluation of hemp (*Cannabis sativa* L.) as an industrial crop: a review. *Environ Sci Poll Res.* https://doi.org/10.1007/s11356-021-16264-5
Nichols PGH et al. (2012) Temperate pasture legumes in Australia – their history, current use and future prospects. *Crop Pasture Sci* 63:691–725.
Noor M, Naveed Ur R, Ajmal J, Fahad S, Muhammad A, Fazli W, Saud S, Hassan S (2020) Climate change and costal plant lives. In: Fahad S, Hasanuzzaman M, Alam M, Ullah H, Saeed M, Khan AK, Adnan M (eds) *Environment, climate, plant and vegetation growth.* Springer Publ Ltd, Springer Nature Switzerland AG. Part of Springer Nature, pp. 93–108. https://doi.org/10.1007/978-3-030-49732-3
Othman Y, Al-Karaki G, Al-Tawaha A et al. (2006) Variation in germination and ion uptake in barley genotypes under salinity conditions. *World J Agric Sci* 2(1):11–15.
Othman Y, VanLeeuwen D, Heerema R et al. (2014) Midday stem water potential values needed to maintain photosynthesis and gas exchange established for pecans. *J Am Soc Hortic Sci* 139(5):537–546.

Pang J, Megan H, Hans L et al. (2018) Phosphorus acquisition and utilization in crop legumes under global change. *Curr Opin Plant Biol* 45:1–7.

Parvin S, Uddin S, Fitzgerald GJ et al. (2019) Free air CO_2 enrichment (FACE) improves water use efficiency and moderates drought effect on N-2 fixation of *Pisum sativum* L. *Plant Soil* 436(1–2):587–606.

Podgorska-Lesiak M, Sobkowicz P (2013) Prevention of pea lodging by intercropping barley with peas at different nitrogen fertilization levels. *Field Crops Res* 149:95–104.

Pooniya V, Choudhary AK, Dass A et al. (2015) Improved crop management practices for sustainable pulse production: An Indian perspective. *Indian J Agric Sci* 85(6):747–758.

Purushothaman R et al. (2013) Root anatomical traits and their possible contribution to drought tolerance in grain legumes. *Plant Prod Sci* 16(1):1–8.

Qamar-Uz Z, Zubair A, Muhammad Y, Muhammad ZI, Abdul K, Fahad S, Safder B, Ramzani PMA, Muhammad N (2017) Zinc biofortification in rice: leveraging agriculture to moderate hidden hunger in developing countries. *Arch Agron Soil Sci* 64:147–161. https://doi.org/10.1080/03650340.2017.1338343

Rashid M, Qaiser H, Khalid SK, Mohammad I-W, Zhang A, Muhammad A, Shahzada SI, Rukhsanda A, Ghulam AS, Shahzada MM, Sarosh A, Muhammad FQ (2020) Prospects of biochar in alkaline soils to mitigate climate change. In: Fahad S, Hasanuzzaman M, Alam M, Ullah H, Saeed M, Khan AK, Adnan M (eds) *Environment, climate, plant and vegetation growth.* Springer Publ Ltd, Springer Nature Switzerland AG. Part of Springer Nature, pp. 133–150. https://doi.org/10.1007/978-3-030-49732-3

Raveneau M, Coste F, Moreau-Valancogne P et al. (2011) Pea and bean germination and seedling responses to temperature and water potential. *Seed Sci Res* 21(3):205–213.

Rehman M, Fahad S, Saleem MH, Hafeez M, Rahman MH, Liu F, Deng G (2020) Red light optimized physiological traits and enhanced the growth of ramie (*Boehmeria nivea* L.). *Photosynthetica* 58(4):922–931.

Sadam M, Muhammad TQ, Ghulam M, Muhammad SK, Faiz AJ (2020) Role of biotechnology in climate resilient agriculture. In: Fahad S, Hasanuzzaman M, Alam M, Ullah H, Saeed M, Khan AK, Adnan M (eds) *Environment, climate, plant and vegetation growth.* Springer Publ Ltd, Springer Nature Switzerland AG. Part of Springer Nature, pp. 339–366. https://doi.org/10.1007/978-3-030-49732-3

Safi UK, Ullah F, Mehmood S, Fahad S, Ahmad Rahi A, Althobaiti F et al. (2021) Antimicrobial, antioxidant and cytotoxic properties of *Chenopodium glaucum* L. *PLoS ONE* 16(10):e0255502. https://doi.org/10.1371/journal.pone.0255502

Sahrish N, Shakeel A, Ghulam A, Zartash F, Sajjad H, Mukhtar A, Muhammad AK, Ahmad K, Fahad S, Wajid N, Sezai E, Carol Jo W, Gerrit H (2022) Modeling the impact of climate warming on potato phenology. *Eur J AgroN* 132:126404.

Sajid H, Jie H, Jing H, Shakeel A, Satyabrata N, Sumera A, Awais S, Chunquan Z, Lianfeng Z, Xiaochuang C, Qianyu J, Junhua Z (2020) Rice production under climate change: adaptations and mitigating strategies. In: Fahad S, Hasanuzzaman M, Alam M, Ullah H, Saeed M, Khan AK, Adnan M (eds) *Environment, climate, plant and vegetation growth.* Springer Publ Ltd, Springer Nature Switzerland AG. Part of Springer Nature, pp. 659–686. https://doi.org/10.1007/978-3-030-49732-3

Sajjad H, Muhammad M, Ashfaq A, Waseem A, Hafiz MH, Mazhar A, Nasir M, Asad A, Hafiz UF, Syeda RS, Fahad S, Depeng W, Wajid N (2019) Using GIS tools to detect the land use/land cover changes during forty years in Lodhran district of Pakistan. *Environ Sci Pollut Res.* https://doi.org/10.1007/s11356-019-06072-3

Saleem MH, Fahad S, Adnan M, Mohsin A, Muhammad SR, Muhammad K, Qurban A, Inas AH, Parashuram B, Mubassir A, Reem MH (2020a) Foliar application of gibberellic acid endorsed phytoextraction of copper and alleviates oxidative stress in jute (*Corchorus capsularis* L.) plant grown in highly copper-contaminated soil of China. *Environ Sci Pollut Res.* https://doi.org/10.1007/s11356-020-09764-3

Saleem MH, Fahad S, Shahid UK, Mairaj D, Abid U, Ayman ELS, Akbar H, Analía L, Lijun L (2020c) Copper-induced oxidative stress, initiation of antioxidants and phytoremediation potential of flax (*Linum usitatissimum* L.) seedlings grown under the mixing of two different soils of China. *Environ Sci Poll Res.* https://doi.org/10.1007/s11356-019-07264-7

Saleem MH, Rehman M, Fahad S, Tung SA, Iqbal N, Hassan A, Ayub A, Wahid MA, Shaukat S, Liu L, Deng G (2020b) Leaf gas exchange, oxidative stress, and physiological attributes of rapeseed (*Brassica napus* L.) grown under different light-emitting diodes. *Photosynthetica* 58(3):836–845.

Saman S, Amna B, Bani A, Muhammad TQ, Rana MA, Muhammad SK (2020) QTL mapping for abiotic stresses in cereals. In: Fahad S, Hasanuzzaman M, Alam M, Ullah H, Saeed M, Khan AK, Adnan M (eds) *Environment, climate, plant and vegetation growth.* Springer Publ Ltd, Springer Nature Switzerland AG. Part of Springer Nature, pp. 229–252. https://doi.org/10.1007/978-3-030-49732-3

Saud S, Chen Y, Fahad S, Hussain S, Na L, Xin L, Alhussien SA (2016) Silicate application increases the photosynthesis and its associated metabolic activities in Kentucky bluegrass under drought stress and post-drought recovery. *Environ Sci Pollut Res* 23(17):17647–17655. https://doi.org/10.1007/s11356-016-6957-x

Saud S, Chen Y, Long B, Fahad S, Sadiq A (2013) The different impact on the growth of cool season turf grass under the various conditions on salinity and drought stress. *Int J Agric Sci Res* 3:77–84.

Saud S, Fahad S, Cui G, Chen Y, Anwar S (2020) Determining nitrogen isotopes discrimination under drought stress on enzymatic activities, nitrogen isotope abundance and water contents of Kentucky bluegrass. *Sci Rep* 10:6415. https://doi.org/10.1038/s41598-020-63548-w

Saud S, Fahad S, Yajun C, Ihsan MZ, Hammad HM, Nasim W, Amanullah Jr, Arif M, Alharby H (2017) Effects of nitrogen supply on water stress and recovery mechanisms in Kentucky bluegrass plants. *Front Plant Sci* 8:983. https://doi.org/10.3389/fpls.2017.00983

Saud S, Li X, Chen Y, Zhang L, Fahad S, Hussain S, Sadiq A, Chen Y (2014) Silicon application increases drought tolerance of Kentucky bluegrass by improving plant water relations and morph physiological functions. *SciWorld J* 2014:1–10. https://doi.org/10.1155/2014/368694

Senol C (2020) The effects of climate change on human behaviors. In: Fahad S, Hasanuzzaman M, Alam M, Ullah H, Saeed M, Khan AK, Adnan M (eds) *Environment, climate, plant and vegetation growth.* Springer Publ Ltd, Springer Nature Switzerland AG. Part of Springer Nature, pp. 577–590. https://doi.org/10.1007/978-3-030-49732-3

Serraj R, Drevon J (1998) Effects of salinity and nitrogen source on growth and nitrogen fixation in alfalfa. *J Plant Nutr* 21(9):805–1818.

Shafi MI, Adnan M, Fahad S, Fazli W, Ahsan K, Zhen Y, Subhan D, Zafar-ul-Hye M, Martin B, Rahul D (2020) Application of single superphosphate with humic acid improves the growth, yield and phosphorus uptake of wheat (*Triticum aestivum* L.) in calcareous soil. *Agronomy* 10:1224. https://doi.org/10.3390/agronomy10091224

Shah F, Lixiao N, Kehui C, Tariq S, Wei W, Chang C, Liyang Z, Farhan A, Fahad S, Huang J (2013) Rice grain yield and component responses to near 2°C of warming. *Field Crop Res* 157:98–110.

Shaharoona B, Imran M, Arshad M et al. (2011) Manipulation of ethylene synthesis in roots through bacterial ACC deaminase for improving nodulation in legumes. *Crit Rev Plant Sci* 30:279–291.

Siddique KHM et al. (2000) Adaptation of chickpea (*Cicer arietinum* L.) and faba bean (*Vicia faba* L.) to Australia. In: Knight R (eds) *Linking research and marketing opportunities for pulses in the 21st century.* Current Plant Science and Biotechnology in Agriculture, vol. 34. Springer, Dordrecht, pp. 289–303.

Sidra K, Javed I, Subhan D, Allah B, Syed IUSB, Fatma B, Khaled DA, Fahad S, Omaima N, Ali TKZ, Rahul D (2021) Physio-chemical characterization of indigenous agricultural waste materials for the development of potting media. *J Saudi Soc Agric Sci.* https://doi.org/10.1016/j.sjbs.2021.08.058

Signorelli S, Corpas F, Rodriguez-Ruiz M et al. (2019) Drought stress triggers the accumulation of NO and SNOs in cortical cells of *Lotus japonicus* L. roots and the nitration of proteins with relevant metabolic function. *Environ Exp Bot* 161:228–241.

Singh SK, Reddy VR (2017) Potassium starvation limits soybean growth more than the photosynthetic processes across CO_2 levels. *Front Plant Sci* 8:991.

Sprent J, Gehlot H (2010) Nodulated legumes in arid and semi-arid environments: are they important? *Plant Ecol Divers* 3(3):211–219.

Subhan D, Zafar-ul-Hye M, Fahad S, Saud S, Martin B, Tereza H, Rahul D (2020a) Drought stress alleviation by ACC deaminase producing *Achromobacter xylosoxidans* and *Enterobacter cloacae*, with and without timber waste biochar in maize. *Sustainability* 12(6286). https://doi.org/10.3390/su12156286

Sulieman S, Tran L (2015) Phosphorus homeostasis in legume nodules as an adaptive strategy to phosphorus deficiency. *Plant Sci* 239:36–43.

Sulpanjani A, Yang MS, Tawaha ARM et al. (2005) Effect of magnesium application on yield, mineral contents and active components of *Chrysanthemum coronarium* L. under hydroponics conditions. *Biosci Res* 2(2):73–79.

Supanjani A, Tawaha AM, Min Yang MS et al. (2005) Role of calcium in yield and medicinal quality of *Chrysanthemum coronarium* L. *J Agron* 4(3):188–192.

Subhan D, Zafar-ul-Hye M, Fahad S, Saud S, Martin B, Tereza H, Rahul D (2020b) Drought stress alleviation by ACC deaminase producing *Achromobacter xylosoxidans* and *Enterobacter cloacae*, with and without timber waste biochar in maize. *Sustainability* 12(6286). https://doi.org/10.3390/su12156286

Taub D (2010) Effects of rising atmospheric concentrations of carbon dioxide on plants. *Nat Educ Knowl* 3(10):21.

Tausz M, Tausz-Posch S, Norton RM et al. (2013) Understanding crop physiology to select breeding targets and improve crop management under increasing atmospheric CO_2 concentrations. *Environ Exp Bot* 88:71–80.

Tausz-Posch S, Seneweera S, Norton R et al. (2012) Can a wheat cultivar with high transpiration efficiency maintain its yield advantage over a near-isogenic cultivar under elevated CO_2? *Field Crops Res* 133:160–166.

Tawaha AM, Singh VP, Turk MA et al. (2003) A review on growth, yield components and yield of barley as influenced by genotypes, herbicides and fertilizer application. *Res Crop* 4(1):1–9.

Tawaha AM, Turk MA (2002) Lentil (*Lens culinaris* Medic.) productivity as influenced by rate and method of phosphate placement in a Mediterranean environment. *Acta Agron Hung* 50(2):197–201.

Tawaha AM, Turk MA (2004) Field pea seeding management for semi-arid Mediterranean conditions. *J Agron Crop Sci* 190:86–92.

Toker C, Yadav SS (2010) Legumes cultivars for stress environments. In: Yadav S, Redden R (eds) *Climate change and management of cool season grain legume crops*. Springer, Dordrecht, pp. 351–376.

Trytsman M et al. (2016) Diversity and biogeographical patterns of legumes (Leguminosae) indigenous to southern Africa. *PhytoKeys* 70:53–96.

Turk MA, Shatnawi MK, Tawaha AM (2003b) Inhibitory effects of aqueous extracts of black mustard on germination and growth of alfalfa. *Weed Biol Manag* 3(1):37–40.

Turk MA, Tawaha AM, Lee KD (2004) Seed germination and seedling growth of three lentil cultivars under moisture stress. *Asian J Plant Sci* 3(3):394–397.

Turk MA, Tawaha AM, Samara N (2003a) Effects of seeding rate and date, and phosphorus application on growth and yield of narbon vetch (*Vicia narbonensis*). *Agronomie* 23:1–4.

Turk MA, Tawaha AM (2001) Common vetch (*Vicia sativa* L.) productivity as influenced by rate and method of phosphate fertilization in a Mediterranean environment. *Agric Mediterranea* 131:108–111.

Turk MA, Tawaha AM (2002b) Impact of seeding rate, seeding date and method of phosphorous application in faba bean (*Vicia faba* L. minor) in the absence of moisture stress. *Biotechnol Agron Soc Environ* 6(3):171–178.

Turk MA, Tawaha AM (2002a) Onion (*Allium cepa* L.) as influenced by rate and method of phosphorus placement. *Crop Res* 23(1):105–107.

Ullah H, Khan WU, Alam M, Khalil IH, Adhikari KH, Shahwar D, Jamal Y, Jan I, Adnan M (2016) Assessment of G × E interaction and heritability for simplification of selection in spring wheat genotypes. *Can J Plant Sci* 96(3):1021–1025.

Ullah H, Rashid A, Liu G et al. (2018) Perceptions of mountainous people on climate change, livelihood practices and climatic shocks: a case study of Swat District, Pakistan. *Urban Clim* 26:244–257.

Unsar Naeem-U, Muhammad R, Syed HMB, Asad S, Mirza AQ, Naeem I, Muhammad HR, Fahad S, Shafqat S (2020) Insect pests of cotton crop and management under climate change scenarios. In: Fahad S, Hasanuzzaman M, Alam M, Ullah H, Saeed M, Khan AK, Adnan M (eds) *Environment, climate, plant and vegetation growth*. Springer Publ Ltd, Springer Nature Switzerland AG. Part of Springer Nature, pp. 367–396. https://doi.org/10.1007/978-3-030-49732-3

Upadhyaya HD et al. (2010) Phenotyping chickpeas and pigeon peas for adaptation to drought. In: *Drought phenotyping in crops: from theory to practice*. Generation Challenge Programme, pp. 347–355.

Valentine A, Kleinert A, Benedito V (2017). Adaptive strategies for nitrogen metabolism in phosphate deficient legume nodules. *Plant Sci* 256:46–52.

Valliyodan B, Ye H, Song L et al. (2017) Genetic diversity and genomic strategies for improving drought and waterlogging tolerance in soybeans. *J Exp Bot* 68(8):1835–1849.

Wahid F, Fahad S, Subhan D, Adnan M, Zhen Y, Saud S, Manzer HS, Martin B, Tereza H, Rahul D (2020) Sustainable management with mycorrhizae and phosphate solubilizing bacteria for enhanced phosphorus uptake in calcareous soils. *Agriculture* 10(334). https://doi.org/10.3390/agriculture10080334

Wajid N, Ashfaq A, Asad A, Muhammad T, Muhammad A, Muhammad S, Khawar J, Ghulam MS, Syeda RS, Hafiz MH, Muhammad IAR, Muhammad ZH, Muhammad HR, Veysel T, Fahad S, Suad S, Aziz K, Shahzad A (2017) Radiation efficiency and nitrogen fertilizer impacts on sunflower crop in contrasting environments of Punjab. *Pakistan Environ Sci Pollut Res* 25:1822–1836. https://doi.org/10.1007/s11356-017-0592-z

Wang D, Grieve C, Suarez D (2005) Composition of irrigation water salinity affects growth characteristics and uptake of selenium and salt ions by soybean. *J Plant Nutr* 28(6):1073–1088.

Wilson RG, Sbatella GM (2014) Integrating irrigation, tillage, and herbicides for weed control in dry bean. *Weed Technol* 28:479–485.

Wu C, Kehui C, She T, Ganghua L, Shaohua W, Fahad S, Lixiao N, Jianliang H, Shaobing P, Yanfeng D (2020) Intensified pollination and fertilization ameliorate heat injury in rice (*Oryza sativa* L.) during the flowering stage. *Field Crops Res* 252:107795.

Wu C, Tang S, Li G, Wang S, Fahad S, Ding Y (2019) Roles of phytohormone changes in the grain yield of rice plants exposed to heat: a review. *PeerJ* 7:e7792. https://doi.org/10.7717/peerj.7792

Yadav SS, McNeil DL, Redden R et al. (2010) *Climate change and management of cool season grain legume crops*. Springer, Dordrecht Heidelberg London New York, p. 460, ISBN 978-90-481-3708-4 e-ISBN 978-90-481-3709-1, https://doi.org/10.1007/978-90-481-3709-1.

Yang F, Feng L, Liu Q et al. (2018) Effect of interactions between light intensity and red-to-far-red ratio on the photosynthesis of soybean leaves under shade condition. *Environ Exp Bot* 150:79–87.

Yang Z, Zhang Z, Zhang T, Fahad S, Cui K, Nie L, Peng S, Huang J (2017) The effect of season-long temperature increases on rice cultivars grown in the central and southern regions of China. *Front Plant Sci* 8:1908. https://doi.org/10.3389/fpls.2017.01908

Zafar-ul-Hye M, Muhammad TH, Muhammad A, Fahad S, Martin B, Tereza D, Rahul D, Subhan D (2020b) Potential role of compost mixed biochar with rhizobacteria in mitigating lead toxicity in spinach. *Scientific Rep* 10:12159. https://doi.org/10.1038/s41598-020-69183-9

Zafar-ul-Hye M, Muhammad N, Subhan D, Fahad S, Rahul D, Mazhar A, Ashfaq AR, Martin B, Ji ˇrí H, Zahid HT, Muhammad N (2020a) Alleviation of cadmium adverse effects by improving nutrients uptake in bitter gourd through cadmium tolerant rhizobacteria. *Environments* 7(54). https://doi.org/10.3390/environments7080054

Zafar-ul-Hye M, Akbar MN, Iftikhar Y, Abbas M, Zahid A, Fahad S, Datta R, Ali M, Elgorban AM, Ansari MJ et al. (2021) Rhizobacteria inoculation and caffeic acid alleviated drought stress in lentil plants. *Sustain* 13:9603. https://doi.org/10.3390/su13179603

Zahida Z, Hafiz FB, Zulfiqar AS, Ghulam MS, Fahad S, Muhammad RA, Hafiz MH, Wajid N, Muhammad S (2017) Effect of water management and silicon on germination, growth, phosphorus and arsenic uptake in rice. *Ecotoxicol Environ Saf* 144:11–18.

Zahir SM, Zheng HG, Ala Ud D, Amjad A, Ata Ur R, Kashif J, Shah F, Saud S, Adnan M, Fazli W, Saud A, Manzer HS, Shamsher A, Wajid N, Hafiz MH, Fahad S (2021) Synthesis of silver nanoparticles using *Plantago lanceolata* extract and assessing their antibacterial and antioxidant activities. *Sci Rep* 11:20754.

Zia-Ur-Rehman M (2020) Environment, climate change and biodiversity. In: Fahad S, Hasanuzzaman M, Alam M, Ullah H, Saeed M, Khan AK, Adnan M (eds) *Environment, climate, plant and vegetation growth*. Springer Publ Ltd, Springer Nature Switzerland AG. Part of Springer Nature, pp. 473–502. https://doi.org/10.1007/978-3-030-49732-3

12

Plant-Insect Interaction: Plant Defence Barrier vs Counter Defence Mechanism of Insects

Gulnaz Afzal, Farooq Ahmad, Nargis Naz, Sayeda Sadaf Zehra, and Muhammad Abdul Ghafar
The Islamia University of Bahawalpur, Bahawalpur, Pakistan

CONTENTS

DOI: 10.1201/9781003286417-12

12.1 Introduction

In natural environment, plants and insects belong to diversified groups, but they have a dynamic association between them that may be responsible to bring continual changes in their existence with the passage of time. It has been evident that they both have extensive range of niches from natural to manmade and desert to Antarctic ecosystems. This evolutionary arm race has a history of almost 400 million years (Labandeira 2013; Schmelz 2015), when flowering plants started to get evolved into land plants and insects followed them. Since that time, the phytophagous insects are required to be nourished by plants (Hopkins 1916), and also adopted dynamic strategies as their counter-defence (Bruce 2015). First, the coevolution processes between insect and plant communities were highlighted by Darwin (1899). Then, this concept of both organisms drew the attention of Ehrlich and Raven (1964). Their theory reflects the complex ecological interactions on some peculiar aspects between plants and insects. Both have developed different mechanistic strategies to interact and cope with each other's defence systems.

Plants play their role of being a source of energy and also serve as nesting sites for insects. On the other hand, all insects may act as guardians and provide fertilizers for plants (Hogenhout and Bos 2011; Mithöfer and Boland 2012; Smith and Clement 2012). However, in the same environment both organisms undergo selection pressure for their best survival. Usually, insect herbivores are separated into two groups, i.e. generalist (polyphagous) and specialists (monophagous and oligophagous). The generalists depend on several families of host plants for food and are more tolerant against various plant defence mechanisms, while specialists feed on plants of the same family with specific range of releasing defensive chemicals from host plants that may function as feeding stimulants and restrict the tolerance capacity of these species. Plants defence against herbivores' attack can be constituted in two ways i.e., directly and indirectly by affecting host plant preference and reproductive success and through other natural enemies of the insect pests, respectively. Plants have adapted some particular morphological structures, act as toxic chemicals (phenols, alkaloids, anthocyanins, quinones, and terpenoids) and repellent on the herbivores. They not only have the potential to recognize the herbivore challenge but can keep their metabolism adaptive to be modulated with respect to its need and time (Mithöfer and Boland 2012). Moreover, a few mechanical characteristics (spines, hairs, thorns, trichomes, and thicker leaves) of plants can also kill or alter the insect's biology.

It is known that plant-insect association is spatially, temporally, and ecologically complex, which has always been challenging to investigate by ecologists (Becerra and Venable 1999; Mitter et al. 1991; Thompson 1994). It is evident that with due course of time, these interactions are becoming more intense due to herbivore pressure, signifying plant defence strategies against this pressure and of course Anthropocene impacts (Burkepile and Parker 2017; Javed 2018).

Therefore, more studies are needed to be explored, in terms of insect pest management for best crop yield by knowing the insect maternal ecology. The major aim of this chapter is to explore the defensive mechanisms of one group and to know the other's ability to cope with the noxious cues in response.

12.2 Types of Plant-Insect Interactions

Plant and insect interactions have evolved in response to selective pressures caused by their physical environment and other interactive organisms. This background provides many types of interactions; few of them can be separated on the bases of their effects including harmful, beneficial, and neutral with other species (Abrahamson et al. 1989). These distinguishing interactions are still unclear due to their continual changes with time and space.

a. **Mutualism (+,+):** The interaction between different organisms in which both get benefits from each other is called *mutualism*. Plants also show mutualistic behaviour by being dependent on insects for pollination, and in turn providing food and shelter to insects. For example, flower plants/pollinators, plants/ants, plant guarding, and plant seed scattering.
b. **Antagonistic herbivory (+,–):** In this relation one is benefitted and the other is harmed. This interaction includes phytophagy in which insects eat plants and plants suffer tissue loss, low survival, and reproduction, and insectivory by plants (carnivorous plants). This interaction can cause severe threat to plants and damage the crop yield.
c. **Antagonistic insectivory (+,–):** This is reverse interaction to antagonistic herbivory. This is another type of predation in which plants eat insects for nitrogen nutrients.
d. **Commensalism (+, 0):** In this relationship, one group get benefits but the other is not harmed. For example, insects can take some products from plants without harming the plant like Monarch butterfly larvae seek shelter from milkweeds to store cardiac glycosides for their defence (Tan et al. 2018).

12.3 Plant Defence Barriers

During the long history of plant-insect coevolution, plants have established intimate relations with insects relating to intricate defensive approaches which are substantial to be dealt with herbivores attack. These approaches can be divided into two groups. First is the preformed or constitutive defence (before insect's attack preexisted physical and chemical responses) while second is inducible defence (response activation upon insect attack). Constitutive defence is more specific, and it may sustain less metabolic costs of plant defence, so does the inducible defence by keeping itself off until insect attack is noticed (Dicke and Poecke 2002). Inducible defence can be categorized into direct and indirect defences. Basically, plant defence works on tolerance as well as resistance strategies. Usually, plants show some expressions through their specific morphological traits that reduce or prevent the damage caused by insects. In direct defence, all plant traits are preformed that increase resistance against insect herbivory by affecting the physiology and behaviour of insects while indirect defence traits do not have ability to show direct impact upon the attackers without being encouraged by exogenous interventions, e.g. carnivores (Hagenbucher et al. 2013; Xu et al. 2015; Adnan et al. 2016; Zhuang et al. 2012). Although, plants do not have nervous system like animals, they can voluntarily sense and respond to the insect attack through vibrant and intricate signalling pathways. These pathways may be structural, chemical, and mechanical, which activate the direct and indirect defence responses (War et al. 2012).

12.3.1 Direct Defence

In *direct defence*, host plant traits are more vulnerable towards insect attacks (Kessler and Baldwin 2002). Generally, direct plant defences against insects are categorized to include malnutrition, physical damage, hindering chemical pathways, and reducing preference of the attacking insect. There are several different kinds of factors which have additive as well as synergistic impact against specific type of insects. These factors/barriers may have different impacts on same species at one time or may be changed with others on some other time (Blau et al. 1978). Scientists have revealed that antinutrition (including limited food supply and reduced nutrient values) followed by toxicity (including physical damage and chemical disruption) are specific plant traits that have greater influence on herbivorous insects (Divol et al. 2007; Goussain et al. 2005; Pechan et al. 2002; Salunke et al. 2005; Shah et al. 2015). In plant-insect systems, different defence strategies by plants and their correlation with insect's physiological processes in counter-defence are intricate and hard to explore. However, a preliminary review of some morphological and biochemical components involved in host plant resistance is being discussed here.

12.3.2 Morphological Features

Physical barriers are the first line of plant defence in response to insect pests. Whole body of plants display various kinds of obstructions against herbivory. These structural traits (morphological and anatomical) have been raised in simple to highly complex structures, which ultimately help in preventing the insect's feeding in various ways from thickening of cell wall to apparent protuberance of plant (Agrawal et al. 2009; He et al. 2011; Mazid et al. 2011; Strauss and Agrawal 1999).

12.3.3 Waxes

Plant cuticle layer produces wax as a major component, which can protect aerial parts from biotic and abiotic stressors. The slipperiness on plant surface is amplified due to waxes, which reduce the grab of insect herbivores and avert them from feeding or ovipositing on the leaf surfaces. For example, waxy layer of *Eucalyptus globulus* leaves are responsible for the decrease in the adherence of psyllids attachment (Walters 2010). Similarly, waxy surface of *Triticale* plant acts as an antifeedant and antiprobing barrier against grain aphid *Sitobion avenae*.

12.3.4 Tissue Hardness

Tissue hardness in plants directly affects the insect's mouth parts, not allowing them to penetrate, pierce, and suck through it. It may help in preventing the feeding load by increasing the mandibular wear in biting-chewing insects. For example, tough surfaced leaves (sclerophylly) reduce the palatability and digestibility of its tissues (War et al. 2012; Altaf et al. 2016). Similarly, some plants have become so hard with age that they can successfully reduce the diffusion of stylet of *Parabemisia myricae* (Hemiptera, Aleyrodidae). In addition, this tissues hardness is due to deposition of different chemicals with the passage of time. This toughness may lead to mechanical confrontation among species. For example, females prefer young leaves than hard and old leaves to feed and live (Fernandes 1994; Keeping and Kvedaras 2008).

12.3.5 Pubescence

Pubescence is another structural trait of plants, which represents as epidermal outgrowths termed as *Trichomes*. They consist of a hairy layer extending throughout the epidermis of different plant parts including leaves, stem, and fruits only. These are reported in various forms like branched, unbranched glandular, hooked, spiral, and stellate forms (Chamarthi et al. 2011). They also show density-dependent effects on feeding, ovipositional responses, and larval nutrition of insects (Levin 1973). In addition, more dense trichomes have ability to sense and interfere with arthropods' movement on their bodies and can generate auto-defence secondary metabolites in a very short time against attackers (War et al. 2012). Some herbaceous plants are known as more pubescent with young leaves as compared to older ones (Tian et al. 2012). For example, trichomes with hooked shape of *Passiflora adenopoda* (Passifloraceae) demonstrate a specific defence against specific class of insects (Cardoso 2008).

12.3.6 Latex

Latex is a sticky emulsion which is a combination of organic compounds formed by some plants and held under internal pressure in special cells called lacticifers. When injured by herbivores, the canal system present in different plant families like Asteraceae, Euphorbiaceae, Asclepiadaceae, and Apocynaceae exude latex, which serves to trap and kill the insects. For example, milkweed's latex becomes thicker when it is subjected to direct air and ultimately kills insect larvae (Fernandes 1994). Latex may include more secondary metabolites in number and concentrations as compared to leaves. Many of these compounds have toxic or antinutritive effects, while others enhance the stickiness of the latex (Agrawal and Konno 2009). Such compounds are terpenoids, cardenolides, phenolics, alkaloids, and a range of proteins like proteases in species of *Ficus* and *Carica papaya* and proteinase inhibitors (Mithöfer and Boland 2012). It has been essential for the encounter of all types of boring, mining, and chewing insects. The latex of plant *Calotropis procera* is reported as a source of promising insecticidal proteins against *Callosobruchus maculates* (Ramos et al. 2010; Ali et al. 2016). Many specialist insect herbivores are modified to feed on plants that exude latex. This is frequently done either by destroying the lactifer routes in proximal area by vein-cutting and trenching or by rising physiological adaptations. For instance, the larvae of the monarch butterfly feeding on Apocynaceae plants are modified to cardenolides by a single amino acid alteration (Konno 2011).

12.3.7 Resins

The resin-based defences are well established in conifers. Basically, resins are a combination of monoterpenes, diterpenes, and sesquiterpenes, and can be collected in specialized resin-secreted ducts in some plants (Howe and Schaller 2008). Some conifer resins act as barriers to prevent bark beetle attack and are also lethal to its eggs, larval, and adult stages (Rosenthal and Berenbaum 1991). Some monoterpenes become highly volatile in air and help insects to be trapped and solidified at wound place.

12.3.8 Thigmonasty

Thigmonastic movements in some plants take place in response to touch, shaking, thermal or electrical stimulus and also act as defence line against plants. The leaves of the sensitive plant *Mimosa pudica*, also known as touch-me-not plant, close leaves rapidly in response to such external stimuli. This response is not a growth movement but turgor movements caused by osmotic pressure of the pulvini cells at the base of leaves (Mondal 2013). This response is extended throughout the plant via both electrical and chemical ways. This electrical impulse acts on a plant hormone which makes the water migrate from the cells of pulvinus to the intercellular spaces of the other half of pulvinus. This water movement causes the pulvinus to lose its firmness making the leaf to fold. Likewise, all the pulvini lose firmness and become limp due to which all the leaves above them crumple and collapse. After folding of leaves, water usually diffuses back into same cells of pulvinus from which it left, and the leaf returns to its novel position in 15–30 minutes (Braam 2005; Tabasum 2017). The hasty folding of the leaflets may provide to scare away potential predators and give the appearance of a less voluminous meal.

12.3.9 Mimicry and Camouflage

According to these mechanisms, organisms show adaptive resemblance with a mimic organism or tend to mask their emergence with their surroundings. The woody vine *Boquila trifoliolata* mimics or set its leaves in specific orientations that help them to avoid attack of some weevils and leaf beetles (Gianoli and Carrasco-Urra 2014). Similarly, the bracts of a woodland plant *Monotropsis odorata* function as camouflage with its surrounding in response to avoid herbivores. These are brown-coloured that resembles the leaf litter (Hund 2017). Heliconiines larvae are important defoliating agents of *Passiflora* spp. A number of *Passiflora* species have a number of distinctive structures on their leaves. These structures mimic the occurrence of *Heliconius* butterflies' yellow eggs on their leaves. Females are mostly reluctant to lay their eggs on the same plants where already butterfly eggs are present. This is due to the fact that larvae of many *Heliconius* feed on congeneric eggs and larvae (Williams and Gilbert 1981).

12.4 Plant Secondary Metabolites

Despite being generally categorized, there are some specific defence approaches, which are varied and more complex among different environmental plant species. Different insects may use different kinds of damaging mechanisms either on same tissues or on different tissues of same plant. In response of these attacks, plants have to develop counter-defence mechanisms with different tactics. It is correlated with other biological processes. For example, plant secondary metabolites, also known as *bioactive specialized organic compounds*, are toxic to herbivores and act as defence compounds (Wittstock and Gershenzon 2002). These chemical compounds cannot affect the overall reproduction, growth, and development of plants but can decrease palatability of its tissues for insects. Plants have the ability to produce a wide range of secondary metabolites as byproducts of primary metabolites synthesis (Herbert 1989). Few of them can carry out different physiological activities within plants including storage and transportation of other significant chemicals, attraction of insects for pollination, seed dispersal for reproduction and also play their role in UV protection (Wink 1999). However, their major role is to serve as defence barriers against herbivores and pathogen attacks (Bennet and Wallsgrove 1994; Mao et al. 2007; Theis and Lerdau 2003). There are almost 50,000 plant secondary metabolites which are considered to be a small fraction of overall existing in nature (De Luca and St Pierre 2000). After paying a heavy cost, every plant produces not only a little amount, but also an exclusive combination of these defensive compounds. For example, cyclic hydroxamic acid is produced exclusively in cereal crops (Niemeyer 1988), glycosylated triterprenes and steroids are produced in dicotyledonous plants but rare in grass species (Marston and Hostettman 1996). Despite the enormous secondary metabolites and massive amount of literature regarding their defensive potential, particular defensive functions are still unidentified. However, advanced technologies like mass spectrometry, gene expression analysis, and high throughput sequencing have made it quite easier and cost-effective to profile them accordingly. Moreover, study of secondary metabolites may help to identify the genes and enzymes involved during their processes of biosystnesis. Plant secondary metabolites can be separated into three chemically distinct groups, viz. terpenes, phenolics, and N and S containing compounds (Mazid et al. 2011).

12.4.1 Terpenes

Terpeniods are being considered among the highest group of plant secondary metabolites with diversified chemical structures (Yazaki et al. 2017). These compounds are derivatives of five-carbon isoprene units (Singh and Sharma 2015). The terpenes insecticidal action of plant defence is not only due to toxic and antifeedant properties but they can also cope with biotic and abiotic stresses by mixing with volatile blends (Chen et al. 2011). Azadirechtin (triterpene) is one of the main potent feeding deterrents to many insects, which exerts various toxic effects and inhibits egg maturation (Rosenthal and Berenbaum 1991).

12.4.2 Plant Phenolics

Plant phenols are included in most general and frequently occurring compounds along with other secondary metabolites. Plant phenols represent one of the most frequent and general group of defensive compounds, which play an antagonistic role to insects, pathogens, and their competing plants (Rani and Jyothsna 2010). In addition, they are cytotoxic in nature and can reduce the reactive oxygen species (ROS), which results in activation of different cyclic pathways leading to enzymes activation for defence reprisal (Maffei et al. 2007). Phenolic are aromatic ring compounds with one (phenol) or more (polyphenol) hydroxyl substituents. Some important phenolic defence compounds are coumarin, furano-coumarins, lignins, flavonoids, isoflavonoids, and tannins.

12.4.3 N and S Containing Compounds

Sulphur compounds comprise of glutathione, glucosinolates, defensins, phytoalexins, thionins, and allinin. Nitrogen compounds contain cyanogenic glucosides, alkaloids, and nonprotein amino acids. Glucosinolates are included in sulphur- and nitrogen-containing plant secondary metabolites, which are more frequent in family Brassicaceae, an increasing level resulted in a decrease of damage by generalist herbivores (Redovniković et al. 2008). Sinalbin (glucosinolate) is found in high amount in young cotyledenous plant and serves as flea or beetle *Phyllotreta cruciferae* repellent (Hopkins et al. 2009).

12.5 Antinutritional or Digestive Proteins

Plants can also protect themselves by producing proteins that reduce the nutrient cost to the attacking insect or cause physical damage to the insect digestive tract. These defensive proteins can be categorized into following classes.

12.5.1 Alpha Amylase Inhibitor

These interact with α-amylase activity by inhibiting the digestion of glycogen and starch in insects (de Sales et al. 2012). For instance, a 50% activity of α-amylase inhibitor of *Vigna unguiculata* can be inhibited by *Callosobruchus maculates* larvae. Triticale-α amylase inhibitor shows a strongest inhibitory action on gut of *Eurygaster integriceps* (Mehrabadi et al. 2011).

12.5.2 Chitinase

Chitin is the foremost constituent of the insect cuticle membrane, and the enzyme chitinase is used as a pest management tool to damage peritrophic membrane of insect alimentary canal (Chandrasekaran et al. 2014). Development of Colorado potato beetle is inhibited by poplar chitinase in transgenic tomato (Sharma et al. 2011).

12.5.3 Lectin

Lectin is one of the particular entomotoxic proteins, which falls in the nearby plant species having group of carbohydrate binding proteins or lectins (Vandenborre et al. 2011). When lectins come into contact with the glycoprotein lining of the insect's intestine, they inhibit the assimilation of nutrients. First lectin to which anti-insect properties were ascribed on the origin of its poisonous effect was the larvae of bruchid beetle *Callosobruchus maculatus* (Peumans and Van Damme 1995).

12.5.4 Polyphenol Oxidases

Polyphenol Oxidases (PPOs) are omnipresent copper-containing antinutritive enzymes, which oxidize molecular oxygen into orthodiphenolic compounds to highly reactive quinines. PPO-generated quinones additionally react with amino acids and reduce their availability. It causes classic browning of plant extracts and damaged plant tissues. Overexpression of PPO genes in tomato and hybrid aspen has found greater resistance as compared to its normal state, and silencing of PPOs leads to increased vulnerability to insect herbivory (Araji et al. 2014).

12.5.5 Proteinase Inhibitor

Proteinase inhibitors can alter the digestive system of herbivores by acting as antimetabolic proteins. By reducing the enzyme proteases in insect's guts, they may cause a reduction in the accessibility of essential amino acids for their normal growth and development. Pepstatin is an influential and physically powerful inhibitor of aspartyl proteases which is well known to inhibit proteolysis of midgut enzymes present in Colorado potato beetle, *Leptinotarsa decemlineata* (Habib 2007).

12.6 Indirect Defence Response

Indirect defences comprise plant traits that indirectly save themselves and also serve as attractants to natural enemies of herbivore insects (Chen 2008). The plants catch the attention to nourish or save other organisms to reduce enemy pressure. It enhances the chance of attracting the natural enemies of herbivores by producing or emitting the exclusive blend of volatiles, extra floral nectar, food bodies, and nesting sites (Clavijo McCormick et al. 2012).

12.6.1 Volatiles

Plant leaves normally emit small amount of chemicals called *volatiles*. After herbivore's attack, plant recognizes herbivore-associated elicitors and releases more volatiles. The qualitative features of all volatiles not only vary with the plant but with insect species as well. They can attract both natural parasites and predatory enemies of insects (Paré and Tumlinson 1999). For instance, when different wasps like *Cotesia marginiventris* are approaching for feeding on maize, they are killed by armyworm, *Spodoptera exigua* (Kessler and Heil 2011). Moreover, volatiles can also provide underground defence against herbivory (Howe and Schaller 2008).

12.6.2 Extrafloral Nectaries

Extrafloral nectaries (EFNs) are very common, nonpollinating organs of more than 100 families of plants. They are engaged to recruit the predatory and parasitoid insects against plant defences (Gish et al. 2016). These structures are high on metabolic cost and frequently found in young or developing leaves. They secrete abundant sugars which attract various ants. Besides ants, these nectaries attract other arthropods including Araneae (jumping spiders), Diptera (tachinid flies), Coleoptera, and Hymenoptera (ichneumonid, braconid, and chalcidoid wasps), which can reduce herbivores (So 2004). EFNs consist mainly of monosaccharides and disaccharides, amino acids, carbohydrates, lipids, and proteins. Their production is associated with herbivore intensity. Examples of plant bearing EFNs include cotton species and fruit trees of the genus *Prunus* such as almond, plum, peach, and cherry (Heil 2015).

12.6.3 Nesting Spaces

Plants offer nesting space for insects, and insects defend plants against herbivores. For example, some ant species take refuge at nesting site of Myrmecophyte plants in hollow stems. Likewise, tree *Cordia alliodora* can reduce herbivory by promoting *Azteca pittieri* ant-colony. Another species of the same genus of ant colonies, *Azteca chartifex*, mostly select *Goupia glabra* (Goupiaceae) trees to make their nests and also protect the plant leaves from biotic stressors (Dejean et al. 2008).

12.7 Types of Herbivore Insects Based on Feeding Behaviour

Herbivory has evolved with both positive and negative impacts on plants over the evolutionary period. Herbivory is also called as phytophagy that can be subdivided into two major categories of insects like generalists and specialists. The insects which have diverse feeding and not limited to specific group of plants are known as *generalists*, while the insects that have specific feeding choice and are limited to specific diet or plants group are called as *specialists*. Generally, we can say generalists insects are polyphagous (relying upon several plant species) in nature and specialists are monophagous (depending upon single species). Both types of herbivore insects are able to interact with plants in various distinct ways. However, it has been well argued by different scientists and ecologists that specialist herbivores have distinct and predictable association with their hosts. Due to specialized nature where they have lost the ability to choose many hosts, they have adapted with more tolerance ability by manipulating host plants for their fitness. On the other hand, generalists have evolved with very general mechanisms to tolerate plant defences and also have ability to manipulate their host plants in a much conserved plant pathway. These evidences have been a mainstay of coevolutionary studies over the years (Krieger et al. 1971).

12.8 Resistance Pattern of Herbivore Insects

Insects are dependent on plants for food and also have coevolved in response to plants defence mechanisms as offensive traits (Karban and Agrawal 2002). On the other hand, plants always try to save their resources for their own growth and development, by restraining the insect's attacks through different mechanistic approaches. Association between both organisms results in reciprocal evolutionary change. Plants recognize their attackers and initiate defence responses through physical and biochemical methods or induce a counter-attack (Futuyma and Slatkin 1983). It has been reported that coevolution between both organisms is due to a driving force called *speciation* (Ehrlich and Raven 1964). A competitive individual forces its competitor to enhance its performance through selection, and this ultimately results in evolutionary arms race between the two individuals. Insects evolve is response to plant defences because plants must enhance their competitive strength first due to insect's competitive success (Barnett and Hansen 1996). However, insect herbivores have developed complementary adaptations in response to every defensive pattern of the host plant as discussed below.

12.8.1 Mechanical Resistance Pattern

Insects have adopted various physical structures for the easy approach and consumption of plant tissues. Mostly, insect feeding depends on some specific foraging methods and also on favourite food materials (Bernays 1991). For instance, herbivores, specifically Sphinx Moth, possess low crowned teeth for soft leaves, as well as snipping mandibles for hard grasses (Bernays and Janzen 1988). Furthermore, feeding adaptations among insects are associated with their nature of diet. Grasshoppers have evolved with powerful jaws to graze on relatively hard grasses as compared to soft diets. Few Lepidopteran species' guts have also been well adapted with the nature of intaking feed (Thompson 1992).

12.8.2 Biochemical Resistance Pattern

Herbivores are adaptive to generate enzymes to detoxify and decrease the effectiveness of secondary metabolic products manufactured by host plants. Among these important enzymes Mixed Function Oxidases (MFOs), and Cytochrome P450 oxidases and proteases have specifically been connected to counter-defence mechanism of insects (Feyereisen 1999; Jongsma and Bolter 1997; Snydcr and Glcndinning 1996). Herbivores also produce salivary enzymes, for example, glucose oxidase; a component of *Helicoverpa zea*, activates the induced defences in tobacco and aphid saliva and serves as barrier between the respective species and the plant cells (Felton and Eichenseer 1999).

12.8.3 Behavioural Resistance Pattern

Insects attack on plants spatiotemporally and indulge in selective feeding to avoid plant defences For example, in early season the winter moth feeds on oak leaves to get more nutritive food as compared to the late season when the amount of tannins are reduced in the tree (Feeny 1970). Insects are well trained in plant selection; they prefer to eat at toxin-free areas or window feeding. According to geophagy behaviour, some individuals consume large amounts of toxic materials, then they can mitigate its effects by eating clay or other minerals. Another big behavioural adaptation of insects is to deal with chemicals of insecticides and pesticides over time, minimizing the rate of insect's evolution in feeding mechanisms.

12.8.4 Microbial Symbionts Resistance Pattern

Microsymbionts (bacteria, fungi) residing in insect's body can help them to digest cellulose, which cannot be easily broken down due to its complex structure. They also allow insects to eat those plants which usually are not supposed to be eaten without detoxifying by plant secondary metabolites. Moreover, symbionts not only help is producing detoxification enzymes but they may use alleleo-chemicals as their source of carbon, for instance, fungal symbiont *Lasioderma serricorne*. It may also assist in weakening the host plant defence by ingesting the whole plant materials. Sometimes damaged hosts are more preferable and successful sites for herbivores feeding. As an example, some beetles introduce some specific fungal strains into trees before eating (Dowd 1991).

12.8.5 Host Manipulation Resistance

Insects are able to make or manipulate their host plants according to their needs. They can change their microhabitat where they feed and live to counter the plant defences. Many caterpillar families such as Pyralidae and Ctenuchidae roll the leaves of *Psychotria horizontalis* around an expanding bud and then eat. This activity helps them to protect their food from environmental constraints and also help them to maintain its nutritive value for a long time (Sagers 1992). Insects protect their microhabitat by making gall on plants, which may acts as shelter and source of food for herbivore plants.

12.9 Tolerance Strategies of Host Plants

Tolerance is the capability of plants which confer some resistance characters, to alleviate the antieffects of herbivory (Strauss and Agrawal 1999). It has been confirmed by genetic analysis that tolerance in host plants has been inherited through natural selection. It is likely that they have evolved as an adaptive defence mechanism against herbivores (Fornoni 2011; Strauss and Zangerl 2002).

Many essential plant factors, such as growth rate, change phytohormone, photosynthetic rates, nutrient distribution and uptake, storage capability, activation of dormant meristems, changes in plant phenology and plant structural design can affect the capacity of plants to tolerate damage (Duan et al. 2014; Rosenthal and Kotanen 1994). The compensation of tolerance is that it is a plant response and does not affect reproduction, growth, or development and insect behaviour by itself. Therefore, it does not undergo selection pressure on the insect population for new biotypes. It has none of the deleterious effects on natural enemies. Furthermore, it does not decrease their prey populations (Reese et al. 1994).

12.9.1 Signalling in Plants

After insect's attack, numerous signalling transduction pathways are able to detect and transduce the signals, which ultimately lead to shifts in gene expression and finally ends with the production of various chemicals for their defence (Chen 2008; Duan et al. 2014). Plants constantly interact with the peripheral world through plasma membrane. An external stimulus changes plasma membrane potential (*Vm*) intracellularly, which is followed by a rapid electric signal passing through the whole plant. Immediately after this signal, the free cytosolic Ca^{2+} concentration increases at the injury site, which is usually much lower in healthy cells. This increased amount of Ca^{2+} is recognized by calcium-binding proteins, which can turn on many calcium-dependent protein kinases. These enzymes (kinases) are responsible for controlling the function of several stress-responsive genes and ultimately show the response of stress tolerance in plants (Tuteja and Mahajan 2007). For example, mitogen-activated protein kinase (MAPK) activity can be examined at the feeding site and on distant locations of the damaged leaf (Schuman and Baldwin 2016). Some components are able to elicit induced systemic resistance like systemin, peptides, oligogalacturonides (OGAs) (Ferrari et al. 2013), and jasmonates (Fürstenberg-Hägg et al. 2013; Huang et al. 2017). The kinase signalling induces production of jasmonic acid, which plays a significant function in systemic signalling. Jasmonates (JA) are transported throughout the plant and trigger the defence response, both in wounded and unwounded sites through transcriptional control of specific genes (Zhu-Salzman et al. 2004). It has been evident that signalling transduction works with some phytohormones either antagonistically or synergistically. For instance, salicylic acid and ethylene react with jasmonic acid and vice versa.

12.9.2 Photosynthetic Activity

It has been reported that plants' tolerance against herbivore damage can be retained by enhancing the photosynthetic rate at damaged sites (Trumble et al. 1993). However, photosynthetic rate may normally be enhanced via so many other pathways like presence of rubisco enzyme in leaves as well as growing age of plants (Stowe et al. 2000). So, evidence of increased photosynthetic rate is not sufficient to prove that it has damage-tolerant capacity where other secondary metabolites are involved too (Tiffin 2000).

Furthermore, the idea that increase in photosynthetic rate is able to increase resistance instead of tolerance is still debatable (Trumble et al. 1993).

12.9.3 Reallocation of Resources in Plants

Plants accumulate and redistribute its reserves and resources in different organs during their life cycle and also in stressful conditions (Deepranjan et al. 2021; Haider et al. 2021; Huang et al. 2021; Ikram et al. 2021; Jabborova et al. 2021; Khadim et al. 2021a, 2021b; Manzer et al. 2021; Muzammal et al. 2021; Abdul et al. 2021a, 2021b; Ashfaq et al. 2021; Amjad et al. 2021; Atif et al. 2021; Athar et al. 2021; Adnan et al. 2018a, 2018b, 2019; Akram et al. 2018a, 2018b; Aziz et al. 2017a, 2017b; Chang et al. 2021; Chen et al. 2021; Emre et al. 2021; Habib Ur et al. 2017; Hafiz et al. 2016, 2018, 2019, 2020a, 2020b; Ghulam et al. 2021; Guofu et al. 2021; Hafeez et al. 2021; Khan et al. 2021; Kamarn et al. 2017; Muhammad et al. 2019; Safi et al. 2021; Sajjad et al. 2019; Saud et al. 2013, 2014, 2016, 2017, 2020; Shah et al. 2013; Qamar-Uz et al. 2017; Hamza et al. 2021; Irfan et al. 2021; Wajid et al. 2017; Yang et al. 2017; Zahida et al. 2017; Depeng et al. 2018; Hussain et al. 2020; Shafi et al. 2020; Wahid et al. 2020; Subhan et al. 2020; Zafar-ul-Hye et al. 2020a, 2020b; Zafar-ul-Hye et al. 2021; Adnan et al. 2020; Ilyas et al. 2020; Saleem et al. 2020a, 2020b, 2020c; Rehman et al. 2020; Farhat et al. 2020; Wu et al. 2020; Mubeen et al. 2020; Farhana et al. 2020; Jan et al. 2019; Wu et al. 2019; Ahmad et al. 2019; Baseer et al. 2019; Tariq et al. 2018; Fahad and Bano 2012; Fahad et al. 2013, 2014a, 2014b, 2015a, 2015b, 2016a, 2016b, 2016c, 2016d, 2017, 2018a, 2018b, 2019a, 2019b, 2020, 2021a, 2021b, 2021c, 2021d, 2021e, 2021f; Hesham and Fahad 2020; Iqra et al. 2020; Akbar et al. 2020; Mahar et al. 2020; Noor et al. 2020; Bayram et al. 2020; Amanullah and Fahad 2017, 2018a, 2018b; Amanullah et al. 2020, 2021; Rashid et al. 2020; Arif et al. 2020; Amir et al. 2020; Saman et al. 2020; Muhammad et al. 2020; Md Jakir and Allah 2020; Mahmood Ul et al. 2021; Farah et al. 2020; Sadam et al. 2020; Unsar et al. 2020; Fazli et al. 2020; Md. Enamul et al. 2020; Gopakumar et al. 2020; Zia-Ur-Rehman 2020; Ayman et al. 2020; Mohammad et al. 2020a, 2020b; Senol 2020; Amjad et al. 2020; Ibrar et al. 2020; Sajid et al. 2020; Muhammad et al. 2021; Sidra et al. 2021; Zahir et al. 2021; Sahrish et al. 2022).

Upon insect attack, resources can be reallocated to tolerate the damage. For instance, in case of plant *Centaurea maculosa*, nitrogen flow can be shifted to other parts of plants like shoots after being attacked by insect species of *Agapeta zoegana*. Plants have ability to translocate nitrogen almost twice to the shoot even as root grazing condensed total nitrogen uptake by 30%–50% (Schultz et al. 2013). One more example is that the 'Sugars' can also be shifted from infested green parts to the nonaffected roots, i.e. *Manduca sexta*–infested *Nicotiana attenuata* plants. Hence, at the time of need all rescued substances can easily be remobilized within the plant to help in building new above-ground plant parts (Mithöfer and Boland 2012; Tiffin 2000).

12.9.4 Phenological Change

Differential changes in plant phenology are a widespread mechanism of tolerance. Herbivore damage can cause phenology delays (i.e., plant growth, flowering, and fruit production), which may lead to change its tolerance capacity (Tiffin 2000). These phenological delays are due to certain reasons, i.e. shorter life span or season of reproduction. It is reported that the faster recovering genotypes experience shorter delays in production and are more tolerant (Pilson 2000).

12.9.5 Plant Architecture after Damage

This type of tolerance mechanism is constitutive in nature and mostly relies on the morphology of plants upon herbivore attack and damage. A sound architecture of plants (i.e., roots-to-shoots ratios, number of stems and their rigidity, and vasculature system) acts in various ways to enhance the tolerance at damage site (Marquis 1996). The vascular system of plant allows sectoriality and manages to transport resources from undamaged to damaged parts of plants in response to resistance. A very high root-to-shoot ratio may help to absorb more nutrients after herbivory. Similarly, rigid stem structures did not allow plants to collapse after severe damage (Marquis 1996).

12.9.6 Plants Regrowth Following Herbivory

Compensatory growth following herbivory has been proven to be an indicator of tolerance in plants. Apical meristem damage (AMD) is one of the most important mechanisms of tolerance, which has been studied more extensively (Haukioja and Koricheva 2000; Marquis 1996; Suwa and Maherali 2008; Trumble et al. 1993; Wise and Abrahamson 2007). Plant meristems have been observed as highly cell proliferating units and are also enriched with nutrition than other tissues. After damage it may activate the growth of axillary meristems with multiple branching to compensate the damage (Wise and Abrahamson 2007).

12.10 Conclusion and Future Perspectives

Globally, insects and plants play a pivotal role in agriculture. Both are associated with each other dynamically and compete for their best survival and coexistence. Plants have evolved very effective defence mechanisms against herbivore pests. However, these mechanisms are susceptible to counter-defence adaptation. Therefore, an accurate and exact understanding about these interactions is required to be gathered to develop or adopt better pest management techniques. The counter-defence mechanism by herbivore insects to host plant defence is very complicated and has become more challenging to explore new plant varieties with resistance to insects. Both competitors do not show visible tug of war apparently, but internally a number of mechanical, chemical, and other defensive adaptations are working together silently. Normally, phytophagous insects use their sensory genes, proteins, and insect detoxifying enzymes to cope with toxic materials of plants produced as secondary byproducts or other defensive tactics. Hereafter, the stronger participant wins and makes itself stronger for reproduction. Concisely, in the light of above discussion, it is evident that host plants and insect pests have coevolved to tolerate each other's defence strategies. Knowing about plant defence and mechanism of insects counter-adaptations will be a basic step to setting new challenging goals for sustainable pest management programmes. Further, the exact and robust information of genome sequencing studies of both insect and plant species can provide us better information about the highly dynamic and ever-evolving insect-plant interactions.

REFERENCES

Abdul S, Muhammad AA, Shabir H, Hesham AEE, Sajjad H, Niaz A, Abdul G, Sayyed RZ, Fahad S, Subhan D, Rahul D (2021a) Zinc nutrition and arbuscular mycorrhizal symbiosis effects on maize (*Zea mays* L.) growth and productivity. *J Saudi Society Agricultural Sci* https://doi.org/10.1016/j.sjbs.2021.06.096

Abdul S, Muhammad AA, Subhan D, Niaz A, Fahad S, Rahul D, Mohammad JA, Omaima N, Muhammad HR, Bernard RG (2021b) Effect of arbuscular mycorrhizal fungi on the physiological functioning of maizeunder zinc-deficient soils. *Sci Rep* 11:18468.

Abrahamson WG, Sattler JF, McCrea KD, Weis AE (1989) Variation in selection pressures on the goldenrod gall fly and the competitive interactions of its natural enemies. *Oecologia* 79:15–22.

Adnan M, Fahad S, Muhammad Z, Shahen S, Ishaq AM, Subhan D, Zafar-ul-Hye M, Martin LB, Raja MMN, Beena S, Saud S, Imran A, Zhen Y, Martin B, Jiri H, Rahul D (2020) Coupling phosphate-solubilizing bacteria with phosphorus supplements improve maize phosphorus acquisition and growth under lime induced salinity stress. *Plants* 9(900). https://doi.org/10.3390/plants9070900

Adnan M, Fahad S, Khan IA, Saeed M, Ihsan MZ, Saud S, Riaz M, Wang D, Wu C (2019) Integration of poultry manure and phosphate solubilizing bacteria improved availability of Ca bound P in calcareous soils. *3 Biotech* 9(10):368.

Adnan M, Shah Z, Sharif M, Rahman H (2018b). Liming induces carbon dioxide (CO_2) emission in PSB inoculated alkaline soil supplemented with different phosphorus sources. *Environ Sci Poll Res* 25(10):9501–9509.

Adnan M, Zahir S, Fahad S, Arif M, Mukhtar A, Imtiaz AK, Ishaq AM, Abdul B, Hidayat U, Muhammad A, Inayat-Ur R, Saud S, Muhammad ZI, Yousaf J, Amanullah, Hafiz MH, Wajid N (2018a) Phosphate-solubilizing bacteria nullify the antagonistic effect of soil calcification on bioavailability of phosphorus in alkaline soils. *Sci Rep* 8:4339. https://doi.org/10.e1038/s41598-018-22653-7

Adnan M, Shah Z, Ullah H, Khan B, Arshad M, Mian IA, Khan GA, Alam M, Basir A, Rahman IU, Ali M, Khan WU (2016) Yield response of wheat to nitrogen and potassium fertilization. *Pure Appl Bio* 5(4):868–875.

Agrawal AA, Fishbein M, Jetter R, Salminen JP, Goldstein JB, Freitag AE, Sparks JP (2009) Phylogenetic ecology of leaf surface traits in the milkweeds (Asclepias spp.) Chemistry, ecophysiology, and insect behavior. *New Phytol* 183:848–867.

Agrawal AA, Konno K (2009) Latex A model for understanding mechanisms, ecology, and evolution of plant defense against herbivory. *Annu Rev Ecol Evol Syst* 40:311–331.

Ahmad S, Kamran M, Ding R, Meng X, Wang H, Ahmad I, Fahad S, Han Q (2019) Exogenous melatonin confers drought stress by promoting plant growth, photosynthetic capacity and antioxidant defense system of maize seedlings. *PeerJ* 7:e7793. http://doi.org/10.7e717/peerj.7793

Akbar H, Timothy JK, Jagadish T, Golam M, Apurbo KC, Muhammad F, Rajan B, Fahad S, Hasanuzzaman M (2020) Agricultural land degradation: processes and problems undermining future food security. In: Fahad S, Hasanuzzaman M, Alam M, Ullah H, Saeed M, Khan AK, Adnan M (eds) *Environment, climate, plant and vegetation growth*. Springer Publ Ltd, Springer Nature Switzerland AG. Part of Springer Nature, pp. 17–62. https://doi.org/10.1007/978-3-030-49732-3

Akram R, Turan V, Hammad HM, Ahmad S, Hussain S, Hasnain A, Maqbool MM, Rehmani MIA, Rasool A, Masood N, Mahmood F, Mubeen M, Sultana SR, Fahad S, Amanet K, Saleem M, Abbas Y, Akhtar HM, Waseem F, Murtaza R, Amin A, Zahoor SA, ul Din MS, Nasim W (2018a) Fate of organic and inorganic pollutants in paddy soils. In: Hashmi MZ, Varma A (eds) *Environmental pollution of paddy soils, soil biology*. Springer International Publishing Ag, Cham, Switzerland, pp. 197–214.

Akram R, Turan V, Wahid A, Ijaz M, Shahid MA, Kaleem S, Hafeez A, Maqbool MM, Chaudhary HJ, Munis, MFH, Mubeen M, Sadiq N, Murtaza R, Kazmi DH, Ali S, Khan N, Sultana SR, Fahad S, Amin A, Nasim W (2018b) Paddy land pollutants and their role in climate change. In: Hashmi MZ, Varma A (eds) *Environmental pollution of paddy soils, soil biology*. Springer International Publishing Ag, Cham, Switzerland, pp. 113–124.

Ali J, Muhammad H, Ullah I, Rashid JA, Adnan M, Ali M, Ahmad W, Rehman A, Khan J (2016) Mango seed germination in different media at different depth. *J Nat Sci Res* 6(1):56–59.

Altaf M, Raziq F, Khan I, Hussain H, Shah B, Ullah W, Naeem A, Adnan M, Junaid K, Shah SRA, Attaullah, Iqbal M (2016) Study on the response of different maize cultivars to various inoculum levels of *Bipolarismaydis*(Y. Nisik & C. Miyake) shoemaker under field conditions. *J Ento Zool Stud* 4(2):533–537.

Amanullah, Muhammad I, Haider N, Shah K, Manzoor A, Asim M, Saif U, Izhar A, Fahad S, Adnan M et al. (2021) Integrated foliar nutrients application improve wheat (*Triticum aestivum* L.) productivity under calcareous soils in drylands. *Commun Soil Sci Plant Anal.* https://doi.org/10.1080/00103624.2021.1956521

Amanullah, Fahad S (eds) (2017) Rice - technology and production. IntechOpen Croatia 2017. http://dx.doi.org/10.5772/64480

Amanullah, Fahad S (eds) (2018a) Corn - Production and human Health in changing climate. IntechOpen United Kingdom 2018. http://dx.doi.org/10.5772/intechopen.74074

Amanullah, Fahad S (eds) (2018b) Nitrogen in agriculture – updates. IntechOpen Croatia 2018. http://dx.doi.org/10.5772/65846

Amanullah, Shah K, Imran, Hamdan AK, Muhammad A, Abdel RA, Muhammad A, Fahad S, Azizullah S, Brajendra P (2020) Effects of climate change on irrigation water quality. In: Fahad S, Hasanuzzaman M, Alam M, Ullah H, Saeed M, Khan AK, Adnan M (eds) *Environment, climate, plant and vegetation growth*. Springer Publ Ltd, Springer Nature Switzerland AG. Part of Springer Nature, pp. 123–132. https://doi.org/10.1007/978-3-030-49732-3

Amir M, Muhammad A, Allah B, Sevgi Ç, Haroon ZK, Muhammad A, Emre A (2020) Bio fortification under climate change: the fight between quality and quantity. In: Fahad S, Hasanuzzaman M, Alam M, Ullah H, Saeed M, Khan AK, Adnan M (eds) *Environment, climate, plant and vegetation growth*. Springer Publ Ltd, Springer Nature Switzerland AG. Part of Springer Nature, pp. 173–228. https://doi.org/10.1007/978-3-030-49732-3

Amjad I, Muhammad H, Farooq S, Anwar H (2020) Role of plant bioactives in sustainable agriculture. In: Fahad S, Hasanuzzaman M, Alam M, Ullah H, Saeed M, Khan AK, Adnan M (eds) *Environment, climate, plant and vegetation growth*. Springer Publ Ltd, Springer Nature Switzerland AG. Part of Springer Nature, pp. 591–606. https://doi.org/10.1007/978-3-030-49732-3

Amjad SF, Mansoora N, Din IU, Khalid IR, Jatoi GH, Murtaza G, Yaseen S, Naz M, Danish S, Fahad S et al. (2021) Application of zinc fertilizer and mycorrhizal inoculation on physio-biochemical parameters of wheat grown under water-stressed environment. *Sustainability* 13:11007. https://doi.org/10.3390/su131911007

Araji S, Grammer TA, Gertzen R, Anderson SD, Mikulic-Petkovsek M, Veberic R, Phu ML, Solar A, Leslie CA, Dandekar AM, Escobar MA (2014) Novel roles for the polyphenol oxidase enzyme in secondary metabolism and the regulation of cell death in walnut. *Plant Physiol* 164:1191–1203.

Arif M, Talha J, Muhammad R, Fahad S, Muhammad A, Amanullah, Kawsar A, Ishaq AM, Bushra K, Fahd R (2020) Biochar; a remedy for climate change. In: Fahad S, Hasanuzzaman M, Alam M, Ullah H, Saeed M, Khan AK, Adnan M (eds) *Environment, climate, plant and vegetation growth.* Springer Publ Ltd, Springer Nature Switzerland AG. Part of Springer Nature, pp. 151–172. https://doi.org/10.1007/978-3-030-49732-3

Ashfaq AR, Uzma Y, Niaz A, Muhammad AA, Fahad S, Haider S, Tayebeh Z, Subhan D, Süleyman T, Hesham AELE, Pramila T, Jamal MA, Sulaiman AA, Rahul D (2021) Toxicity of Cadmium and nickel in the context of applied activated carbon biochar for improvement in soil fertility. *Saudi Soc Agric Sci.* https://doi.org/10.1016/j.sjbs.2021.09.035

Athar M, Masood IA, Sana S, Ahmed M, Xiukang W, Sajid F, Sher AK, Habib A, Faran M, Zafar H, Farhana G, Fahad S (2021) Bio-diesel production of sunflower through sulphur management in a semi-arid subtropical environment. *Environ Sci Pollution Res.* https://doi.org/10.1007/s11356-021-16688-z

Atif B, Hesham A, Fahad S (2021) Biochar coupling with phosphorus fertilization modifies antioxidant activity, osmolyte accumulation and reactive oxygen species synthesis in the leaves and xylem sap of rice cultivars under high-temperature stress. *Physiol Mol Biol Plants.* https://doi.org/10.1007/s12298-021-01062-7

Ayman ELS, Hossain A, Barutçular C, Iqbal MA, Islam MS, Fahad S, Sytar O, Çig F, Meena RS, Erman M (2020) Consequences of salinity stress on the quality of crops and its mitigation strategies for sustainable crop production: an outlook of arid and semi-arid regions. In: Fahad S, Hasanuzzaman M, Alam M, Ullah H, Saeed M, Khan AK, Adnan M (eds) *Environment, climate, plant and vegetation growth.* Springer Publ Ltd, Springer Nature Switzerland AG. Part of Springer Nature, pp. 503–534. https://doi.org/10.1007/978-3-030-49732-3

Aziz K, Daniel KYT, Fazal M, Muhammad ZA, Farooq S, Fan W, Fahad S, Ruiyang Z (2017a) Nitrogen nutrition in cotton and control strategies for greenhouse gas emissions: a review. *Environ Sci Pollut Res* 24:23471–23487. https://doi.org/10.1007/s11356-017-0131-y

Aziz K, Daniel KYT, Muhammad ZA, Honghai L, Shahbaz AT, Mir A, Fahad S (2017b) Nitrogen fertility and abiotic stresses management in cotton crop: a review. *Environ Sci Pollut Res* 24:14551–14566. https://doi.org/10.1007/s11356-017-8920-x

Barnett WP, Hansen MT (1996) The red queen in organizational evolution. *Strat Manage J* 17:139–157.

Baseer M, Adnan M, Fazal M, Fahad S, Muhammad S, Fazli W, Muhammad A, Jr. Amanullah, Depeng W, Saud S, Muhammad N, Muhammad Z, Fazli S, Beena S, Mian AR, Ishaq AM (2019) Substituting urea by organic wastes for improving maize yield in alkaline soil. *J Plant Nutr.* https://doi.org/10.1080/01904167.2019.1659344

Bayram AY, Seher Ö, Nazlican A (2020) Climate change forecasting and modeling for the year of 2050. In: Fahad S, Hasanuzzaman M, Alam M, Ullah H, Saeed M, Khan AK, Adnan M (eds) *Environment, climate, plant and vegetation growth.* Springer Publ Ltd, Springer Nature Switzerland AG. Part of Springer Nature, pp. 109–122. https://doi.org/10.1007/978-3-030-49732-3

Becerra JX, Venable DL (1999) Macroevolution of insect-plant associations The relevance of host biogeography to host affiliation. *Proc Natl Acad Sci U S A* 96:12626–12631.

Bennet RN, Wallsgrove RM (1994) Secondary metabolites in plant defence mechanisms. *New Phytol* 127: 617–633.

Bernays EA (1991) Evolution of insect morphology in relation to plants. *Philos Trans R Soc London B* 333:257–264.

Bernays EA, Janzen DH (1988) Saturniid and sphingid caterpillars two ways to eat leaves. *Ecology* 69:1153–1160.

Blau PA, Feeny P, Contardo L, Robson DS (1978) Allylglucosinolate and herbivorous caterpillars A contrast in toxicity and tolerance. *Science* 200:1296–1298.

Braam J (2005) In touch Plant responses to mechanical stimuli. *New Phytol* 165:373–389.

Bruce TJA (2015) Interplay between insects and plants Dynamic and complex interactions that have coevolved over millions of years but act in milliseconds. *J Exp Bot* 66:455–465.

Burkepile DE, Parker JD (2017) Recent advances in plant-herbivore interactions. F1000Research 6.

Cardoso MZ (2008) Herbivore handling of a plant's trichome The case of *Heliconius charithonia* (L.) (Lepidoptera Nymphalidae) and *Passiflora lobata* (Killip) Hutch. (Passifloraceae). *Neotrop Entomol* 37:247–252.

Chamarthi SK, Sharma HC, Vijay PM, Narasu ML (2011) Leaf surface chemistry of sorghum seedlings influencing expression of resistance to sorghum shoot fly, *Atherigona soccata*. *J Plant Biochem Biotechnol* 20:211–216.

Chandrasekaran R, Revathi K, Thanigaivel A, Kirubakaran SA, Senthil-Nathan S (2014) Bacillus subtilis chitinase identified by matrix-assisted laser desorption/ionization time-of flight/time of flight mass spectrometry has insecticidal activity against Spodoptera litura Fab. *Pestic Biochem Physiol* 116:1–12.

Chang W, Qiujuan J, Evgenios A, Haitao L, Gezi L, Jingjing Z, Fahad S, Ying J (2021) Hormetic effects of zinc on growth and antioxidant defense system of wheat plants. *Sci Total Environ.* https://doi.org/10.1016/j.scitotenv.2021.150992

Chen F, Tholl D, Bohlmann J, Pichersky E (2011) The family of terpene synthases in plants A mid-size family of genes for specialized metabolism that is highly diversified throughout the kingdom. *Plant J* 66:212–229.

Chen M-S (2008) Inducible direct plant defense against insect herbivores a review. *Insect Sci* 15:101–114.

Chen Y, Guo Z, Dong L, Fu Z, Zheng Q, Zhang G, Qin L, Sun X, Shi Z, Fahad S, Xie F, Saud S (2021) Turf performance and physiological responses of native Poa species to summer stress in Northeast China. *PeerJ* 9:e12252 http://doi.org/10.7717/peerj.12252

Clavijo McCormick A, Unsicker SB, Gershenzon J (2012) The specificity of herbivore-induced plant volatiles in attracting herbivore enemies. *Trends Plant Sci* 17:303–310.

Darwin, Charles. 1899. *The various contrivances by which orchids are fertilized by insects.* 2nd ed. London: John Murray. https://www.oxfordbibliographies.com/view/document/obo-9780199830060/obo-9780199830060-0193.xml#obo-9780199830060-0193-bibItem-0001

De Luca V, St Pierre B (2000) The cell and developmental biology of alkaloid biosynthesis. *Trends Plant Sci* 5:168–173.

de Sales PM, de Souza PM, Simeoni LA, Magalhães PDO, Silveira D (2012) α-amylase inhibitors A review of raw material and isolated compounds from plant source. *J Pharmacy Pharmaceutical Sci* 10.

Deepranjan S, Ardith S, Siva D, Sonam S, Manoj P, Amitava R, Sayyed RZ, Abdul G, Mohammad JA, Subhan D, Fahad S, Rahul D (2021) Optimizing nutrient use efficiency, productivity, energetics, and economics of red cabbage following mineral fertilization and biopriming with compatible rhizosphere microbes. *Sci Rep* 11:15680.

Dejean A, Grangier J, Leroy C, Orivel J, Gibernau M (2008) Nest site selection and induced response in a dominant arboreal ant species. *Naturwissenschaften* 95:885–889.

Depeng W, Fahad S, Saud S, Muhammad K, Aziz K, Mohammad NK, Hafiz MH, Wajid N (2018) Morphological acclimation to agronomic manipulation in leaf dispersion and orientation to promote "Ideotype" breeding: evidence from 3D visual modeling of "super" rice (*Oryza sativa* L.). *Plant Physiol Biochem.* 135:499–510 https://doi.org/10.1016/j.plaphy.2018.11.010

Dicke M, Poecke RMPV (2002) Signaling in plant-insect interactions signal transduction in direct and indirect plant defence. In Scheel C, Wasternack D (Hrsg.), *Plant signal transduction.* Oxford University Press, pp. 289–316.

Divol F, Vilaine F, Thibivilliers S, Kusiak C, Sauge MH, Dinant S (2007) Involvement of the xyloglucan endotransglycosylase/hydrolases encoded by celery XTH1 and Arabidopsis XTH33 in the phloem response to aphids. *Plant Cell Environ* 30:187–201.

Dowd P (1991) Symbiont-mediated detoxification in insect herbivores. In Barbosa V, Krischik A, Jones C (Editors), *Microbial mediation of plant – herbivore interactions.* Wiley & Sons, New York, pp. 411–440.

Duan C, Yu J, Bai J, Zhu Z, Wang X (2014) Induced defense responses in rice plants against small brown planthopper infestation. *Crop J* 2:55–62.

Ehrlich PR, Raven PH (1964) Butterflies and Plants A Study in Coevolution. *Evolution* 18:586–608.

Emre B, Ömer SU, Martín LB, Andre D, Fahad S, Rahul D, Muhammad Z-ul-H, Ghulam SH, Subhan D (2021) Studying soil erosion by evaluating changes in physico-chemical properties of soils under different land-use types. *J Saudi Soc Agric Sci* 20:190–197.

Fahad S, Abdul B, Adnan M (eds) (2018b) Global wheat production. IntechOpen United Kingdom 2018. http://dx.doi.org/10.5772/intechopen.72559

Fahad S, Bajwa AA, Nazir U, Anjum SA, Farooq A, Zohaib A, Sadia S, Nasim W, Adkins S, Saud S, Ihsan MZ, Alharby H, Wu C, Wang D, Huang J (2017) Crop production under drought and heat stress: plant responses and management options. *Front Plant Sci* 8:1147. https://doi.org/10.3389/fpls.2017.01147

Fahad S, Bano A (2012) Effect of salicylic acid on physiological and biochemical characterization of maize grown in saline area. *Pak J Bot* 44:1433–1438.

Fahad S, Chen Y, Saud S, Wang K, Xiong D, Chen C, Wu C, Shah F, Nie L, Huang J (2013) Ultraviolet radiation effect on photosynthetic pigments, biochemical attributes, antioxidant enzyme activity and hormonal contents of wheat. *J Food Agri Environ* 11(3&4):1635–1641.

Fahad S, Hussain S, Bano A, Saud S, Hassan S, Shan D, Khan FA, Khan F, Chen Y, Wu C, Tabassum MA, Chun MX, Afzal M, Jan A, Jan MT, Huang J (2014a) Potential role of phytohormones and plant growth-promoting rhizobacteria in abiotic stresses: consequences for changing environment. *Environ Sci Pollut Res* 22(7):4907–4921. https://doi.org/10.1007/s11356-014-3754-2

Fahad S, Hussain S, Matloob A, Khan FA, Khaliq A, Saud S, Hassan S, Shan D, Khan F, Ullah N, Faiq M, Khan MR, Tareen AK, Khan A, Ullah A, Ullah N, Huang J (2014b) Phytohormones and plant responses to salinity stress: a review. *Plant Growth Regul* 75(2):391– 404. https://doi.org/10.1007/s10725-014-0013-y

Fahad S, Hussain S, Saud S, Hassan S, Chauhan BS, Khan F et al (2016a) Responses of rapid viscoanalyzer profile and other rice grain qualities to exogenously applied plant growth regulators under high day and high night temperatures. *PLoS One* 11(7):e0159590. https://doi.org/10.1371/journal.pone.0159590

Fahad S, Hussain S, Saud S, Hassan S, Ihsan Z, Shah AN, Wu C, Yousaf M, Nasim W, Alharby H, Alghabari F, Huang J (2016c) Exogenously applied plant growth regulators enhance the morphophysiological growth and yield of rice under high temperature. *Front Plant Sci* 7:1250. https://doi.org/10.3389/fpls.2016.01250

Fahad S, Hussain S, Saud S, Hassan S, Tanveer M, Ihsan MZ, Shah AN, Ullah A, Nasrullah KF, Ullah S, Alharby HNW, Wu C, Huang J (2016d) A combined application of biochar and phosphorus alleviates heat-induced adversities on physiological, agronomical and quality attributes of rice. *Plant Physiol Biochem* 103:191–198.

Fahad S, Hussain S, Saud S, Khan F, Hassan S, Jr, Nasim W, Arif M, Wang F, Huang J (2016b) Exogenously applied plant growth regulators affect heat-stressed rice pollens. *J Agron Crop Sci* 202:139–150.

Fahad S, Hussain S, Saud S, Tanveer M, Bajwa AA, Hassan S, Shah AN, Ullah A, Wu C, Khan FA, Shah F, Ullah S, Chen Y, Huang J (2015a) A biochar application protects rice pollen from high-temperature stress. *Plant Physiol Biochem* 96:281–287.

Fahad S, Muhammad ZI, Abdul K, Ihsanullah D, Saud S, Saleh A, Wajid N, Muhammad A, Imtiaz AK, Chao W, Depeng W, Jianliang H (2018a) Consequences of high temperature under changing climate optima for rice pollen characteristics-concepts and perspectives. *Archives Agron Soil Sci.* https://doi.org/10.1080/03650340.2018.1443213

Fahad S, Nie L, Chen Y, Wu C, Xiong D, Saud S, Hongyan L, Cui K, Huang J (2015b) Crop plant hormones and environmental stress. *Sustain Agric Rev* 15:371–400.

Fahad S, Saud S, Yajun C, Chao W, Depeng W (eds) (2021f) Abiotic stress in plants. IntechOpen United Kingdom 2021. http://dx.doi.org/10.5772/intechopen.91549

Fahad S, Hasanuzzaman M, Alam M, Ullah H, Saeed M, Ali Khan I, Adnan M. (Eds.) (2020) *Environment, climate, plant and vegetation growth.* Springer Nature Switzerland AG. https://doi.org/10.1007/978-3-030-49732-3

Fahad S, Adnan M, Hassan S, Saud S, Hussain S, Wu C, Wang D, Hakeem KR, Alharby HF, Turan V, Khan MA, Huang J (2019b) Rice responses and tolerance to high temperature. In: Hasanuzzaman M, Fujita M, Nahar K, Biswas JK (eds) *Advances in rice research for abiotic stress tolerance.* Woodhead Publ Ltd, Abington, Cambridge, UK, pp. 201–224.

Fahad S, Rehman A, Shahzad B, Tanveer M, Saud S, Kamran M, Ihtisham M, Khan SU, Turan V, Rahman MHU (2019a) Rice responses and tolerance to metal/metalloid toxicity. In: Hasanuzzaman M, Fujita M, Nahar K, Biswas JK (eds) *Advances in rice research for abiotic stress tolerance.* Woodhead Publ Ltd, Abington, Cambridge, UK, pp. 299–312.

Fahad S, Sönmez O, Saud S, Wang D, Wu C, Adnan M, Arif M, Amanullah (eds) (2021e) Engineering tolerance in crop plants against abiotic stress. In: *Footprints of climate variability on plant diversity*, 1st edn. CRC Press, Boca Raton.

Fahad S, Sonmez O, Saud S, Wang D, Wu C, Adnan M, Turan V (eds) (2021b) Climate change and plants: biodiversity, growth and interactions. In: *Footprints of climate variability on plant diversity*, 1st edn. CRC Press, Boca Raton.

Fahad S, Sonmez O, Saud S, Wang D, Wu C, Adnan M, Turan V (eds) (2021c) Developing climate resilient crops: improving global food security and safety. In: *Footprints of climate variability on plant diversity*, 1st edn. CRC Press, Boca Raton.

Fahad S, Sönmez O, Saud S, Wang D, Wu C, Adnan M, Turan V (eds) (2021a) Plant growth regulators for climate-smart agriculture. In: *Footprints of climate variability on plant diversity*, 1st edn. CRC Press, Boca Raton.

Fahad S, Sönmez O, Turan V, Adnan M, Saud S, Wu C, Wang D (eds) (2021d) Sustainable soil and land management and climate change. In: *Footprints of climate variability on plant diversity*, 1st edn. CRC Press, Boca Raton.

Farah R, Muhammad R, Muhammad SA, Tahira Y, Muhammad AA, Maryam A, Shafaqat A, Rashid M, Muhammad R, Qaiser H, Afia Z, Muhammad AA, Muhammad A, Fahad S (2020) Alternative and non-conventional soil and crop management strategies for increasing water use efficiency. In: Fahad S, Hasanuzzaman M, Alam M, Ullah H, Saeed M, Khan AK, Adnan M (eds) *Environment, climate, plant and vegetation growth*. Springer Publ Ltd, Springer Nature Switzerland AG. Part of Springer Nature, pp. 323–338. https://doi.org/10.1007/978-3-030-49732-3

Farhana G, Ishfaq A, Muhammad A, Dawood J, Fahad S, Xiuling L, Depeng W, Muhammad F, Muhammad F, Syed AS (2020) Use of crop growth model to simulate the impact of climate change on yield of various wheat cultivars under different agro-environmental conditions in Khyber Pakhtunkhwa, Pakistan. *Arabian J Geosci* 13:112. https://doi.org/10.1007/s12517-020-5118-1

Farhat A, Hafiz MH, Wajid I, Aitazaz AF, Hafiz FB, Zahida Z, Fahad S, Wajid F, Artemi C (2020) A review of soil carbon dynamics resulting from agricultural practices. *J Environ Manage* 268:110319.

Fazli W, Muhmmad S, Amjad A, Fahad S, Muhammad A, Muhammad N, Ishaq AM, Imtiaz AK, Mukhtar A, Muhammad S, Muhammad I, Rafi U, Haroon I, Muhammad A (2020) Plant-microbes interactions and functions in changing climate. In: Fahad S, Hasanuzzaman M, Alam M, Ullah H, Saeed M, Khan AK, Adnan M (eds) *Environment, climate, plant and vegetation growth*. Springer Publ Ltd, Springer Nature Switzerland AG. Part of Springer Nature, pp. 397–420. https://doi.org/10.1007/978-3-030-49732-3

Feeny P (1970) Seasonal changes in Oak Leaf Tannins and nutrients as a cause of spring feeding by winter moth caterpillars. *Ecology* 51:565–581.

Felton GW, Eichenseer H (1999) Herbivore saliva and its effect on plant defense against herbivores and pathogens. In Agrawal A, Tuzun S, Bent E (Hrsg.), *Induced plant defenses against pathogens and herbivores. Biochemistry, ecology and agriculture*. American Phytopathologial Society, St. Paul, MN, pp. 19–36.

Fernandes GW (1994) Plant mechanical defenses against insect herbivory. *Rev Bras Entomol* 38:421–433.

Ferrari S, Savatin DV, Sicilia F, Gramegna G, Cervone F, De Lorenzo G (2013) Oligogalacturonides plant damage-associated molecular patterns and regulators of growth and development. *Frontiers Plant Sci* 4:1–9.

Feyereisen R (1999) Insect P450 enzymes. *Annu Rev Entomol* 44:507–533.

Fornoni J (2011) Ecological and evolutionary implications of plant tolerance to herbivory. *Funct Ecol* 25: 399–407.

Fürstenberg-Hägg J, Zagrobelny M, Bak S (2013) Plant defense against insect herbivores. *Int J Mol Sci* 14:10242–10297.

Futuyma DJ, Slatkin M (1983) Coevolution. Sinauer Associates, 555 pp.

Ghulam M, Muhammad AA, Donald LS, Sajid M, Muhammad FQ, Niaz A, Ateeq Ur R, Shakeel A, Sajjad H, Muhammad A, Summia M, Aqib HAK, Fahad S, Rahul D, Mazhar I, Timothy DS (2021) Formalin fumigation and steaming of various composts differentially influence the nutrient release, growth and yield of muskmelon (*Cucumis melo* L.). *Sci Rep* 11:21057.

Gianoli E, Carrasco-Urra F (2014) Leaf mimicry in a climbing plant protects against herbivory. *Curr Biol* 24:357–359.

Gish M, Mescher MC, De Moraes CM (2016) Mechanical defenses of plant extrafloral nectaries against herbivory. *Commun Integr Biol* 9.

Gopakumar L, Bernard NO, Donato V (2020) Soil microarthropods and nutrient cycling. In: Fahad S, Hasanuzzaman M, Alam M, Ullah H, Saeed M, Khan AK, Adnan M (eds) *Environment, climate, plant and vegetation growth*. Springer Publ Ltd, Springer Nature Switzerland AG. Part of Springer Nature, pp. 453–472. https://doi.org/10.1007/978-3-030-49732-3

Goussain MM, Prado E, Moraes JC (2005) Effect of silicon applied to wheat plants on the biology and probing behaviour of the greenbug *Schizaphis graminum* (Rond.) (Hemiptera: Aphididae). *Neotrop Entomol* 34:807–813.
Guofu L, Zhenjian B, Fahad S, Guowen C, Zhixin X, Hao G, Dandan L, Yulong L, Bing L, Guoxu J, Saud S (2021) Compositional and structural changes in soil microbial communities in response to straw mulching and plant revegetation in an abandoned artificial pasture in Northeast China. *Glob Ecol Conserv* 31:e01871.
Habib KM (2007) Plant protease inhibitors a defense strategy in plants. *Biotechnol Mol Biol Rev* 2:68–85.
Habib Ur R, Ashfaq A, Aftab W, Manzoor H, Fahd R, Wajid I, Md. Aminul I, Vakhtang S, Muhammad A, Asmat U, Abdul W, Syeda RS, Shah S, Shahbaz K, Fahad S, Manzoor H, Saddam H, Wajid N (2017) Application of CSM-CROPGRO-cotton model for cultivars and optimum planting dates: evaluation in changing semi-arid climate. *Field Crops Res.* http://dx.doi.org/10.1016/j.fcr.2017.07.007
Hafeez M, Farman U, Muhammad MK, Xiaowei L, Zhijun Z, Sakhawat S, Muhammad I, Mohammed AA, Mandela F-G, Nicolas D, Muzammal R, Fahad S, Yaobin L (2021) Metabolic-based insecticide resistance mechanism and ecofriendly approaches for controlling of beet armyworm *Spodoptera exigua*: a review. *Environ Sci Pollution Res.* https://doi.org/10.1007/s11356-021-16974-w
Hafiz MH, Abdul K, Farhat A, Wajid F, Fahad S, Muhammad A, Ghulam MS, Wajid N, Muhammad M, Hafiz FB (2020b) Comparative effects of organic and inorganic fertilizers on soil organic carbon and wheat productivity under arid region. *Commun Soil Sci Plant Anal.* https://doi.org/10.1080/00103624.2020.1763385
Hafiz MH, Farhat A, Ashfaq A, Hafiz FB, Wajid F, Carol Jo W, Fahad S, Gerrit H (2020a) Predicting kernel growth of maize under controlled water and nitrogen applications. *Int J Plant Prod.* https://doi.org/10.1007/s42106-020-00110-8
Hafiz MH, Farhat A, Shafqat S, Fahad S, Artemi C, Wajid F, Chaves CB, Wajid N, Muhammad M, Hafiz FB (2018) Offsetting land degradation through nitrogen and water management during maize cultivation under arid conditions. *Land Degrad Dev* 1–10. https://doi.org/10.1002/ldr.2933
Hafiz MH, Muhammad A, Farhat A, Hafiz FB, Saeed AQ, Muhammad M, Fahad S, Muhammad A (2019) Environmental factors affecting the frequency of road traffic accidents: a case study of sub-urban area of Pakistan. *Environ Sci Pollut Res.* https://doi.org/10.1007/s11356-019-04752-8
Hafiz MH, Wajid F, Farhat A, Fahad S, Shafqat S, Wajid N, Hafiz FB (2016) Maize plant nitrogen uptake dynamics at limited irrigation water and nitrogen. *Environ Sci Pollut Res* 24(3):2549–2557. https://doi.org/10.1007/s11356-016-8031-0
Hagenbucher S, Olson DM, Ruberson JR, Wäckers FL, Romeis J (2013) Resistance mechanisms against arthropod herbivores in cotton and their interactions with natural enemies. *Crit Rev Plant Sci* 32:458–482.
Haider SA, Lalarukh I, Amjad SF, Mansoora N, Naz M, Naeem M, Bukhari SA, Shahbaz M, Ali SA, Marfo TD, Subhan D, Rahul D, Fahad S (2021) Drought stress alleviation by potassium-nitrate-containing chitosan/montmorillonite microparticles confers changes in *Spinacia oleracea* L. *Sustainability* 13:9903. https://doi.org/10.3390/su13179903
Hamza SM, Xiukang W, Sajjad A, Sadia Z, Muhammad N, Adnan M, Fahad S et al. (2021) Interactive effects of gibberellic acid and NPK on morpho-physio-biochemical traits and organic acid exudation pattern in coriander (*Coriandrum sativum* L.) grown in soil artificially spiked with boron. *Plant Physiol Biochem* 167(2021):884–900.
Haukioja E, Koricheva J (2000) Tolerance to herbivory in woody vs. herbaceous plants. *Evol Ecol* 14:551–562.
He J, Chen F, Chen S, Lv G, Deng Y, Fang W, Liu Z, Guan Z, He C (2011) Chrysanthemum leaf epidermal surface morphology and antioxidant and defense enzyme activity in response to aphid infestation. *J Plant Physiol* 168:687–693.
Heil M (2015) Extrafloral nectar at the plant-insect interface a spotlight on chemical ecology, phenotypic plasticity, and food webs. *Annu Rev Entomol* 60:213–232.
Herbert RBRB (1989) *The biosynthesis of secondary metabolites.* Chapman and Hall, 231 pp.
Hesham FA and Fahad S (2020) Melatonin application enhances biochar efficiency for drought tolerance in maize varieties: modifications in physio-biochemical machinery. *Agron J* 112(4):1–22.
Hogenhout SA, Bos JIB (2011) Effector proteins that modulate plant-insect interactions. *Curr Opin Plant Biol* 14:422–428.
Hopkins AD (1916) Economic investigations of the scolytid bark and timber beetles of North America, pp. 353.
Hopkins RJ, van Dam NM, van Loon JJA (2009) Role of glucosinolates in insect-plant relationships and multitrophic interactions. *Annu Rev Entomol* 54:57–83.

Howe GA, Schaller A (2008) Direct defenses in plants and their induction by wounding and insect herbivores. In Schaller A (Hrsg.), *Induced plant resistance to herbivory*. Springer, Dordrecht, The Netherlands, pp. 7–29.

Huang H, Liu B, Liu L, Song S (2017) Jasmonate action in plant growth and development. *J Exp Bot* 68:1349–1359.

Huang L-Y, Li X-X, Zhang Y-B, Fahad S, Wang F (2021) dep1 improves rice grain yield and nitrogen use efficiency simultaneously by enhancing nitrogen and dry matter translocation. *J Integrative Agri*. https://doi.org/10.1016/S2095-3119(21)63795-4

Hund R (2017) When camouflage is a plant's best protection. EurekAlert.

Hussain MA, Fahad S, Rahat S, Muhammad FJ, Muhammad M, Qasid A, Ali A, Husain A, Nooral A, Babatope SA, Changbao S, Liya G, Ibrar A, Zhanmei J, Juncai H (2020) Multifunctional role of brassinosteroid and its analogues in plants. *Plant Growth Regul*. https://doi.org/10.1007/s10725-020-00647-8

Ibrar K, Aneela R, Khola Z, Urooba N, Sana B, Rabia S, Ishtiaq H, Mujaddad R, Salvatore M (2020) Microbes and environment: global warming reverting the frozen zombies. In: Fahad S, Hasanuzzaman M, Alam M, Ullah H, Saeed M, Khan AK, Adnan M (eds) *Environment, climate, plant and vegetation growth*. Springer Publ Ltd, Springer Nature Switzerland AG. Part of Springer Nature, pp. 607–634. https://doi.org/10.1007/978-3-030-49732-3

Ikram U, Khadim D, Muhammad T, Muhammad S, Fahad S (2021) Gibberellic acid and urease inhibitor optimize nitrogen uptake and yield of maize at varying nitrogen levels under changing climate. *Environ Sci Pollution Res*. https://doi.org/10.1007/s11356-021-16049-w

Ilyas M, Mohammad N, Nadeem K, Ali H, Aamir HK, Kashif H, Fahad S, Aziz K, Abid U (2020) Drought tolerance strategies in plants: a mechanistic approach. *J Plant Growth Regul*. https://doi.org/10.1007/s00344-020-10174-5

Iqra M, Amna B, Shakeel I, Fatima K, Sehrish L, Hamza A, Fahad S (2020) Carbon cycle in response to global warming. In: Fahad S, Hasanuzzaman M, Alam M, Ullah H, Saeed M, Khan AK, Adnan M (eds) *Environment, climate, plant and vegetation growth*. Springer Publ Ltd, Springer Nature Switzerland AG. Part of Springer Nature, pp. 1–16. https://doi.org/10.1007/978-3-030-49732-3

Irfan M, Muhammad M, Muhammad JK, Khadim MD, Dost M, Ishaq AM, Waqas A, Fahad S, Saud S et al. (2021) Heavy metals immobilization and improvement in maize (*Zea mays* L.) growth amended with biochar and compost. *Sci Rep* 11:18416.

Jabborova D, Sulaymanov K, Sayyed RZ, Alotaibi SH, Enakiev Y, Azimov A, Jabbarov Z, Ansari MJ, Fahad S, Danish S et al. (2021) Mineral fertilizers improves the quality of turmeric and soil. *Sustainability* 13:9437. https://doi.org/10.3390/su13169437

Jan M, Muhammad AH, Adnan NS, Muhammad Y, Javaid I, Xiuling L, Depeng W, Fahad S (2019) Modulation in growth, gas exchange, and antioxidant activities of salt-stressed rice (*Oryza sativa* L.) genotypes by zinc fertilization. *Arabian J Geosci* 12:775 https://doi.org/10.1007/s12517-019-4939-2

Jongsma MA, Bolter C (1997) The adaptation of insects to plant protease inhibitors. *J Insect Physiol* 43:885–893.

Javed MW (2018) Dynamicity of plant-insect interaction, defense communism and key-prospects. *J Agric Res* 3:000188.

Kamarn M, Wenwen C, Irshad A, Xiangping M, Xudong Z, Wennan S, Junzhi C, Shakeel A, Fahad S, Qingfang H, Tiening L (2017) Effect of paclobutrazol, a potential growth regulator on stalk mechanical strength, lignin accumulation and its relation with lodging resistance of maize. *Plant Growth Regul* 84:317–332. https://doi.org/10.1007/s10725-017-0342-8

Karban R, Agrawal AA (2002) Herbivore offense. *Annu Rev Ecol Syst* 33:641–664.

Keeping MG, Kvedaras OL (2008) Silicon as a plant defence against insect herbivory Response to Massey, Ennos and Hartley. *J Anim Ecol* 77:631–633.

Kessler A, Baldwin IT (2002) Plant responses to insect herbivory the emerging molecular analysis. *Annu Rev Plant Biol* 53:299–328.

Kessler A, Heil M (2011) The multiple faces of indirect defences and their agents of natural selection. *Funct Ecol* 25:348–357.

Khadim D, Fahad S, Jahangir MMR, Iqbal M, Syed SA, Shah AK, Ishaq AM, Rahul D et al. (2021a) Biochar and urease inhibitor mitigate NH_3 and N_2O emissions and improve wheat yield in a urea fertilized alkaline soil. *Sci Rep* 11:17413.

Khadim D, Saif R, Fahad S, Syed SA, Shah AK et al. (2021b) Influence of variable biochar concentration on yield-scaled nitrous oxide emissions, wheat yield and nitrogen use efficiency. *Sci Rep* 11:16774.

Khan MMH, Niaz A, Umber G, Muqarrab A, Muhammad AA, Muhammad I, Shabir H, Shah F, Vibhor A, Shams HA-H, Reham A, Syed MBA, Nadiyah MA, Ali TKZ, Subhan D, Rahul D (2021) Synchronization of Boron application methods and rates is environmentally friendly approach to improve quality attributes of *Mangifera indica* L. On sustainable basis. *Saudi J Bio Sci.* https://doi.org/10.1016/j.sjbs.2021.10.036

Konno K (2011) Plant latex and other exudates as plant defense systems Roles of various defense chemicals and proteins contained therein. *Phytochemistry* 72:1510–1530.

Krieger RI, Feeny PP, Wilkinson CF (1971) Detoxication enzymes in the guts of caterpillars an evolutionary answer to plant defenses? *Science* 172:579–580.

Labandeira CC (2013) A paleobiologic perspective on plant-insect interactions. *Curr Opin Plant Biol* 16: 414–421.

Levin DA (1973) The role of trichomes in plant defense. *Quart Rev Biol* 48:3–15.

Maffei ME, Mithöfer A, Boland W (2007) Insects feeding on plants rapid signals and responses preceding the induction of phytochemical release. *Phytochemistry* 68:2946–2959.

Mahar A, Amjad A, Altaf HL, Fazli W, Ronghua L, Muhammad A, Fahad S, Muhammad A, Rafiullah, Imtiaz AK, Zengqiang Z (2020) Promising technologies for Cd-contaminated soils: drawbacks and possibilities. In: Fahad S, Hasanuzzaman M, Alam M, Ullah H, Saeed M, Khan AK, Adnan M (eds) *Environment, climate, plant and vegetation growth.* Springer Publ Ltd, Springer Nature Switzerland AG. Part of Springer Nature, pp. 63–92. https://doi.org/10.1007/978-3-030-49732-3

Mahmood Ul H, Tassaduq R, Chandni I, Adnan A, Muhammad A, Muhammad MA, Muhammad H-Ur-R, Mehmood AN, Alam S, Fahad S (2021) Linking plants functioning to adaptive responses under heat stress conditions: a mechanistic review. *J Plant Growth Regul.* https://doi.org/10.1007/s00344-021-10493-1

Manzer HS, Saud A, Soumya M, Abdullah AA, Qasi DA, Bander MAM, Hayssam MA, Hazem MK, Fahad S, Vishnu DR, Om PN (2021) Molybdenum and hydrogen sulfide synergistically mitigate arsenic toxicity by modulating defense system, nitrogen and cysteine assimilation in faba bean (*Vicia faba* L.) seedlings. *Environ Pollut* 290:117953.

Mao J, Burt AJ, Ramputh AI, Simmonds J, Cass L, Hubbard K, Miller S, Altosaar I, Arnason JT (2007) Diverted secondary metabolism and improved resistance to European corn borer (*Ostrinia nubilalis*) in maize (*Zea mays* L.) transformed with wheat oxalate oxidase. *J Agric Food Chem* 55:2582–2589.

Marquis RJ (1996) Plant architecture, sectoriality and plant tolerance to herbivores. *Vegetatio* 127:85–97.

Marston K, Hostettman A (1996) Saponins chemistry and pharmacology of natural products. *J Am Chem Soc* 118:8509–8509.

Mazid M, Khan TA, Mohammad F (2011) Role of secondary metabolites in defense mechanisms of plants. *Biol Med* 3:232–249.

Md Jakir H, Allah B (2020) Development and applications of transplastomic plants; a way towards eco-friendly agriculture. In: Fahad S, Hasanuzzaman M, Alam M, Ullah H, Saeed M, Khan AK, Adnan M (eds) *Environment, climate, plant and vegetation growth.* Springer Publ Ltd, Springer Nature Switzerland AG. Part of Springer Nature, pp. 285–322. https://doi.org/10.1007/978-3-030-49732-3

Md. Enamul H, Shoeb AZM, Mallik AH, Fahad S, Kamruzzaman MM, Akib J, Nayyer S, Mehedi KM, Swati AS, Md Yeamin A, Most SS (2020) Measuring vulnerability to environmental hazards: qualitative to quantitative. In: Fahad S, Hasanuzzaman M, Alam M, Ullah H, Saeed M, Khan AK, Adnan M (eds) *Environment, climate, plant and vegetation growth.* Springer Publ Ltd, Springer Nature Switzerland AG. Part of Springer Nature, pp. 421–452. https://doi.org/10.1007/978-3-030-49732-3

Mehrabadi M, Bandani AR, Saadati F, Mahmudvand M (2011) Inhibition of Sunn pest, Eurygaster integriceps, α-amylases by α-amylase inhibitors (T-αAI) from Triticale. *J Agric Sci Technol* 10:179.

Mithöfer A, Boland W (2012) Plant defense against herbivores chemical aspects. *Annu Rev Plant Biol* 63: 431–450.

Mitter C, Farrell B, Futuyma DJ (1991) Phylogenetic studies of insect-plant interactions Insights into the genesis of diversity. *Trends Ecol Evol* 6:290–293.

Mohammad IA-W, Abdelazeem S, Munir A, Khalid E, Adel RAU (2020b) Extent of climate change in Saudi Arabia and its impacts on agriculture: a case study from Qassim Region. In: Fahad S, Hasanuzzaman M, Alam M, Ullah H, Saeed M, Khan AK, Adnan M (eds) *Environment, climate, plant and vegetation growth.* Springer Publ Ltd, Springer Nature Switzerland AG. Part of Springer Nature, pp. 635–658. https://doi.org/10.1007/978-3-030-49732-3

Mohammad IA-W, Ahmad M, Usman ARA, Akanji M, Rafique MI (2020a) Advances in pyrolytic technologies with improved carbon capture and storage to combat climate change. In: Fahad S, Hasanuzzaman

M, Alam M, Ullah H, Saeed M, Khan AK, Adnan M (eds) *Environment, climate, plant and vegetation growth*. Springer Publ Ltd, Springer Nature Switzerland AG. Part of Springer Nature, pp. 535–576. https://doi.org/10.1007/978-3-030-49732-3

Mondal AK (2013) What makes plants 'sleep'? *Sci Repor* 50(7):45–48.

Mubeen M, Ashfaq A, Hafiz MH, Muhammad A, Hafiz UF, Mazhar S, Muhammad SD, Asad A, Amjed A, Fahad S, Wajid N (2020) Evaluating the climate change impact on water use efficiency of cotton-wheat in semi-arid conditions using DSSAT model. *J Water Clim Change*. https://doi.org/10.2166/wcc.2019.179/622035/jwc2019179.pdf

Muhammad N, Muqarrab A, Khurram S, Fiaz A, Fahim N, Muhammad A, Shazia A, Omaima N, Sulaiman AA, Fahad S, Subhan D, Rahul D (2021) Kaolin and Jasmonic acid improved cotton productivity under water stress conditions. *J Saudi Society Agricultural Sci* 28:6606–6614. https://doi.org/10.1016/j.sjbs.2021.07.043

Muhammad TQ, Amna F, Amna B, Barira Z, Xitong Z, Ling-Ling C (2020) Effectiveness of conventional crop improvement strategies vs. omics. In: Fahad S, Hasanuzzaman M, Alam M, Ullah H, Saeed M, Khan AK, Adnan M (eds) *Environment, climate, plant and vegetation growth*. Springer Publ Ltd, Springer Nature Switzerland AG. Part of Springer Nature, pp. 253–284. https://doi.org/10.1007/978-3-030-49732-3

Muhammad Z, Abdul MK, Abdul MS, Kenneth BM, Muhammad S, Shahen S, Ibadullah J, Fahad S (2019) Performance of *Aeluropus lagopoides* (mangrove grass) ecotypes, a potential turfgrass, under high saline conditions. *Environ Sci Pollut Res*. https://doi.org/10.1007/s11356-019-04838-3

Muzammal R, Fahad S, Guanghui D, Xia C, Yang Y, Kailei T, Lijun L, Fei-Hu L, Gang D (2021) Evaluation of hemp (*Cannabis sativa* L.) as an industrial crop: a review. *Environ Sci Pollution Res*. https://doi.org/10.1007/s11356-021-16264-5

Niemeyer HM (1988) Hydroxamic acids (4-hydroxy-1,4-benzoxazin-3-ones), defence chemicals in the gramineae. *Phytochem* 27:3349–3358.

Noor M, Naveed Ur R, Ajmal J, Fahad S, Muhammad A, Fazli W, Saud S, Hassan S (2020) Climate change and costal plant lives. In: Fahad S, Hasanuzzaman M, Alam M, Ullah H, Saeed M, Khan AK, Adnan M (eds) *Environment, climate, plant and vegetation growth*. Springer Publ Ltd, Springer Nature Switzerland AG. Part of Springer Nature, pp. 93–108. https://doi.org/10.1007/978-3-030-49732-3

Paré PW, Tumlinson JH (1999) Plant volatiles as a defense against insect herbivores. *Plant Physiol* 121:325–332.

Pechan T, Cohen A, Williams WP, Luthe DS (2002) Insect feeding mobilizes a unique plant defense protease that disrupts the peritrophic matrix of caterpillars. *Proc Natl Acad Sci U S A* 99:13319–13323.

Peumans WJ, Van Damme EJ (1995) Lectins as plant defense proteins. *Plant Physiol* 109:347–352.

Pilson D (2000) Herbivory and natural selection on flowering phenology in wild sunflower, *Helianthus annuus*. *Oecologia* 122:72–82.

Qamar-Uz Z, Zubair A, Muhammad Y, Muhammad ZI, Abdul K, Fahad S, Safder B, Ramzani PMA, Muhammad N (2017) Zinc biofortification in rice: leveraging agriculture to moderate hidden hunger in developing countries. *Arch Agron Soil Sci* 64:147–161. https://doi.org/10.1080/03650340.2017.1338343

Ramos MV, Grangeiro TB, Freire EA, Sales MP, Souza DP, Araújo ES, Freitas CDT (2010) The defensive role of latex in plants Detrimental effects on insects. *Arthropod Plant Interact* 4:57–67.

Rani PU, Jyothsna Y (2010) Biochemical and enzymatic changes in rice plants as a mechanism of defense. *Acta Physiol Plant* 32:695–701.

Rashid M, Qaiser H, Khalid SK, Mohammad I-W, Zhang A, Muhammad A, Shahzada SI, Rukhsanda A, Ghulam AS, Shahzada MM, Sarosh A, Muhammad FQ (2020) Prospects of biochar in alkaline soils to mitigate climate change. In: Fahad S, Hasanuzzaman M, Alam M, Ullah H, Saeed M, Khan AK, Adnan M (eds) *Environment, climate, plant and vegetation growth*. Springer Publ Ltd, Springer Nature Switzerland AG. Part of Springer Nature, pp. 133–150. https://doi.org/10.1007/978-3-030-49732-3

Redovniković IR, Glivetić T, Delonga K, Vorkapić-Furač J (2008) Glucosinolates and their potential role in plant. *Period Biol* 110:297–309.

Reese JC, Schwenke JR, Lamont PS, Zehr DD (1994) Importance and quantification of plant tolerance in crop pest management programs for aphids greenbug resistance in Sorghum. *J Agric Entomol* 11:255–270.

Rehman M, Fahad S, Saleem MH, Hafeez M, Rahman MH, Liu F, Deng G (2020) Red light optimized physiological traits and enhanced the growth of ramie (*Boehmeria nivea* L.). *Photosynthetica* 58(4):922–931.

Rosenthal GA, Berenbaum MR (1991) *Herbivores: their interactions with secondary plant metabolites*, Vol. 1, The chemical participants, 308 pp.

Rosenthal JP, Kotanen PM (1994) Terrestrial plant tolerance to herbivory. *Trends Ecol Evol* 9:145–148.

Sadam M, Muhammad TQ, Ghulam M, Muhammad SK, Faiz AJ (2020) Role of biotechnology in climate resilient agriculture. In: Fahad S, Hasanuzzaman M, Alam M, Ullah H, Saeed M, Khan AK, Adnan M (eds) *Environment, climate, plant and vegetation growth*. Springer Publ Ltd, Springer Nature Switzerland AG. Part of Springer Nature, pp. 339–366. https://doi.org/10.1007/978-3-030-49732-3

Safi UK, Ullah F, Mehmood S, Fahad S, Ahmad Rahi A, Althobaiti F, et al. (2021) Antimicrobial, antioxidant and cytotoxic properties of *Chenopodium glaucum* L. *PLoS ONE* 16(10):e0255502. https://doi.org/10.1371/journal.pone.0255502

Sagers CL (1992) Manipulation of host plant quality herbivores keep leaves in the dark. *Funct Ecol* 6:741–743.

Sahrish N, Shakeel A, Ghulam A, Zartash F, Sajjad H, Mukhtar A, Muhammad AK, Ahmad K, Fahad S, Wajid N, Sezai E, Carol Jo W, Gerrit H (2022) Modeling the impact of climate warming on potato phenology. *Eur J AgroN* 132:126404.

Sajid H, Jie H, Jing H, Shakeel A, Satyabrata N, Sumera A, Awais S, Chunquan Z, Lianfeng Z, Xiaochuang C, Qianyu J, Junhua Z (2020) Rice production under climate change: adaptations and mitigating strategies. In: Fahad S, Hasanuzzaman M, Alam M, Ullah H, Saeed M, Khan AK, Adnan M (eds) *Environment, climate, plant and vegetation growth*. Springer Publ Ltd, Springer Nature Switzerland AG. Part of Springer Nature, pp. 659–686. https://doi.org/10.1007/978-3-030-49732-3

Sajjad H, Muhammad M, Ashfaq A, Waseem A, Hafiz MH, Mazhar A, Nasir M, Asad A, Hafiz UF, Syeda RS, Fahad S, Depeng W, Wajid N (2019) Using GIS tools to detect the land use/land cover changes during forty years in Lodhran district of Pakistan. *Environ Sci Pollut Res*. https://doi.org/10.1007/s11356-019-06072-3

Saleem MH, Fahad S, Adnan M, Mohsin A, Muhammad SR, Muhammad K, Qurban A, Inas AH, Parashuram B, Mubassir A, Reem MH (2020a) Foliar application of gibberellic acid endorsed phytoextraction of copper and alleviates oxidative stress in jute (*Corchorus capsularis* L.) plant grown in highly copper-contaminated soil of China. *Environ Sci Pollution Res*. https://doi.org/10.1007/s11356-020-09764-3

Saleem MH, Fahad S, Shahid UK, Mairaj D, Abid U, Ayman ELS, Akbar H, Analía L, Lijun L (2020c) Copper-induced oxidative stress, initiation of antioxidants and phytoremediation potential of flax (*Linum usitatissimum* L.) seedlings grown under the mixing of two different soils of China. *Environ Sci Poll Res*. https://doi.org/10.1007/s11356-019-07264-7

Saleem MH, Rehman M, Fahad S, Tung SA, Iqbal N, Hassan A, Ayub A, Wahid MA, Shaukat S, Liu L, Deng G (2020b) Leaf gas exchange, oxidative stress, and physiological attributes of rapeseed (*Brassica napus* L.) grown under different light-emitting diodes. *Photosynthetica* 58(3):836–845.

Salunke BK, Kotkar HM, Mendki PS, Upasani SM, Maheshwari VL (2005) Efficacy of flavonoids in controlling *Callosobruchus chinensis* (L.) (Coleoptera: Bruchidae), a post-harvest pest of grain legumes. *Crop Prot* 24:888–893.

Saman S, Amna B, Bani A, Muhammad TQ, Rana MA, Muhammad SK (2020) QTL mapping for abiotic stresses in cereals. In: Fahad S, Hasanuzzaman M, Alam M, Ullah H, Saeed M, Khan AK, Adnan M (eds) *Environment, climate, plant and vegetation growth*. Springer Publ Ltd, Springer Nature Switzerland AG. Part of Springer Nature, pp. 229–252. https://doi.org/10.1007/978-3-030-49732-3

Saud S, Chen Y, Fahad S, Hussain S, Na L, Xin L, Alhussien SA (2016) Silicate application increases the photosynthesis and its associated metabolic activities in Kentucky bluegrass under drought stress and post-drought recovery. *Environ Sci Pollut Res* 23(17):17647–17655. https://doi.org/10.1007/s11356-016-6957-x

Saud S, Chen Y, Long B, Fahad S, Sadiq A (2013) The different impact on the growth of cool season turf grass under the various conditions on salinity and drought stress. *Int J Agric Sci Res* 3:77–84.

Saud S, Fahad S, Cui G, Chen Y, Anwar S (2020) Determining nitrogen isotopes discrimination under drought stress on enzymatic activities, nitrogen isotope abundance and water contents of Kentucky bluegrass. *Sci Rep* 10:6415. https://doi.org/10.1038/s41598-020-63548-w

Saud S, Fahad S, Yajun C, Ihsan MZ, Hammad HM, Nasim W, Amanullah Jr, Arif M, Alharby H (2017) Effects of nitrogen supply on water stress and recovery mechanisms in Kentucky bluegrass plants. *Front Plant Sci* 8:983. https://doi.org/10.3389/fpls.2017.00983

Saud S, Li X, Chen Y, Zhang L, Fahad S, Hussain S, Sadiq A, Chen Y (2014) Silicon application increases drought tolerance of Kentucky bluegrass by improving plant water relations and morph physiological functions. *SciWorld J* 2014:1–10. https://doi.org/10.1155/2014/368694

Schmelz EA (2015) Impacts of insect oral secretions on defoliation-induced plant defense. *Curr Opin Insect Sci* 9:7–15.

Schultz JC, Appel HM, Ferrieri AP, Arnold TM (2013) Flexible resource allocation during plant defense responses. *Front Plant Sci* 324:1–11.

Schuman MC, Baldwin IT (2016) The layers of plant responses to insect herbivores. *Annu Rev Entomol* 61:373–394.

Senol C (2020) The effects of climate change on human behaviors. In: Fahad S, Hasanuzzaman M, Alam M, Ullah H, Saeed M, Khan AK, Adnan M (eds) *Environment, climate, plant and vegetation growth.* Springer Publ Ltd, Springer Nature Switzerland AG. Part of Springer Nature, pp. 577–590. https://doi.org/10.1007/978-3-030-49732-3

Shafi MI, Adnan M, Fahad S, Fazli W, Ahsan K, Zhen Y, Subhan D, Zafar-ul-Hye M, Martin B, Rahul D (2020) Application of single superphosphate with humic acid improves the growth, yield and phosphorus uptake of wheat (*Triticum aestivum* L.) in calcareous soil. *Agronomy* 10:1224. https://doi.org/10.3390/agronomy10091224

Shah F, Lixiao N, Kehui C, Tariq S, Wei W, Chang C, Liyang Z, Farhan A, Fahad S, Huang J (2013) Rice grain yield and component responses to near 2°C of warming. *Field Crop Res* 157:98–110.

Shah SAR, Khan SA, Junaid K, Sattar S, Zaman M, Saleem N, Adnan M (2015) Screening of mustard genotypes for antixenosis and multiplication against mustard aphid, *Lipaphiserysimi*(Kalt) (Aphididae: Homoptera). *J Ento Zool Stud* 3(6):84–87.

Sharma N, Sharma KP, Gaur RK, Gupta VK (2011) Role of chitinase in plant defense. *Asian J Biochem* 6:29–37.

Sidra K, Javed I, Subhan D, Allah B, Syed IUSB, Fatma B, Khaled DA, Fahad S, Omaima N, Ali TKZ, Rahul D (2021) Physio-chemical characterization of indigenous agricultural waste materials for the development of potting media. *J Saudi Society Agricultural Sci.* https://doi.org/10.1016/j.sjbs.2021.08.058

Singh B, Sharma RA (2015) Plant terpenes defense responses, phylogenetic analysis, regulation and clinical applications. *3 Biotech* 5:129–151.

Smith CM, Clement SL (2012) Molecular bases of plant resistance to arthropods. *Annu Rev Entomol* 57:309–328.

Snydcr MJ, Glcndinning JI (1996) Causal connection between detoxification enzyme activity and consumption of a toxic plant compound. *J Comp Physiol: A Sens Neural Behav Physiol* 179:255–261.

So ML (2004) The occurrence of extrafloral nectaries in Hong Kong plants. *Bot Bull Acad Sinica* 45:237–245.

Stowe KA, Marquis RJ, Hochwender CG, Simms EL (2000) The evolutionary ecology of tolerance to consumer damage. *Annu Rev Ecol Syst* 31:565–595.

Strauss S, Zangerl A (2002) Plant-Insect Interactions in Terrestrial Ecosystems. In Pellmyr MHAO (eds) *Plant-animal interactions an evolutionary approach.* Blackwell Publishing Company, pp. 77–106.

Strauss SY, Agrawal AA (1999) The ecology and evolution of plant tolerance to herbivory. *Trends Ecol Evol* 14:179–185.

Subhan D, Zafar-ul-Hye M, Fahad S, Saud S, Martin B, Tereza H, Rahul D (2020) Drought stress alleviation by ACC deaminase producing *Achromobacter xylosoxidans* and *Enterobacter cloacae*, with and without timber waste biochar in maize. *Sustainability* 12(6286). https://doi.org/10.3390/su12156286

Suwa T, Maherali H (2008) Influence of nutrient availability on the mechanisms of tolerance to herbivory in an annual grass, *Avena barbata* (Poaceae). *Am J Bot* 95:434–440.

Tabasum (2017) Directional response of plants to the touch of an object thigmotropism. Thigmotropism.

Tan WH, Tao L, Hoang KM, Hunter MD, de Roode JC (2018) The effects of milkweed induced defense on parasite resistance in monarch butterflies, Danaus plexippus. *J Chem Ecol* 183:848–867.

Tariq M, Ahmad S, Fahad S, Abbas G, Hussain S, Fatima Z, Nasim W, Mubeen M, Rehman MH, Khan MA, Adnan M (2018) The impact of climate warming and crop management on phenology of sunflower-based cropping systems in Punjab, Pakistan. *Agri Forest Met* 256:270–282.

Theis N, Lerdau M (2003) The evolution of function in plant secondary metabolites. *Int J Plant Sci* 164:S93–S102.

Thompson DB (1992) Consumption rates and the evolution of diet-induced plasticity in the head morphology of *Melanoplus femurrubrum* (Orthoptera: Acrididae). *Oecologia* 89:204–213.

Thompson JN (1994) The coevolutionary process. 376 pp.

Tian D, Tooker J, Peiffer M, Chung SH, Felton GW (2012) Role of trichomes in defense against herbivores comparison of herbivore response to woolly and hairless trichome mutants in tomato (*Solanum lycopersicum*). *Planta* 234:1053–1066.

Tiffin P (2000) Mechanisms of tolerance to herbivore damage What do we know? *Evol Ecol* 14:523–536.

Trumble JT, Kolodny-Hirsch DM, Ting IP (1993) Plant compensation for arthropod herbivory. *Ann Rev Entomol* 38(38):93–119.

Tuteja N, Mahajan S (2007) Calcium signaling network in plants: an overview. *Plant Signal Behav* 2:79–85.

Unsar Naeem-U, Muhammad R, Syed HMB, Asad S, Mirza AQ, Naeem I, Muhammad HR, Fahad S, Shafqat S (2020) Insect pests of cotton crop and management under climate change scenarios. In: Fahad S, Hasanuzzaman M, Alam M, Ullah H, Saeed M, Khan AK, Adnan M (eds) *Environment, climate, plant and vegetation growth.* Springer Publ Ltd, Springer Nature Switzerland AG. Part of Springer Nature, pp. 367–396. https://doi.org/10.1007/978-3-030-49732-3

Vandenborre G, Smagghe G, Van Damme EJM (2011) Plant lectins as defense proteins against phytophagous insects. *Phytochemistry* 72:1538–1550.

Wahid F, Fahad S, Subhan D, Adnan M, Zhen Y, Saud S, Manzer HS, Martin B, Tereza H, Rahul D (2020) Sustainable management with mycorrhizae and phosphate solubilizing bacteria for enhanced phosphorus uptake in calcareous soils. *Agriculture* 10(334). https://doi.org/10.3390/agriculture10080334

Wajid N, Ashfaq A, Asad A, Muhammad T, Muhammad A, Muhammad S, Khawar J, Ghulam MS, Syeda RS, Hafiz MH, Muhammad IAR, Muhammad ZH, Muhammad HR, Veysel T, Fahad S, Suad S, Aziz K, Shahzad A (2017) Radiation efficiency and nitrogen fertilizer impacts on sunflower crop in contrasting environments of Punjab. *Pakistan Environ Sci Pollut Res* 25:1822–1836. https://doi.org/10.1007/s11356-017-0592-z

Walters DR (2010) *Plant defense: warding off attack by pathogens, herbivores and parasitic plants*, Wiley-Blackwell, pp. 30–31.

War AR, Paulraj MG, Ahmad T, Buhroo AA, Hussain B, Ignacimuthu S, Sharma HC (2012) Mechanisms of plant defense against insect herbivores. *Plant Signal Behav* 7:1306–1320.

Williams KS, Gilbert LE (1981) Insects as selective agents on plant vegetative morphology Egg mimicry reduces egg laying by butterflies. *Science* 212:467–469.

Wink M (1999) Biochemistry of plant secondary metabolism. *Ann Plant Rev* 2:358.

Wise MJ, Abrahamson WG (2007) Effects of resource availability on tolerance of herbivory: a review and assessment of three opposing models. *Am Nat* 169:443–454.

Wittstock U, Gershenzon J (2002) Constitutive plant toxins and their role in defense against herbivores and pathogens. *Curr Opin Plant Biol* 5:300–307.

Wu C, Kehui C, She T, Ganghua L, Shaohua W, Fahad S, Lixiao N, Jianliang H, Shaobing P, Yanfeng D (2020) Intensified pollination and fertilization ameliorate heat injury in rice (*Oryza sativa* L.) during the flowering stage. *Field Crops Res* 252:107795.

Wu C, Tang S, Li G, Wang S, Fahad S, Ding Y (2019) Roles of phytohormone changes in the grain yield of rice plants exposed to heat: a review. *PeerJ* 7:e7792. https://doi.org/10.7717/peerj.7792

Xu S, Zhou W, Pottinger S, Baldwin IT (2015) Herbivore associated elicitor-induced defences are highly specific among closely related Nicotiana species. *BMC Plant Biol* 15:1–13.

Yang Z, Zhang Z, Zhang T, Fahad S, Cui K, Nie L, Peng S, Huang J (2017) The effect of season-long temperature increases on rice cultivars grown in the central and southern regions of China. *Front Plant Sci* 8:1908. https://doi.org/10.3389/fpls.2017.01908

Yazaki K, Arimura GI, Ohnishi T (2017) 'Hidden' terpenoids in plants their biosynthesis, localization and ecological roles. *Plant Cell Physiol* 58:1615–1621.

Zafar-ul-Hye M, Muhammad TH, Muhammad A, Fahad S, Martin B, Tereza D, Rahul D, Subhan D (2020b) Potential role of compost mixed biochar with rhizobacteria in mitigating lead toxicity in spinach. *Scientific Rep* 10:12159. https://doi.org/10.1038/s41598-020-69183-9

Zafar-ul-Hye M, Muhammad N, Subhan D, Fahad S, Rahul D, Mazhar A, Ashfaq AR, Martin B, Ji˘rí H, Zahid HT, Muhammad N (2020a) Alleviation of cadmium adverse effects by improving nutrients uptake in bitter gourd through cadmium tolerant rhizobacteria. *Environments* 7(54). https://doi.org/10.3390/environments7080054

Zafar-ul-Hye M, Akbar MN, Iftikhar Y, Abbas M, Zahid A, Fahad S, Datta R, Ali M, Elgorban AM, Ansari MJ et al. (2021) Rhizobacteria inoculation and caffeic acid alleviated drought stress in lentil plants. *Sustain* 13:9603. https://doi.org/10.3390/su13179603

Zahida Z, Hafiz FB, Zulfiqar AS, Ghulam MS, Fahad S, Muhammad RA, Hafiz MH, Wajid N, Muhammad S (2017) Effect of water management and silicon on germination, growth, phosphorus and arsenic uptake in rice. *Ecotoxicol Environ Saf* 144:11–18.

Zahir SM, Zheng HG, Ala Ud D, Amjad A, Ata Ur R, Kashif J, Shah F, Saud S, Adnan M, Fazli W, Saud A, Manzer HS, Shamsher A, Wajid N, Hafiz MH, Fahad S (2021) Synthesis of silver nanoparticles using *Plantago lanceolata* extract and assessing their antibacterial and antioxidant activities. *Sci Rep* 11:20754.

Zhuang X, Fiesselmann A, Zhao N, Chen H, Frey M, Chen F (2012) Biosynthesis and emission of insect herbivory-induced volatile indole in rice. *Phytochemistry* 73:15–22.

Zhu-Salzman K, Salzman RA, Ahn JE, Koiwa H (2004) Transcriptional regulation of sorghum defense determinants against a phloem-feeding aphid. *Plant Physiol* 134:420–431.

Zia-Ur-Rehman M (2020) Environment, climate change and biodiversity. In: Fahad S, Hasanuzzaman M, Alam M, Ullah H, Saeed M, Khan AK, Adnan M (eds) *Environment, climate, plant and vegetation growth.* Springer Publ Ltd, Springer Nature Switzerland AG. Part of Springer Nature, pp. 473–502. https://doi.org/10.1007/978-3-030-49732-3

Index

J

K

L

M

N

O

P

R

S

T

U

V

W